Fermented Meat Products
Health Aspects

Food Biology Series

Fermented Meat Products
Health Aspects

Editor
Nevijo Zdolec
University of Zagreb, Faculty of Veterinary Medicine
Department of Hygiene
Technology and Food Safety
Zagreb, Croatia

CRC Press is an imprint of the
Taylor & Francis Group, an **informa** business
A SCIENCE PUBLISHERS BOOK

CRC Press
Taylor & Francis Group
6000 Broken Sound Parkway NW, Suite 300
Boca Raton, FL 33487-2742

© 2017 by Taylor & Francis Group, LLC
CRC Press is an imprint of Taylor & Francis Group, an Informa business

No claim to original U.S. Government works

Printed on acid-free paper
Version Date: 20160622

International Standard Book Number-13: 978-1-4987-3304-5 (Hardback)

This book contains information obtained from authentic and highly regarded sources. Reasonable efforts have been made to publish reliable data and information, but the author and publisher cannot assume responsibility for the validity of all materials or the consequences of their use. The authors and publishers have attempted to trace the copyright holders of all material reproduced in this publication and apologize to copyright holders if permission to publish in this form has not been obtained. If any copyright material has not been acknowledged please write and let us know so we may rectify in any future reprint.

Except as permitted under U.S. Copyright Law, no part of this book may be reprinted, reproduced, transmitted, or utilized in any form by any electronic, mechanical, or other means, now known or hereafter invented, including photocopying, microfilming, and recording, or in any information storage or retrieval system, without written permission from the publishers.

For permission to photocopy or use material electronically from this work, please access www.copyright.com (http://www.copyright.com/) or contact the Copyright Clearance Center, Inc. (CCC), 222 Rosewood Drive, Danvers, MA 01923, 978-750-8400. CCC is a not-for-profit organization that provides licenses and registration for a variety of users. For organizations that have been granted a photocopy license by the CCC, a separate system of payment has been arranged.

Trademark Notice: Product or corporate names may be trademarks or registered trademarks, and are used only for identification and explanation without intent to infringe.

Preface to the Series

Food is the essential source of nutrients (such as carbohydrates, proteins, fats, vitamins, and minerals) for all living organisms to sustain life. A large part of daily human efforts is concentrated on food production, processing, packaging and marketing, product development, preservation, storage, and ensuring food safety and quality. It is obvious therefore, our food supply chain can contain microorganisms that interact with the food, thereby interfering in the ecology of food substrates. The microbe-food interaction can be mostly beneficial (as in the case of many fermented foods such as cheese, butter, sausage, etc.) or in some cases, it is detrimental (spoilage of food, mycotoxin, etc.). The *Food Biology* series aims at bringing all these aspects of microbe-food interactions in form of topical volumes, covering food microbiology, food mycology, biochemistry, microbial ecology, food biotechnology and bio-processing, new food product developments with microbial interventions, food nutrification with nutraceuticals, food authenticity, food origin traceability, and food science and technology. Special emphasis is laid on new molecular techniques relevant to food biology research or to monitoring and assessing food safety and quality, multiple hurdle food preservation techniques, as well as new interventions in biotechnological applications in food processing and development.

The series is broadly broken up into food fermentation, food safety and hygiene, food authenticity and traceability, microbial interventions in food bio-processing and food additive development, sensory science, molecular diagnostic methods in detecting food borne pathogens and food policy, etc. Leading international authorities with background in academia, research, industry and government have been drawn into the series either as authors or as editors. The series will be a useful reference resource base in food microbiology, biochemistry, biotechnology, food science and technology for researchers, teachers, students and food science and technology practitioners.

Ramesh C. Ray
Series Editor

Preface

Fermented meat products are traditionally and commercially most the valuable meat products. As a subject of research, they are an inexhaustible source of new knowledge becoming available through development of new technologies. Fermented meat products can be observed from various aspects: technological, chemical, biochemical, microbiological, toxicological, nutritional and alike, always in connection with certain public health topics.

This book focuses exactly on health aspects of fermented meat products, presented by eminent researchers and experts of various profiles – veterinarians, microbiologists, food technologists, molecular biologists, analytical chemists, biotechnologists, and nutritionists. The complexity of production and health aspects of fermented meat products require such multidisciplinary approach providing the reader with a deep insight into potential risks and their control. The book does not aim to criticize meat products but just the opposite – to point out to the complexity of prevention and control of potential hazards/risks in the production which greatly contributes to a higher total value of fermented meat products.

The introductory chapters present the specific types of fermented meat products and the basic principles of contemporary production. Basic technological operations are presented in the context of their impact on the products safety, hurdle technologies as well as risks of the traditional production. Two chapters are dedicated to health aspects of fermented meat products originating from wild game and small ruminants. The following chapters in the book concern microbiology and biotechnology, including microbial ecology, autochthonous starter cultures, biopreservation, probiotics, molecular analytics and microbiological risks in form of pathogenic microorganisms, resistance to antibiotics and microbial spoilage. The next set of chapters is related to chemical and toxicological topics such as chemical hazards, particularly biogenic amines, the composition of products and their impact on health, particularly fats and fat oxidation in fermented meat products. Finally, there are chapters about self-control in the production, pre-requisite programs, HACCP and official controls. Contents of the majority of chapters are not restricted strictly to fermented meat products but also describe general principles applicable in the context of other foodstuffs, i.e. the entire production chain from field to table. This book is a result of collaborative efforts of a number of distinguished authors with international reputation from renowned institutions and its editing has been a great challenge and pleasure for me. My special thanks go to Dr Ramesh C. Ray who invited me to edit this book under the Book Series "Food Biology". I trust the book will find its way to a wide audience and contribute in the area of science and practice of fermented meat products.

Nevijo Zdolec
Editor

Contents

Preface to the Series	*v*
Preface	*vii*
List of Contributors	*xi*

1. **Fermented Meat Products — An Overview**   **1**
   Friedrich-Karl Lücke

2. **Current Status of Fermented Meat Production**   **15**
   Herbert W. Ockerman and *Lopa Basu*

3. **Technology of Fermented Meat Products**   **27**
   Helga Medić

4. **Traditional Production of Fermented Meats and Related Risk**   49
   Hirokazu Oiki, Hirokazu Kimura and *Nevijo Zdolec*

5. **Game Meat Fermented Products — Food Safety Aspects**   58
   Peter Paulsen, Kashif Nauman, Friedrich Bauer, Alessandra Avagnina and *Frans J.M. Smulders*

6. **Sheep and Goat Fermented Meat Products— Health Aspects**   78
   Alfredo Teixeira, Sandra Rodrigues, Teresa Dias and *Letícia Estevinho*

7. **Hurdle Technologies in Fermented Meat Production**   95
   Josef Kameník

8. **Microbial Ecology of Fermented Sausages and Dry-cured Meats**   127
   Bojana Danilović and *Dragiša Savić*

9. **Application of Molecular Methods in Fermented Meat Microbiota: Biotechnological and Food Safety Benefits**   167
   Maria Grazia Bonomo, Caterina Cafaro and *Giovanni Salzano*

x   *Fermented Meat Products: Health Aspects*

10. **Foodborne Pathogens of Fermented Meat Products**   **196**
    Spiros Paramithiotis and Eleftherios H. Drosinos

11. **Protective Cultures and Bacteriocins in Fermented Meats**   **228**
    Maria João Fraqueza, Luis Patarata and Andrea Lauková

12. **Autochthonous Starter Cultures**   **270**
    Jadranka Frece and Ksenija Markov

13. **Probiotics in Fermented Meat Products**   **294**
    Rodrigo J. Nova, George Botsaris and Fabiola Cerda-Leal

14. **Antimicrobial Resistance of Lactic Acid Bacteria in Fermented Meat Products**   **319**
    Nevijo Zdolec, Slavica Vesković-Moračanin, Ivana Filipović and Vesna Dobranić

15. **Microbial Spoilage of Fermented Meat Products**   **343**
    Spiros Paramithiotis and Eleftherios H. Drosinos

16. **Chemical and Sensorial Properties of Fermented Meat Products**   **359**
    Tanja Bogdanović, Jelka Pleadin, Nada Vahčić and Sandra Petričević

17. **Fermented Meats Composition— Health and Nutrition Aspects**   **389**
    Peter Popelka

18. **Chemical Hazards in Fermented Meats**   **417**
    Jelka Pleadin and Tanja Bogdanović

19. **Biogenic Amines in Fermented Meat Products**   **450**
    José M. Lorenzo, Daniel Franco and Javier Carballo

20. **Fat Content of Dry-cured Sausages and its Effect on Chemical, Physical, Textural and Sensory Properties**   **474**
    José M. Lorenzo, Daniel Franco and Javier Carballo

21. **Lipid Oxidation of Fermented Meat Products**   **488**
    Slavomír Marcinčák

22. **HACCP in Fermented Meat Production**   **512**
    Igor Tomašević and Ilija Djekić

23. **Official Controls of Raw Meat Fermented Sausage Production**   **535**
    Milorad Radakovic and Slim Dinsdale

Index   *553*

List of Contributors

Avagnina, Alessandra
Institute of Meat Hygiene,
Meat Technology and Food Sciences
University of Veterinary Medicine Vienna
Veterinaerplatz 1, A 1210 Vienna, Austria
Tel. +431250773301
Fax +431250773390
Email: anarvik@infinito.it

Basu, Lopa
University of Kentucky
38 Charles E Barnhart Bldg
Lexington, KY 40546, USA
Tel: 6148432749
Email: lopa.basu@uky.edu

Bauer, Friedrich
Institute of Meat Hygiene
Meat Technology and Food Sciences
University of Veterinary Medicine Vienna
Veterinaerplatz 1, A 1210 Vienna, Austria
Tel: +431250773302
Fax +431250773390
Email: friedrich.bauer@vetmeduni.ac.at

Bogdanović, Tanja
Croatian Veterinary Institute
Regional Department Split
Laboratory for Analytical Chemistry and
Residues, Poljička cesta 33
21000 Split, Croatia
Tel: +38521789705
Email: t.bogdanovic.vzs@veinst.hr

Bonomo, Maria Grazia
Dipartimento di Scienze
Università degli Studi della Basilicata
Viale dell'Ateneo Lucano 10
85100 Potenza, Italy
Tel: +390971205690
Email: mariagrazia.bonomo@unibas.it

Botsaris, George
Department of Agricultural Sciences
Biotechnology and Food Science
Cyprus University of Technology
Limassol, Cyprus
Tel: +35725002582
Email: george.botsaris@cut.ac.cy

Cafaro, Caterina
Dipartimento di Scienze
Università degli Studi della Basilicata
Viale dell'Ateneo Lucano 10
85100 Potenza, Italy
Tel: +390971205690
Email: caterina.cafaro@unibas.it

Carballo, Javier
Área de Tecnología de los Alimentos
Facultad de Ciencias de Ourense
Universidad de Vigo
32004 Ourense, Spain
Tel: +34988387052
Fax: +34988387001
Email: carbatec@uvigo.es

Cerda-Leal, Fabiola
Departamento de Ingeniería en Alimentos
Universidad del Bío-Bío
Campus Chillán
Avenida Andrés Bello
S/N, Chillán, Chile
Tel: + 56422463249
Email: fcerda@ubiobio.cl

Danilović, Bojana
Faculty of Technology
Bulevar Oslobodjenja 124
16000 Leskovac
Serbia
Tel: +38116247203
Email: danilovic@tf.ni.ac.rs

Dias, Teresa
School of Agriculture of Polytechnique
Institute of Bragança (ESA-IPB)
Apartado 172, Bragança
Portugal
Tel: +351273303298
Email: tdias@ipb.pt

Dinsdale, Slim
FoodSafetyExperts Limited
Maltyard, Brooke Road, Shotesham
Norwich NR15 1XL UK
Tel: +441508550007
Cell: +447768662188
Fax: +448704601150
Email: dinsdale@foodsafetyexperts.co.uk

Djekić, Ilija
Food Safety and Quality Management
Department, University of Belgrade –
Faculty of Agriculture, Nemanjina 6
11080 Belgrade
Republic of Serbia
Tel: +381112615315/427
Cell: +38165 5127848
Email: idjekic@agrif.bg.ac.rs

Dobranić, Vesna
University of Zagreb
Faculty of Veterinary Medicine
Department of Hygiene
Technology and Food Safety
Heinzelova 55, 10000 Zagreb
Croatia
Tel: +38512390191
Email: vdobranic@vef.hr

Drosinos, Eleftherios H.
Laboratory of Food Quality Control and
Hygiene, Department of Food Science and
Human Nutrition
Agricultural University of Athens
Iera Odos 75, GR-11855 Athens
Greece
Tel: +302105294713
Fax: +302105294683
Email: ehd@aua.gr

Estevinho, Letícia
School of Agriculture of Polytechnique
Institute of Bragança (ESA-IPB)
Apartado 172, Bragança
Mountain Research Centre (CIMO)
Portugal
Tel: + 351273303221
Fax + 351273325405
Email: leticia@ipb.pt

Filipović, Ivana
University of Zagreb
Faculty of Veterinary Medicine
Department of Hygiene
Technology and Food Safety
Heinzelova 55, 10000 Zagreb, Croatia
Tel: +38512390192
Email: ivanaf@vef.hr

Franco, Daniel
Centro Tecnológico de la Carne de Galicia
Rua Galicia no. 4
Parque Tecnológico de Galicia
San Cibrao das Viñas
32900 Ourense
Spain
Tel: +34988548277
Email: danielfranco@ceteca.net

Fraqueza, Maria João
CIISA, Faculty of Veterinary Medicine
University of Lisbon
Avenida da Universidade Técnica
Pólo Universitário do Alto da Ajuda
1300-477 Lisbon
Portugal
Tel: +351213652884
Email: mjoaofraqueza@fmv.ulisboa.pt

Frece, Jadranka
Laboratory for General Microbiology
and Food Microbiology, Faculty of Food
Technology and Biotechnology University
of Zagreb, Pierottijeva 6,10000 Zagreb,
Croatia
Tel: +38514605284
Fax. + 38514836424
Email: jfrece@pbf.hr

Kameník, Josef
Department of Meat Hygiene and
Technology, Faculty of Veterinary Hygiene
and Ecology, University of Veterinary and
Pharmaceutical Sciences Brno
Palackého tř. 1/3, CZ-61242 Brno
Czech Republic,
Tel: + 420 604 220851
Email: kamenikj@vfu.cz

Kimura, Hirokazu
Department of Life Science
Shokei University, 2-6-78 Kuhonji
Chuo-ku, Kumamoto 862-8678
Japan
Tel: +81-96-362-2011(452)
FAX: +81-96-363-2975
Email:kimura@shokei-gakuen.ac.jp

List of Contributors

Lauková, Andrea
Institute of Animal Physiology
Slovak Academy of Sciences
Laboratory of Animal Microbiology,
Šoltésovej 4-6, 040 01 Košice
Slovakia
Tel: + 421557922964
Email: laukova@saske.sk

Lorenzo, José M.
Centro Tecnológico de la Carne de Galicia
Rua Galicia no. 4, Parque Tecnológico de
Galicia, San Cibrao das Viñas
32900 Ourense, Spain
Tel: +34988548277
Email: jmlorenzo@ceteca.net

Lücke, Friedrich-Karl
Fulda University of Applied Sciences
P. O. Box 2254, 36012 Fulda
Germany
Tel: +49 661 9640 376
Fax +49 661 9640 399
Email: Friedrich-karl.luecke@he.hs-fulda.de

Marcinčák, Slavomír
University of Veterinary Medicine and
Pharmacy in Košice, Department of Food
Hygiene and Technology
Komenského 73, 041 81 Košice, Slovakia
Tel: +421915984756
Email: slavomir.marcincak@uvlf.sk

Markov, Ksenija
Laboratory for General Microbiology and
Food Microbiology
Faculty of Food Technology and
Biotechnology University of Zagreb
Pierottijeva 6, 10000 Zagreb
Croatia
Tel: +38514605284
Email: kmarko@pbf.hr

Medić, Helga
University of Zagreb, Faculty of Food
Technology and Biotechnology
10000 Zagreb, Pierottijeva 6
Croatia
Tel: + 38514605126
Fax:+385 1 4605 072
Email: hmedic@pbf.hr

Nauman, Kashif
Department of Meat Technology
University of Veterinary and Animal
Sciences, Lahore
Pakistan
Tel: +431250773301
Email: drkashif@uvas.edu.pk

Nova, Rodrigo J.
School of Veterinary Medicine and Science
Sutton Bonington Campus
University of Nottingham
LE12 5RD, United Kingdom
Tel: 00441159516391
Email: rodrigo.nova@nottingham.ac.uk

Ockerman, Herbert W.
The Ohio State University
Plumb Hall 230A, 2027 Coffey Rd
Columbus, OH 43210, USA
Tel: 6142924317
Email: ockerman.2@osu.edu

Oiki, Hirokazu
Department of Biochemistry and Applied
Chemistry, Kurume National College of
Technology, 1-1-1 Komorino, Kurume
Fukuoka 830-8555, Japan
Tel: +81942359407
Fax: +81942359400
Email:oiki@kurume-nct.ac.jp

Paramithiotis, Spiros
Laboratory of Food Quality Control and
Hygiene, Department of Food Science and
Human Nutrition
Agricultural University of Athens
Iera Odos 75, GR-11855 Athens, Greece
Tel: +302105294705
Fax: +302105294683
Email: sdp@aua.gr

Patarata, Luis
Veterinary and Animal Research Centre
(CECAV),
Universidade de Trás-os-Montes e Alto
Douro, 5001-801 Vila Real, Portugal
Tel: +351259350539
Email: lpatarat@utad.pt

Paulsen, Peter
Institute of Meat Hygiene
Meat Technology and Food Sciences
University of Veterinary Medicine Vienna
Veterinaerplatz 1, A 1210 Vienna, Austria
Tel: +431250773318
Fax +431250773390
Email: peter.paulsen@vetmeduni.ac.at

Petričević, Sandra
Croatian Veterinary Institute
Regional Department Split
Laboratory for Analytical
Chemistry and Residues
Poljička cesta 33, 21000 Split, Croatia
Tel: + 38521789705
Email: petricevic.vzs@veinst.hr

Pleadin, Jelka
Croatian Veterinary Institute
Laboratory for Analytical Chemistry
Savska cesta 143, 10000 Zagreb
Croatia
Tel: + 38516123626
Email: pleadin@veinst.hr

Popelka, Peter
Department of Food Hygiene and
Technology, University of Veterinary
Medicine and Pharmacy in Košice
Komenského 73, 041 81 Košice
Slovakia
Email: peter.popelka@uvlf.sk

Radakovic, Milorad
University of Cambridge
Department of Veterinary Medicine
Madingley Road,
Cambridge CB3 0ES, UK
Tel: +44 1223 765048
Cell: +44 7917 228664
Email: mr412@cam.ac.uk

Rodrigues, Sandra
School of Agriculture of Polytechnique
Institute of Bragança (ESA-IPB)
Apartado 172, Bragança, Portugal
Tel: + 351273303221
Email: srodrigues@ipb.pt

Salzano, Giovanni
Dipartimento di Scienze
Università degli Studi della Basilicata
Viale dell'Ateneo Lucano 10
85100 Potenza, Italy
Tel: + 390971205690
Email: giovanni.salzano@unibas.it

Savić, Dragiša
Faculty of Technology, Bulevar
Oslobodjenja 124,16000 Leskovac
Serbia
Tel: +38116247203
Email: savic@tf.ni.ac.rs

Smulders, Frans J.M.
Institute of Meat Hygiene
Meat Technology and Food Sciences
University of Veterinary Medicine Vienna
Veterinaerplatz 1, A 1210 Vienna
Tel. +431250773301
Fax +431250773390
Email: frans.smulders@vetmeduni.ac.at

Teixeira, Alfredo
School of Agriculture of Polytechnique
Institute of Bragança (ESA-IPB)
Apartado 172, Bragança, Portugal
Tel: + 351273303206
Email: teixeira@ipb.pt

Tomašević, Igor
Animal Source Food Technology Department
University of Belgrade
Faculty of Agriculture
Nemanjina 6, 11080 Belgrade
Republic of Serbia
Tel: + 381112615315/272
Cell: + 381604299998
Email: tbigor@agrif.bg.ac.rs

Vahčić, Nada
Faculty of Food Technology and
Biotechnology, University of Zagreb
Pierottijeva 6, 10000 Zagreb
Tel: + 38514605 277
Email: nvahcic@pbf.hr

Vesković-Moračanin, Slavica
Institute of Meat Hygiene and Technology
Kaćanskog 13, 11000 Belgrade, Serbia
Tel: +381112650655
Email:slavica@inmesbgd.com

Zdolec, Nevijo
University of Zagreb, Faculty of Veterinary
Medicine, Department of Hygiene
Technology and Food Safety
Heinzelova 55, 10000 Zagreb, Croatia
Tel: + 385 12 390 199
Email: nzdolec@vef.hr

Chapter 1

Fermented Meat Products—An Overview

Friedrich-Karl Lücke

1 INTRODUCTION

Fermented meats are meat products that owe, at least partially, their characteristic properties to the activity of microorganisms. They may be subdivided into fermented sausages (made from comminuted meat) and meat products prepared by salting/curing entire muscles or cuts, followed by an ageing period in which enzymes (mostly meat proteases) bring about tenderness and flavor. In this chapter, such products are referred to as "ripened meats" (see also Toldrá 2015). Bacon and dried meats such as jerky and biltong are not considered here since there is little, if any, effect of microbial or tissue enzymes on their characteristics.

There are hundreds of different fermented meats. For example, the DOOR list of the European Union (http://ec.europa.eu/agriculture/quality/door/list.html) contains (as of July 2015), in the category Meat products, cooked, salted, smoked, etc., 117 entries as "Protected Geographic Indication" (PGI) and 34 entries as "Protected Designation of Origin" (PDO). Of these, the majority are fermented and/or ripened raw meat products, predominantly from the Mediterranean countries.

This chapter provides an overview on the types of fermented meats and on factors affecting their quality. It is not the intention of this chapter to describe all types of raw fermented and/or ripened meat products. Rather, they are classified by using a limited set of variables, and some

---

For Correspondence: Fulda University of Applied Sciences, P.O. Box 2254, 36012 Fulda, Germany. Tel: +49 661 9640 376, Fax +49 661 9640 399, Email: Friedrich-karl.luecke@he.hs-fulda.de

examples for each category are given. For a more detailed account of the various types of fermented meats, their specific characteristics and their history, the reader is referred to the books edited by Toldrá (2015) and by Campbell-Platt and Cook (1995).

Salting and fermentation developed as a method to preserve meat in times before refrigeration became available. Hence, most traditional raw meat products are stable at ambient temperatures (after a salting/drying process). Some traditional sausages with short ripening time and high moisture content are normally cooked before consumption. Semi-dry fermented sausages mainly appeared on the market once starter cultures were introduced and the need for saving costs and reducing the fermentation time increased.

Ripened meat products prepared by salting/curing entire muscles or cuts are often referred to as "raw hams" because pork hams are nowadays most widely used as raw material. However, depending on regional traditions, other muscles and cuts (e.g. shoulder, *M. longissimus dorsi*) from various animals are used, too.

Quality parameters both for fermented sausages and ripened meats include:

- colour (affected by the type and amount of lean meat, curing agents, ageing time, pH, nitrate reductase activity);
- texture (affected by the type and amount of lean meat, method of comminution, acidification, drying and ageing time, weight loss);
- flavour (affected by the type and amount of lean meat, spices, acidification, ageing time);
- nutritional value (salt and fat content, fatty acid spectrum);
- safety (affected by levels of pathogens and contaminants in raw material and processing/ripening environment; lactic acid bacteria and fermentable carbohydrates in recipe; fermentation temperature and time; drying);
- shelf life (affected by composition of fatty tissue; spoilage flora in processing environment; lactic acid bacteria and fermentable carbohydrates in recipe; salting and/or fermentation time and temperature, relative humidity and time; drying).

Moreover, consumers have an increasing interest in the quality of production processes, namely, ecology (low input of energy, low output of waste and greenhouse gases etc.), animal welfare, and corporate social responsibility.

2 FERMENTED SAUSAGES

The following variables affect the characteristics of fermented sausages:

Fermented Meat Products—An Overview

- Animal species and breeds for the lean meat component: these include pigs, cattle, small ruminants, poultry, horse, reindeer and game.
- Animal species for the fat component: pork back fat is preferred, especially for aged, high-quality products. The feeding regime is also important. If religious tradition prohibits the use of pork, fatty tissues from ruminants is used, such as sheep tail fat for Turkish sucuk (Kilic 2009).
- Cuts/muscles used. These vary in the percentage and composition of adherent fat and connective tissue, and pH value.
- Method for comminution of the meat: mincer or cutter.
- Degree of comminution of the meat: coarse, medium, or fine.
- Non-meat bulk ingredients from dairy (e.g. casein, milk powder) or plant origin. For example, potato flour and pre-cooked cereals may be used as extenders and binders, and some vegetables, especially paprika, for colour and taste.
- Spices (pepper, garlic, mustard seeds etc.).
- Additives; these include
  - salt: sodium chloride
  - curing agents (nitrite and/or nitrate) and adjuncts (ascorbate, isoascorbate)
  - fermentable sugars of different types and ingoing amounts: glucose, sucrose, lactose, starch hydrolysates etc. A sugar derivative, glucono-δ-lactone (GdL), is hydrolyzed in the sausages shortly after stuffing, to give gluconic acid which is subsequently fermented into lactic and acetic acid. GdL is sometimes used as a chemical acidulant
  - starter microorganisms (lactic acid bacteria, catalase-positive cocci, moulds or "house flora"). Various traditional fermentations still rely on the "house flora" but in order to standardize products and processes, use of commercial starter cultures is widespread, especially in industrial production
  - other (e.g. polyphosphates, citrate, antioxidants).
- Type of casing (natural, semi-synthetic, synthetic), with different vapour permeability.
- Diameter of casing (may range from 1 to 15 cm).
- Fermentation conditions (time, temperature, relative humidity, air velocity).
- Surface treatment during or after fermentation: smoke, surface flora or none.
- Ageing (time, temperature, relative humidity, air velocity).

- Treatment of final product: slicing, packaging (material, gas atmosphere).
- Handling by the consumer: some products are cooked before consumption.

Table 1 shows the main categories for fermented sausages. Legally, classification is based on parameters such as moisture-protein ratio (e.g. in the USA), weight loss (e.g. in Austria), and lean muscle tissue (BEFFE; e.g. in Germany). Hence, the categories "dry", "semi-dry", "spreadable (ripened)", and "spreadable (unripened)" are used in Table 1. However, there is no general clear, uniform distinction between dry and semi-dry fermented sausages in many countries, especially with traditional "artisanal" products: Many products having the same name can be prepared either as "dry" or "semi-dry" sausages, or, occasionally, even as unfermented hot-smoked product, and details, if any, are given in the small print part of the label. In Germany, for example, the "Leitsätze" (Codes of Practice; Anonymous 2014a) categorize fermented sausages either as "sliceable" ("schnittfest") or "spreadable" ("streichfähig"). Within these categories, products are classified by the minimum content of muscle protein (total protein nitrogen × 6.25, minus collagen protein; German acronym "BEFFE"). Common formulations (with about 14% initial protein content and 2.5% ingoing sodium chloride) with a weight loss of about 25% result in a moisture/protein ration of 2.0 or below and a water activity of 0.91 or below. Such sausages can be classified as "dry" and microbiologically stable without refrigeration (Lücke 2015).

With respect to safety and stability, raw fermented and/or ripened meats may be divided into two categories, depending on whether or not they support growth of *Listeria monocytogenes*. According to Regulation (EC) No. 2073/2005 (Anonymous 2014b), up to 100 cells of *L. monocytogenes* can be tolerated if a food has a pH $\leq$ 4.4, or a water activity (a_w) value $\leq$ 0.92, or a combination of pH $\leq$ 5.0 and a_w value $\leq$ 0.94. If protected from undesired mould growth (e.g. by smoking, modified atmosphere packaging), sausages meeting this requirement are usually stable at 15°C or even without refrigeration. This applies to most, if not all, dry and semi-dry fermented sausages and many dry-aged hams while for meats not meeting this requirement, it is up to the manufacturer to provide scientific evidence that *L. monocytogenes* does not grow in his product.

2.1 European Dry and Semi-dry Sausages

The terms "Northern European Technology" and "Mediterranean Technology" have been coined by Flores (1997) and Demeyer et al. (2000). Briefly, "Northern-style sausages" have lower pH values (usually below 5.0 after fermentation) and are smoked after fermentation. Smoking meats

Table 1 Classification of fermented sausages

Category	Final water activity	Total production time (fermentation + drying/ageing, if applicable)	Surface treatment	Fermentation temperature	Drying loss	Moisture/protein	pH of final product	Example	Reference[3]
Dry	<0.91	> 3 weeks[1]	Smoke or none	Traditional: <18°C;	25-40%	<2.0	Variable, generally 5.0-5.3	German-style salami; US-style pepperoni	(1)
Dry	<0.91	> 4 weeks	Mould growth	Other: 18-25°C[2]	25-40%	<2.0	Variable, 5.5-6.0	Italian salami; saucisse/saucisson (France); téliszalámi (Hungary)	(2), (3), (4), (5)
Turkish style	<0.90	> 2 weeks	None	20-25°C	25-40%	<2.0	Variable, generally 4.8-5.5	Sucuk (Turkey)	(6)
South-East Asian style	ca. 0.96	4-7 days	None	30-35°C	<10%	ca. 3.0	4.0-4.5	Nham (Thailand); Nemchua (Vietnam)	(7), (8)
Semi-dry, European style	0.91-0.94	10-20 days	Usually smoke	20-27°C	15-20%	2-3	< 5.2	Various Northern-style European sausages	(9)
Semi-dry, US style	0.92-0.94	7-14 days	Smoke	25-43°C	10-15%	>2.3	<5.0	Summer sausage (USA)	(10)
Spreadable, ripened	0.94-0.96	ca. 7 days	Smoke or none	18-25°C	<10%	>3	5.0-5.4	"Teewurst" (Germany), „sopréssa" (Italy), „sobrasada de Mallorca" (Spain)	(11)
Spreadable, not ripened	0.95-0.97	1-2 days	None	18-25°C	<10%	>3	5.3-5.6	Merguez (Maghreb); „frische Mettwurst" (Germany)	(9), (12)

[1]Shorter with very thin sausages such as some snack sausages; [2]US-style pepperoni: >30°C;
[3](1) Holck et al. 2015; (2) Cocolin et al. 2009; (3) Di Cagno et al., 2008; (4) Incze 1986; (5) Thévenot et al. 2007); (6) Kilic 2009; (7) Tran et al. 2011; (8) Chen et al. 2015; (9) Lücke 2015; (10) Maddock 2015; (11) Benkerroum 2013

6   *Fermented Meat Products: Health Aspects*

is more common in regions where climatic conditions make it difficult to "air-dry" the meat products, e.g. in maritime or mountain regions with much rainfall and where firewood is easily available (see e.g. Holck et al. 2015). The manufacturing process of Mediterranean-style sausages is slower and results in products with low final water activity (and long shelf life at hot ambient temperatures) while pH values are usually 5.0–5.3 after fermentation and rise again during subsequent drying and ageing, especially if moulds and yeasts grow on the surface.

2.1.1 Dry and Semi-dry Sausages, "Northern European Style"

Most dry sausages typical for Northern and Eastern Europe are smoked, and mould-ripened sausages are uncommon. Traditionally, the sausages were prepared in winter and ripened at temperatures at 15°C or below for extended periods, so that they could be stored with little or no refrigeration for up to one year. By smoking, the sausage surface was protected from mould growth (especially in rainy climate) and rancidity development (Holck et al. 2015). However, so-called "air dried" sausages are common in Central Germany where specially designed loam-coated ripening chambers ("Wurstekammer") provided a "low-tech" method to control the humidity (Lücke and Vogeley 2012). Traditional sausages were mostly made from pork (or, especially in mountain areas such as in Norway, from sheep meat), by use of a mincer to comminute the meat. The reason for this is that pigs can be slaughtered on farms more easily than cattle, and mincers, unlike cutters, can also comminute warm meat. For traditional product, nitrate and only low amounts of sugar (0.3 percent) are used, and starter cultures are rarely added. Nowadays, however, most of the fermented dry and semi-dry sausages in Northern, Central and Eastern Europe are prepared from chilled and frozen raw materials (pork and beef, the proportion between which depending on availability and desired colour) in the cutter, with 0.5% or more sugar, a combination of lactic acid bacteria and catalase-positive cocci as starters, and nitrite and ascorbate as curing agents. Fermentation is usually at 20–27°C, followed by a drying process at 10–15°C. The overall production time is generally about two to three weeks for thin dry sausages (or even less for thin snack sausages) and semi-dry sausages, and four weeks or longer for dry sausages of larger diameter (>60 mm). Certain sausages (e.g. "Landjäger", "Kantwurst") common in Southern Germany and Austria are pressed into moulds before fermentation. Categorization of German (Anonymous 2014a) and Austrian (Anonymous 2015) types of dry sausages is based on lean meat content, degree of comminution, and diameter, as determined by the casing used. In these countries, the term "salami" is used as a synonym for various dry sausage types of

Fermented Meat Products—An Overview 7

intermediate particle size, refers to various dry sausage types irrespective of surface treatment.

A review on dry sausages common in Scandinavia and Eastern Europe is given by Holck et al. (2015). As many designations indicate, there was a strong influence of German traditions. Differences are mainly caused by availability of different raw materials (such as lamb and mutton in Norway), by use of non-meat ingredients (such as potatoes and cereals in some Swedish sausages), paprika in the Danube region ("kolbász", "petrovská klobása", "kulen"; Kozačinski et al. 2006, Ikonić et al. 2010, Vuković et al. 2014), and spices. A special case is genuine Hungarian salami ("téliszalámi", meaning "winter salami") which is cold-smoked during the initial 2 weeks of ripening, and subsequently dried and aged under conditions supporting mold growth on the surface (Incze 1986).

2.1.2 Dry Sausages, Mediterranean Style

In Italy, there are various types of "salami" sausages. Nine varieties have PGI or PDO status. All of them are dry pork sausages with a mold cover. They differ in the amount of sugar added (and therefore, in the pH after fermentation), the composition of the spice mixture, the degree of comminution, the diameter of the casings, and the ripening regime (Cocolin et al. 2009, Di Cagno et al. 2008,Tabanelli et al. 2013). Some are fermented at low temperatures (15°C) but there is a trend to raise the temperature and use lactic starter cultures.

In Spain (Lorenzo et al. 2015), various types of fermented sausages are specific for the region. Many of these are made with paprika. Most fermented sausages are of the "Chorizo" (or "Chouriço" in Portugal) type. Such sausages are also common in Latin America. "Chorizo" sausages have in common that they are heavily spiced and coloured by paprika and mostly garlic, and they are neither smoked nor have a mold layer. Traditionally, they were made without curing agents. On the other hand, the varieties "Androlla" and "Botello" are typical for the maritime North West of Spain. They are smoked semi-dry or dry sausages made from cheaper cuts of pork (such as rib meat and jowls), and which are normally cooked before consumption. They also resemble certain Portuguese sausages such as "Alheira" (Ferreira et al. 2006). "Fuet" (diameter 34–38 mm; popular in Catalunya) and "Salchichón" (resembling French-type "Saucisson") are mold-ripened types. The latter has a larger diameter and usually contains beef, too.

In France, unsmoked dry and semi-dry sausages with a surface microflora predominate. "Saucisses" usually have diameters of about 40, and "Saucissons" of about 60 mm (Thévenot et al. 2005). Accordingly, the final pH and weight losses differ. Especially artisanal products may cover a wide range of pH and a_w values (Rason et al. 2007).

2.2 US-type Fermented Sausages

Even though fermented sausages in the United States originated from European-type processes, there are major differences. These may be due to the fact that there is still no compulsory inspection of pig carcasses for trichinae, but various legal requirements regarding the reduction of pathogens during ripening. Generally speaking, the regulatory approach is different in a way that production processes and controls are specified in more detail by USDA-FSIS rules. Accordingly, many of the sausages received a final treatment for pathogen inactivation (Maddock 2015). Moreover, use of elevated fermentation temperatures (>30°C) is common.

2.3 Asian-type Dry Sausages

In South-East Asia, (pork) sausages are usually prepared with the addition of pre-cooked ingredients (pork rinds, sometimes precooked rice), and high amounts of fermentable sugar. They are fermented at ambient temperatures (>30°C), and not dried extensively. Hence, they are stabilized by low pH only, and have a rather short shelf life. Typical examples are "Nham" (Thailand; Paukatong and Kunawasen 2001) and "Nemchua" (Vietnam; Tran et al. 2011, Chen et al. 2015). Chinese sausages ("Lupcheong") are preserved mainly by drying (Leistner 2000).

2.4 Sucuk

Sucuk (Turkey) and related products in the Middle East and Maghreb is a dry sausage made entirely from beef or water buffalo meat, and (typically) sheep tail fat (Kilic 2009). It is fermented at 20–25°C and subsequently dried to water activity values usually below 0.90.

2.5 Spreadable, Ripened Sausages

Only in Germany, spreadable, ripened sausages make up a major share of fermented sausages. Common varieties include "Teewurst", "Streichmettwurst" and "Braunschweiger". They are fermented to pH values around 5.3, resulting in cohesion between the meat and fat particles, but preparation of the mix is done in a way to keep the desired soft texture. For similar purpose, vegetable oils are partially replacing pork back fat in the formulation. Comparable products are Italian sausages of the "Sopréssa" or "Soppressata" (Coppola et al. 1997) and "Ciauscolo" type (Aquilanti et al. 2007), and "Sobrasada de Mallorca" (Rosselló et al. 1995) which undergo a longer ageing process.

Fermented Meat Products—An Overview

2.6 Fresh Sausages

German-style "Frische Mettwurst" and Maghreb-type "Merguez" (Benkerroum 2013) are fermented only briefly, just enough to ensure a pH drop to below 5.6 and cohesion between the meat and fat particles. They must be stored refrigerated and have only a short shelf life. In Germany, where many people eat raw minced meat, they are often eaten raw while merguez and many other varieties are normally cooked before consumption.

3 RIPENED MEATS

The following variables affect the characteristics of ripened meats:

- Animal species and breeds for the meat component: pigs, cattle, small ruminants, poultry, reindeer, game etc., depending on local availability and cultural traditions.
- Cuts/muscles used. These vary in the percentage of adherent fat and connective tissue, and pH value.
- Spices (for certain varieties).
- Additives; these include:
  - salt (coarse or fine)
  - curing agents (nitrite and nitrate) and adjuncts (ascorbate, isoascorbate). Most hams are cured by using nitrate or a combination of nitrite and nitrate.
  - (rarely) starter cultures (catalase-positive cocci or "house flora" in recycled brines)
  - (occasionally) carbohydrates.
- Salting/curing conditions: Salt (and curing agents) may be applied by rubbing the meat surface (dry salting/curing), by immersing the meat into brines (brine salting/curing), or by injection of brine. Some hams (e.g. "WestfälischerSchinken") are first dry-salted, followed by salting in brine. To prevent growth of psychrotrophic pathogens such as non-proteolytic strains of *Clostridium botulinum* in the interior of the meat, temperatures during salting/curing and the subsequent salt equilibration are below 5°C, and time must be sufficient to guarantee a water activity below 0.96 throughout the meat. Nevertheless, there is some variation in the salting/curing and salt equilibration regime.
- Surface treatment during or after fermentation: Some hams receive no surface treatment, others, especially in Germany and Northern Europe, are smoked, others develop a surface flora similar to

10 *Fermented Meat Products: Health Aspects*

mould-ripened sausages, and others are covered with a layer of paste based on lard (e. g. Parma hams) or "Çemen". The latter is a paste containing fenugreek, hot red pepper and garlic, and is used for manufacture of Turkish-style "Pastırma" (Kilic 2009).

- Drying and ageing: For the production of high-quality products stable at ambient temperature, the hams are dried and aged extensively (up to one year). The rate of weight loss and tenderization (by meat proteases) is affected by time, temperature, relative humidity, and air velocity.
- Treatment of final product: slicing, packaging.

Table 2 gives an overview on major types of ripened meats.

Table 2 Classification of ripened meats

Category	Total production time (salting, drying/ageing)	Surface treatment	Drying loss	Examples
Pork ham/ legs	4–7 months	Smoke	10–15%	Westfälischer Schinken, Holsteiner Katenschinken (Germany)
	> 2 months	Smoke	25%	Schwarzwälder Schinken (Germany)
	> 6 months	None or lard	25–30%	Prosciutto di Parma, Prosciutto di San Daniele (Italy); Jamón Serrano (Spain); Jambon de Bayonne (France); Kraški pršut (Slovenia), Istarski pršut (Croatia)
	> 4 months	None or smoke	>20%	Country ham (USA)
Pork neck	5–6 months	Smoke or none	40%	Coppa (Italy)
Lamb	2–9 months	None or smoke	28–42%	Fenalår (Norway)
Beef muscle (silverside)	> 3 months	None	40%	Bündnerfleisch (Switzerland); Bresaola (Italy)
Beef (various cuts)	ca. 1 month	Cover with spicy paste	25–40%	Pastırma/basturma/ pastrami (Turkey, Egypt, Maghreb countries)

Source: Compiled from information in the book edited by Toldrá 2015

3.1 Pork Hams

In Germany, most pork hams are smoked and not dried extensively. In the production of "Westfälischer Schinken", the bone is not removed before salting, and "Schwarzwälder Schinken" and "Holsteiner Katenschinken" are produced by using special smoking methods. Similar smoked products prevail in Northern and Eastern Europe. In analogy to dry sausages, Mediterranean-style raw hams are not smoked but dried and aged for extended periods. Accordingly, they lose more weight and are more tender and aromatic due to extended proteolysis. Typical examples are given in Table 2 and by Estévez et al. (2015).

Some regional specialties owe their characteristics to special pig breeds and special feeding regimes. Examples are Iberian ham (Spain) and Noir de Bigorreham (south-west France; see Estévez at al. 2015).

3.2 Products from Ruminants: Pastirma and Related Products, BündnerFleisch, Fenalår

"Pastirma" (basterma, basturma, pastrami) is produced from beef (sometimes sheep) muscles by using a special salting, pressing and drying regime (Kilic 2009, Benkerroum 2013), and the resulting product is covered with a paste consisting of hot pepper, garlic and fenugreek, named "Çemen" in Turkish. The finished product has a water activity of 0.85 to 0.90, and moisture of 35 to 52 percent (Leistner 2000). Hence, a good-quality pastirma is stable at ambient temperature.

"Bündnerfleisch" and "Bresaola" are typical products from Graubünden (Switzerland) and adjacent regions of Italy, respectively. They are prepared from beef silverside ("Unterschale") by a salting and pressing procedure, and lose about 40 percent weight during drying.

"Fenalår" and some similar products common in Scandinavia are produced from lamb cuts. For "Fenalår", lamb or sheep legs are salted and dried for up to nine months (Håseth et al. 2015).

4 CONCLUSION

Fermented and ripening meats are very diverse, and a part of the regional and national culture (see Leroy et al. 2013). Any industrialization and standardization of their production should take care to maintain this diversity (Settanni and Moschetti 2014). Control measures must be risk-based and should not put small manufactures at a disadvantage.

Key Words: *Fermented sausages, ripened meats, hams, varieties, specialties*

REFERENCES

Anonymous. 2014a. Leitsätze für Fleisch und Fleischerzeugnisse, status 8 January 2010. Bundesanzeiger Vol. 16, p. 336ff, available from http://www.bmelv.de/SharedDocs/Downloads/Ernaehrung/Lebensmittelbuch/LeitsaetzeFleisch.pdf?__blob=publicationFile; last accessed 3 August 2015

Anonymous. 2014b. Commission Regulation (EC) no. 2073/2005 of 15 November 2005 on microbiological criteria for foodstuffs. 2005. Official Journal of the EU L338:1–26. Last modified by Commission Regulation (EU) No 209/2013 of 7 March 2014. Official Journal of the EU L 68, pp. 19-23.

Anonymous. 2015. Österreichisches Lebensmittelbuch (Codex Alimentarius Austriacus), Chapter B 14 - Fleisch und Fleischerzeugnisse. BundesministeriumfürGesundheit, Wien.Last modified 27 January 2015. Available from http://www.lebensmittelbuch.at/fleisch-und-fleischerzeugnisse; last accessed 3 August 2015

Aquilanti, L., S. Santarelli, G. Silvestri, A. Osimani, A. Petruzzelli and F. Clementi. 2007. The microbial ecology of a typical Italian salami during its natural fermentation. Int. J. Food Microbiol. 120: 136–145.

Benkerroum, N. 2013. Traditional fermented foods of North African countries: Technology and food safety challenges with regard to microbiological risks. Compr. Rev. Food Sci. F. 12: 54–89.

Campbell-Platt, G. and P.E. Cook. 1995. Fermented meats. Blackie Academic and Professional, London.

Chen, M.-J., R-J. Tu and S.-Y. Wang. 2015. Asian products. Chapter 37, pp. 321–327. In: F. Toldrá (ed.). Handbook of fermented meat and poultry, 2nd edition. Wiley-Blackwell, Chichester, UK.

Cocolin, L., P. Dolci, K. Rantsiou, R. Urso, C. Cantoni and G. Comi. 2009. Lactic acid bacteria ecology of three traditional fermented sausages produced in the North of Italy as determined by molecular methods. Meat Sci. 82: 125–132.

Coppola, R.M., R. Iorizzo, R. Saotta, E. Sorrentino and L. Grazia. 1997. Characterization of micrococci and staphylococci isolated from soppressata-molisana, a Southern Italy fermented sausage. Food Microbiol. 14: 47–53.

Demeyer, D., M. Raemaekers, A. Rizzo, A. Holck, A. De Smedt, B. ten Brink, B. Hagen, C. Montel, E. Zanardi, E. Murbrekk, F. Leroy, F. Vandendriessche, K. Lorentsen, K. Venema, L. Sunesen, L. Stahnke, L. De Vuyst, R Talon, R. Chizzolini and S. Eerola. 2000. Control of bioflavour and safety in fermented sausages: First results of a European project. Food Res. Int. 33: 171–180.

Di Cagno, R., C.C. Lòpez, R. Tofalo, G. Gallo, M. De Angelis, A. Paparella, W.P. Hammes and M. Gobbetti. 2008. Comparison of the compositional, microbiological, biochemical and volatile profile characteristics of three Italian PDO fermented sausages. Meat Sci. 79: 224–235.

Estévez, M., S. Ventanas, D. Morcuende and J. Ventanas. 2015. Mediterranean products. Chapter 42, pp. 361–369. In: F. Toldrá (ed.). Handbook of fermented meat and poultry, 2nd edition.Wiley-Blackwell, Chichester, UK.

Ferreira, V., J. Barbosa, S. Vendeiro, A. Mota, F. Silva, M.J. Monteiro, T. Hogg and P.A. Gibbs. 2006. Chemical and microbiological characterization of alheira: A typical Portuguese fermented sausage with particular reference to factors relating to food safety. Meat Sci. 73: 570–575.

Flores, J. 1997. Mediterranean vs. northern European meat products. Processing technologies and main differences. Food Chem. 59: 505–510.

Håseth, T.T., G. Thorkelsson, E. Puolanne and M.S. Sidhu. 2007. North European products. Chapter 36, pp. 371–376. *In*: F. Toldrá (ed.). Handbook of fermented meat and poultry, 2nd edition. Wiley-Blackwell, Chichester, UK.

Holck, A., E. Heir, T.C. Johannessen and L. Axelsson. 2015. Northern European Products. Chapter 36, pp. 313–320. *In*: F. Toldrá (ed.). Handbook of fermented meat and poultry, 2nd edition.Wiley-Blackwell, Chichester, UK.

Ikonić, P., L.S. Petrović, T.A. Tasić, N.R. Džinić, M.R. Jokanović and V.M. Tomović. 2010. Physicochemical, biochemical and sensory properties for the characterization of Petrovská Klobása (traditional fermented sausage). APTEFF 41: 19–31.

Incze, K. 1986. Technologie und Mikrobiologie der ungarischen Salami. Tradition und Gegenwart (Technology and microbiology of Hungarian salami. Tradition and present status). Fleischwirtschaft 66: 1305, 1308, 1311.

Kilic, B. 2009. Current trends in traditional Turkish meat products and cuisine. LWT - Food Sci. Technol. 42: 1581–1589.

Kozačinski, L., N. Zdolec, M. Hadžiosmanović, Ž. Cvrtila, I. Filipović and T. Majić. 2006. Microbial flora of the Croatian traditional fermented sausage. Arch. Lebensmittelhyg. 57: 141–147.

Leroy, F., A. Geyzen, M. Janssens and P. Scholliers. 2013. Meat fermentation at the crossroads of innovation and tradition: A historical outlook. Trends Food Sci. Tech. 31: 130–137.

Leistner, L. 2000. Use of combined preservative factors in food of developing countries. pp. 294–314. *In*: B.M. Lund, T.C. Baird-Parker and G.W. Gould (eds.). The microbiological safety and food quality. Aspen Publication, Gaithersburg, MD.

Lorenzo, J.M., S. Martínez and J. Carballo. 2015. Microbiological and biochemical characteristics of Spanish fermented sausages. pp. 55–72. *In*: V.R. Rai and J.A. Bai (eds.). Beneficial microbes in fermented and functional foods. CRC Press, Boca Raton.

Lücke, F.-K. 2015. European products. Chapter 33. pp. 287–292. *In*: F. Toldrá, (ed.). Handbook of fermented meat and poultry, 2nd edition,. Wiley-Blackwell, Chichester, UK.

Lücke, F.-K. and I. Vogeley. 2012. Traditional 'air-dried' fermented sausages from Central Germany. Food Microbiol. 29: 242–246.

Maddock, R. 2015. US Products – Dry sausage. Chapter 34, pp. 295–299. *In:* F. Toldrá (ed.). Handbook of fermented meat and poultry, 2nd edition. Wiley-Blackwell, Chichester, UK.

Paukatong, K.V. and S. Kunawasen. 2001. The Hazard Analysis and Critical Control Point (HACCP) generic model for the production of Thai fermented pork sausage (nham). Berl. Munch. Tierarztl. Wschr. 114: 327–330.

Rason, J., A. Laguet, P. Berge, E. Dufour and A. Lebecque. 2007. Investigations of the physicochemical and sensory homogeneity of traditional French dry sausages. Meat Sci. 75: 359–550.

Rosselló, C., J.I. Barbas, A. Berna and N. López. 1995. Microbial and chemical changes in Sobrasada during ripening. Meat Sci. 40: 379–385.

Settanni, L. and G. Moschetti. 2014. New trends in technology and identity of traditional dairy and fermented meat production processes. Trends Food Sci. Tech. 37: 51–58.

Tabanelli, G., F. Coloretti, C. Chiavari, L. Grazia, R. Lanciotti and F. Gardini. 2013. Effects of starter cultures and fermentation climate on the properties of two types of typical Italian dry fermented sausages produced under industrial conditions. Food Control 26: 416–426.

Thévenot, D., M-L. Delignette-Muller, S. Christieans and C. Vernozy-Rozand. 2005. Fate of *Listeria monocytogenes* in experimentally contaminated French sausages. Int. J. Food Microbiol. 101: 189–200.

Toldrá, F. (ed.). 2015. Handbook of fermented meat and poultry, 2nd edition. Wiley-Blackwell, Chichester, UK.

Tran, K.T.M., B.K. May, P.M. Smooker, T.T. Van and P.J. Coloe. 2011. Distribution and genetic diversity of lactic acid bacteria from traditional fermented sausage. Food Res. Int. 44: 338–343.

Vuković, I., D. Vasilev and S. Sačić. 2014. Mikroflora und Qualität traditioneller Rohwurst Lemeški Kulen. Fleischwirtschaft 94, 8: 114–118.

Chapter 2

Current Status of Fermented Meat Production

Herbert W. Ockerman* and Lopa Basu

1 INTRODUCTION

Fermentation science is known as zymology and this system is an ancient technique of preserving food and beverages. Yeast ferments sugar to alcohol and it has been postulated that this preservation was used in the B.C. era. In just the food area there are in excess of 100 different food items that are preserved by fermentation (http://en.wikipedia. org/wiki/List_of_fermented_foods#Fermented_foods). This type of food preservation is conducted in most areas of the world not only for its keeping ability but also for its flavor. Meat fermentation is a biological preservation method that requires little energy, and results in unique and distinctive meat properties such as microbiological safety, flavor and palatability, color, tenderness, and a host of other desirable attributes of this specialized meat item. Lowering of pH could also be accomplished by simply adding acid (e.g. vinegar, lemon juice, etc.); however this does not have the same desirable flavor as that produced in fermented products. This difference is speculated to be caused by other flavor desirable microbial metabolites which are formed during fermentation in addition to the production of lactic acid.

---

*For Correspondence. The Ohio State University, Plumb Hall 230A, 2027 Coffey Rd, Columbus, OH 43210, Tel: 6142924317, Email: ockerman.2@osu.edu

Changes from raw meat to a fermented product can be caused by "cultured" or "wild" microorganism (wild inoculation is sometimes called "back slop" since often part of a previous production batch is added to the current production unit without knowing if the previous batch was desirable). The lactic acid bacteria (LAB) consume sugar and carbohydrates and convert it into lactic acid which lowers the pH. This fermentation is a biological system, and is influenced by many production environmental conditions that need to be controlled to produce a consistent product. Some of these factors would include a consistent inoculum (impossible with a "wild inoculum"), consistent fresh, low-contaminate raw material, a strict sanitation production system, and control of time, temperature, and humidity during production, followed by a consistent smoke cycle and volume, and use of appropriate and approved additives. Lactic acid, which accounts for most of the antimicrobial properties of fermented meats, originates from LAB conversion of glycogen reserves in the carcass tissues and from the sugar added during product formulation. A desirable and safe fermentation product is the outcome of acidulation caused by lactic acid production along with lowering of water activity (a_w). Safety is also aided by the addition of salt (curing) and drying which lowers the water activity and often cold smoking which adds flavor, reduce microbial numbers on the surface and retards surface oxidation due to phenols in the smoke. Heating sometimes accompanies smoking (hot smoke) which reduces internal bacterial growth (both desirable LAB and undesirable bacteria).

LAB cultures ("natural" or "wild") and controlled fermentations involve the metabolism of carbohydrates. LAB growth must be understood to produce a safe, uniform and marketable product Most commercial starter cultures today consist of lactobacilli and/or micrococci, selected for their metabolic activity production of lactic acid and other metabolites, which often improves the flavor profile. The reduction of pH and the lowering of water activity are both microbial hurdles that aid in producing a safe product. Fermented sausages often have a long storage life due to lactic acid production by LAB organisms in the early stages of storage. Added salt, nitrite, and/or nitrate, low pH, and drying which also reduces the water activity, assist in long term preservation. Production and composition figures for fermented meat products are difficult to obtain, particularly, since many of these fermented products are local products and are produced in rural areas and consumed locally and quantities are not recorded. The limited number of references available on fermentation would suggest that the production and consumption is sizeable.

2 DEFINITIONS OF A FEW FERMENTED PRODUCTS

The characteristics and types of a few fermented products can be found in Tables 1 and 2. Guidelines proposed in the U.S. (American Meat Institute 1982, Hui et al. 2004) for making fermented dry or semi-dry sausages include a definition of dry sausage as chopped or ground meat products

Table 1 Characteristics of different types of fermented sausages

Type of sausage	Characteristics
Dry; long ripening, e.g. dry or hard salami (types – Arles, Genoa, Hungarian and Milano salami, Italian salami) Saucission, pepperoni (Italian for bell pepper) and these products are shelf stable	Chopped and ground meat (often beef/pork)
	Commercial starter culture or back inoculum
	U.S. fermentation temperature 15–40°C for 1–5 d. Most contain nitrite – red color
	Not smoked or lightly smoked, highly spiced
	U.S. Bacterial action reduces pH to 4.7–5.3 (0.5–1.0% lactic acid, total acidity 1.3% which facilitates drying by denaturing protein resulting in a firm texture), moisture protein ratio <2.3 : 1, moisture loss 20–50%, Moisture level <35%
	European bacterial action reduces pH 5.3–5.6 for a more mild taste than U.S., and processing at 22–26°C for 12–14 wk
	Dried to remove 20–50% of moisture; contains 20–45% moisture, fat 39%, protein 21%, salt 4.2%, a_w 0.85–0.86. yield 64%
	Moisture protein ratio no greater than 2.3 to 1.0
	Less tangy taste than semi-dry
Semi-dry; sliceable, e.g. summer sausage (Spain – include salchichón, chorizo, and types of embutido)., Holsteiner, Cervelat (Zervelat), Tuhringer, Chorizos, refrigerate, usually shelf stable	Chopped or ground meat
	Bacterial action reduces pH to 4.7–5.3 (lactic acid 0.5–1.3%, total acidity 1%), processing time 1–4 wk
	Dried to remove 8–30% of moisture by heat; contains 30–50% moisture, fat 24%, protein 21%, salt 3.3%, a_w 0.92–0.94, and yield 90%
	Usually packaged after fermentation/heating
	Generally smoked during fermentation
	Moisture protein ratio no greater than 2.3–3.7 to 1.0, No mold
Moist; undried; spreadable e.g. Teewurst (Rügenwalde-style Teewurst; Mettwurst, Frishe, Braunschweiger	Contains 34–60% moisture, production time 3–5 days
	Weight loss ~ 10%, a_w 0.95–0.96
	Usually smoked and cooked
	Highly perishable, refrigerate, consume in 1–2 days, No mold

Source: Modified from American Meat Institute 1982, Gilliland 1985, Campbell-Platt and Cook 1995, Doyle et al. 1997, Farnworth 2003, Ockerman and Basu 2014; http://kb.osu.edu/dspace/handle/1811/45275

Table 2 Types of fermented sausages and areas of production

Type	Area	Production	Sausages	Sensory
Dry fermented, mixed culture	Southern and Eastern Europe, psychrotrophic LAB, optimum growth 10-15°C	40-60 mm, pork, 2-6 mm particle, nitrate, starters, fungi, ferment 10-24°C, 3 d, ripen 10-18°C, 3-6 wk, weight loss higher than 30%, low water activity, higher pH, 5.2 to 5.8, lower lactate, 17 mmol/100g of dry matter	Italian Salami	Fruity, sweet odor,
			Spanish Salchichon, Chorizo (usually no starter)	medium buttery, sour and pungent, more mature
			French Saucisson, 9 mg/g of lactate/g	
	Northern Europe, LAB (*Lb. plantarum* or *Pediococcus*) and Micrococcaceae (*St. carnosus* or *Micrococcus*)	90 mm, pork/beef, 1-2 mm particle, nitrite, starters, smoked, ferment 20-32°C, 2-5 d, rapid acidification to below 5, smoking ripen 2-3 wk, weight loss higher than 20%, lower pH, 4.8 to 4.9, higher lactate 20-21 mmol/100g of dry matter	German Salami, 15 mg/g lactate/g	Buttery, sour odor, low levels of spice and fruity notes, more acid
			Hungarian Salami	More acid
			Nordic Salami	More acid
Mold ripened	Europe, U.S.	Usually with starters	French Salami	Slightly musty, fruity, sweet odor,
		Usually with starters, bowl chopper	German Salami	medium buttery, sour and pungent, more mature
		Without starters, bowl chopper	Hungarian Salami	
		Usually with starters, ground	Italian Salami	
		Usually with starters	California Salami	
Semi-dry	U.S.	Usually with starters	Summer Sausage, M/P ratio 2.0-3.7:1.0	Moist
			Thuringer	
			Beef sticks	
Dry		Usually with *Pd. acidilactici*	Beef sticks	Dry, Compact
		Usually with *Pd. acidilactici*	Pepperoni, M/P ratio 1.6:1.0	

Source: Hui et al. 2004, Schmidt and Berger 1998, Demeyer et al. 2000, Stahnke et al. 1999, Ockerman and Basu 2014; http://kb.osu.edu/dspace/handle/1811/45275

Current Status of Fermented Meat Production

that due to bacterial action, reaches (LAB "wild" or starter culture) a pH of 5.3 or less. The drying removes 20 to 50% of the moisture resulting in moisture to protein ratio (M/P) of no greater than 2.3 to 1.0. Dry salami (U.S.) has an M/P ratio of 1.9 to 1, pepperoni 1.6 to 1, and jerky 0.75 to 1.

Semi-dry sausages are similar except that they have 15 to 20% loss of moisture during processing. Therefore, semi-dry sausages also have a softer texture and a different flavor profile than dry sausages. Because of the higher moisture content, semi-dry sausages are more susceptible to spoilage and are usually fermented to a lower pH (which increases shelf stability), to produce a very tangy flavor. Semi-dry products are generally sold after fermentation (pH of 5.3 or less). These are heated, and do not go through a drying process (water activity is usually 0.86 or higher). They are also usually smoked during the fermentation cycle and have a maximum pH of 5.3 in less than 24 hr. If the semi-dry sausage has a pH of 5.0 or less and a moisture protein ratio of 3.1 to 1 or less, it is considered to be shelf stable but most semi-dry products require refrigeration (2°C). In Europe, fermented meat with a pH of 5.2 and a water activity of 0.95 or less is considered shelf stable (USDA FSIS Food Labeling Policy Manual).

To decrease the pH (below 5.0) with limited drying, the U.S. semi-dry products are often fermented rapidly (12 hr or less) at a relatively high temperature (32–46°C). In Europe, fermentation is slower (24 hr or more) at a lower temperature and results in a higher pH. These differences in speed of fermentation and final pH result in products having different flavors.

3 FERMENTED PRODUCT INGREDIENTS

3.1 Raw Meat Used to Produce Fermented Products

Beef, mechanically separated beef (up to ~ 5%), pork, lamb, chicken, mechanically separated chicken (up to ~ 10%), duck, water buffalo, horse, donkey, reindeer, gazelle, porcupine, whale, fish, rabbit, by-products and other tissue from a variety of species are used to make fermented meat products. Fermented meat is often divided into two groups; products made from whole pieces of meat, such as hams, and products made from meat chopped into small pieces, such as various sausage types.

The predominant bacteria that appear in fresh meat are typically Gram-negative, oxidase-positive, aerobic rods of psychrotrophic pseudomonads along with psychrotrophic *Enterobacteriaceae*, small numbers of LAB, and other Gram-positive bacteria. The lactic and other Gram-positive bacteria become the dominant flora if oxygen is excluded and is encouraged during the fermentation stages. Since the production of fermented meat depends on microorganism growth, it is essential that these products are hygienically processed and chilled prior to use and maintained under refrigeration prior and during the curing operation.

3.2 Starter Cultures Usually Used for Fermented Products

Traditionally, fermented products depend on a "wild inoculum", which usually do not conform to any specific species but are usually related to *Lactobacillus plantarum*. However, other species such as *Lb. casei* and *Lb. leichmanii*, as well as many others, have been isolated from traditionally fermented meat products (Anon, 1978). In the U.S., *Lb. plantarum, Pediococcus pentosaceus*, or *Pd. acidilactici* are the most commonly used starter cultures. In Europe, the most used starter cultures include *Lb. sakei, Lb. plantarum, Pd. pentosaceus, Staphylococcus xylosus, St. carnosus*, and to a lesser extent *Micrococcus* spp. Reliance on natural flora results in products with inconsistent quality. The advantage of a starter culture is that the same microorganisms can be used repeatedly which reduces the variation of the finished product, and a larger number of organisms can be added. Currently, combined starter cultures are available in which one organism produces lactic acid (e.g. *lactobacilli*) and another improves desirable flavors (*Micrococcaceae, Lb. brevis, Lb. buchneri*). This translates into a lot of very good and acceptable product and almost no undesirable fermented product. However, very little excellent product is produced since most starter cultures are a combination of just a few species of microorganisms and they cannot produce as balanced a flavor as sometimes can be obtained when many species are included ("wild"). But "wild" inoculation also produces a lot of junk.

Particularly in the south European countries dry sausages are inoculated with atoxinogenic yeast and molds to produce products with specific flavor notes. This is done by dipping or spraying. Mold cultures tend to suppress natural molds and consequently, reduce the risk of mycotoxins. Due to the extended ripening and drying for these products, the final pH is usually higher (pH >5.5), even if the pH was lower after fermentation, because molds can utilize lactic acid and produce ammonia. This requires the final water activity to be low enough for preservation.

3.3 Other Ingredients Used in Fermented Meat Products

3.3.1 Salt

Salt is the major additive in fermented meat products. It is added in levels of 2 to 4% (2% minimum for desired bind and up to 3% will not retard fermentation) that will allow LAB to grow and will inhibit several unwanted microorganisms.

Current Status of Fermented Meat Production

3.3.2 Nitrite

Nitrite is used at 80 to 240 mg/kg for antibacterial, color, and antioxidant purposes. Nitrate and nitrite are often used in combination, but nitrate is usually not necessary except as a reservoir for nitrite, which could be useful in long term processing. Nitrite is also a hurdle, which inhibits bacterial growth and retards *Salmonellae* multiplication. Fermentation can be produced with only salt but there is a greater microbial risk if no nitrite is used.

3.3.3 Simple Sugars

Sugars such as glucose (dextrose, a minimum of 0.5%, 0.75% is often recommended), is the fermentation substrate that can be readily utilized by all LAB. The quantity of sugar influences the rate and extent of acidulation, and also contributes to flavor, texture, and product yield properties. The amount of dextrose added will directly influence the final product pH and additional sugar will not decrease pH further since bacterial cultures cannot grow in excess acid.

3.3.4 Spices

Spices are often used in fermented meat products (e.g. black, red, and white pepper, cardamom, mustard, all spice, paprika, nutmeg, ginger, mace, cinnamon, garlic), and various combinations are often included in the fermented meat formula. Spices are used for flavor, anti-oxidant properties, and to stimulate growth of lactic bacteria.

3.3.5 Sodium Ascorbate

Sodium ascorbate (also in U.S., sodium erythorbate) or ascorbic acid (also in U.S. erythorbic acid) is used for improvement and stability of color and retardation of oxidation.

4 PROCESSING OF FERMENTED PRODUCTS

Formulations are numerous even for products with the same name and some are held in strict security. Examples of formulations and processing procedures can be found in Komarik et al. (1974), Rust (1976), Ockerman (1989), Campbell-Platt and Cook (1995), and Klettner and Baumgartner (1980). Time, temperature, humidity, and smoke are also variables that control the quality of the final product (http://www.meatsandsausages. com/sausage-types/fermented-sausage). Type of casings available include natural (cleaned) casings, artificial casings (made from cotton linters), collagen casing, fibrous casings (synthetic, inedible) for use in the smokehouse or cooker, and they are sometimes netted, pre-stuck (pin pricked) to allow for better smoke penetration and elimination of air pockets, and cloth bags.

4.1 Environment

Temperature, time, and relative humidity combinations are quite variable in industrial productions. In general, the higher the fermentation temperature and water activity, the faster the lactic acid production. In Europe, the fermentation temperatures range from 5 to 26°C, with lower temperatures used in the Mediterranean area and higher temperatures in northern Europe. In the U.S. semidried products are usually fermented with slowly rising temperatures to over 35°C to shorten the fermentation time, which is frequently 12 hr or less. In the U.S. half-dried summer sausage is fermented three days at 7°C, three days at 27 to 41°C, and two days at 10°C, then heated to 58°C for 4 to 8 hours. Smoking depends on tradition and product type in the area the product is produced and can vary from no smoke to heavy smoke. If smoked, it is used to contribute flavor and to retard surface bacteria, molds and yeast and acts as a surface antioxidant. Many semi-dry sausages are heated after fermentation and/or smoking and this often increases the pH. Often, an internal temperature of 58.3°C is used. European drying temperatures of 14°C, and 78 to 88% relative humidities are often used. Air velocity of 1 m/s is also used. This results in a final water activity of < 0.90 in the finished fermented product (Klettner and Baumgartner 1980).

5 COMPOSITION OF FERMENTED MEAT PRODUCTS

Composition varies widely due to the number of fermented product types. Products with the same name vary considerably when made in different countries or even in the same country or even made by the same company in different locations. The major factors influencing composition of the finished product is naturally the composition and ratio of raw materials used. Other factors that are critical include processing procedures utilized, and the quantity of drying. Since most meat composition data are usually expressed on a percentage basis, the dryer (less moisture) the sausage the higher will be the other individual ingredients in the finished product. A few selected examples of average composition can be found in Table 3.

Current Status of Fermented Meat Production

Table 3 Nutritive composition of examples of fermented meat products

Properties	Cervelat		Salami; Pork, Beef;	Pepperoni; Pork, Beef
	Soft	Dry	Dry	Dry
Moisture	48.4%	29.4%		30.5%
Calorie (kcal)	307	451	424	466
Protein	18.6%	24.6%	23.1%	20.3%
Fat	25.5%	37.6%	34.8%	40.3%
–Mono-unsaturated	13.0 g/100g	–	17 g/100g	19 g/kg
–Poly-unsaturated	1.2 g/100g	–	3.2 g/100g	2.6 g/100g
–Saturated	12.0 g/100g		12.4 g/100g	16.1 g/100g
Ash	6.8%	6.7%	5.5%	4.8%
Fiber	0	0	–	1.5%
Carbohydrate	1.6%	1.7%	–	5 g/100g
Sugar	0.85%	–	–	0.7%
Calcium	11mg/100g	14 mg/100g	–	21 mg/110g
Iron	2.8 mg/100g	2.7 mg/100g	1.,5 mg/100g	1.4 mg/100g
Magnesium	14.0 mg/100g	–	17.8 mg/100g	18 mg/100g
Phosphorus	214 mg/100g	294 mg/100g	143 mg/100g	176 mg/100g
Potassium	260 mg/100g	–	379 mg/100g	315 mg/100g
Sodium	1242 mg/100g	–	1881 mg/100g	1788 mgt/100g
Zinc	2.6 mg/100g	–	3.3 mg/100g	2.7 mg/100g
Copper	0.15 mg/100g	–	0.07 mg/100g	0.07 mg/100g
Manganese	–	–	0.4 mg/100g	0.3 mg/100g
Selenium	20.3 mg/100g	–	–	21.8 µ/100g
Vitamin C (ascorbic acid)	16.6 mg/100g	–	–	0.7 mg/100g
Thiamin (B$_1$)	0.3 mg/100g	–	0.6 mg/100g	0.5 mg/100g
Riboflavin (B$_2$)	0.2 mg/100g	–	0.3 mg/100g	0.2 mg/100g
Niacin (B$_3$)	5.5 mg/100g	–	4.9 mg/100g	5.4 mg/100g
Pantothenic Acid (B$_5$)	–	–	1.1 mg/100g	0.6 mg/100g
Vitamin B$_6$	0.26 mg/100g	–	0.5 mg/100g	0.4 mg/100g
Folate	2 DFE/100g	–	–	6 DFE/100g
Vitamin B$_{12}$	–	–	1.9 mg/100g	1.6 mg/100g
Vitamin E (alpha-tocopherol)	5.5 mcg/100g	–	–	0.3 µ/100g
Cholesterol	75 g/100g	–	78 mg/100g	118 mg/100g

– = data not reported

Source: Calorie-Counter.net 2003, Nutrition Info 2005, United States Department of Agriculture 1963, National Livestock and Meat Board 1984, Ockerman and Basu 2014, http://kb.osu.edu/dspace/handle/1811/45275

6 CONCLUSION

Health and spoilage will be the center focus of research in the future and a better understanding of the fermentation mechanisms will help tremendously. Hazard Analysis Critical Control Point (HACCP) and hurdle technology control are becoming commonplace in the food industry and both are particularly important in fermented sausage. Easily controlled fermentation chambers are helpful and gaining in popularity to accurately control the environmental conditions during processing.

Genetic determinants and transfer mechanisms will be utilized to develop superior LAB strains. These genetically modified microorganisms have the potential to improve food safety and quality, but they must be evaluated on a case-by-case basis. Genetic manipulation to "tailor-made" cultures to be used as starters and fungal starter cultures are being evaluated for elimination of toxin production. Regulation of fermentation including proteolytic and lipolytic activity is critical. Bio-preservative properties such as production of natural antimicrobial compounds for elimination of pathogens are also being investigated.

Muscle enzyme activity of a carcass and its influence on meat quality has been demonstrated (Toldra and Flores 2000, Russo et al. 2002). Therefore, enzymes and genes can be used to select breeding animals and raw material for production of fermented sausages. For more value added fermented products, muscle protease, muscle and fat lipase (activities) and spice influence on fermented sausage flavor (Fournaud 1978, Claeys et al. 2000) will be evaluated. If these can be altered, a flavor modification is probably possible. New product developments with new ingredients will evolve. For example, making fermented product from ostrich (Bohme et al. 1996), carp (Arslan et al. 2001), and utilization of olive oil (Muguerza et al. 2001) have been suggested. Acceptance of fermented meat products in additional international food items seems to be gaining in popularity such as increased consumption of pizza containing fermented meat. With the past progress and the possible future developments, the future looks bright for fermented meat products.

Key words: *Meat, Chicken, Beef, Pork, Lamb, Sausage, Lactic acid bacteria, Reduced pH, Formulation, Fermented, Semi-dried, Dried.*

REFERENCES

American Meat Institute. 1982. Good manufacturing practices, fermented dry and semi-dry sausages. Washington DC: American Meat Institute AMI. Washington, DC

Anon. 1978. Some aspects of dry sausage manufacture. Fleischwirtschaft 58: 748–749.

Arslan, A., A.H. Dincoglu and Z. Gonulalan. 2001. Fermented Cyrinus carpio L. sausage. Turk. J. Vet. Anim. Sci. 25: 667–673.

Bohme, H.M., F.D. Mellett, L.M.T. Dicks and D.S. Basson. 1996. Production of salami from ostrich meat with strains of *Lactobacillus sakei*, *Lactobacillus curvatus* and Micrococcus. Meat Sci. 44: 173–180.

Campbell-Platt, G. and P.E. Cook. 1995. Fermented Meats. Blackie Academic and Professional. London, UK.

Claeys, E., S. De Smet, D. Demeyer, R. Geers and N. Buys. 2000. Effect of rate of pH decline on muscle enzyme activities in two pig lines. Meat Sci. 57: 257–263.

Demeyer, D., M. Raemaekers, A. Rizzo, A. Holck, A. De Smedt, B. ten Brink, B. Hagen, C. Montel, E. Zanardi, E. Murbrekk, F. Leroy, F. Vandendriessche, K. Lorentsen, K. Venema, L. Sunesen, L. Stahnke, L. De Vuyst, R Talon, R. Chizzolini and S. Eerola. 2000. Control of bioflavour and safety in fermented sausages: First results of a European project. Food Res. Int. 33: 171–180.

Doyle, M.P., L.R. Deuchatand and T.J. Montville. 1997. Food microbiology. Washington DC: American Society for Microbiology.

Farnworth, E.R. 2003. Handbook of fermented functional foods. CRC Press, Boca Raton, USA.

Fournaud, J. 1978. La microbiologie du saucisson sec. L'Alimentation et la vie. 64: 382–392.

Gilliland, S.E. 1985. Bacterial starter cultures for foods. CRC Press Inc., Boca Raton, USA.

Hui, Y.H., L.M. Goddik, A.S. Hansen,W.K. Nip, P.S. Stanfield and F. Toldrá. 2004. Handbook of food and beverage fermentation technology. Marcel Dekker, Inc. New York.

Klettner, P.G. and P.A. Baumgartner. 1980. The technology of raw dry sausage manufacture. Food Technol. Austral. 32: 380–384.

Komarik, S.L., D.K. Tresslerand and L. Long. 1974. Food products formulary: The AVI Publishing Company. Westport Connecticut.

Muguerza, E., O. Gomeno, D. Ansorena, J.G. Bloukas and I. Astiasaran. 2001. Effect of replacing pork back fat with pre-emulsified olive oil on lipid fraction and sensory quality of Chorizo de Pamplona. Meat Sci. 59: 251–258.

Ockerman, H.W. 1989. Sausage and processed meat formulations. New York: Van Nostrand Reinhold.

Ockerman, H.W. and L. Basu. 2014. Production and consumption of fermented meat products. pp. 7–12. *In:* F. Toldrá (ed.). Handbook of fermented meat and poultry. 2nd edition. Blackwell Publishing. Hoboken, NJ.

Russo, V., L. Fontanesi, R. Davoli, L.N. Costa, M. Cagnazzo, L. Buttazoni, R. Virgili and M. Yerle. 2002. Investigation of genes for meat quality in dry-cured ham production: the porcine cathepsin B (CTSB) and cystatin B (CSTB) genes. Anim. Genet. 33: 123–131.

Rust, R.E. 1976. Sausage and processed meat manufacturing. American Meat Institute. Washington, DC.

Schmidt, S. and R.G. Berger. 1998. Microbial formed aroma compounds during the maturation of dry fermented sausages. Adv. Food Sci. 20: 144–152.

Stahnke, L.H., L.O. Sunesen and A.D.E. Smedt. 1999. Sensory characteristics of European dried fermented sausages and the correlation to volatile profile. pp. 559–566. Thirteenth Forum for Applied Biotechnology. Med. Fac. Landbouw. Univ. Wageningen, Netherlands.

Toldrá, T. and M. Flores 2000. The use of muscle enzymes as predictors of pork meat quality. Food Chem. 69: 387–395.

Websites

http://www.meatsandsausages.com/sausage-types/fermented-sausage 2014

http://kb.osu.edu/dspace/handle/1811/45275

http://www.meatsandsausages.com/sausage-types/fermented-sausage

http://en.wikipedia.org/wiki/List_of_fermented_foods#Fermented_foods

Ockerman H.W. and L. Basu 2010. Fermented Meat Products: Production and Consumption. 2010-04-29, http://hdl.handle.net/1811/45275

Chapter 3

Technology of Fermented Meat Products

Helga Medić

1 INTRODUCTION

Drying and fermentation of meat is one of the oldest ways for meat preservation and these two processes cannot be separated in practice. Also, thousands of years ago, it was found that the shelf life of meats was rather extended if the meat was mixed with salt and aromatic herbs and then dried (Ordonez and de la Hoz 2007). Nowadays, the basic principles have not changed and a number of different fermented meat products are produced worldwide. The most famous products arising for these technologies are dry-fermented sausages and dry-cured hams. They have undergone a long ripening (aging) process, where complicated biochemical, proteolytic and lipolytic modifications take place, and which in turn are responsible for the distinctive flavors of these products. Dry-meat products provide a large added value to the producer because they are regarded as high quality products. The quality of dry-meat products is related to both the quality of raw materials (meat and fat tissue) and the control of complex biochemical reactions which take place during processing (Gandemer 2002).

Dry-fermented sausage is a comminuted meat product, consisting of meat, fat, additives, starter cultures (optional) and different spices. There is a variety of fermented meat products in the world because

---

For Correspondence: University of Zagreb, Faculty of Food Technology and Biotechnology, 10000 Zagreb, Pierottijeva 6, Croatia, Tel: + 385 1 4605 126, Fax: +385 1 4605 072, Email: hmedic@pbf.hr

28   *Fermented Meat Products: Health Aspects*

of regional preferences, different cultures, environmental variations, processing technology and other factors but for all of them processing requires dehydration and fermentation. Dry-cured ham is a traditional meat product with a strong presence in markets in the Mediterranean area. Despite the important differences in the processing and sensory characteristics of dry-cured hams between countries, all production technologies have two operations in common: salting and drying. The processing technology for the production of dry-fermented sausage and dry-cured ham in a relation to food safety will be discussed in this chapter.

2 PROCESSING TECHNOLOGY OF DRY-FERMENTED SAUSAGES

The production of dry-fermented sausages consists of meat and fat selection, chopping of them, addition of food additives, spices, curing agents (salt, nitrate, nitrite, others) and starter cultures, mixing and stuffing into casings. The sausages are fermented and dried with optional smoking phase and further subjected to ripening.

2.1 Selection of Raw Materials

The meat used for processing is usually pork, which can be mixed with beef. Other meats like lamb, cow, goat, horse and donkey can be used also. The composition of meat varies with anatomical region and age of an animal. Meat from adult animals and firm fat with high melting point and low content of polyunsaturated fatty acids (PUFA) is preferred. That meat has lower water content and higher fat content which supports drying and has higher myoglobin content important for a color of a product. In most products pork backfat is used as it can be minced to apparently defined particles, has low content of PUFA and remains stable without pronounced rancidity (Heinz and Hautzinger 2007).

It must be pointed out that only raw meat materials of excellent hygienic quality should be used for the processing, as these kind of products are never subjected to heat during the production. Contamination of meat should be avoided at any stage of production from slaughtering the animals, boning and cutting of meat to processing of sausages. The bacteria count of meat should be less than 10^4 per gram and the pH value of meat should be between 5.5 and 5.8. *Salmonella* spp., *Campylobacter* spp., *Staphylococcus aureus*, *Listeria monocytogenes* and verotoxigenic strains of *Escherichia coli* are the most important pathogens in relation to meat products and their absence in raw material is crucial

Technology of Fermented Meat Products

for the production of safe meat products. Some of the above mentioned pathogens have an ability to survive in meat products with low pH and water activity (a_w) and effective control strategies should be applied during the processing.

Measures to control pathogens during processing include: hygienic design and construction of food facilities and equipment, prerequisite programs, standard sanitation procedures, implementation of Hazard Analysis Critical Control Point (HACCP) which comprise personnel hygiene and education and others. Hygienic design principles for equipment and facilities should enable greater control over the risk of contamination of ready-to-eat (RTE) meat products during processing. They permit appropriate maintenance, cleaning and disinfection of facilities and equipment. The European Hygienic Engineering and Design Group published guideline documents that apply to the meat industry including design of equipment for open processing (Seward 2007, Koning 2008). It is crucial because in meat-processing plants, all surfaces and materials may present potential harborage niches for microorganisms without hygienic design of the equipment, if surfaces are difficult to clean and disinfect and sanitation procedures are inadequate. These bacteria can colonize and occupy different surfaces for finally adapt and tolerate sanitation procedures. In any cases and once developed, biofilms are a significant potential source of contamination of processed meat-products, which may lead to important risks for consumer health upon consumption (Giaouris et al. 2014). It has been well documented that biofilms are more resistant to environmental challenges than are their planktonic counterparts in suspension and can be attached to biotic or abiotic surfaces, even to stainless steel (Bae et al. 2012).

By the end of the 20th century, HACCP had been mandated, implemented and was in routine operation at every meat company involved in international trade (Jenson and Sumner 2012). HACCP system addresses hazards in the production that their elimination or reduction to acceptable levels is essential for producing safe food (Asefa et al. 2011). Under this system, potential hazards in the entire process are determined, critical points are identified, where such hazards can be controlled and preventive measures taken. This can assure basic food safety during the processing of meat products. The specific operations and HACCP system for the production of fermented meat products will be further discussed in detail in later chapters.

Parasites like *Trichinella spiralis*, *Cysticercus bovis* and *Toxoplasma gondii* may be present in meat. They do not grow in food and therefore, control is focused on destroying the parasites and/or preventing their introduction. Adequate cooking destroys parasites but in this type of products which does not undergo any heat treatment during processing, they are biological hazards (Hui 2012, Robertson et al. 2014). Apart from cooking, freezing, curing and drying may destroy parasites in meat but freezing temperature and time requirements have to be established for specific parasite, the effectiveness of curing depends on a closely monitored combination of salt concentration, temperature and time. In salting procedures, a level of a_w below 0.92 in meat products may be adequate to destroy *Trichinella* larvae (Table 1). Although irradiation methods for destruction of *Trichinella* larvae are approved in some jurisdictions, it may be some time before irradiated pork becomes widely acceptable (Gajadhar et al. 2009).

Chemical hazards associated with meat involve veterinary drugs and growth promoters' residues and environmental contaminants. Veterinary drugs have been used in animals like therapeutic, prophylactic and growth promoting agents. Their residues may be present in meat which can lead to adverse effects on human health. The presence of these substances in food and their monitoring are regulated in European Union by different directives. Environmental contaminants such as dioxins, organophosphorus and organochlorine pesticides, polychlorinated biphenyls, heavy metals and mycotoxins produces by molds (Asefa et al. 2011), among others, may be present in meat through contaminated feed eaten by animals. The reasons for feed contamination vary from inadequate control of ingredients and processing to growth of molds in feed grains and meals. These contaminants are distributed worldwide and are difficult to control.

2.2 Chopping of Meat and Fat

The degree of chopping depends on the type of sausage and can vary from 2–5 mm for moderately to 6–12 mm for coarsely chopped sausages. The meat should be chilled or slightly frozen and the fat should be frozen to achieve a clean cut and to prevent smearing.

Chopping can be done in meat grinder or in a bowl cutter. When only meat grinding equipment is used after grinding the chopped meat and fat are transferred to vacuum mixer to create a sausage batter with all additives (salt, nitrates, nitrites, sugars, starter cultures, spices etc.). Mixing should be sufficient to uniformly distribute ingredients. If a bowl cutter is used for the production the large pieces of lean meat is firstly chopped and if starter cultures are used, they are added. Then the frozen fat along with spices and sugars are joined and chopping is continued till the desired particle size of the fat is achieved. Afterwards the minced chilled meat is added and at the end the curing salt is supplemented to the sausage batter (Ordonez and de la Hoz 2007).

Commercial starter cultures usually consist of bacteria belonging to the genera *Lactobacillus*, *Pediococcus*, *Staphylococcus* and *Micrococcus*.

Technology of Fermented Meat Products

Table 1 The a_w limits for the microbial growth

Bacteria	Yeasts	Molds	Parasite	a_w
E. coli				0.99
Streptococcus fecalis			Cysticercus bovis (+ NaCl)	0.98
Vibrio metschnikovii				0.97
Pseudomonas fluorescens				0.97
Clostridium botulinum				0.97
Campylobacter spp.				0.97
Shigella			.	0.97
Yersinia enterocolitica				0.97
Clostridium perfringens				0.96
Bacillus cereus				0.96
Bacillus subtilis				0.95
Salmonella newport			Toxoplasma (pH ≤ 5.3) (pH5.3)	0.95
Enterobacter aerogenes				0.94
Microbacterium				0.94
Vibrio parahaemoliticus				0.94
Lactobacillus viridescens	Schizosaccharomyces	Rhisopus		0.93
		Mucor		0.93
	Rodotorula		Trichinella	0.92
Micrococcus roseus	Pichia			0.91
Anaerobic Staphylococcus	Saccharomyces			0.91
Lactobacillus				0.90
Pediococcus	Hansenula			0.90
	Candida	Aspergillus niger		0.88
Staphylococcus aureus toxin		Debaryomyces		0.88
	Torulopsis	Cladosporium		0.87
Staphylococcus aureus growth	Torulaspora	Paecilomyces		0.86
Listeria monocytogenes				0.83
		Penicillium		0.80
		Aspergillus achraceus		0.80
Halophillic bacteria				0.75
		Aspergillus glaucus		0.72
		Chrysosporium fastidum		0.70
Zygosaccharomyces rouxii		Monascus bisporus		0.60

Sources: FAO 2001, Rödel 2001, Anonymous 2006, Gajadhar et al. 2009, Zukál and Incze 2010

They produce lactic acid, develop ripening flavor, inhibit undesired microorganisms and they are generally safe in terms of consumers' health (Bassi et al. 2015, Bonomo et al. 2009). The use of *Lactobacillus* spp. result in rapid acidification with low pH values and the use of *Pediococcus* spp. lead to slower and milder acidification, while *Staphylococcus* spp. cause fast reduction of nitrite, stable color and reduced risk of rancidity. *Micrococcus* spp. has nitrate reducing activity and support color and aroma of the product. Different mixtures of bacterial strains are usually used to achieve specific results (Heinz and Hautzinger 2007). Lactic acid bacteria can also be an alternative to chemical additives and act as extra hurdle for microorganisms because they can exert antagonism through competition for nutrients and production of antimicrobial substances (Aymerich et al. 2008). Molds and yeasts are aerobic fungi and may be inoculated to the surface of the sausages either by spraying on the product or dipping the product into solution. Frequently chosen mold is from genus *Penicillium* (Tabanelli et al. 2012) and yeast from genus *Debaryomyces*. Added molds are white or grey-white in color while unwanted molds are often black, green or yellowish (Feiner 2006). Addition of molds has several advantages during the production process. With a mold layer on the sausage drying is more balanced, thus reducing frequency of drying failure. It slows down moisture loss in the final product and because of its light protection, discoloration and rancidity is retarded (Incze 2010). On the other hand, unwanted molds can have adverse effect to human health due to production of mycotoxins. Their growth on the surface of the sausage is connected with too high relative humidity (RH) in the drying chambers.

2.3   Addition of Salt, Curing Agents and Sugars

One of the primary reasons for adding salt to meat is providing preservation, mainly to preserve meat from microbial spoilage (Barat and Toldrá 2011). Salt exerts important functions in the sausages like lowering the a_w value of the sausage by absorbing water, which presents initial hurdle for bacteria. Also, salt gives a typical flavor to meat products. Salt-soluble proteins are extracted from chopped meat in presence of salt and these solubilized or gelatinous proteins bind meat and fat particles. The result is an increasingly firm structure with continuous drying and ripening of the products. The amount of salt added to sausages should be > 25 g/kg and the salt content in the final product will be higher due to the reduction in water content (Heinz and Hautzinger 2007).

Food safety and mass manufacturing are dependent upon safe and effective means to cure and preserve meats. Nitrite remains the most effective curing agent for prevention of food spoilage and bacterial

Technology of Fermented Meat Products   33

contamination. Nitrites are used as powerful antimicrobials (inhibition of *Clostridium botulinum*), they contribute to color through the formation of nitrosylmyoglobin, improve oxidative stability of lipids and contribute to typical flavor of the meat products (Skibsted 2011, Parthasarathy and Bryan 2012). Nitrate has no preservative effect but is a reservoir of nitrite. They should be reduced to nitrites by bacteria. Nitrites can be reduced to nitric oxide, which is a nitrosating agent and may react with secondary amines and produce potent carcinogenic nitrosamines. The amount of N-nitrosamines in fermented sausages and dry-cured hams depends on a variety of factors like the amount of added nitrites, processing conditions and presence of antioxidants (Honikel 2008). The antioxidants usually used in the production of fermented sausages are ascorbic acid, ascorbate and erythorbate. Hygienic quality of raw materials used for production of dry-fermented sausages plays a great role in formation of biogenic amines. High amounts of proteins in meat and the proteolytic activity during ripening provide free amino acids as the precursors for decarboxylase activity of microflora. Therefore, the meat used should be of high hygienic quality and all technological steps should be properly controlled (Kalač 2006).

Added spices influence aroma of the meat products and majority of dry-fermented sausages contain pepper and garlic. Commonly used are also coriander, paprika, nutmeg, mace and chilli. The contamination of spices and herbs with microorganisms is often and usually decontamination process is applied before using them in meat products. On the other hand, some spices have strong antimicrobial activity against foodborne pathogens (Ceylan and Fung 2004, Shan et al. 2007).

Certain additives may help in acidification like glucono-δ-lactone (GDL), citric acid and sugars.

The bacterial breakdown of added sugars results in the accumulation of lactic acid, acidification and development of typical flavor. Often a mixture of different sugars is used because dextrose and fructose support an early drop in pH values and the breakdown of lactose and maltose is slower.

Vacuum stuffing is the next step in processing. The casings can be natural or artificial but must be permeable for water-vapor and smoke. This operation is carried out in vacuum in order to exclude oxygen from the mixture, to prevent discoloration of meat, development of undesirable flavors and reduction of the shelf life of the sausage.

2.4   Fermentation

During fermentation, drying and ripening, the temperature, air velocity and relative humidity are adjusted and gradually decreased. Directly

34   *Fermented Meat Products: Health Aspects*

after stuffing, the temperature of the sausage is below zero. The inclusion of a tempering period of three hours at moderate room temperature before the sausages are transferred to the drying/ripening chamber is advisable, to allow moisture release from the sausages and to initiate the fermentation. A high relative humidity at the beginning of the drying, which keeps sausage casings wet and soft, and the gradual lowering of the air humidity in the advanced stages of the process are the key factors to enable the migration of moisture from the interior to the outer layer of the sausage. In the first phase of drying, the red cured meat color is built up and the curing color progresses from the center to the outer region of sausage.

The temperatures are initially kept at 22–24°C and are slowly reduced to 18°C. The relative humidity decreases gradually from typical values of 92–94% on the first day to 82–84% and air movement of 0.8–0.5 m/s before the sausages are transferred to the ripening/storage room. If the humidity is kept too high, excessive surface moisture is retained usually resulting in increased bacterial growth on the surface, thus forming a slimy layer. If humidity is reduced too fast especially in the early stages of the process, a hard and dry crust is formed at the outer layer of the sausage. This crust is unable to adjust to the reducing diameter caused by continuous loss of moisture and as a result cracks will appear in the center of the product (Heinz and Hautzinger 2007).

2.5   Smoking

Smoke is a complex mix of high and low molecular weight components. Due to its formaldehyde content and phenolic components smoking has antibacterial and fungicide effects (Flores 1997). Sausages are smoked from several hours to several weeks depending on their diameter and a type of product. Cold smoke (temperature below 25°C) is applied to impart flavor, aroma, and color to meat products, as well as to preserve them. The traditional method of direct smoking is still used today in the production but in the past few decades, commercial, natural wood smoke flavorings, or liquid smokes have become more popular throughout the world as a convenient and consistent way to add smoke flavor and color to meat products without using a smokehouse. Wood smoke is generated by the controlled combustion of wood. During pyrolysis, hemicellulose is most readily degraded followed by cellulose and finally lignin. Hemicellulose degradation yields primarily lactones and furans, while pyrolysis of cellulose produces aliphatic acids and aldehydes. The thermal degradation of lignin gives rise to the most important class of smoke flavor compounds, the phenols. These include guaiacol, phenol, 4-methylguaiacol, and syringol. The particular composition of the various types of woods is an important factor affecting the volatile composition

Technology of Fermented Meat Products

of the smoke (Cadwallader 2007). Smoke may contain some health-hazardous compounds like polycyclic aromatic hydrocarbons but their content is highly variable and depends on many factors such as smoking method, the type of smoke generator, the type of wood (Hitzel et al. 2013) and the temperature of the process. They are generated during pyrolysis of wood at the temperatures above 500°C.

2.6 Drying

During drying, water is removed as vapor from the sausages. The activity of microorganisms decreases because the portion of water they can utilize decreases. As the result, the shelf life of the product increases, the mass and volume of product decreases, the texture of product will be harder and aroma compounds develop during longer processes. The preservation effect of drying is achieved by lowering water activity in the sausages. The pressure of water vapor will be constant in the closed space around the material. This pressure is lower or equal to vapor pressure above pure water at the same temperature. The numerical expression of this is the water activity (a_w), and it ranges from 0.00 – 1.00. Microorganisms have optimum and minimum levels of a_w for growth and Gram-negative bacteria are generally more sensitive to low a_w than Gram-positive bacteria. Microorganisms cannot grow below their specific minimal values (Table 1) (Zukál and Incze 2010). Due to a_w, which ranges from 0.85 to 0.91, dry-fermented sausages exhibit shelf stability and can be kept without refrigeration. The typical lower a_w values of these products is achieved by air-drying and by smoking (if applied) (Vignolo et al. 2010).

During ripening the temperature is maintained at <16°C at a relative humidity of 75–78% to ensure controlled bacterial fermentation and gradual dehydration resulting in lowering of pH value and moisture content in finished dry-fermented sausages. The duration of the drying/ripening process mainly depends on the diameter of sausages and type of sugars and starter cultures used (Heinz and Hautzinger 2007). Controlling temperature and drying rate plays an important role in food safety by inhibiting the growth of undesired microbes, but this control is needed during ripening also for a consistent sensory quality. As a result of a_w -reduction, less resistant microorganisms gradually disappear, yet less sensitive ones may survive for a longer period of time. In the final product of long-ripened dried sausages, spoilage microflora has practically no chance to grow and cause deterioration (Incze 2012). The composition of dry-fermented sausage is constantly changing during the ripening process due to complex biochemical reactions and a wide range of compounds are generated. Amino acids, amines and long chain fatty acids contribute to taste and short chain fatty acids contribute to taste and/or odor of the products. Other compounds become aromatic substances (aldehydes, ketones, secondary alcohols). All these compounds are responsible for the aroma of the final product. They can be of diverse origin: some are result of microbial metabolism, some arise from endogenous lipolysis and proteolysis, and others from nitric oxide reactions, lipid autooxidation and finally, some originate from spices (Marušić et al. 2014).

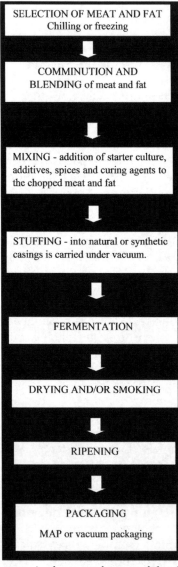

Figure 1 Flow diagram in the manufacture of dry-fermented sausages.

Technology of Fermented Meat Products

3 PROCESSING TECHNOLOGY OF DRY–CURED HAM

Dry-cured ham is generally considered to be a typical Mediterranean product, although it is also manufactured outside Southern Europe. The Mediterranean types of dry-cured ham are characterized by dry salting and a long ripening period (up to 24 months). On the contrary, the Northern Europe types of dry-cured ham are characterized by smoking followed by a short ripening period. Despite the important differences in the processing and sensory characteristics of dry-cured hams between countries, all production technologies have two operations in common: salting and drying. The main differences between dry-cured hams are due to specific raw material (different breeds, weight, or trimming), specific production steps (e.g., different salting systems, smoking, etc.), and specific air conditions and duration of the drying and ripening processes. The use of chilling chambers or artificial dryers, have improved the process standardization and, consequently, the control of the safety and quality of the final product (Gou et al. 2012). Dry-cured hams are generally very stable and robust as well as being safe and great tasting products which can be produced without major difficulty if a few important rules are followed during the production (Feiner 2006).

3.1 Raw Material Selection and Handling

Dry-cured hams are produced from the hind legs of pork carcasses. The characteristics of the raw ham depend on genetics, pig production (feeding, age, season, etc.), preslaughter treatments, slaughter conditions and ham handling. Raw hams weight usually between 9–16 kg. Only meat of a good microbiological quality should be used (10^2–10^4 cfu per gram of product) with a pH value between 5.5 and 5.8 and meat should be well chilled (0–2°C) prior to salting. If using frozen meat, a growth of bacteria should be avoided during thawing (proper temperature and RH) (Feiner 2006). Fresh ham should not have visible signs of any defects like cuts into meat during boning, cuts and hanging parts of muscle and rind and also dark, firm and dry (DFD) meat should not be processed. High pH value of DFD meat can prompt the development of undesirable microorganisms. Visible connective tissue should be removed during trimming for faster penetration of salt into the meat. Raw hams can be presented in different ways: with or without hoof, rind can be partially or completely removed. The aitch bone can be eliminated, which allows faster salt penetration and water losses. Hazards connected with raw materials are described previously in the section of dry-fermented sausages production.

3.2 Salting

From the food safety aspect, salting phase is most critical in the production of a dry-cured ham, so the low temperature should be maintained over the entire process. Before salting, residual blood from the whole ham, particularly of the femoral artery should be rubbed and squeezed out. Hams are firstly salted at low temperature (up to 4°C) and are subsequently salted after 10–15 d and this continues for another 20–50 days. Salt should be applied at around 35 g/kg of product through dry salting. The maximum permitted level for the addition of nitrite should be obeyed and a temperature below 5°C and a RH of around 85% should be maintained. During the initial steps of salting, hurdles such as a high level of salt, the presence of nitrite (optional), processing meat with a low bacteria count and storing salted meat at temperatures below 5°C avoids microbiological spoilage.

Equalization of curing additives is carried out at a temperature below 5°C and an RH of around 80% until a_w value of 0.95 is obtained in all parts of meat, which is also important for product safety.

3.3 Fermentation and Smoking

Following equalization, meat is to be held at 22–24°C for around 24–48 hours to start fermentation and enzyme activity. Application of cold smoke (with some of the hams) should be performed at around 20–25°C and an RH of 80%. Products should be further dried at 16–18°C and around 75–80% RH for a few days. Among different components of smoke, phenols and carboxylic acids are the primary active antimicrobial and antifungal compounds.

3.4 Drying and Ripening

Final drying and storage of product is carried out at 12–15°C and 72–75% RH; the loss in drying is between 25 and 40%. During drying, important processes such as weight loss occur, which contributes to firmness of the product and enhances sliceability. A more intense salting usually leads to higher dehydration and, therefore, harder texture of the final product. It also influences proteolytic (Laureati et al. 2014) and inhibits lipolytic activity. The later can be influenced also by using frozen/thawed hams in the production of dry-cured hams (Flores et al. 2009). Flavor development is another major occurrence during drying and ripening. The degradation of lipids and proteins occur leading to generation of free fatty acids and low-weight nitrogen containing molecules. Free fatty acids are subsequently degraded through oxidative reactions and the denaturation and degradation of proteins by proteases causes loss of protein functionality and solubility (Feiner 2006, Estévez et al. 2007).

The aroma is perhaps the most important quality parameter and it is markedly affected by the raw material and the production process. In the case of dry-cured hams, the aroma is due to the presence of many volatile compounds, most of them produced by chemical and enzymatic mechanisms during the post-mortem process (Flores et al. 1997). The main biochemical reactions involved in the generation of these compounds are lipolysis and proteolysis that produce a wide range of volatiles and precursors (Toldrá 1998). Muscle and adipose tissue lipids are subject to intense lipolysis generating free fatty acids by the action of lipases (lysosomal acid lipase, acid phospholipase and adipose tissue lipase) that are transformed to volatiles as a result of oxidation. The most of the volatile compounds of the dry-cured ham is formed by oxidation of fatty acids through a process of auto-oxidation and beta-oxidation (Gandemer 2002, Harkouss et al. 2015). Muscle proteins undergo an intense proteolysis resulting in a great number of small peptides and high amounts of free amino acids. The enzymes responsible of these changes are proteinases (cathepsins, calpains, peptidases and aminopeptidases). The combination of small peptides and free amino acids contribute to the characteristic flavor of dry-cured ham (Aristoy and Toldrá 1995). Non-volatile components, amino acids and peptides are the active ingredients of flavors that have a major impact on the final flavor of the ham. Approximately 240 volatile compounds have been identified in dry-cured hams (Flores et al. 1998, Marušić et al. 2011). These volatile compounds are: aldehydes, alcohols, hydrocarbons, pyrazines, ketones, esters, lactones, furans, sulfur and chloride compounds and carboxylic acids (Buscailhon et al. 1994). Some volatile compounds such as pyrazines, sulfur compounds and branched aldehydes have significant effects on flavor. They can be formed by amino acids degradation, or their formation may depend on the processing conditions (Flores et al. 1997). The final flavor of dry-cured ham depends on the specific aroma for each particular volatile compound (Estévez et al. 2007).

3.5 Extended Ripening

In some cases, hams are submitted to ripening cellars at mild temperatures for up to 24 months at temperatures 10–20°C and RH of 65 to 82%. This long ripening time allows the development of exquisite and intense flavors generated through further chemical and enzymatic reactions. Large amounts of free amino acids are generated because of very intensive proteolysis phenomena and in most cases, tyrosine crystals, which are quite insoluble, may be observed as white spots on the cut surface (Toldrá and Aristoy 2010). These crystals do not pose any negative effect to human health but consumers complain about the presence of white crystals on dry-cured ham. The origin of tyrosine is enzymatic and a muscle protease activity may determine the concentration of free tyrosine. The crystallization process may also be influenced by structural changes brought about during freezing and thawing because the number of tyrosine crystals was higher in hams originated of frozen than of refrigerated raw hams (Arnau et al. 1994, Banón et al. 1999).

At or below a_w of 0.89, the two main risks from a microbiological point of view, *Salmonella* spp. as well as *S. aureus*, are well controlled and neither of these can grow or produce toxin any longer.

Shelf stable final products have water activity lower than 0.89 while some dry-cured products have a_w value around 0.91–0.92 and are therefore not shelf stable without being stored at temperatures below 4°C. The reason for not obtaining a_w of 0.89 or below is that there is a significant difference in weight loss required to produce a product with a_w of 0.89 compared with that with an a_w of 0.92. Products exhibiting a_w of 0.91–0.92 contain considerably more moisture than fully dried products, displaying a_w of 0.89 or below. Main difference is that characteristic flavor and texture of a cured dried meat product are missing (Feiner 2006).

3.6 Packaging

Whole products are shelf stable but deboned, cut and sliced products are generally modified atmosphere packed (MAP) as well as packed under vacuum and stored at temperatures below 4°C to ensure product safety. MAP is commonly used technology to inhibit the growth of microorganisms. Gases inhibit microorganisms by their direct toxic effect that can inhibit growth and proliferation (carbon dioxide) and this mechanism is dependent upon the physico-chemical properties of the gas and its interaction with the aqueous and lipid phases of the food. A second inhibitory mechanism is achieved by modifying composition of the gasses, which has indirect inhibitory effect through altering the competitive environment. Controlled atmosphere and modified atmosphere packaging of certain foods can significantly extend their shelf life. The inhibitory effects of CO_2 increase with decreasing temperature due to its increased solubility at lower temperatures. Nitrogen, being an inert gas, has no direct antimicrobial properties. It is typically used to displace oxygen in the food package either alone or in combination with CO_2, thus having an indirect inhibitory effect on aerobic microorganisms. Proposed gas mixture in MAP of cured meats is 20–50% of CO_2, 0% of O_2 and 50–80% of N_2. Another important parameter is barrier properties of the package, which have a major effect on the microbial growth by influencing the time in which the selected modified atmosphere gases remain in contact with the product and the rate at which oxygen enters the package (FDA 2001).

4 SELECTED SAFETY AND QUALITY ISSUES IN FERMENTED MEAT PROCESSING

4.1 Oxidation of Lipids and Proteins

Lipid oxidation is one of the main causes of deterioration in the quality of meat during storage and processing (Gandemer 2002). The main process of lipid oxidation in meat and meat products is a chemical process named autooxidation. Ground meat has greater autooxidation potential than whole cuts because the grinding process incorporates oxygen, mixes reactive components, and increases surface area as a result of particle size reduction. Also, some lipid compounds like phospholipids are very susceptible to autooxidation due to its high content in PUFA. Flavorless hydroperoxides are the primary products of oxidation, while the secondary products of oxidation can contribute to off flavors, color deterioration, and potential generation of toxic compounds (Faustman et al. 2010). Lipid and protein oxidation may also be induced by hydrogen peroxide generated by certain bacteria during meat fermentation. Some products of lipid oxidation may be chronic toxicants but the low levels (usually present in sausages) are far from any acute toxicity. Cholesterol oxidation may occur resulting in formation of cholesterol oxides which are considered as prejudicial for health but no cholesterol oxides have been detected after heating of pork sausages. Studies made on European sausages (based on in vivo tests with laboratory animals) revealed that the content of cholesterol oxides were quite below toxic. Sulfur amino acids of proteins are those more susceptible to oxidation by peroxide reagents, thus cystine is oxidized to sulfinic and cysteic acids, whereas methionine is oxidized to methionine sulfoxide and methionine sulfone (Gallego et al. 2015). Homolanthionine sulfoxide is the main product of homocysteine oxidation. Peptides like the reduced glutathione can also be oxidized by hydrogen peroxide and most of the cysteine in the glutathione is oxidized to the monoxide or dioxide forms (Toldrá and Reig 2007). The formation of carbonyl compounds from specific amino acids (like lysine, arginine and proline) is one of the most remarkable measurable changes in oxidized proteins and has been commonly used as indicator of protein oxidation in food (Utrera et al. 2011).

4.2 Lipid Content: Dry-Cured Ham

Lipid content and fatty acid profile of different dry-cured hams are shown in Table 2. Fat content is one of the most important quality traits of cured hams (the acceptability of cured hams increases with the fat content). Fat stimulates the saliva secretion and contributes directly to juiciness

Table 2 Lipid content and fatty acid profile (percentage of total fatty acids) of the different dry-cured hams

Ham (feeding system)	Fat[*]	SFA	MUFA	PUFA	PUFA/SFA	n-6/ n-3
1. Iberian (acorn)	11.28	32.51	59.37	8.12	0.25	12.1
2. Iberian (acorn)	9.51 (19.24)	33.97	54.60	11.43	0.33	18.1
3. Iberian (commercial feed)	5.47	35.15	51.39	13.44	0.38	31.2
4. Iberian (acorn)	(17.23)	34.9	57.2	6.76	0.19	9.4
5. Serrano (commercial feed)	4.8 (18.8)	32.7	52.7	10.2	0.31	16.2
6. Serrano (commercial feed)	3.5	33.4	55.6	11.0	0.33	12.7
7. Bayonne (commercial feed)	2.6	36.4	52.9	10.7	0.29	14.1
8. Bayonne (commercial feed)	3.5	36.52	47.49	15.3	0.42	29.6
9. Corsican (chestnut)	12.3	35.0	53.8	11.2	0.32	–
10. Corsican (commercial feed)	5.3	34.9	55.4	9.7	0.28	8.7
11. Parma (commercial feed)	3.57	35.99	54.04	8.59	0.23	39.9
12. Parma (commercial feed)	(18.4)	37.9	52.0	9.9	0.26	–
13. San Danielle (commercial feed)	(23.0)	38.5	51.9	9.6	0.25	–
14. Cinta Senese (acorn)	–	33.26	51.35	15.38	0.46	14.2
15. Jinhua (commercial feed)	–	37.10	46.63	14.24	0.38	–
16. Iberian (feed with MUFA)	7.08	34.79	54.21	11.00	0.38	28.20
17. Serrano (feed with corn oil)	–	32.60	47.77	19.58	0.60	8.2
18. Serrano (feed with n-3 and MUFA)	–	33.06	47.40	19.98	0.60	1.97
19. Parma (feed with corn oil)	–	31.82	50.20	17.83	0.56	20.5
20. Parma (feed with rapeseed oil)	–	30.42	54.62	15.26	0.46	12.3
21. Parma (commercial feed)	–	35.99	54.04	9.36	0.26	26.94
22. Parma (supplem. with CLA)	–	38.99	53.08	7.33	0.18	29.39
23. Istrian ham (commercial feed)	13.84	38.94	53.54	7.51	0.19	16.58
24. Dalmatian ham (commercial feed)	13.85	41.42	50.60	7.93	0.19	14.72
25. Teruel (based on cereals)	24.2	41.5	48.7	8.71	0.21	10.12
26. Dehesa (acorn)	20.7	34.9	57.2	6.76	0.19	9.36

Sources: Fernández et al. 2007, Jiménez-Colmenero et al. 2010, Marušić et al. 2013.

*Expressed as g IMF/100 g of the muscle Biceps femoris: or g fat/100 g of the lean fraction of the slice with remaining intermuscular and 5 mm of subcutaneous fat (in bracket).; SFA- saturated fatty acids; MUFA- monounsaturated fatty acids; PUFA- polyunsaturated fatty acids.

Technology of Fermented Meat Products 43

by coating parts of the mouth acting as a lubricant agent. Intramuscular fat (IMF) also plays an important role in the perception of the texture of dry-cured hams, particularly in juiciness because the contribution of moisture is limited since these products are strongly dehydrated (Lorido et al. 2015).

The high amount of intramuscular fat is one of the most relevant quality aspects in muscles. IMF reduces the shear force during chewing, making easier muscle fiber separation, and hence, improving the juiciness and tenderness sensation (Fuentes et al. 2014). The highest amount of IMF is determined in Spanish Tereul dry-cured ham. Dry-cured ham presents different fat levels due to different pig breeds (genetics) and feeding of animals (Jiménez-Colmenero et al. 2010). The content of PUFA is best in Serrano ham which is produced from pigs fed with diets rich in n-3 PUFA and MUFA. Fat from meat and meat products, such as dry-cured ham, is not considered among the "healthy" fats, since it contains cholesterol, large amounts of saturated fatty acids and very low levels of n-3 PUFA. Saturated fatty acids are correlated with increased risk of cardiovascular disease while a high intake of monounsaturated and n-3 polyunsaturated fatty acids has been shown to have an inverse effect. The high MUFA percentage in hams indicates their suitability for healthier diets, since MUFA (and PUFA) rich diets decrease cholesterol levels in blood and are related to a low incidence of cardiovascular diseases (Fernández et al. 2007).

4.3 Microbiological-Toxicological Hazards: Dry-Cured Ham

Dry-cured hams usually present low bacterial counts due to limiting factors, such as its high salt content, the use of nitrate (reservoir or nitrite) and the progressive reduction in water activity. Nitrite is a powerful inhibitor of the growth of *Clostridium botulinum*, yet some types of dry-cured hams do not contain additives but only salt. Negative aspects of the addition of nitrites as well as of PAH were discussed in previous section. Also, safety aspects of processing and measures to control pathogens during processing were analyzed previously under the production of dry-fermented sausages.

The natural flora of ham is composed of certain lactic acid bacteria, such as *Lb. sakei*, *Lb. curvatus* and *Pd. pentosaceus*, but the counts are below 10^4/g. Other bacteria, such as *St. xylosus*, have nitrate reductase activity, which is an important enzyme for the reduction of nitrate to nitrite. Amines levels are usually low or even negligible in dry-cured hams, but these levels could rise in case of spoilage with microorganisms with decarboxylase activity (Toldrá and Aristoy 2010). The environmental conditions throughout ripening phase favor growth of a fungal population on the surface of hams, which could include a number of potentially toxigenic molds like *Aspergillus ochraceus, Penicillium nordicum* and *Penicillium verrucosum*. The occurrence of toxigenic molds usually prevails at the drying and ripening stages of production and therefore it is an important critical point at which potential mycotoxigenic food safety hazards can emerge (Asefa et al. 2011). According to the results of one recent research, the toxins produced by molds were isolated from the surface of the hams and were detected even in deep sections of ham (Rodríguez et al. 2012). To prevent the growth of undesirable molds, it is suggested to use the non-toxigenic strains, as starters. In this case, the health hazard due to the mycotoxins presence can be minimized (Comi et al. 2004). A drying process that lowers the water activity of the products faster to a level near to 0.9 minimizes the risk of the production of toxic secondary metabolites and mycotoxins production by toxigenic molds was not observed at a water activity of 0.9 (Asefa et al. 2011).

5   CONCLUSION

Dry-fermented sausages and dry-cured hams are considered ready-to-eat products. To produce safe dry meat products the equipment used during each processing step should be hygienically designed must be clean and HACCP must be implemented at all stages of production of meat and meat product as well as distribution. Also quantitative microbial risk assessment should be carried out on processing lines and training of personnel in food industry coupled with education of consumers on hygienic handling of food. Raw materials used in production must be properly chilled with optimal pH of meat below 5.8 and should exhibit low bacteria counts. Low temperatures should be maintained during salting of raw hams and preparing sausage batters. During the processing control of actual concentration of additives, control of temperature and RH during drying and ripening are essential for product safety. Final products are microbiologically safe when have water activity below 0.89 and pH below 5.2 (for sausages).

Some recently employed technics and technologies that tend to improve safety and quality of meat products are carcass decontamination, high pressure processing of boned dry-cured hams which has showed some promising results in the microbial stability of vacuum packed products, the use of new generation of starter cultures which produce antimicrobial compounds contributes to safety of dry-fermented sausages, improvements in composition of product in terms of reduction of salt, nitrites, fat, saturated fatty acids and addition of functional ingredients and improved packaging systems for meat and meat products.

Technology of Fermented Meat Products

Recent researches on functional meat compounds like bioactive peptides, taurine, coenzyme Q10, carnosine and anserine revealed some beneficial impacts of these meat compounds on human health and well-being. At the end, we may conclude that the constant development of meat-processing technologies will continue bringing increased quality and safety of meat products.

Key words: dry-fermented sausage, dry-cured ham, production technology, lipid content, health aspects of production.

REFERENCES

Anonymous. 2006. Technologie spielt Schlüsselrolle. Fleischwirtschaft 1: 43–44.

Aristoy, M.C. and F. Toldrá. 1995. Isolation of flavor peptides from raw pork meat and dry-cured ham. Dev. Food Sci. 37: 1323–1344.

Arnau, J., P. Gou and L. Guerrero. 1994. The effects of freezing, meat pH and storage temperature on the formation of white film and tyrosine crystals in dry-cured hams. J. Sci. Food Agric. 66: 219–282.

Asefa, D.T., C.F. Kure, R.O. Gjerde, S. Langsrud, M.K. Omer, T. Nesbakken, and I. Skaar. 2011. A HACCP plan for mycotoxigenic hazards associated with dry-cured meat production processes. Food Control 22: 831–837.

Aymerich, T., P.A. Picouet and J.M. Monfort. 2008. Decontamination technologies for meat products. Meat Sci. 78: 114–129.

Bae, Y.M., S.Y. Baek and S.Y. Lee. 2012. Resistance of pathogenic bacteria on the surface of stainless steel depending on attachment form and efficacy of chemical sanitizers. Int. J. Food Microbiol. 153: 465–473.

Banón, S., J.M. Cayuela, M.V. Granados and M.D. Garrido. 1999. Pre-cure freezing affects proteolysis in dry-cured hams. Meat Sci. 51: 11–16.

Barat, J.M. and F. Toldrá. 2011. Reducing salt in processed meat products. pp. 331–345. *In*: J.P. Kerry and J.F. Kerry (eds.) Processed Meats. Improving safety, nutrition and quality. Woodhead Publishing Limited, Cambridge.

Bassi, D., E. Puglisi and P.S. Cocconcelli. 2015. Comparing natural and selected starter cultures in meat and cheese fermentations. Curr. Opin. Food Sci. 2: 118–122.

Bonomo, M.G., A. Ricciardi, T. Zotta, M.A. Sico and G. Salzano. 2009. Technological and safety characterization of coagulase-negative staphylococci from traditionally fermented sausages of Basilicata region (Southern Italy). Meat Sci. 83: 15–23.

Buscailhon, S., J.L. Berdagué, J. Bousset, M. Cornet, G. Gandemer, C. Touraille and G. Monin. 1994. Relations between compositional traits and sensory qualities of French dry-cured ham. Meat Sci. 37: 229–243.

Cadwallader, K.R. 2007. Wood smoke flavor. pp. 201–210. *In*: L.M.L. Nollet (ed.). Handbook of meat, poultry and seafood quality. Blackwell Publishing Professional, Ames, Iowa.

Ceylan, E. and D.Y.C. Fung. 2004. Antimicrobial activity of spices. J. Rapid Meth. Aut. Mic. 12: 1–55.

Comi, G., S. Orlic, S. Redzepovic, R. Urso and L. Iacumin. 2004. Moulds isolated from Istrian dried ham at the pre-ripening and ripening level. Int. J. Food Microbiol. 96: 29–34.

Estévez, M., D. Marcuende, J. Ventanas and S. Ventanas. 2007. Mediterranean products. pp. 393–406. In: F. Toldrá (ed.). Handbook of fermented meat and poultry. Blackwell Publishing Professional, Ames, Iowa.

Faustman, C., Q. Sun, R. Mancini and S.P. Suman. 2010. Myoglobin and lipid oxidation interactions: Mechanistic bases and control. Meat Sci. 86: 86–94.

FDA (US Food and Drug Administration). 2001. Factors that influence microbial growth. Evaluation & definition of potentially hazardous foods. A report of the Institute of Food Technologists for the Food and Drug Administration of the U.S. Department of Health and Human Services.

Feiner, G. 2006. Meat products handbook. Practical science and technology. CRC Press, Boca Raton, Florida.

Fernández, M., J.A Ordoñez, I. Cambero, C. Santos, C. Pin and C. de la Hoz. 2007. Fatty acid compositions of selected varieties of Spanish dry ham related to their nutritional implications. Food Chem. 101: 107–112.

Flores, J. 1997. Mediterranean vs northern European meat products. Processing technologies and main differences. Food Chem. 59: 505–510.

Flores, M., M.C. Aristoy, T. Antequera, J.M. Barat and F. Toldrá. 2009. Effect of prefreezing hams on endogenous enzyme activity during the processing of Iberian dry-cured hams. Meat Sci. 82: 241–246.

Flores, M., C.C. Grimm, F. Toldrá and A.M. Spanier. 1997. Correlations of sensory and volatile compounds of Spanish Serrano dry-cured hams as a function of two processing times. J. Agric. Food Chem. 45: 2178–2186.

Flores, M., A.M. Spanier and F. Toldrá. 1998. Flavour analysis of dry-cured ham. pp. 320–341. In: F. Shahidi, (ed.). Flavor of meat, meat products & seafood, 2nd ed. Blackie Academic & Professional, London.

Fuentes, V., M. Estévez, N. Grèbol, J. Ventanas and S. Ventanas. 2014. Application of time–intensity method to assess the sensory properties of Iberian dry-cured ham: effect of fat content and high-pressure treatment. Eur. Food Res. Technol. 238: 397–408.

Gandemer, G. 2002. Lipids in muscles and adipose tissues, changes during processing and sensory properties of meat products. Meat Sci. 62: 309–321.

Gajadhar, A.A., E. Pozio, H.R. Gamble, K. Nöckler, C. Maddox-Hyttel, L.B. Forbes, I. Vallée, P. Rossi, A. Marinculić and P. Boireaur. 2009. Trichinella diagnostics and control: Mandatory and best practices for ensuring food safety. Vet. Parasitol. 159: 197–205.

Gallego, M., L. Mora, M.C. Aristoy and F. Toldrá. 2015. Evidence of peptide oxidation from major myofibrillar proteins in dry-cured ham. Food Chem. 187: 230–235.

Giaouris, E., E. Heir, M. Hébraud, N. Chorianopoulos, S. Langsrud, T. Møretrø, O. Habimana, M. Desvaux, S. Renier and G.J. Nychas. 2014. Attachment and biofilm formation by foodborne bacteria in meat processing environments: Causes, implications, role of bacterial interactions and control by alternative novel methods. Meat Sci. 97: 98–309.

Gou, P., J. Arnau, N. Garcia-Gil, E. Fulladosa and X. Serra. 2012. Dry-cured ham. pp. 673–688. In: Y.H. Hui (ed.). Handbook of meat and meat processing. CRC Press, Boca Raton.

Technology of Fermented Meat Products

Harkouss, R., T. Astruc, A. Lebert, P. Gatellier, O. Loison, H.S.S. Portanguen, E. Parafita and P.S. Mirade. 2015. Quantitative study of the relationships among proteolysis, lipid oxidation, structure and texture throughout the dry-cured ham process. Food Chem. 166: 522–530.

Heinz, G. and P. Hautzinger. 2007. Meat processing technology. For small-to-medium scale producers. FAO. Bangkok.

Hitzel, A., M. Pöhlmann, F. Schwägele, K. Speer and W. Jira. 2013. Polycyclic aromatic hydrocarbons (PAH) and phenolic substances in meat products smoked with different types of wood and smoking spices. Food Chem. 139: 955–962.

Honikel, K.O. 2008. The use and control of nitrate and nitrite for the processing of meat products. Meat Sci. 78: 68–76.

Hui, Y.H. 2012. Hazard Analysis and Critical Control Point. pp. 741–768. In: Y.H. Hui (ed.). Handbook of meat and meat processing. CRC Press, Boca Raton.

Incze, K. 2010. Mold-Ripened Sausages. pp. 363–378. In: F. Toldrá (ed.). Handbook of meat processing. Blackwell Publishing Professional, Ames, Iowa.

Incze, K. 2012. Mold-ripened sausages. pp. 647–662. In: Y.H. Hui (ed.). Handbook of meat and meat processing. CRC Press, Boca Raton.

Jenson, I. and J. Sumner. 2012. Performance standards and meat safety — Developments and direction. Meat Sci. 92: 260–266.

Jiménez-Colmenero, J. Ventanas and F. Toldrá. 2010. Nutritional composition of dry-cured ham and its role in a healthy diet. Meat Sci. 84: 585–593.

Kalač, P. 2006. Biologically active polyamines in beef, pork and meat products: A review. Meat Sci. 73: 1–11.

Koning, C.J. 2008. Improving the hygienic design of packaging equipment. pp. 228–238. In: H.L.M. Lelieveld, M.A. Mostert and J. Holan (eds.). Handbook of hygiene control in the food industry. CRC Press, Boca Raton.

Laureati, M., S. Buratti, G. Giovanelli and M. Corazzin. 2014. Characterization and differentiation of Italian Parma, San Daniele and Toscano dry-cured hams: A multi-disciplinary approach. Meat Sci. 96: 288–294.

Lorido, L., M. Estévez, J. Ventanas and S. Ventanas. 2015. Salt and intramuscular fat modulate dynamic perception of flavour and texture in dry-cured hams. Meat Sci. 107: 39–48.

Marušić, N., M. Petrović, S. Vidaček, T. Petrak and H. Medić. 2011. Characterization of Istrian dry-cured ham by means of physical and chemical analyses and volatile compounds. Meat Sci. 88: 786–790.

Marušić, N., M. Petrović, S. Vidaček, T. Janči, T. Petrak and H. Medić. 2013. Fat content and fatty acid composition in Istrian and Dalmatian dry-cured ham. MESO: The first Croatian meat journal 15(4): 307–313.

Marušić, N., S. Vidaček, T. Janči, T. Petrak and H. Medić. 2014. Determination of volatile compounds and quality parameters of traditional Istrian dry-cured ham. Meat Sci. 96: 1409–1416.

Ordonez, J.A. and L. de la Hoz. 2007. Mediterranean products. pp. 333–348. In: F. Toldrá (ed.). Handbook of fermented meat and poultry. Blackwell Publishing Professional, Ames, Iowa.

Parthasarathy, D.P. and N.S. Bryan. 2012. Sodium nitrite: The "cure" for nitric oxide insufficiency. Meat Sci. 92: 274–279.

Robertson, L.J., H. Sprong, Y. R. Ortega, J.W.B. van der Giessen and R. Fayer. 2014. Impacts of globalisation on foodborne parasites. Trends Parasitol. 30: 37–52.

Rodríguez, A., M. Rodríguez, A. Martín, J. Delgado and J. J. Córdoba. 2012. Presence of ochratoxin A on the surface of dry-cured Iberian ham after initial fungal growth in the drying stage. Meat Sci. 92: 728–734.

Rödel, W. 2001. Water activity and its measurement in food. pp. 453–483. *In*: E. Kress-Rogers and C.J.B. Brimelow (eds.). Instrumentation and sensors for the food industry. CRC Press, Boca Raton.

Seward, S. 2007. Sanitary design of ready-to-eat meat and poultry processing equipment and facilities. Trends Food Sci. Tech. 18: S108–S111.

Shan, B., Y.Z. Cai, J.D. Brooks and H. Corke. 2007. The in vitro antibacterial activity of dietary spice and medicinal herb extracts. Int. J. Food Microbiol. 117: 112–119.

Skibsted, L.H. 2011. Nitric oxide and quality and safety of muscle based foods. Nitric Oxide 24: 176–183.

Tabanelli, G., F. Coloretti, C. Chiavari, L. Grazia, R. Lanciotti and F. Gardini. 2012. Effects of starter cultures and fermentation climate on the properties of two types of typical Italian dry fermented sausages produced under industrial conditions. Food Control 26: 416–426.

Toldrá, F. 1998. Proteolysis and lypolisis in flavour development of dry-cured meat products. Meat Sci. 49: 101–110.

Toldrá, F. and M. Reig. 2007. Chemical origin toxic compounds. pp. 469–476. *In*: F. Toldrá (ed.). Handbook of fermented meat and poultry. Blackwell Publishing Professional, Ames, Iowa.

Toldrá, F. and M.C. Aristoy. 2010. Dry-cured ham. pp. 351–362. *In*: F. Toldrá (ed.). Handbook of meat processing. Blackwell Publishing Professional, Ames, Iowa.

Utrera, M., D. Morcuende, J.G. Rodríguez-Carpena and M. Estévez. 2011. Fluorescent HPLC for the detection of specific protein oxidation carbonyls– α-aminoadipic and γ-glutamic semialdehydes – in meat systems. Meat Sci. 89: 500–506.

Vignolo, G., C. Fontana and S. Fadda. 2010. Semidry and dry fermented sausages. pp. 379–398. *In*: F. Toldrá (ed.). Handbook of meat processing. Blackwell Publishing Professional, Ames, Iowa.

Zukál, E. and K. Incze. 2010. Drying. pp. 219–229. *In*: F. Toldrá (ed.). Handbook of meat processing. Blackwell Publishing Professional Ames, Iowa.

Chapter 4

Traditional Production of Fermented Meats and Related Risk

Hirokazu Oiki, Hirokazu Kimura and Nevijo Zdolec*

1 INTRODUCTION

Autochthonous fermented meat products occupy an important place in a wide range of foodstuffs available on regional, and possibly, international markets. They are, at the same time, of immense value as regards the preservation of culture heritage and national identity. Congruently to modern trends in the production of foodstuffs of animal origin, implementation of and control of adherence to good manufacturing and hygienic practice make the basis for the improvement of production and subsistence of the autochthonous products on the demanding market.

Regional food specialties, such as dry-fermented meat products and sausages and other food products of animal origin, have been traditionally manufactured in rural households and family farms. The production of some autochthonous and traditional products is not standardized or accompanied by the corresponding regulations (Kozačinski et al. 2008). On the other hand, many kinds of these products are nowadays produced under controlled conditions in small manufacturing facilities. Such production often develops into artisan production. In fact, excellent acceptance of that kind of meat products resulted in industrial scale production i.e. under the controlled conditions in order to ensure the

---

*For Correspondence: University of Zagreb, Faculty of Veterinary Medicine, Department of Hygiene, Technology and Food Safety, Heinzelova 55, 10000 Zagreb, Croatia, Tel: ++385 12 390 199, Email:nzdolec@vef.hr

permanent supply of the market with products of equable quality. During recent years, traditional production of fermented meat products is more precisely regulated by international or national rules and many products and traditional technologies are now recognized under the mark of "Protected Geographic Indication" (PGI) or "Protected Designation of Origin" (PDO). Very important regulatory approach in the protection of traditional meat products is flexibility, meaning the use of traditional tools (e.g. wood), lower structural requirements for small-producing facilities or milder hygienic requirements. Of course, the flexibility approach should not be misused, and food safety must be assured.

General measures of human health protection imply primarily the assurance of food safety and sanitary-technical and hygienic conditions for the production of traditional fermented meats. As regards veterinary-sanitary conditions of production of autochthonous meat products, of utmost importance is the microbiological and technological quality of raw material (Kozačinski et al. 2008). Some countries and regions allow the slaughtering of animals (e.g. pigs) in households, but food products produced from these animals are only intended for private consumption, not for the market. The quality of meat in traditional slaughtering procedures could be sometimes compromised by poor welfare conditions. However, welfare rules and regulations must be strict followed in any kind of slaughtering conditions, including households. Despite quality, the safety of meat could be also compromised by poor hygienic procedures in traditional production, as well as technological mistakes and climatic impact during the meat fermentation/maturation and storage. In this chapter some microbiological and technological aspects of traditional production of fermented meat production will be summarized, in order to assess the potential hazards for consumers.

2 TRADITIONAL PRODUCTIONS: RAW HAM AND FERMENTED SAUSAGES

The production of ready to eat meat products such as dry-cured ham has to be carefully performed under the condition of traditional production. Most important factors that may threaten the safety of product are hygienic handling procedures and natural environmental conditions (temperature, moisture). Usually, meat intended for ham processing is left in cool conditions for 24 hours and then salted. The dry curing method involves rubbing salt directly on to the surface of meat parts to remove water by exosmosis. Moreover, the total water content of the meat decreases due to the presence of salt in the meat. Water activity decreases by salt curing, resulting in the inhibition of bacterial growth. The addition of herbs and spices, along with salt, imparts greater flavor and taste to raw hams. In addition, certain herbs exert strong antimicrobial effects. After salt curing of raw ham, the salt is removed by running water in case of high salt concentrations. The water used for removing this salt must be sterile in order to prevent contamination of the raw ham. Generally, salt removal is carried out using running water whose volume is 5 times the weight of the ham being handled.

The second most common salt curing method used in traditional production is using of brine, i.e. by dipping meat (ham) in salt water. The salting treatment is carried out by sequentially immersing the meat in salt water of at least two different concentrations. The salt solution used for salt curing must be sterile to prevent raw ham from becoming contaminated with harmful bacteria. The salt water residue obtained after salt curing is a potential hotbed for the growth of harmful bacteria, and should therefore be removed immediately. After the removal of salt, some ham types are cold-smoked. Smoking results in a further decrease in water activity, thereby enhancing the preservation of product. Final production phase is drying for several months in specific conditions (cold, dark rooms) depending on ham type. In general, low safety hazards can occur in the production of dry cured meats, however the basic hygienic and technological rules should be followed.

Traditional fermented sausages are produced worldwide by different methods and techniques which are closely related to regional climate conditions. The growth of microorganisms associated with raw material/batter contamination is controlled by the dry, cold climate (winter season); however, great attention is required in the high temperature and humid districts (e.g. Asia). In the case of sausage, particular attention is focused on the prevention of contamination of the meat surface, because the process of mincing meat is associated with a higher risk of microbial contamination. Basic procedures of fermented sausages production are presented elsewhere in this book, however, traditional production may be accompanied with some additional hazards which can be effectively controlled in large-scale sausage production.

3 MICROBIOLOGY AND TRADITIONAL FERMENTED MEAT PRODUCTION

Ancient methods of food preservation, such as salting, fermentation, smoking, drying etc., have not been abandoned, but are still employed in the production of meat products. The improvement of industrial processing methods has resulted in the development of different additional protective methods, which have the same aim as in the early days of food preservation i.e. prevention of spoilage, inhibition of growth of spoilage microorganisms and pathogens or prolongation of the product's shelf life (Zdolec 2007).

As known, fermentation causes unfavorable conditions for growth and multiplication of undesirable microbiota. However, occasional findings of pathogenic microorganisms and other harmful substances in final products are indicators of non-hygienic manufacturing conditions, neglects in technological process and inadequate control and inspection both in the manufacture and sale of fermented products. The microbiological stability of dry-fermented sausages is based on physicochemical changes during the ripening mainly lowering the pH, increasing of salt content and decreasing of water activity. Besides that, the microbiological interactions in the filling are of great importance too, with particular regard antimicrobial mechanisms of lactic acid bacteria (Zdolec et al. 2007).

Fermentation is one of the oldest processes of meat preservation, which depends on the biological activity of lactic acid bacteria, (LAB) i.e. on the production of different metabolites capable of suppressing the growth of undesirable microbial flora (Ross et al. 2002, Hutkins 2006). Knowledge about the role of microorganisms in the food fermentation process dates long ago, so that the individual strains, considered fit from the technological and hygienic point of view, have been introduced in the production of fermented foodstuffs in form of starter cultures. It is well known that the microbiological and chemical succession during the maturation of fermented sausages depends on many factors, such as type of sausage, raw material used, additives, microclimatic conditions and duration of ripening process. Natural fermentation performed by indigenous microbial flora could result in technological failures and non-uniform quality of products. For this reason, implementation of technologically acceptable, meat-adapted and safe strains of LAB can contribute to suppressing of natural microbial flora and enhancing of the sensory profile of fermented sausages (Zdolec et al. 2008).

3.1 Bacterial Contaminants

The quality and safety of naturally fermented meat products relies on indigenous ("wild") microbiota. For this reason the traditional production of fermented meat products must be accompanied with stricter hygienic procedures and basic technological requirements (cold storage and meat maturing, proper salting and spicing, brining, temperature conditions during the ripening, storage during the shelf-life etc.).

Microbiological quality of raw meat, fat and other ingredients used in traditional home-made fermented sausages is expected to be at lower level than in meat-facilities with implemented food safety and quality standards. Zdolec et al. (2007) reported that pork meat, fat and sausage filling in household condition was contaminated by enterobacteria (3.47–3.54 log cfu/g), enterococci (2.00–4.43 log cfu/g), sulfite-reducing clostridia (1–2 log cfu/g) and St. aureus (2.6–3.47 log cfu/g). Enterobacteria, St. aureus

Traditional Production of Fermented Meats and Related Risk

and clostridia were detectable in fermented sausages by cultured methods until day 60, 33 and 8, respectively. The population of enterococci, as potentially hazardous microbiota, was stabile during the ripening; however their number was reduced by 1.7 log compared to initial counts. In any case, traditionally fermented sausages at the end of production were in compliance with comparable microbiological standards/criteria used in controlled sausage production. The basic physico-chemical changes in naturally fermented sausages are much slower than in controlled conditions (microclimate chambers), thus microbiological stability and appropriate pH reduction will occur after several weeks (Zdolec et al. 2007).

3.2 Molds and Mycotoxins in Traditional Fermented Meat Production

Molds are microorganism with strong historical and technological impact on sensorial properties of different kinds of fermented meat products. In natural production of fermented meat product, the specific microbiological-toxicological problem could arise by development of toxogenic molds. Hygienic and microclimate conditions during the ripening of fermented sausages in households usually support a surface growth of molds. Some technological failures, such as casing damaging, could additionally support the growth of toxogenic molds (Pleadin et al. 2015a). Markov et al. (2013) reported that dry-cured fermented meat products from individual producers (home-made products) are mainly contaminated with *Penicillium* spp. and most frequently detected mycotoxin was ochratoxin A. Pleadin et al. (2015b) reported that the maximal observed ochratoxin A level in the traditional fermented sausages and hams was around 5 to 10 times higher than the maximal recommended level (1 µg/kg) stipulated for pork products in some EU countries (Table 1). The health aspects of mycotoxins in fermented meat products are further discussed in chapter 18 of this book.

Table 1 Mycotoxin content in selected traditionally fermented meat products

Product	Mycotoxin	Determined amount	Reference
Game sausages, fermented dry-meat products	Ochratoxin A Aflatoxin B1	up to 7.83 µg/kg up to 3.00 µg/kg	Markov et al. 2013
Dry-fermented sausage "Slavonski Kulen"	Ochratoxin A	up to 17.00 µg/kg (damaged casings)	Pleadin et al. 2015a
Ham Dry-fermented sausages	Ochratoxin A	up to 9.95 µg/kg up to 5.10 µg/kg	Pleadin et al. 2015b
"Dalmatinski pršut" "Kraški pršut"	Ochratoxin A	up to 2.75 µg/kg up to 2.86 µg/kg	Pleadin et al. 2014

3.3 Antimicrobial Resistance in Indigenous Microbiota

Lactic acid bacteria (LAB) and coagulase-negative staphylococci (CoNS) are the main bacterial groups included in intrinsic microbiota of naturally fermented meat products. Their safety aspects have been evaluated, particularly in relation to transfer of antimicrobial resistance determinants to foodborne pathogens or closely related fermentative bacteria. The resistance of LAB from fermented meat products is discussed elsewhere in this book (Zdolec et al., Chapter 14). The resistance of CoNS seems to be equally important, despite a common opinion on low presence of safety hazards in CoNS from fermented meat products (Resch et al. 2008). In this respect, frequent isolation and even the dominance of opportunistic pathogen *St. epidermidis* in spontaneously fermented sausages is of particular significance (Marty et al. 2012, Zdolec et al. 2013). It is known that natural site of *St. epidermidis* is human and animal skin, and thus its finding in home-made fermented sausages is usual (mixing the sausage ingredients by hands). Although *St. epidermidis* was not involved in food poisoning outbreaks yet, its presence in food could be of public health importance as carrier of transmissible resistance genes. Zdolec et al. (2013) reported the dominance of *St. epidermidis* among CoNS in naturally fermented sausages produced in households, and frequent presence of tetracycline resistance genes *tet*K and *tet*M. The majority of strains harbored phenotypic resistance to erythromycin, tetracycline and ampicillin. Martin et al. (2006) reported the highest rate of resistance in *St. epidermidis* and *St. warneri* strains to ampicillin and erythromycin. Even et al. (2010) found that 69% of multiresistant strains isolated from food/clinical specimens belonged to *St. epidermidis* species with dominant resistance to erythromycin, tetracycline and penicillins. The *tet*K and *tet*M genes were also reported as most prevalent antibiotic resistance genes in CoNS from the production chain of swine meat commodities where *St. epidermidis* was dominant species (Simeoni et al. 2008). Although *St. epidermidis* is not classical food poisoning bacteria, its presence in food, including fermented meat products, and food-production chain could be of public health significance due to a possible spreading of antimicrobial resistance determinants.

4 OTHER HEALTH ISSUES IN TRADITIONAL FERMENTED MEAT PRODUCTION

Some additional practical problems and public health issues may arise during the fattening (domestic pigs) and meat processing in households and family farms. As already mentioned, slaughtering of pigs is allowed at farm-level, if welfare rules are followed and meat/meat products are not intended for public consumption (market). So, the responsibility for animal health and meat/meat products safety intended for private consumption relies on pig owner. One of the well-known omissions with public-health consequences is *Trichinella* poisoning which is still present in some pork-producing regions. The farm owner is obliged to test the samples for *Trichinella* by authorized organization/laboratory. However, the practice is sometimes different; if several pigs are slaughtered simultaneously (during winter season), the owner doesn't take all samples for *Trichinella* testing (reasons are saving the money, or previous negative findings of *Trichinella* etc.). It is hard to expect that meat from different pigs slaughtered at the same day will be separated and traced in households. Therefore, the *Trichinella* poisoning is quite possible when un-tested pork is mixed with *Trichinella*-free (tested) pork. Fermented meat products are ideal vehicle of *Trichinella* poisoning, as *Trichinella* is not totally destroyed by fermentation, curing, smoking or drying (Gamble 1997). The second possible mistake is making fermented sausages by mixing pork with un-tested wild boar meat which is more likely to be *Trichinella* invaded. Further huge mistake can be done during the preparation of home-made fermented sausages, by trying a taste of raw sausage mixture before stuffing (salinity, spiciness), when *Trichinella* testing is still in progress (if tested). Generally, the presence of *Trichinella* and other multicellular parasites in raw sausages is extremely rare under well-established inspection system, and sporadic cases of trichinelosis could be traced back to unlawful slaughtering (Lücke and Zangerl 2014).

Finally, very frequent practical problem in home-made fermented meat production is a pest infestation of products. The most important pests involved in deterioration of home-cured meat products are flies *Calliphora vomitoria* and *Piophila casei*, beetle *Dermestes ladarius*, mites *Tyrophagus casei, Tyrophagus putrescentiae* (Karolyi 2009). Their invasion (larvae) into fresh hams, prosciutto hams or other cured meats can even jeopardize the safety of product which must be rejected as unfit for human consumption.

5 CONCLUSION

Despite usually low hygienic practice in natural production of fermented sausages and cured fermented meats, this kind of production should be supported because of preserving regional products and heritage. Nowadays, family farms are often transferred to artisanal production that must be compliant with food safety and quality requirements and approved by competent authorities. Technological, hygienic and other requirements are not strict such as in industrial production. Nevertheless, the quality/safety of raw material intended for fermented

56 Fermented Meat Products: Health Aspects

meat production is remarkably improved; by compulsory slaughtering of farm animals in approved slaughterhouses under the continuous veterinary controls. Still, the safety of traditional home-made fermented meat products intended for private consumption rely on conscience of farmers during animal fattening, meat processing and controls.

Key words: *home-made fermented meat products, salt cured meat, fermented sausages, bacteria, parasites, mycotoxins, antimicrobial resistance, biogenic amines, tradition, heritage*

REFERENCES

Even, S., S. Leroy, C. Charlier, N.B. Zakour, J.P. Chacornac, I. Lebert, E. Jamet, M.H. Desmonts, E. Coton, S. Pochet, P.Y. Donnio, M. Gautier, R. Talon and Y. Le loir. 2010. Low occurrence of safety hazards in coagulase negative staphylococci isolated from fermented foodstuffs. Int. J. Food Microbiol. 139: 87–95.

Gamble, H.R. 1997. Parasites associated with pork and pork products. Rev. Sci. Tech. Off. Int. Epiz. 16: 496–506.

Hutkins, R.W. 2006. Meat fermentation. pp. 207–232. *In*: R.W. Hutkins (ed.). Microbiology and technology of fermented foods, Blackwell Publishing.

Karolyi, D. 2009. Rotting, pests and mistakes in prosciutto ham. Meso 11: 134–143. (in Croatian).

Kozačinski, L., M. Hadžiosmanović, Ž. Cvrtila Fleck, N. Zdolec, I. Filipović and Z. Kozačinski. 2008. Quality of dry and garlic sausages from individual households. Meso 10: 74–80.

Lücke, F.-K. and P. Zangerl. 2014. Food safety challenges associated with traditional foods in German-speaking regions. Food Control 43: 217–230.

Markov, K., J. Pleadin, M. Bevardi, N. Vahčić, D. Sokolić-Mihalek and J. Frece. 2013. Natural occurrence of aflatoxin B1, ochratoxin A and citrinin in Croatian fermented meat products. Food Control 34: 312–317.

Martin, B., M. Garriga, M. Hugas, S. Bover-Cid, M.T. Veciana-Nogués and T. Aymerich. 2006. Molecular, technological and safety characterization of Gram-positive catalase-positive cocci from slightly fermented sausages. Int. J. Food Microbiol. 107: 148–158.

Marty, E., J. Buchs, E. Eugster-Meier, C. Lacroix and L. Meile. 2012. Identification of staphylococci and dominant lactic acid bacteria in spontaneously fermented Swiss meat products using PCR-RFLP. Food Microbiol. 29: 157–166.

Pleadin, J., L. Demšar, T. Polak and D. Kovačević. 2014. Levels of aflatoxin B_1 and ochratoxin A in Croatian and Slovenian traditional meat products. Meso 16: 516–521.

Pleadin, J., D. Kovačević and N. Perši. 2015a. Ochratoxin A contamination of the autochthonous dry-cured meat product "Slavonski Kulen" during a six-month production process. Food Control. 57: 377–384.

Pleadin, J., M. Malenica Staver, N. Vahčić, D. Kovačević, S. Milone, L. Saftić and G. Scortichini. 2015b. Survey of aflatoxin B_1 and ochratoxin A occurrence in traditional meat products coming from Croatian households and markets. Food Control 52: 71–77.

Resch, M., V. Nagel and C. Hertel. 2008. Antibiotic resistance of coagulase-negative staphylococci associated with food and used in starter cultures. Int. J. Food Microbiol. 127: 99–104.

Ross, R.P., S. Morgan and C. Hill. 2002. Preservation and fermentation: past, present and future. Int. J. Food Microbiol. 79: 3-16.

Simeoni, D., L. Rizzotti, P. Cocconcelli, S. Gazzola, F. Dellaglio and S. Torriani. 2008. Antibiotic resistance genes and identification of staphylococci collected from the production chain of swine meat commodities. Food Microbiol. 25: 196–201.

Zdolec, N. 2007. Influence of protective cultures and bacteriocins on safety and quality of fermented sausages. PhD Thesis, University of Zagreb.

Zdolec, N., M. Hadžiosmanović, L. Kozačinski, Ž. Cvrtila, I. Filipović, K. Leskovar, N. Vragović and D. Budimir. 2007. Home-made fermented sausages – microbiological quality. Meso 9: 318–324.

Zdolec, N., M. Hadžiosmanović, L. Kozačinski, Ž. Cvrtila, I. Filipović, M. Škrivanko and K. Leskovar. 2008. Microbial and physicochemical succession in fermented sausages produced with bacteriocinogenic culture of *Lactobacillus sakei* and semi-purified bacteriocin mesenterocin. Meat Sci. 80: 480–487.

Zdolec, N., I. Račić, A. Vujnović, M. Zdelar-Tuk, K. Matanović, I. Filipović, V. Dobranić, Ž. Cvetnić and S. Špičić. 2013. Antimicrobial resistance of coagulase-negative staphylococci isolated from spontaneously fermented sausages. Food Technol. Biotechnol. 51: 240–246.

Chapter 5

Game Meat Fermented Products—Food Safety Aspects

Peter Paulsen*, Kashif Nauman, Friedrich Bauer,
Alessandra Avagnina and Frans J.M. Smulders

1 INTRODUCTION

Meat from wild animals has been presumably the first source of animal protein for human nutrition (Ahl et al. 2002). In poor economies, rural population still has to rely on this "cheap" source of protein in the form of bushmeat (Awaiwanont et al. 2014, Kamins et al. 2014), but meat from "exotic" or unusual species has a potential to contribute to the growing demand for sustainably produced food (Hoffman and Cawthorn 2013). It is conceivable that since ancient times attempts have been made to preserve the nutritional qualities of such meat for an extended period of time by drying, smoking and fermentation. Nowadays, consumers concerned about "healthy and natural" meat with low environmental footprint consider game meat as an alternative to meat from farmed animals (Hoffman and Wiklund 2006, Hoffman and Cawthorn 2012), and according to the "palaeo diet" concept (Cordain 2011, Cordain and Friel 2012), humans might be better adapted to the nutrient composition provided by meat from wild game than to that from domesticated animals. Fermented meat products fit well in this line of thinking, since

---

*For Correspondence: Institute of Meat Hygiene, Meat Technology and Food Sciences, University of Veterinary Medicine Vienna, Veterinaerplatz 1, A 1210 Vienna, Tel: +43 1 25077 3318, Fax: +43 1 25077 3390, Email: peter.paulsen@vetmeduni.ac.at

Game Meat Fermented Products—Food Safety Aspects

fermentation is one of the most natural food preserving processes. Meat producers can fulfil the demand for fermented meats from game either by adapting conventional meat product formulations, including nitrate/ nitrite as curing agents, the use of commercial starter cultures and of artificial casings for instance, or by adhering to minimizing additives, e.g. by relying on spontaneous fermentation or in-house flora, or by omission of curing agents.

General issues on meat processing technology, e.g. food additives and starter bacteria, are discussed elsewhere in this book, whereas this chapter will deal with peculiarities of game meat products, according to species and mode of keeping and harvesting (wild/free-living vs. farmed). A focus will be on wild ungulates, and a separate section will deal with ostrich. The systematic approach will start with intrinsic properties of meat and fat tissues which can have impact on fermentation and curing processes, namely pH and haem iron content of muscle tissue, fatty acid pattern and (boar) taint of fat tissue. Microbiota either present in the tissues of the live animal or contaminating the tissue surfaces during killing, eviscerating and cutting of game carcasses will be discussed. Physical hazards introduced by hunting activities [i.e. bone fragments as typical foreign objects, and fragments of bullet/shot embedded in edible tissues that can release metal (ions) and, thus constitute a chemical hazard] will not be addressed specifically. In sum, conditions rendering fermented game meat products "unsafe" will be discussed, including those hazardous to human health and those unfit for consumption as well. Biological agents without zoonotic evidence will not be taken into account.

2 WHAT IS "GAME MEAT"?

"Game meat" is often not clearly defined. For example, European Union food hygiene legislation distinguishes "wild" from "farmed" game, but leaves the definition, which animal species are covered, largely to the member states (Anonymous 2004). In detail, "wild game" comprises "wild ungulates and lagomorphs (in this context: rabbits, hares and rodents), as well as other land mammals that are hunted for human consumption and are considered to be wild game under the applicable law in the member state concerned, including mammals living in enclosed territory under conditions of freedom similar to those of wild game, and wild birds that are hunted for human consumption". The EU definition of farmed game comprises farmed ratites and farmed land mammals other than domestic animals. Among "wild game", the regulation distinguishes "small wild game" (i.e. wild game birds and lagomorphs living freely in the wild) from "large wild game" (i.e. wild land mammals living freely in the wild

60 Fermented Meat Products: Health Aspects

that do not fall within the definition of small wild game). Depending on the region, meat from "exotic species", as crocodiles (Gill 2007) or monitors and primates (Awaiwanont et al. 2014) will be considered as game meat. The term "venison" usually means meat from deer.

The distinction between "wild" and "farmed" becomes complicated when animals are semi-domesticated (e.g. reindeer in Sweden), kept in large parks under extensive conditions (e.g. deer in the UK), or when stocking density in fenced hunting areas is too high, which means that "free living" game is *de facto* restrained and dependent on feeding in the same way as farmed animals.

In essence, meat production from farmed game relies on regular health checks of life animals, treatment of diseases, and on veterinary inspection of animals before and after slaughter ("approved meat chain"), and is far more under control than hunting of free-living wild game. It has been argued, that under such conditions, crowding of animals and chronic social stress may occur more frequently and thus, transmission of zoonotic – enteric bacteria is more likely to occur (Gill 2007). Harvesting procedures may require transportation of animals. Likewise, night cropping with artificial illumination, use of helicopters for herding animals or for shooting them from the helicopter etc. can be applied on ranched or semi-domesticated game. Depending on the actual setting, such practices can exert stress and strain on the animal, which is sometimes reflected in injuries or depletion of energy depots in the muscle (Stevenson-Barry et al. 1999, Hoffman 2000, Van Schalkwyk et al. 2011a, Wiklund and Smulders 2011). Consequently, the ultimate pH of muscles will remain high, i.e. 5.8–6.2 or above, in other words, indicate a "dark-firm-dry" condition (Lawrie and Ledward 2006). This condition is associated with shorter shelf-life of the fresh meat since the proteolytic microflora is favoured instead of the bacteria splitting carbohydrates (Newton and Gill 1981). The possibility of high ultimate pH in meat should be critically considered in fermented meats production (Wirth et al. 1990).

Regulations on hunting of free-living wild game and placing such game meat on the market differ between countries or even regions. Whereas strict rules apply when meat from wild game is traded between countries or with the EU, supply to local retailers or to final consumers and finally the private domestic consumption are less strictly regulated. Health checks of the animals, microbiological contamination from inexpert shots/killing or deficiencies during evisceration, transport and cooling are major issues. Some examples on how the game meat chain works in these less or not formalized settings are given by Fettinger et al. (2011) for Europe and Bekker et al. (2011) for South Africa. Animal welfare issues of killing farmed game have been evaluated by EFSA (2006), and the microbiological risk profiles associated with farmed game been evaluated later by the same organization (EFSA 2013). As regards wild game, microbiological profiles in specific primary production settings have been reported by e.g. Atanassova et al. (2008), Deutz et al. (2006), Avagnina et al. (2012), and, in a food chain approach, reviewed by Paulsen (2011).

3   INTRINSIC PROPERTIES OF GAME MEAT WITH IMPACT ON SAFETY OF FERMENTED MEATS

Intrinsic properties which may have an impact of safety and quality of fermented meat from wild game are pH, haem (iron) content (Irschik et al. 2012), and fatty acid profiles. Although the latter are usually presented in a positive context in terms essential fatty acids and of omega3 – omega6 fatty acids ratio (Valencak and Gamsjäger 2014), concerns have been raised if tissues rich in polyunsaturated fatty acids should be processed into dry fermented products with long shelf-life (Paulsen et al. 2011). In addition to these intrinsic properties, killing, evisceration, and cooling can cause sensory deviations of bacterial ("haut gout") or endogenous nature (referred to as "stifling maturation"; Bauer et al. 2014).

3.1   Fermented Meats from Ungulates

A detailed review of the various biological hazards reported in meat from wild or farmed game is beyond the scope of this contribution. Also, only a selection of works dealing with fermented game meat products could be considered, with a focus on sausages, and food safety related characteristics. Quality traits of dry sausage from venison or wild boar meat have been studied by Soriano et al. (2006). Food quality and safety traits of dry-cured ham products were not considered. For these products, the interested reader may refer to Morgante et al. (2003) and Paleari et al. (2002). As regards fermented sausages technology, game meat products and products from domestic animals often do not differ markedly, also because very often pork fat is used in game meat sausage production.

3.1.1   Wild Boar, Europe

Intrinsic Properties. In Europe, abundance of wild boars is increasing and thus, more wild boar meat is going to be placed on the market (Quaresma et al. 2011, Sales and Kotrba 2013). In addition, it was reported that a substantial fraction of the deboned meat (13%) is processed into sausages or pâté (Winkelmayer and Paulsen 2008). Quaresma et al. (2011) reported fatty acid profiles similar to pork, but the wild boars in the study received additional feeding. Meat composition and health aspects of fatty acid profiles have been reviewed by Razmaité et al. (2012), Sales and Kotrba (2013) and Strazdina et al. (2014), and other specific quality issues by Bauer et al. (2014). Wild boar is regarded as high pH meat with limited shelf life (Boers et al. 1994). Ultimate pH has been reported to range from 5.43 to 6.08 (extreme values), with lower values for thigh and longissimus, but higher average values around 5.6–5.9 for small distal leg muscles and shoulder (Koomen 2014).

Biological Hazards of Concern. In 2007, Gill reviewed the microbiological condition of game meat and concluded that *Salmonella*, *Campylobacter*, *Yersinia* and *Listeria* are generally infrequently isolated from wild boar. More recent studies reviewed by Paulsen et al. (2012) and reports from various European countries (Wacheck et al. 2010, Vieria-Pinto et al. 2011, Conedera et al. 2014) indicate that more emphasis needs to be put on *Salmonella* in free-living wild boar. Among the meatborne parasites, not only *Trichinella* and *Toxoplasma* are of concern, since in the 1990ies, mesocercarial stages of the trematode *Alaria* have been detected more frequently in wild boar samples in Germany (Große and Wüste 2006), and, later in other European countries. A review of Möhl et al. (2009) collated evidence that *Alaria* mesocercariae could be transmitted via meat. As regards *Trichinella* in wildlife, both biosecurity measures and testing of wild boar carcasses need to be observed (Buncic and Mirilovic 2011).

Hepatitis E virus is one of the emerging zoonotic agents (Kumar et al. 2013); in particular the genotypes 3 and 4 can be transmitted from animals to humans. In Japan, it was detected in deer and wild boar in 2004 (Sonoda et al. 2004), and subsequently, transmission to humans via consumption of wild boar meat was described (Li et al. 2005). In the following years, virus or antibodies against the virus were detected in European countries, e.g. Germany (Schielke et al. 2009), Italy (Martelli et al. 2008, Martinelli et al. 2015), France (Carpentier et al. 2012) and Spain (Kukielka et al. 2015). However, the actual extent of exposure of consumers of products from wild boar meat remains to be evaluated.

In 2013, EFSA published an assessment which biological hazards should be considered when meat inspection is performed on farmed game (EFSA 2013). The assessment was conducted for Europe, but considered reports from all over the world. *Bacillus cereus*, *Clostridium botulinum*, *Clostridium perfringens*, *Listeria monocytogenes*, *Staphylococcus aureus*, and more specifically, *Campylobacter* spp., *Salmonella* spp., pathogenic VTEC, *Yersinia enterocolitica*, *Toxoplasma gondii*, *Trichinella* spp., and Hepatitis E virus were considered in a first stage. The outcome of the assessment was *Salmonella* spp., *Toxoplasma gondii* are of high priority for meat inspection, *Y. enterocolitica* and pathogenic verotoxigenic *Escherichia coli* (VTEC) and *Trichinella* spp. were low priority hazards, the last because of currently applied controls. The priority of *Campylobacter* spp. and HEV in farmed wild boar could not be determined due to a lack of data.

3.1.2 Wild Ruminants, Europe

Intrinsic Properties. Detailed data exist for semi-domesticated or farmed wild deer species, and have been reviewed by Wiklund and Smulders (2011) and Wiklund (2014). Ultimate pH exceeding 6.2, rendering a DFD condition with negative impact on microbial stability is a major issue. In particular, social stress or the mode of animal herding, transporting and culling can lead to energy depletion, whereas in non-stressed farmed ruminants ultimate pH values should be below 5.8. In principle, the same holds true for hunted wild ruminants (Winkelmayer et al. 2004, Paulsen et al. 2005, Hofbauer et al. 2006, Sauvala et al. 2015). However, wild ruminants that had been chased by dogs or wounded animals that tried to escape the hunters or dogs had higher ultimate pH, sometimes exceeding 6.2 (Deutz et al. 2006, Sauvala et al. 2015).

Biological Hazards of Concern. Compared to wild boar, bacterial pathogens of primary concern are verotoxinogenic *E. coli* of various serotypes (Thoms 1999, Lahti et al. 2001, Lehmann et al. 2006, Miko et al. 2009), for which meatborne transmission from deer products has been proven (Keene et al. 1997, Rabatsky-Ehr et al. 2002). Most studies focus on fecal samples rather than on meat, but fecal contamination of game meat can occur during shooting or evisceration of large game (Paulsen 2011).

In wild roe deer in Italy, none of 124 samples in 2007–2008 tested positive for VTEC; however *stx-/eae+* strains were found (Magnino et al. 2011). Caprioli et al. (1991) found no VT-producing *E. coli* in fecal samples from 46 red and 13 roe deer. A Belgian study from 2008/2009 reported VTEC in 15 of 133 red and roe deer fecal samples. Twelve isolates carried the *stx2* gene; however, none of the isolates with either *stx1* or *stx2* tested also positive for *eae* (Bardiau et al. 2010). In a Norwegian study (Lillehaug et al. 2005) two *stx+/eae-* isolates were recovered from fecal samples from 135 red and 206 roe deer. A Likewise, no O157-VTEC were found in a Swedish study (Wahlström et al. 2003). However, a German study demonstrated in 46 of 140 VTEC isolated from game both *stx2* and *eae* gene (Miko et al. 2009). Sánchez et al. (2009) reported a frequency of more than 50% VTEC in roe deer in Spain. *Stx2* gene was common, but the combination *stx2* and *eae* was rare.

Yersinia pseudotuberculosis is a recognized pathogen in farmed deer (Böhm et al. 2007), but considerable few information is available on *Y. enterocolitica* (Gill 2007). This bacterium has been detected in feces samples of wild deer from 19/249, 15/132 and 8/170 samples in Italy (Magnino et al. 2011), Switzerland (Joutsen et al. 2013) and Norway (Aschfalk et al. 2008), respectively. Avagnina et al. (2012) detected *Y. enterocolitica* on 1/61 roe deer carcasses, and 1/56 red deer carcasses in Italy, whereas in a game meat cutting plant in Austria, 16% of roe deer

64   *Fermented Meat Products: Health Aspects*

carcasses were positive for this bacterium (Paulsen et al. 2003). A higher prevalence was reported for meat cuts in Germany (Bucher et al. 2008); however, the meat species were not reported. Carriage of *Salmonella* or contamination of meat from wild ruminants is a very rare event (Paulsen et al. 2012).

As regards farmed deer, EFSA (2013) ranked *Toxoplasma gondii* as high-priority hazard, *Y. enterocolitica* and *Y. pseudotuberculosis* as low-priority in farmed deer, whereas for *Campylobacter* spp., *Salmonella* spp., pathogenic VTEC and Hepatitis E virus (HEV) in farmed deer the priority was not determined due to insufficient data.

3.1.3   *Fermented Sausages from Meat from Wild Game, Europe*

Dried Sausages. Findings of pathogenic *E. coli* and *L. monocytogenes* in dried fermented sausages from game meat have occasionally caused product recalls of in Europe under RASFF scheme. There are, however, few studies on these issues.

Food safety issues of an Italian fermented sausage "luganeghe" have been dealt with by Armani et al. (2014) and Lucchini et al. (2014). This sausage contains meat from wild ruminants or wild boar (chamois, red deer, roe deer), and 40–60% backfat from domestic pork, and sodium nitrite plus potassium nitrate as curing agents. The sausage is kept at 16–20°C for one week, and matured at 11–12°C for 2–5 weeks. Armani et al. (2014) studied 51 samples of luganeghe from four different producers and observed pronounced differences in pH and water activity (a_w) between producers (most probably because of different maturation periods), but, within one company, not between sausages produced from the different meat species, indicating that either there were no species-specific differences between the meat types used or such differences were overridden by processing technology. The initial pH of the sausage mass was 5.8–5.9, and in the finished product it varied from 5.4 to 6.5, with corresponding a_w values of 0.77 to 0.97. The same group of authors (Lucchini et al. 2014) studied the prevalence of *L. monocytogenes* in this type of sausage. The pathogen was detected in fresh game meat used for the production of the sausage, in fresh sausage as well as in the finished mature sausage. *L. monocytogenes* was detected in 14 of 59 samples, other *Listeria* species in 31 of the 59 samples. However, in all samples counts of *Listeria* were <10 cfu/g. As regards the finished sausage, a_w was 0.85 and pH was 5.7. As the a_w did not exceed 0.92, the product is considered not to support growth of *L. monocytogenes* according to Regulation (EC) No. 2075/2005. According to this regulation, ready-to-eat foods with either a pH ≤ 4.4 or an a_w ≤ 0.92 or the combination of pH ≤ 5.0 and a_w ≤ 0.94 will not support growth of *L. monocytogenes* during their shelf life (Anonymous 2005). As regards fermented sausages, it should not be relied upon a low pH, but rather on low a_w. Likewise, Capita et al. (2006) report pH values >5, but below 5.6 in salchichon and chorizo from deer on the Spanish market.

A study on the effect of meat and fat selection and application of starter cultures on the formation of biogenic amines and polyamines and on fat oxidation in dried fermented sausages was conducted by Paulsen et al. (2011). Sausage batches were produced from 7 kg meat and 3 kg fat, plus 2.8% nitrite curing salt (NaCl with 0.6% sodium nitrite). Eight different batches were prepared, reflecting a three-factorial design: (a) starter culture (*Lactobacillus sakei, Staphylococcus carnosus, Staphylococcus xylosus, Pediococcus pentosaceus, Candida famata*): present/absent; (b) meat selection: shoulder vs. hind leg from wild boar; (c) fat selection: pork backfat vs. fat from wild boar. Batches were ripened/dried until a weight loss of 32% was recorded (between days 28–35 after manufacture). Batches produced with a bacterial starter culture were generally preferred by taste panels, and had significantly lower concentrations of TBARS (<1.5 mg malondialdehyde/kg) and peroxide values and lower concentrations of cadaverine (<50 mg/kg), histamine (<10 mg/kg) and putrescine (<60 mg/kg). Batches produced from shoulder muscles contained significantly higher concentrations of cadaverine, histamine and putrescine; however, the levels would not be of health concern for the average consumer. TBARS were highest in batches manufactured with fat tissue from wild boars. These findings should be considered when guides to good practice for the manufacture of game meat products are developed.

Spreadable Fermented Sausages. Presence of *L. monocytogenes* is critical in fresh fermented sausage or in spreadable fermented sausage where virtually no drying and reduction of a_w will occur. This issue was studied by Schimpl et al. (2011), who produced three series consisting of several batches (i.e. all batches within one series were produced from the same meat and fat cuts) of a model soft raw sausage from 60% shoulder muscle of wild boar and 40% backfat from domestic pig, with addition of nitrite curing salt (2.7%), spices and four different commercial starter cultures plus a control without starters. Batter was vacuum-packaged in 90 μm PA/PE film and stored up to seven days at 18°C. The initial pH was 5.7, and dropped to 5.0–5.1 at the end of storage (irrespective of the presence of starter cultures). Numbers of lactic acid bacteria, *Enterobacteriaceae* and *Pseudomonas* were determined by culture methods at the day of manufacture and days 2 and 7 of storage. Average *Enterobacteriaceae* counts were initially 3.7, 5.3 and 3.1 log cfu/g for the three series and decreased to 2.6, 2.8; <2 log cfu/g at day seven in batches manufactured without addition of starter cultures. Corresponding counts of *Pseudomonas* were 3.0, 4.2 and 4.0 at day 0 to 2.4, 4.1 and 2.2 log cfu/g at day seven. Only a slight increase was observed for artificially inoculated *L. monocytogenes* (NCTC 11994), as initial counts of 3.5–3.9 log cfu/g developed to 3.9–4.1 log cfu/g (average values) at day seven. The batches demonstrating the fastest acidification (produced with a starter culture containing *St. xylosus* and *Pd. pentosaceus)* were associated with lowest multiplication of listeriae. Artificially inoculated *St. aureus* ATCC 25923 (5 log cfu/g) did not multiply in all batches produced with starter cultures, but increased by 1.0–1.5 log during seven days in the control (i.e. without starter cultures). As regards inoculated *E. coli* ATCC 25922 (4 log cfu/g), bacterial numbers decreased by 0.5 to 1.5 log during seven days, but this effect was inconsistent between different batches/ series of experiments. As regards behaviour of pathogenic bacteria, the study confirmed findings on "teewurst" from meat of domestic animals (Dourou et al. 2009). Some of the inter-series variations in the behaviour of the tested pathogens might be due to differences in hygienic quality of the wild boar meat used to manufacture the sausages. Although the fresh meat was tested to be initially free from the pathogens under study, different levels of the background flora may have interfered with the results. Batches containing a mix of *Lb. curvatus, St. xylosus, Debaryomyces hansenii* demonstrated a moderate speed of acidification and received significantly best sensory scores. Likewise, these batches demonstrated significantly lowest histamine and putrescine concentrations (averages of 10 and 30 mg/kg at day seven, respectively). Concentrations of cadaverine (<10 mg/kg) and of the polyamines spermidine and spermine (average values 3.0 and 13.0 mg/kg, respectively) remained virtually constant, whereas tyramine concentrations increased to >50 mg/kg fresh matter at day 7. Expectedly, a reduction in numbers of Gram negative bacteria was concurrent with pH decline and comparably lower biogenic amines accumulation, in particular as regards histamine (Hernandez-Jover et al. 1997, Bover-Cid et al. 1999, 2000, 2001). The authors concluded that safety and quality issues of this model fermented sausage from wild boar meat were comparable to those of spreadable raw sausage from pork.

Quite recently, the tenacity of *Alaria alata* mesocercariae in fermented sausages made from infested as well as from naturally contaminated wild boar meat was studied (González-Fuentes et al. 2014). The authors reported that in both nitrite cured salami with starter cultures *(Lb. sakei* and *St. carnosus)* and in a coarse, soft, nitrite cured raw sausage, the parasite was recovered 24 h after manufacture, but was completely inactivated within seven days after manufacture. It was assumed that inactivation is due to the cascade of "hurdles" emerging during fermentation and maturation (Leistner 1992). As regards this specific parasite, deep-freezing of game meat intended for fermented sausage production is a suitable pre-treatment to inactivate this parasite (González-Fuentes et al. 2015). Given suitable time-temperature combinations, such pre-treatment would also serve to inactivate other meatborne pathogens as *Trichinella, Toxoplasma* and

Game Meat Fermented Products—Food Safety Aspects

tapeworm cysts. Details on deep-freezing regimens or on decontamination treatments, as well as on strategies against Hepatitis E virus can be taken from the chapters dealing with meat from domestic animals.

3.1.4 Wild Ruminants, South Africa

Intrinsic Properties. Although there is a large variation between species of wild ruminants ranched in South Africa, they have in common that, under normal conditions, the ultimate pH in the longissimus muscle is around 5.4–5.7 (Hoffman et al. 2011) and, thus, in a range recommended for the manufacture of fermented meats (Wirth et al. 1990). Mode of harvesting can exert stress and cause glycogen depletion, for example, Veary (1991) analyzed the effects of shooting from helicopter, from the ground at day and during night on the ultimate pH of longissimus muscle of springbok (*Antidorcas marsupialis*). The corresponding pH values of 5.67, 6.15 and 5.67 indicated that hunting during the day would be more stressful, most probably because the animals need to be driven to the shooters. This is somehow corroborated by the finding that hunting on the foot during the day caused less disturbance and did not result in higher ultimate pH in impala compared to animals shot in the night (Hoffman and Laubscher 2009). Likewise, Hoffman (2000) reported that, during night-cropping, the ultimate pH of an impala wounded at the first shot and being killed after a 4-minutes flight was significantly higher than that of animals killed with the first shot. Fat content in meat from South African wild ruminants is around 1–2.9% (Hoffman et al. 2011), thus, for the manufacture of fermented sausages, fat from other species or vegetable fats has to be added.

Biological Hazards of Concern. Few studies exist on the microbiological condition of game carcasses, meat cuts or meats in South Africa. Magwedere et al. (2013) did not detect *Salmonella* or VTEC virulence genes in springbok meat from Namibia for export. Further studies are clearly needed to elucidate the role of wild ruminants as carriers of foodborne zoonotic bacteria.

Fermented Sausages from Meat from Wild Ruminants. Van Schalkwyk et al. (2011b) compared salami sausages made from springbok, gemsbok and kudu meat. The fat component was pork backfat, a commercial *Pd. pentosaceus – St. carnosus* starter culture was used and both sodium nitrite and potassium nitrate were added as curing agents. Dextrose was added to favour multiplication of lactic acid bacteria. The sausages were fermented and cold smoked at ca. 22°C for 2 days and then ripened for 21 days, after which period they were stored vacuum-packed at +4°C. The pH of the finished products was around 5.0 for batches made from meat from gemsbok and kudu (and, thus, in the range of a beef product manufactured as a control), but ca. 5.46 when meat from springbok had

68 Fermented Meat Products: Health Aspects

been processed – the latter product received also lowest sensory scores. In a previous work, Todorov et al. (2007) compared anti-listerial activity of lactobacilli in cold-smoked nitrite-cured Italian-style salamis made from various species´ meat (blesbok, springbok; cattle, horse, sheep). *Lactobacillus* strains producing plantaricin 423 or curvacin DF 126 were applied. At day 23 after manufacture, pH was 4.4–4.5 for sausages fermented with *Lb. plantarum* 432 and slightly higher when *Lb. curvatus* had been used. The authors reported no influence of meat species on pH of the product. As regards the anti-listerial activity, no consistent pattern was observed. The authors assumed that inactivation of bacteriocins by natural meat compounds as a meat-species-specific effect on one hand and the production of other antibacterial substances by the inoculated bacteria may have interfered.

In sum, the microbiological profile of meat from wild ruminants in South Africa is not well known. Since high-pH meat or DFD condition should not be expected under normal harvesting conditions, the safety of fermented meats from wild ruminants should controlled by the same measures taken for the processing of beef or mutton.

3.2 Fermented Meats from Birds (Ratites)

3.2.1 Intrinsic Properties of Ostrich Meat

A number of studies have been conducted on the processing of meat from ostriches into fermented sausage. Ostriches are indigenous to Africa, and have been farmed there in the 1800s or even earlier, but since the 1980ies, ostrich farming became increasingly popular in Europe and the US (Mozdiak 2004), and correspondingly, there was increased interest in utilizing not only eggs, leather and feathers, but also meat. Ostrich meat is low in fat (average 1.6%; Paleari et al 1998), which means that fat from other species (or vegetable fats or oils) has to be added in the production of fermented sausage. Ultimate pH is ostrich muscles is often around 6.0 (Morris et al. 1995), this in the upper range recommended for manufacture of fermented sausages (Wirth et al. 1990). In detail, Paleari et al. (1998) report an average pH of vacuum-packed leg muscles of 5.86 ± 0.35, and Sales and Mellet (1996) describing pH in individual leg muscles, found average pH_{24} values from 5.92–6.13. The latter authors point out that in their study, this was not the result of *ante-mortem* stress, but rather the reflection of the glycogenolytic metabolism of the muscle fibers, and should be seen as the normal condition.

3.2.2 Biological Hazards of Concern

In 2013, EFSA published an assessment which biological hazards should be considered when meat inspection is performed on farmed game, including ostriches (EFSA 2013). The assessment was conducted for Europe, but considered reports from all over the world. *Bacillus cereus, Cl. botulinum, Cl. perfringens, L. monocytogenes, St. aureus,* and more specifically, *Campylobacter* spp. and *Salmonella* spp. were considered in a first stage. Few data exist on the prevalence of the latter two pathogens on ostrich meat, and results vary from a few per cent to >40%, depending on region. For details, the reader may refer to EFSA (2013). In his review on the microbiological condition of meat from game, Gill (2007) found no evidence in literature that *E. coli* O157:H7 had been detected on ostrich meat.

3.2.3 Fermented Sausages from Ostrich Meat

Böhme et al. (1996) reported microbiological characteristics, pH and sensory properties of Italian-style salami made from ostrich meat and pork backfat. All sausages were produced with sodium nitrite and ascorbate, and dextrose. Sausages were fermented for four days at 20–22°C, and then ripened at 16–18°C. The authors compared batches with starter cultures (combinations of *Lb. sakei, Lb. curvatus* and *Micrococcus* sp. isolated from Italian salami) with a batch containing no starters, but 1% glucono-δ-lactone (GdL). Expectedly, sausages with GdL exhibited a fast pH drop from ca. 5.8 to 5.2 in the first day of fermentation, whereafter pH levels remained at this level. Initial pH of the sausages was in the range of 6.5–7.0, and declined gradually to pH 4.8–5.5 at day 6. The batch with GdL and no starters was least preferred by the evaluators. A further study (Dicks et al. 2004) explored the application of bacteriocin-producing lactobacilli in this Italian-style salami, with a focus on anti-listerial activity. Whereas the absence or presence of sodium nitrite had no effect on growth of *L. monocytogenes,* the bacteriocins (plantaricin 423 and curvacin DF 126) released from the lactobacilli were able to reduce *Listeria* numbers by 3 log cycles, but the pathogen recovered with 15 h. In a Spanish study on chorizo and salchichon dry sausages it was observed that sausages from ostrich meat had higher pH values (i.e. in the range of 5.5) than that made from deer meat or from pork.

In sum, few data exist on the microbiological profile of meat from ostrich. The high pH of ostrich meat can obviously be lowered by application of starter cultures, and it is conceivable that the safety of fermented ostrich meats is controlled by the same measures taken for the processing of high pH meat from other species.

4 CONCLUSION

The processing of game meat into fermented products is not a new phenomenon. However, the utilization of meat from "new" species or the emergence of zoonotic agents in traditionally processed game meat species require a careful consideration if empirical fermentation procedures or practices used for beef and pork provide an adaequate level of product safety. Comparably few studies exist on safety of fermented game meat products. In a view of increased consumers' demand for such products, more information is clearly needed to assess risks and establish appropriate control measures for fermented game meat products. In some species in general and under some stressful *ante-mortem* conditions in particular, high ultimate pH can occur, which requires starter cultures and carbohydrate addition in order to initiate pH decline in the product. In addition, there are some peculiarities of game meat, as sexual odour, high content of haem iron and fatty acid pattern, which have to be considered when meat and fat are selected for production of fermented sausages.

Key words: *game meat, ruminants, ostrich, wild boar, pH, Listeria monocytogenes, STEC, Hepatitis E, biogenic amines, fat oxidation*

REFERENCES

Ahl, A., D. Nganwa and S. Wilson. 2002. Public health considerations in human consumption of wild game. Annals NYAS 969: 48–50.

Anonymous. 2004. Regulation (EC) No 853/2004 of the European Parliament and of the Council of 29 April 2004 laying down specific hygiene rules for on the hygiene of foodstuffs. O.J. L139/55.

Anonymous. 2005. Commission Regulation (EC) No 2073/2005 of 15 November 2005 on microbiological criteria for foodstuffs. O.J. L338 1–26.

Armani, M., R. Lucchini, E. Novelli, S. Rodas, A. Masiero, J. Minenna, C. Pasolli, L. Lucchesa, U. Zamboni and G. Farina. 2014. Hygienic quality and microflora evolution in typical Italian game meat products. pp. 321–334. *In*: P. Paulsen, A. Bauer and F.J.M. Smulders (eds.). Trends in game meat hygiene: From forest to fork. Wageningen Academic Publishers, Wageningen.

Aschfalk, A., N. Kemper, J.M. Arnemo, V. Veiberg, O. Rosef and H. Neubauer. 2008. Prevalence of *Yersinia* species in healthy free-ranging red deer (*Cervus elaphus*) in Norway. Vet. Rec. 163: 27–28.

Atanassova, V., J. Apelt, F. Reich and G. Klein. 2008. Microbiological quality of freshly shot game in Germany. Meat Sci. 78: 414–419.

Avagnina, A., D. Nucera, M.A. Grassi, E. Ferroglio, A. Dalmasso and T. Civera. 2012. The microbiological conditions of carcasses from large game animals in Italy. Meat Sci. 91: 266–271.

Awaiwanont, N., P. Pongsopawijt and P. Paulsen. 2014. Bushmeat consumption and possible risks to consumers in Thailand. pp. 127–133. *In*: P. Paulsen, A. Bauer and F.J.M. Smulders (eds.). Trends in game meat hygiene: From forest to fork. Wageningen Academic Publishers, Wageningen.

Bardiau, M., F. Gregoire, A. Muylaert, A. Nahayo, J.N. Duprez, J. Mainil and A. Linden. 2010. Enteropathogenic (EPEC), enterohaemorragic (EHEC) and verotoxigenic (VTEC) *Escherichia coli* in wild cervids. J. Appl. Microbiol. 109: 2214–2222.

Bauer, F., A. Bauer and P. Paulsen. 2014. Raw material quality and processing technology issues of meat from wild ungulates. pp. 321–334. *In*: P. Paulsen, A. Bauer and F.J.M. Smulders (eds.). Trends in game meat hygiene: From forest to fork. Wageningen Academic Publishers, Wageningen.

Bekker, L.J., L.C. Hoffman and P.J. Jooste. 2011. Essential food safety management points in the supply chain of game meat in South Africa. pp. 39–65. *In*: P. Paulsen, A. Bauer, R. Winkelmayer, M. Vodnansky and F.J.M. Smulders (eds.). Game meat hygiene in focus: Microbiology, epidemiology, risk analysis and quality assurance. Wageningen Academic Publishers, Wageningen.

Böhm, M., P.L.C. White, J. Chambers, L. Smith and M.R. Hutchings. 2007. Wild deer as a source of infection for livestock and humans in the UK. Vet. J. 174: 260–276.

Böhme, H.M., F.D. Mellett, L.M.T. Dicks and D.S. Basson. 1996. Production of salami from ostrich meat with strains of *Lactobacillus sake, Lactobacillus curvatus* and *Micrococcus* sp. Meat Sci. 44: 173–180.

Boers, R.H., K.E. Dijkmann and G. Wijngaards. 1994. Shelf-life of vacuum-packaged wild boar meat in relation to that of vacuum-packaged pork: Relevance of intrinsic factors. Meat Sci. 37: 91–102.

Bover-Cid., S., M. Izquierdo-Pulido and M.C. Vidal-Carou. 1999. Effect of proteolytic starter cultures of *Staphylococcus* spp. on biogenic amine formation during the ripening of dry fermented sausages. Int. J. Food Microbiol. 46: 95–104.

Bover-Cid, S., M. Izquierdo-Pulido and M.C. Vidal-Carou. 2000. Influence of hygienic quality of raw materials on biogenic amine production during ripening and storage of dry fermented sausages. J. Food Prot. 63: 1544–1550.

Bover-Cid, S., M. Izquierdo-Pulido and M.C. Vidal-Carou. 2001. Effect of the interaction between a low tyramine-producing *Lactobacillus* and proteolytic staphylococci on biogenic amine production during ripening and storage of dry sausages. Int. J. Food Microbiol. 65: 113–123.

Bucher, M., C. Meyer, B. Grötzbach, S. Wacheck, A. Stolle and M. Fredriksson-Ahomaa. 2008. Epidemiological data on pathogenic *Yersinia enterocolitica* in Southern Germany during 2000–2006. Foodborne Pathog. Dis. 5: 273–280.

Buncic, S. and M. Mirilovic. 2011. Trichinellosis in wild and domestic pigs and public health: A Serbian perspective. pp. 143–156. *In*: P. Paulsen, A. Bauer, R. Winkelmayer, M. Vodnansky and F.J.M. Smulders (eds.). Game meat hygiene in focus: Microbiology, epidemiology, risk analysis and quality assurance. Wageningen Academic Publishers, Wageningen.

Capita, R., S. Llorente-Marigómez, M. Prieto and C. Alonso-Calleja. 2006. Microbiological profiles, pH, and titratable acidity of chorizo and salchichón (Two Spanish dry fermented sausages) manufactured with ostrich, deer, or pork meat. J. Food Prot. 69: 1183–1189.

Caprioli, A., G. Donelli, V. Falbo, C. Passi, A. Pagano and A. Mantovani. 1991. Antimicrobial resistance and production of toxins in *Escherichia coli* strains from wild ruminants and the alpine marmot. J. Wildl. Dis. 27: 324–327.

72   *Fermented Meat Products: Health Aspects*

Carpentier, A., H. Chaussade, E. Rigaud, J. Rodriguez, C. Berthault, F. Boue, M. Tognon, A. Touze, N. Garcia-Bonnet, P. Choutet and P. Coursaget. 2012. High hepatitis E virus seroprevalence in forestry workers and wild boars in France. J. Clin. Microbiol. 50: 2888–2893.

Conedera, G., M. Ustulin, L. Barco, M. Bregoli, E. Re and D. Vio. 2014. Outbreak of atypical Salmonella Choleraesuis in North Italy. pp. 151–159. *In*: P. Paulsen, A. Bauer and F.J.M. Smulders (eds.). Trends in game meat hygiene: From forest to fork. Wageningen Academic Publishers, Wageningen.

Cordain, L. 2011. The Paleo Diet: lose weight and get healthy by eating the food you were designed to eat. Revised version. J. Wiley & Sons, Hoboken, NJ.

Cordain, L. and J. Friel. 2012. The Paleo diet for athletes: a nutritional formula for peak athletic performance. Rodale Books, Emmaus, PA.

Deutz, A., F. Völk, P. Pless, H. Fötschl and P. Wagner. 2006. Game meat hygiene aspects of dogging red and roe deer. Arch. Lebensmittelhyg. 57: 197–202.

Dicks, L.M.T., F.D. Mellett and L.C. Hoffman. 2004. Use of bacteriocin-producing starter cultures of *Lactobacillus plantarum* and *Lactobacillus curvatus* in production of ostrich meat salami. Meat Sci. 66: 703–708.

Dourou, D., A.C.S. Porto-Fett, B. Shoyer, J.E. Call, G.J.E. Nychas, E. Illg and J.B. Luchansky. 2009. Behaviour of *Escherichia coli* O157:H7, *Listeria monocytogenes*, and *Salmonella* Typhimurium in teewurst, a raw spreadable sausage. Int. J. Food. Microbiol. 130: 245–250.

EFSA (AHAW Panel). 2006. Opinion of the Scientific Panel on Animal Health and Welfare on a request from the Commission related to the welfare aspects of the main systems of stunning and killing applied to commercially farmed deer, goats, rabbits, ostriches, ducks, geese and quail. EFSA J. 326: 1–18.

EFSA (EFSA Panel on Biological Hazards) 2013. Scientific Opinion on the public health hazards to be covered by inspection of meat from farmed game. EFSA J. 11: 3264.

Fettinger, V., F.J.M. Smulders and P. Paulsen. 2011. Structure and legal framework for the direct local marketing of meat and meat products from wild game in Austria: the Lower Austrian model. pp. 259–266. *In*: P. Paulsen, A. Bauer, R. Winkelmayer, M. Vodnansky and F.J.M. Smulders (eds.). Game meat hygiene in focus: Microbiology, epidemiology, risk analysis and quality assurance. Wageningen Academic Publishers, Wageningen.

Gill, C.O. 2007. Microbiological conditions of meats from large game animals and birds. Meat Sci. 77: 149–160.

González-Fuentes, H., A. Hamedy, E. von Borell, E. Lücker and K. Riehn. 2014. Tenacity of *Alaria alata* mesocercariae in homemade German meat products. Int. J. Food Microbiol. 176: 9–14.

González-Fuentes, H., K. Riehn, M. Koethe, E. von Borell, E. Lücker and A. Hamedy. 2015. Effects of in vitro conditions on the survival of *Alaria alata* mesocercariae. Vet. Parasitol. 113: 2383–2389.

Große, K. and T. Wüste. 2006. Der Dunker'sche Muskelegel. Funde bei der Trichinenuntersuchung mittels Verdauungsverfahren. Fleischwirtschaft 4: 106–108.

Hernandez-Jover, T., M. Izquierdo-Pulido, M.T Veciana-Nogues, A. Marine-Font and M.C. Vidal-Carou. 1997. Effect of starter cultures on biogenic amine formation during fermented sausage production. J. Food Prot. 60: 825–830.

Hofbauer, P., F. Bauerand P. Paulsen. 2006. Saisonale Unterschiede von Gämsenfleisch. Fleischwirtschaft 86: 100–102.

Hoffman, L.C. 2000. Meat quality attributes of night-cropped Impala (*Aepyceros melampus*). S. Afr. J. Anim. Sci. 30: 133–137.

Hoffman, L.C. and D.M. Cawthorn. 2012. What is the role and contribution of meat from wildlife in providing high quality protein for consumption? Animal Frontiers 2: 40–53.

Hoffman, L.C. and D.M. Cawthorn. 2013. Exotic protein sources to meet all needs. Meat Sci. 95: 764–771.

Hoffman, L.C. and L.L. Laubscher. 2009. Comparing the effects on meat quality of convventional hunting and night cropping of impala (*Aepyceros melampus*). S. Afr. J. Wildl. Res. 39: 39–47.

Hoffman, L.C. and E. Wiklund. 2006. Game and venison - meat for the modern consumer. Meat Sci. 74: 197–208.

Hoffman, L.C., S. Van Schalkwyk and M. Muller. 2011. Quality characteristics of blue wildebeest (*Connochaetes taurinus*) meat. S. Afr. J. Wildl. Res. 41: 210–213.

Irschik, I., F. Bauer and P. Paulsen. 2012. Meat quality aspects of roe deer, with regard to the mode of killing (shooting). Arch. Lebensmittelhyg. 63: 115–120.

Joutsen, S., E. Sarno, M. Fredriksson-Ahomaa, N. Cernela and R. Stephan. 2013. Pathogenic *Yersinia enterocolitica* O:3 isolated from a hunted wild alpine ibex in Switzerland. Epidemiol. Infect. 141: 612–617.

Kamins, A., K. Baker, O. Restif, A. Cunningham and J.L.N. Wood. 2014. Emerging risks from bat bushmeat in West Africa. pp. 91–105. *In*: P. Paulsen, A. Bauer and F.J.M. Smulders (eds.). Trends in game meat hygiene: From forest to fork. Wageningen Academic Publishers, Wageningen.

Keene, W.E., E. Sazie, J. Kok, D.H. Rice, D.D. Hancock, V.K. Balan, T. Zhao and M.P. Doyle. 1997. An outbreak of *Escherichia coli* O157:H7 infections traced to jerky made from deer meat. J.A.M.A. 277: 1229–1231.

Koomen, J.I. 2014. Microbial shelf-life of vacuum packed wild boar meat cuts at two different storage temperatures. M.Sc. thesis, University of Utrecht, Utrecht.

Kukielka, D., V. Rodriguez-Prieto, J. Vicente and J.M. Sánchez-Vizcaíno. 2015. Constant Hepatitis E Virus (HEV) circulation in wild boar and red deer in Spain: an increasing concern source of HEV zoonotic transmission. Transbound. Emerg. Dis. 2015 Jan 9. doi: 10.1111/tbed.12311.

Kumar, S., S. Subhadra, B. Singh and B.K. Panda. 2013. Hepatitis E virus: the current scenario. Int. J. Infect. Dis. 17: e228–e233.

Lahti, E., V. Hirvela-Koski and T. Honkanen-Buzalski. 2001. Occurrence of *Escherichia coli* O157 in reindeer (*Rangifer tarandus*). Vet. Rec. 148: 633–634.

Lawrie, R.A. and D.A. Ledward. 2006. Lawrie´s Meat Science. 7th ed. Woodhead, Cambridge.

Leistner, L. 1992. Food preservation by combined methods. Food Res. Int. 25: 151–158.

Lehmann, S., M. Timm, H. Steinrück and P. Gallien. 2006. Detection of STEC in faecal samples of free-ranging wild and in wild meat samples. Fleischwirtschaft 4: 93–96. (In German.)

Li, T.C., K. Chijiwa, N. Sera, T. Ishibashi, Y. Etoh, Y. Shinohara, Y. Kurata, M. Ishida, S. Sakamoto, N. Takeda and T. Miyamura. 2005. Hepatitis E virus transmission from wild boar meat. Emerg. Infect. Dis. 11: 1958–1960.

Lillehaug, B., B. Bergsjo, J. Schau, T. Bruheim, T. Vikoren and K. Handeland. 2005. *Campylobacter* spp., *Salmonella* spp., Verocytotoxic *Escherichia coli* and antibiotic resistance in indicator organisms in wild cervids. Acta Vet. Scand. 46: 23–32.

Lucchini, R., M. Armani, E. Novelli, S. Rodas, A. Masiero, J. Minenna, C. Bacchin, I. Drigo, A. Piovesana, M. Favretti, M. Rocca, U. Zamboni and G. Farina. 2014. *Listeria monocytogenes* in game meat cured sausages. pp. 321–334. *In*: P. Paulsen, A. Bauer and F.J.M. Smulders (eds.). Trends in game meat hygiene: From forest to fork. Wageningen Academic Publishers, Wageningen.

Magnino, S., M. Frasnelli, M. Fabbi, A. Bianchi, M.G. Zanoni, G. Merialdi, M.L. Pacciarini and A. Gaffuri. 2011. The monitoring of selected zoonotic diseases of wildlife in Lombardy and Emilia-Romagna, northern Italy. pp. 223–244. *In*: P. Paulsen, A. Bauer, R. Winkelmayer, M. Vodnansky and F.J.M. Smulders (eds.). Game meat hygiene in focus: Microbiology, epidemiology, risk analysis and quality assurance. Wageningen Academic Publishers, Wageningen.

Magwedere, K., R. Shilangale, R.S. Mbulu, Y. Hemberger, L.C. Hoffman and F. Dziva. 2013. Microbiological quality and potential public health risks of export meat from springbok (*Antidorcas marsupialis*) in Namibia. Meat Sci. 93: 73–78.

Martinelli, N., E. Pavoni, D. Filogari, N. Ferrari, M. Chiari, E. Canelli and G. Lombardi. 2015. Hepatitis E virus in wild boar in the central northern part of Italy. Transbound. Emerg. Dis. 62: 217–222.

Martelli, F., A. Caprioli, M. Zengarini, A. Marata, C. Fiegna, I. Di Bartolo, F.M. Ruggieri, M. Delogu and F. Ostanello. 2008. Detection of hepatitis E virus (HEV) in a demographic managed wild boar (*Sus scrofa scrofa*) population in Italy. Vet. Microbiol. 126: 74–81.

Miko, A., K. Pries, S. Haby, K. Steege, N. Albrecht, G. Krause and L. Beutin. 2009. Assessment of Shiga Toxin-Producing *Escherichia coli* isolates from wildlife meat as potential pathogens for humans. Appl. Environ. Microbiol. 75: 6462–6470.

Möhl, K., K. Große, A. Hamedy, T. Wüste, P. Kabelitz and E. Lücker. 2009. Biology of *Alaria* spp. and human exposition risk to *Alaria* mesocercariae a review. Parasitol. Res. 105: 1–15.

Morgante, M., R. Valusso, P. Pittia, L.A. Volpelli and E. Piasentier. 2003. Quality traits of fallow deer (*Dama dama*) dry-cured hams. Ital. J. Anim. Sci. 2(suppl.1): 557–559.

Morris, C. A., S.D. Harris, S.G. May, T.C. Jackson, D.S. Hale, R.K. Miller, J.T. Keeteon, J.R. Accuff, L.M. Lucia and J.W. Savell. 1995. Ostrich slaughter and fabrication: 1. Slaughter yields of carcasses and effects of electrical stimulation on post-mortem pH. Poult. Sci. 74: 1683–1687.

Mozdiak, P. 2004. Species of meat animals: poultry. pp.1296–1302. *In*: W.K. Jensen, C. Devine and M. Dikeman (eds.). Encyclopedia of Meat Sciences. Elsevier Academic Press, Amsterdam.

Newton, K.G. and C.O. Gill. 1981. The microbiology of DFD fresh meats: A review. Meat Sci. 5: 223–232.

Paleari, M.A., S. Camisasca, G. Beretta, P. Renon, P. Corsico, G. Bertolo and G. Crivelli. 1998. Ostrich meat: physico-chemical characteristics and comparison with turkey and bovine meat. Meat Sci. 48: 205–210.

Paleari, M.A., C. Bersani, M.M. Vittorio and G. Beretta. 2002. Effect of curing and fermentation on the microflora of meat of various animal species. Food Control 13: 195–197.

Paulsen, P. 2011. Hygiene and microbiology of meat from wild game: an Austrian view. pp. 19–37. *In*: P. Paulsen, A. Bauer, R. Winkelmayer, M. Vodnansky and F.J.M. Smulders (eds.). Game meat hygiene in focus: Microbiology, epidemiology, risk analysis and quality assurance. Wageningen Academic Publishers, Wageningen.

Paulsen, P., S. Vali and F. Bauer. 2011. Quality traits of wild boar mould-ripened salami manufactured with different selections of meat and fat tissue, and with and without bacterial starter cultures. Meat Sci. 89: 486–490.

Paulsen, P., F.J.M. Smulders and F. Hilbert. 2012. *Salmonella* in meat from hunted game: A Central European perspective. Food Res. Int. 45: 609–616.

Paulsen, P., F. Hilbert, R. Winkelmayer, S. Mayrhofer, P. Hofbauer and F.J.M. Smulders. 2003. Zur tierärztlichen Fleischuntersuchung von Wild, dargestellt an der Untersuchung von Rehen in Wildfleischbearbeitungsbetrieben. Arch. Lebensmittelhyg. 54: 137–140.

Paulsen, P., R. Winkelmayer, F.J.M. Smulders, F. Bauer and P. Hofbauer. 2005. A note on quality traits of vacuum-packaged meat from roe-deer cut and deboned 12 and 24h post mortem. Fleischwirtschaft 85: 114–117.

Quaresma, M.A.G., S.P. Alves, I. Trigo-Rodrigues, R. Pereira-Silva, N. Santos, J.C.P. Lemos, A.S. Barreto and R.J.B. Bessa. 2011. Nutritional evaluation of the lipid fraction of feral wild boar (*Sus scrofa scrofa*) meat. Meat Sci. 89: 457–461.

Rabatsky-Ehr, T., D. Dingman, R. Marcus, R. Howard, A. Kinney and P. Mshar. 2002. Deer meat as the source for a sporadic case of *Escherichia coli* O157: H7 infection, Connecticut. Emerg. Infect. Dis. 8: 525–527.

Razmaitė, V., G.J. Švirmickas and A. Šiukščius.2012. Effect of weight, sex and hunting period on fatty acid composition of intramuscular and subcutaneous fat from wild boar. Ital. J. Anim. Sci. 11: 174–179.

Sales, J. and R. Kotrba. 2013. Meat from wild boar (*Sus scrofa* L.): A review. Meat Sci. 94: 187–201.

Sales, J. and F.D. Mellett. 1996. Post-mortem pH decline in different ostrich muscles. Meat Sci. 42: 235–238.

Sánchez, S., A. García-Sánchez, R. Martínez, J. Blanco, J.E. Blanco, M. Blanco, G. Dahbi, A. Mora, J. Hermoso de Mendoza, J.M. Alonso and J. Rey. 2009. Detection and characterization of Shiga toxin producing *Escherichia coli* other than *Escherichia coli* O157:H7 in wild ruminants. Vet. J. 180: 384–388.

Sauvala, M., S. Laaksonen, K. Jalava and M. Fredriksson-Ahomaa. 2015. Hunting hygiene, contamination and pH values of Finnish moose. p. 16. *In*: A. Seguino, C. Soare and P. Paulsen (eds.). IRFGMH 2015 conference proceddings. Trends in game meat hygiene: From forest to fork. R(D)SVS, Edinburgh.

Schielke, A., K. Sachs, M. Lierz, B. Appel, A. Jansen and R. Johne. 2009. Detection of hepatitis E virus in wild boars of rural and urban regions in Germany and whole genome characterization of an endemic strain. Virol. J. 6: 58.

Schimpl, A., F. Bauer and P. Paulsen. 2010. Quality aspects of a spreadable raw sausage product manufactured from wild boar meat. Arch. Lebensmittelhyg. 61: 153–159.

Sonoda, H., M. Abe, T. Sugimoto, Y. Sato, M. Bando, E. Fukui, H. Mizuo, M. Takahashi, T. Nishizawa and H. Okamoto. 2004. Prevalence of Hepatitis E Virus (HEV) infection in wild boars and deer and genetic identification of a genotype 3 HEV from a boar in Japan. J. Clin. Microbiol. 42: 5371–5374.

Soriano, A., B. Cruz, L. Gómez, C. Mariscal and A. García Ruiz. 2006. Proteolysis, physicochemical characteristics and free fatty acid composition of dry sausages made with deer (*Cervus elaphus*) or wild boar (*Sus scrofa*) meat: A preliminary study. Food Chem. 96: 173–184.

Stevenson-Barry, J.M., W.J. Carseldine, S.J. Duncan and R.O. Littlejohn. 1999. Incidence of high pH venison: implications for quality. Proc. N. Z. Soc. Anim. Prod. 59: 145–147.

Strazdina, V., A. Jemeljanovs, V. Sterna and D. Ikauniece. 2014. Nutritional characteristics of wild boar meat hunted in Latvia. Proc. FoodBalt. 2014: 32–36.

Thoms, B. 1999. Nachweis von verotoxinbildenden *Escherichia coli* in Rehfleisch. Arch. Lebensmittelhyg. 50: 52–54.

Todorov, S.D., K.S.C. Koep, C.A. Van Reenen, L.C. Hoffman, E. Slinde and L.M.T. Dicks. 2007. Production of salami from beef, horse, mutton, Blesbok (*Damaliscus dorcas phillipsi*) and Springbok (*Antidorcas marsupialis*) with bacteriocinogenic strains of *Lactobacillus plantarum* and *Lactobacillus curvatus*. Meat Sci. 77: 405–412.

Valencak, T. and L. Gamsjäger. 2014. Lipids in tissues of wild game: overall excellent fatty acid composition, even better in free-ranging individuals. pp. 335–344. *In*: P. Paulsen, A. Bauer and F.J.M. Smulders (eds.). Trends in game meat hygiene: From forest to fork. Wageningen Academic Publishers, Wageningen.

Van Schalkwyk, D.L., L.C. Hoffman and L.L. Laubscher. 2011a. Game harvesting procedures and their effect on meat quality: the Africa experience. pp. 67–92. *In*: P. Paulsen, A. Bauer, R. Winkelmayer, M. Vodnansky and F.J.M. Smulders (eds.). Game meat hygiene in focus: Microbiology, epidemiology, risk analysis and quality assurance. Wageningen Academic Publishers, Wageningen.

Van Schalkwyk, D.L., K.W. McMillin, M. Booyse, R.C. Witthuhn and L.C. Hoffman. 2011b. Physico-chemical, microbiological, textural and sensory attributes of matured game salami produced from springbok (*Antidorcas marsupialis*), gemsbok (*Oryx gazella*), kudu (*Tragelaphus strepsiceros*) and zebra (*Equus burchelli*) harvested in Namibia. Meat Sci. 88: 36–44.

Veary, C.M. 1991. The effect of three slaughter methods and ambient temperature on the pH and temperatures in springbok (*Antidorcus marsupialis*) meat. M.Med.Vet.(Hyg.) thesis, Faculty of Veterinary Science, University of Pretoria, South Africa.

Vieria-Pinto, M., L. Morais, C. Caleja, P. Themudo, J. Aranha, C. Torrs, G. Igrejas, P. Poeta and C. Martins. 2011. *Salmonella* spp. in wild boar (*Sus scrofa*): a public health concern. pp. 131–136. *In*: P. Paulsen, A. Bauer, R. Winkelmayer, M. Vodnansky and F.J.M. Smulders (eds.). Game meat hygiene in focus: Microbiology, epidemiology, risk analysis and quality assurance. Wageningen Academic Publishers, Wageningen.

Wacheck, S., M. Fredriksson-Ahomaa, M. König, A. Stolle and R. Stephan. 2010. Wild boars as an important reservoir for foodborne pathogens. Foodborne Pathog. Dis. 7: 307–312.

Wahlström, H., E. Tysén, E. Olsson Engvall, B. Brändström, E. Eriksson, T. Mörner and I. Vagsholm. 2003. Survey of *Campylobacter* species, VTEC O157 and *Salmonella* species in Swedish wildlife. Vet. Rec. 153: 74–80.

Wiklund, E. 2014. Experiences during implementation of a quality label for meat from reindeer (*Rangifer tarandus tarandus*). pp. 295–303. *In*: P. Paulsen, A. Bauer and F.J.M. Smulders (eds.). Trends in game meat hygiene: From forest to fork. Wageningen Academic Publishers, Wageningen.

Wiklund, E. and F.J.M. Smulders. 2011. Muscle biological and biochemical ramifications of farmed game husbandry with focus on deer and reindeer. pp. 297–324. *In*: P. Paulsen, A. Bauer, R. Winkelmayer, M. Vodnansky and F.J.M. Smulders (eds.). Game meat hygiene in focus: Microbiology, epidemiology, risk analysis and quality assurance. Wageningen Academic Publishers, Wageningen.

Winkelmayer, R and P. Paulsen. 2008. Top-Wildbretqualität. Österr. Weidwerk: 46–48.

Winkelmayer, R., P. Hofbauer and P. Paulsen. 2004. Qualität des Rückenmuskels–Qualitätsparameter des Rückenmuskels von Rehen aus dem Voralpengebiet in Österreich. Fleischwirtschaft 84: 88–90.

Wirth, F., L. Leistner and W. Rödel. 1990. Richtwerte der Fleischtechnologie. Deutscher Fachverlag, Frankfurt am Main.

Chapter 6

Sheep and Goat Fermented Meat Products—Health Aspects

Alfredo Teixeira*, Sandra Rodrigues, Teresa Dias and Letícia Estevinho

1 INTRODUCTION

The most common sausages use only pork meat and are ripened for long periods. However, some countries with great tradition of sheep and goat meat consumption have the habit of eating some processed products of these meats. In Mediterranean countries as well as in other parts of the world, the meat from young lamb or kid is very usual and appreciated. These young milk fed animals producing lightweight carcasses are highly appreciated by consumers and are traditionally commercialized as quality brands with protected origin designation (PDO) or protected geographical indication (PGI). The animals that come out of these brands, particularly the heavier and culled ones, have very low consumer acceptability and consequently a low commercial value. Meat from these animals is more suitable to be processed as drought, cured or smoked products (Webb et al. 2005). Value may be added to final products by decreasing costs or improving relative value of the final product (McMillin and Brock 2005). With this goal there are several recently studies in goat and sheep meat processed products: Polpara et al. (2008) studied the quality characteristics of raw and canned goat

---

*For Correspondence: Campus Sta Apolónia Apt 1172, Bragança 5301-855, Portugal. Tel: +351 273 303206; Fax: +351 273 325405. Email: teixeira@ipb.pt

meat in water, brine, oil and Thai curry during storage; Das et al. (2009) studied the effect of different fats on the quality of goat meat patties; Teixeira et al. (2011) studied the effect of salting, air-drying and ageing processes in a new goat meat product "manta"; Teixeira and Rodrigues (2014) refer the high contents of protein of new sheep and goat meat products (sausages and mantas) concluding that both products are balanced products in protein and fat contents particularly unsaturated fat; and Oliveira et al. (2014) evaluated the quality of ewe and goat meat cured product.

In several countries, culled animals are slaughtered and their meat is processed, for example, the Spanish *cecina de castron* (Hierro et al. 2004), the Italian *violin di capra* (Fratianni et al. 2008) or the Brazilian *charque* and *manta* (Madruga and Bresan 2011). Devatkal and Naveena (2010) have recently studied the effect of salt and other products as kinnow and pomegrade on color and oxidative stability of goat meat during refrigerated storage.

So, the use of processes as salting, smoking and ripening to preserve meat products as well as to get special flavors was a practice before the global usage of refrigeration. Particularly, sheep or goat sausages as they are made from chopped lean meat and fat mixed with salt, spices and other ingredients, filled into a casing, normally cleaned intestines of cattle, sheep, goat, pigs or increasingly into artificial casings mainly for products with great uniformity. Although it is not a practice that has been commonly studied, Cosenza et al. (2003) evaluated the quality and consumer acceptability of *cabrito* smoked sausage, using goat meat as the sole meat ingredient. Also in Brazil, particularly in northeast the manufacturing of fermented sausages containing goat meat is an alternative use of meat from old animals (Nassu et al. 2003). Leite et al. (2015) reported the effect of different pork fat levels on the physicochemical properties, fatty acid profile and sensory characteristics in sheep and goat meat sausages. About cured products, Tolentino et al. (2016) reported the microbiological safety and sensory characteristics of sheep and goat's cured legs.

Despite the benefits and the control in processing these meat fermented products, some problems related to food safety and public health can occur. The minced meat used to make sausages tends to carry a relatively high level of microbial contamination and some ingredients as sulphite or metabisulphite could be added to control this. The active agent is the sulphur dioxide that is effective against *Pseudomonas* spp. The high content of salt used for drying and reducing the a_w of these products could result in an excessive sodium intake with negative effect in heart health particularly due to the connection between salt intake and hypertension, and consequently to an increased risk of stroke and premature death from cardiovascular disease.

These products are also considered unhealthy because of their fat content and the use of additives and spices in their formulation. Nowadays the addition of probiotics, particularly to the fermented sausages could promote the health benefits associated with lactic acid bacteria and contribute to the increase in the consumption of such products (Lücke 2000, De Vuyst et al. 2008).

The rediscovering of the traditional sheep and goat meat products associated to a developing of a new generation of fermented meat products as functional foods is an interesting food research field and could be a good strategy to meat industry.

2 TYPES OF FERMENTED SHEEP AND GOAT MEATS (RAW MATERIALS AND ADDITIVES)

The process of fermentation is one of the oldest techniques to preserve meat and meat products. The general principles of the fermentation of dry sausages have been reviewed many times and particularly by Toldrá (2004), Demeyer and Toldrá (2004). Fermented meats are a conserved product resulting from the presence of microbes in meat with salt added. Mixing and grinding meat or different kinds of meat with fats and salt together with spices, herbs and other ingredients produce a product which stability will depend of acidulation from acid lactic production and lowering of a_w as an effect of salt and cured salts addition.

Basically the principle of fermentation is based in the breakdown of carbohydrates presented in the meat mixtures, mainly the lactic acid. The process depends on the action of fermentation bacteria, the contaminating flora naturally present in the raw meat. The low temperatures (less than 20°C) stimulate the growth of desired flora and the fermentation bacteria produce acids resulting in a decline of pH values, and the spoilage conditions become gradually unfavorable. During fermentation and ripening the reduction of a_w also avoid the spoilage. The spontaneous fermentation made the sausages stable at ambient temperatures and improves the sensory quality of the product. During fermentation and ripening lipolysis and proteolysis occurred developing the characteristic flavor of fermented cured sausages.

The industrial production of sheep and goat sausages corresponds to a four different products: fresh, semi-dry, dry and emulsion sausages. Fresh sausages containing goat or sheep meat are not fermented, smoked or cooked. The production period of semi-dry or dry sausages is basically separated in two periods: fermentation followed by drying, depending on the technology i.e. the length of ripening. The emulsion sausages are based on the creation of a meat emulsion of sheep or goat with different pork backfat levels (Leite et al. 2015). The different kinds of meat, sheep, goat or pork with different degrees of mixing with other ingredients or additives and submitted to various temperature and humidity conditions are called fermented sausages and their shelf-life or safety as well as their specific flavour, texture or colour are determined by the acidulation (especially lactic acid produced during the ripening process by various beneficial bacteria such lactobacillus), water activity (a_w), salt, curing and drying. There are great variations in size and type of casing, size of meat particles, fat level, salt, water, type and strength of spicing as well as length of fermentation, curring or smoking phases. Fresh sausages are perishable unless refrigerated and must be cooked before consumption, often by frying or grilling. The fermented products with a low a_w and low pH (< 5.3) associated by vacuum stuffing have a good shelf-life and safety and are not normally cooked before eating and are consumed as cured products.

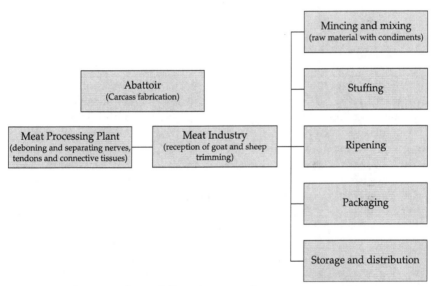

Figure 1 General flow diagram of sausages processing.

The general flow diagram of process technology description with all good manufacturing and hygiene practices is shown in Figure 1. Carcasses were previously deboned and cleaned from nerves, tendons and connective tissues before raw meat was processed at the manufacturing meat industry. Meats used for manufacturing sausages include sheep and goat trimmings. Meats are mincing and mixing with different levels of pork backfat fat (10–30%) salt (2.4%), peppers (0.3%), sugar (0.1%), water and rendimix®. Salt include sodium nitrite ($NaNO_2$) as curing salt (0.4–0.6%) is also added and considered essential because its antibacterial (particularly the growth of *Clostridium botulinum*), antioxidant, color preservation and cured flavouring properties. Other additives could be used such as glucose (0.5–1%), acid ascorbic (0.5–1%) to improve the stability of the red pigment, spices as paprika and garlic.

Raw materials and all additives are mixed and chopping in a mincer or cutter with different knives depending of the particle size of the batter wanted. The batter is then stuffed into pork or synthetic casing, hung and stabilized. The ripening is developed in two stages: firstly in a natural or mostly frequent in air-conditioned climate fermentation chambers at high relative humidity (HR 80–90%); and secondly transferred for another chamber for drying to development the sensory characteristics of a cured fermented product with temperatures between 5–24°C and 75 to 55% HR. In Mediterranean traditionally this ripening is processed in fresh air according the local climatic conditions. Sometimes, depending on consumers' preferences, a brief smoking period could precede the fermentation period.

3 DEVELOPMENT OF SENSORY QUALITY

Small ruminants' meat is traditionally worldwide consumed. Generally, and particularly in Mediterranean countries, the consumer demand is for young or light animals (Risvik 1994, Rodrigues and Teixeira 2010), characterized as tenderer than older or heavier animals (Rodrigues and Teixeira 2009). Meat from older and heavier animals has very low acceptability and market value, due to its hardness, poor structure and, normally, unpleasant taste and aroma. Occasionally, it is consumed in traditional dishes cooked for long time and very seasoned. To use this type of meat some processing must be done to transform it, which can be accomplished by salts or smoke and drying or also by the production of a fresh sausage after grinding, mixing with salts, spices, and other ingredients and casing. In the last years, there have been several studies concerning the incorporation of meat from culled sheep and goats in processed products, as fermented sausages (Nassu et al. 2002a, Cosenza et al. 2003, Pellegrini et al. 2008), demonstrating the possibility of taking advantage of meat usually rejected and transform it into well accepted products by consumers.

All over the world, it is well recognised that fermented sausages have as basic ingredient beef and pork meat. Although less frequent, the production of fermented sausages from other animals' meat follows the same procedure.

Stajić et al. (2011) studied the possibility for the use of goat meat in the production of traditional sucuk (Turkish style dry-fermented sausage) and observed that no significant differences were detected in cut appearance, colour and odour. However, in terms of appearance, texture and taste, evaluated in a 9 points scale from 1 (extremely unacceptable) to 9 (extremely acceptable) assessors gave smaller grades to goat than beef sucuk, but they also refer that those grades were higher than 5. The authors suggest the replacement of goat fat by beef fat, to appease the specific goat flavour, to make the product more acceptable to consumers that may not be used to such flavour.

Lu et al. (2014) made a study to compare the sensory characteristics of fermented, cured sausages made from equivalent muscle groups of beef, pork, and sheep meat, referring that the last had no commercial examples and represented an unexploited opportunity. They used seven replicates of shoulder meat and subcutaneous fat, sausages were made with 64, 29, 4, 2, 0.2, and 0.01 percent of lean meat, fat, NaCl, glucose, sodium pyrophosphate, and lactic culture, respectively. They observed that following anaerobic fermentation (96 h, 30°C), there were no significant species differences in mean texture (hardness, springiness, adhesiveness, cohesiveness), and only minor differences were seen in colour. However, the same authors refer that although not consumer tested, it is argued that consumers would be able to pick a texture difference due to different fat melting point ranges, highest for sheep meat.

Lu et al. (2014) also performed a sensory evaluation to understand if the peculiar sheep meat flavour could be covered or even eliminated to please consumers unused with this type of product. They simulated a very strong characteristic producing a mixed sheep meat and beef sausage, spicing it, or not, with 4-methyloctanoic, 4-methylnonanoic acid, and skatole (5.0, 0.35, and 0.08 mg/kg, respectively). They also, variably added sodium nitrite (at 0.1 g/kg and a garlic/rosemary flavour. Results of Lu et al. (2014), using 60 consumers, were that spiked sheep meat flavour caused an overall significant ($P = 0.003$) decrease from 5.83 to 5.35 in mean liking on a 1–9 scale, but when garlic/rosemary were added an increase ($P < 0.001$) from 5.18 to 6.00 was observed. Nitrite had no effect on liking (5.61 vs. 5.58, $P = 0.82$). Conclusions suggest that "sheep meat flavour could be suppressed to appeal to unhabituated consumers. Commercial examples could thus be made for these consumers, but the mandatory use of the name "mutton" in some markets would adversely affect prospects".

Consumers and processors are concerned about the safety of synthetic food additives, as some products used to mask or improve sensory characteristics can have health implications like synthetic antioxidants. So, a renewed interest in natural antioxidants and its research has increased. The use of natural antioxidants, like rosemary, is well accepted by consumers since it is considered safe, but they have some disadvantages like its cost and its influence in sensory characteristics as colour, aftertaste or off flavours (Brookman 1991, Pokorný 1991). However, the use of additives in fermented sausages can improve sensory characteristics, as registered by Nassu et al. (2003). In a study on using goat meat in processing of fermented sausage, salami type, they observed that the incorporation of rosemary minimized oxidized goat aroma and flavour. Also, Paulos et al. (2015) observed that the use of paprika had influence on the presence and intensity of flavour, spiciness, and off-odour (Figure 2) in sausages made from heavy sheep and goat meat when studying their sensory characteristics. Sausages without paprika presented higher spicy intensity, flavour intensity, and off-flavour than sausages with paprika, which had higher odour intensity and sweetness. Paprika masks the less pleasant sensory characteristics of this type of meat. Related to species, these authors found that goat sausages were harder and more fibrous, while sheep sausages where juicier.

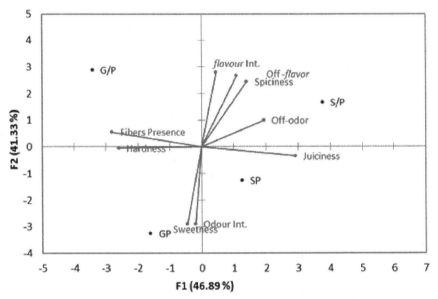

Figure 2 Consensus configuration: joint representation of the correlation between the sensory parameters and their first two dimensions, and groups of sausages sensory analysis. F1 = first principal component of generalized Procrustes analysis (GPA); F2 =second principal component of GPA; SP= sheep with paprika, S/P= sheep without paprika, GP= goat with paprika and G/P= goat without paprika (Paulos et al. 2015).

Besides the effect of additives in sensory characteristics, results of Paulos et al. (2015) show that consumers generally accepted fresh (fermented) sausages made of sheep and goat meat, with an average of 6 in a scale of 10, and no marked preferences were observed for sheep, goat or seasoning, used to mask some unpleasant characteristics as taste, odour or flavour. Figure 3 shows the preference maps obtained by these authors.

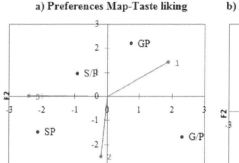
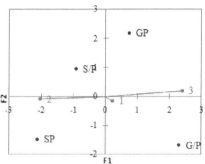
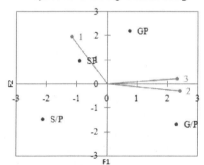
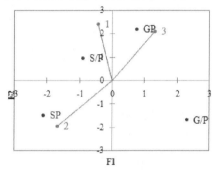

Figure 3 Preferences map for a) taste, b) spiciness, c) texture and overall acceptability; SP = sheep with paprika, S/P = sheep without paprika, GP = goat with paprika and G/P = goat without paprika (Paulos et al. 2015).

Table 1 Mean predicted values for the consumers' evaluation

	\multicolumn{4}{c}{Goat meat sausages}	\multicolumn{4}{c}{Sheep meat sausages}	\multicolumn{3}{c}{Significance}
	G0%	G10%	G30%	SEM	S0%	S10%	S30%	SEM	Species	Fat level	Sp x FL
Taste	5.09[e]	5.75[b]	7.23[a]	0.05	5.88[d]	6.37[b]	6.10[c]	0.05	NS	**	***
Spicy taste	4.83[a]	4.61[a]	4.78[a]	0.55	5.36[a]	4.13[a]	4.02[a]	0.56	NS	NS	NS
Texture	5.18[d]	6.17[c]	7.56[a]	0.06	5.40[d]	6.83[b]	6.27[bd]	0.04	NS	***	***
Overall acceptability	5.12[e]	5.92[c]	7.42[a]	0.06	5.59[d]	6.67[b]	5.95[b]	0.05	***	***	***

SEM – Standard error of the mean. NS - not significant ($P > 0.05$), *$P < 0.05$, **$P < 0.01$, *** $P < 0.001$; G0% goat sausages without pork fat; G10% goat sausages with 10% of pork fat; G30% goat sausages with 30% of pork fat; S0% sheep sausages without pork fat; S10% sheep sausages with 10% of pork fat; S30% sheep sausages with 30% of pork fat.

As already referred, one aspect to consider when processing fermented sausages is the fat addition. With the purpose to verify the effect of the addition of different fat contents (5, 10 and 20 percent) in the sensory acceptance of a goat meat fermented sausage, Nassu et al. (2002a) observed no significant differences for any of all measured sensory attributes (appearance, aroma, taste, texture and global acceptability) using a 1–9 hedonic scale. However, Leite et al. (2015) found a significant effect of the fat content on taste, texture and overall acceptability of sheep and goat sausages, as can be observed in Table 1.

When writing about fermented products, particularly fermented sausages, we must take into account the use of starter cultures. Already in 1970, Everson et al. referred that the right physiologically active starter culture would improve the uniformity of fermented products in terms of flavour, appearance and texture. Sensory evaluation of fermented mutton sausage, using *Pediococcus acidilactici H* and *Lactobacillus plantarum* 27 as starter cultures, had shown acceptable scores after 60 days of storage at 4°C (Wu et al. 1991). To produce dry-fermented sausages, Erkkilä et al. (2001) used starter cultures of *Lb. rhamnosus* strains, as well as *Pd. pentosaceus* and *Lb. plantarum*, reporting their suitability for use as probiotic starter cultures in fermenting dry sausage with respect to flavour profile compared to the commercial starter culture. The use of different starter cultures in processing of goat meat fermented sausages by Nassu et al. (2002b) produced average values between 5.5 and 5.9 for global sensory acceptability, using a 9 points hedonic scale. Global acceptability, aroma, taste and texture mean values presented no significant differences for all treatments, but appearance had the smallest value when the treatment with SPX (*Staphylococcus xylosus* and *Pd. pentosaceus*) culture was used. The authors refer the use of isolated observations from the judges as "rancid", "soap", which can be attributed to the products fat oxidation or even to the lipolytic action of microorganisms present in the used cultures. Using lactic starter cultures of *Lb. casei*, *Lb. plantarum* and *Pd. pentosaceus* Mukherjee et al. (2006) studied the effect of fermentation and drying temperature on the characteristics of goat meat (Black Bengal variety) dry sausages. Results were that the samples fermented at 30°C, followed by drying at 10°C, were the most acceptable samples respecting sensory characteristics as taste, flavour, texture and overall acceptability in a 5 points hedonic scale. The coagulase-negative *Staphylococcus* (CoNS), such as *St. xylosus*, *St. saprophyticus* and *St. carnosus*, are described as the main species that contribute in the formation of colour and flavor in the meat products. In sheep and goat "manta", a new product developed by Teixeira et al. (2011), appreciated by the consumers, *St. xylosus* has been found as dominating CoNS (data not shown).

4 MICROBIOLOGICAL SAFETY, HEALTH BENEFITS AND HAZARDS

4.1 Microbiological Safety

For assess microbiological safety, several procedures and analysis are performed in order to achieve the safety and quality of food involved to safeguard public health and provide assurance on food safety (Centre for Food Safety 2007), but microbiological analysis alone cannot guarantee the safety of food and microbiological criteria should be used to support good hygienic practice (GHP), good manufacturing practices (GMP), good agricultural practices (GAP) and implementation of food safety risk management systems such as hazard analysis and critical control point (HACCP) systems (Health Protection Agency 2009, van Schothorst et al. 2009). The food industry has a duty to ensure that microorganisms are eliminated or minimized to the extent that they cannot cause harm to human health (Anonymous 2004). Therefore microbiological safety plays an important role to be taken both by government and food industry for identifying, assessing and managing risks associated with the consumption of food and drink (Stringer 2005). Even though public authority is of paramount importance in the insurance of consumer's protection, the food industry itself plays a key role in that process, being responsible for handling the product during all the stages of the manufacturing process until its deliver to consumer. For taking on these microbiological roles, the authorities could follow the recommended stepwise by International Commission on Microbiological Specifications for Foods (ICMSF 1997), for the management of microbiological hazards in foods in international trade, applying existing Codex documents in a logical sequence (van Schothorst 1998).

Fermentation and drying has been reported as the oldest methods for food preservation known to mankind consequently the consumption of these products by humans dates from immemorial times (Nassu et al. 2003). Even today fermented foods are still among the most popular type of food consumed due to the fact that these products provides a means for producing safe and well preserved foods.

Recent studies focused on fermented foods, including fermented meat products, proved that these products are an excellent source of microorganisms with probiotic characteristics (Nova et al., Chapter 13 in this book). Schillinger and Lücke (1990) reported that lactic acid bacteria (LAB) enhance the safety of the product through the production of antimicrobial compounds such as lactic acid, acetic acid, hydrogen peroxide, carbon dioxide and bacteriocins. In general way fermented meat products are considered safe due to the reduction in a_w, pH and the

88 *Fermented Meat Products: Health Aspects*

presence of LAB that produce organic acids, mainly lactic acid and acetic acid that are effective antimicrobial agents *i.e.,* could reduce and prevent the growth of several pathogenic food microorganisms and competing for the nutrients (Schillinger and Lücke 1990, Lee 1994, Ferreira et al. 2007). The technology of lactic fermentation could be defined as the fermentation process involving a group of Gram-positive, non-sporing, non-motile, catalase-negative, non-aerobic organisms, which ferment carbohydrates to produce lactic acid as the sole or major organic acid (Oyewole 1997). Studies done by Paleari et al. (2002) observed the inhibitory effect of a_w, pH and the produced LAB on the pathogenic bacteria's during the fermentation process. At the beginning noted the normal flora in the raw materials as the presence of *St. aureus* and coliforms in all the samples and none had salmonella or *L. monocytogenes*. Nevertheless at the final of the fermented process noted an increase of LAB that exerts an antagonistic action on contaminating flora.

4.2 Microbiological Hazards

Wherein the presence of pathogenic microorganisms in food products represents a health hazard to the consumers, therefore, the need of reinforce and urgent implementation of the measures in meat processing industries and also in market points regarding the stability and safety of these meat products along shelf life period are required (Matos et al. 2013). Even that the fermentation inhibits the growth of some pathogenic bacteria's should not be expected to reduce the level of mold (mycotoxins) or bacterial (enterotoxins, botulinum toxin) so that the role of fermentation in reducing mycotoxins in food shows contradictory results.

The next microbiological hazard in fermented meat products are bacteria resistant to antibiotics and their transfer to consumers (Zdolec et al., Chapter 14 in this book). However, due to lower antibiotic pressure in small ruminant husbandry, it could be expected that antimicrobial resistance is of lower importance in production of goat/sheep fermented meat products. In general, the food chain has been associated as one of the main routes for transmission of antibiotic resistant bacteria between animal and human populations (Witte 1997), and in particular, fermented meats that establish direct link between the indigenous animal microflora and the human gastrointestinal tract microflora had been described as a potentially vehicle for horizontal transfer. Several studies *in vitro* and also *in vivo* reported the influence of the natural fermentative and ripening microflora in the dissemination of antibiotic resistance. For exemple the transfer of tetracycline resistance genes from *Lb. plantarum* to *Lactococcus lactis* and *Enterococcus faecalis* (Toomey et al. 2010), of erythromycin resistance genes from *Lb. fermentum* and *Lb. salivarius*, and of tetracycline resistance genes from *Lb. plantarum* and *Lb. brevis* to *E. faecalis* (Nawaz

Sheep and Goat Fermented Meat Products—Health Aspects　　89

et al. 2011) from *Lb. curvatus* and *E. faecalis* to *Lb. curvatus* (Vogel et al. 1992) of tetracycline and erythromycin resistance genes among *E. faecalis* isolates (Cocconcelli et al. 2003), and of tetracycline from *E. faecalis* to *L. monocytogenes* and *L. inocua* (Bertrand et al. 2005). This transfer was also reported in the digestive system of mice by Doucet-Populaire et al. (1991), Gazzola et al. (2012) found that the human isolate *E. faecalis* OG1RF with tetracycline and erythromycin resistances genes on plasmids were able to colonize the meat ecosystem with similar growth kinetics to that of food origin enterococci and to transfer the resistance genes to endogenous microflora (enterococci, pediococci, lactobacilli and staphylococci) present during raw fermented dry sausage ripening. At same time Jahan et al. (2015) confirmed the transfer from food isolates to human-associated *Enterococcus* strains.

In developed countries antibiotics are used especially in the industrialized production of food animals namely in the production of chickens and pigs and calves. On the other hand in Mediterranean area the goats and sheep production is essentially done traditionally without using antibiotics. Therefore the consumption of goat and sheep meat products can be helpful since it can reduce the problems related to the occurrence of antibiotic resistance in the food-chain. Fermented meat products may harbor also chemical hazard which are connected with microbiological activities, and the most relevant are biogenic amines (see Chapter 19 in this book).

In the literature there are not available data, about the presence of biogenic amines in sheep and/or goat fermented products, maybe because these products are not usual. However, due to the high biological value, demand for the sheep and goat meat has increased. Thus, in the last years it increased the effort to develop new products from this raw material. As a result were developed some fermented sausage made from sheep meat (Lu et at. 2015, Leite et al. 2015) and from goat meat (Nassu et al. 2003, Cosenza et al. 2003, Stajić et al. 2011, Leite et al. 2015). On the other hand, Bovolenta et al. (2008) analyzed the physicochemical, microbiological and sensory properties of Pitina, a traditional fermented sheep meat product from Italy.

However, these works are focused principally in the physio-chemical and sensorial characteristics, nevertheless the production principles remain the same; the presence of salt and sugars, and pH falls as lactic acid accumulates, suggesting intensive microbial activity and therefore these products must have the potential for BAs formation.

5　CONCLUSION

Even if sheep and goat are the most widely consumed red meat in the world the sheep and goat fermented products are not so popular.

However sheep and goat fermented product have peculiar sensory quality with high nutritional and nutraceutical characteristics with great demand and very well appreciated in several countries, particularly in Mediterranean area, Middle East, North Africa and Central Asia, as well in specific delicatessen markets, and among Ethnic Groups in Europe and North America. Today also in these products the main consumer concerns are food safety, aspects of health and environmental impact.

Certainly all good practices in manufacturing process and food analysis, to enhance the quality and to improve consumer-safety would be the most important trends in the near future. Research efforts should be addressed to improve the knowledge about the complex food fermentation ecosystems. Such knowledge will allow the determination of critical microbial variables, such as pathogen detection, microbial profiling, determination of survival of starter cultures and pathogens over food manufacturing and ripening, and predicting product shelf life and consequently improving the food quality and safety. Furthermore, it will help understand the relations between microflora and sensory characteristics. Another approach in food safety will be the selection of starter cultures with the ability to produce specific antimicrobial compounds, such as bacteriocin, against pathogenic bacteria or undesirable microorganisms and at same time genetically unable to produce biogenic amines.

Additionally new functional fermented goat and sheep products will be developed. Today a major expansion in functional meat products is related to probiotics cultures mainly LAB or bifidobacteria. However, there is scarce scientific evidence in literature about the positive effect of probiotic meat products on human health. Therefore, more research is required for identifying the strain or strains that produce the greatest health effect as well explore processes that improve the viability of probiotics on final product as well in the body. New functional products will be developed with other advantages than probiotics, such as fermented meat products enriched with different prebiotics.

Key words: *goat, sheep, meat, fermented products*

REFERENCES

Anonymous. 2004. Regulation (EC) 852/2004 of the European Parliament and of the Council of 29 April 2004 on the hygiene of foodstuffs. *Official Journal of the European Communities* 2004; L139: 1–54. Available from: eur-lex.europa. eu/LexUriServ/LexUriServ.do?uri=OJ:L:2004:226:0003:0021:EN:PDF.

Bertrand, S., G. Huys, Y. Marc, K. D'Haene, F. Tardy, M. Vrints, J. Swings and J.M. Collard. 2005. Detection and characterization of tet (M) in tetracycline-

Sheep and Goat Fermented Meat Products—Health Aspects 91

resistant Listeria strains from human andfood-processing origins in Belgium and France. J. Med. Microbiol. 54: 1151–1156.

Bovolenta, S., D. Boscolo, S. Dovier, M. Moragnte, A. Palloti and E. Piasentier. 2008. Effect of pork lard content on the chemical, microbiological and sensory properties of a typical fermented meat product (Pitina) obtained from Alpagota sheep. Meat Sci. 80: 771–779.

Brookman, P. 1991. Antioxidants and consumer acceptance. Food Technol. New Zealand 26: 24–28.

Centre for Food Safety. 2007. Microbiological Guidelines for Ready-to-eat Food. Revised. Food and Environmental Hygiene Department. 66 Queensway, Hong Kong.

Cocconcelli, P.S., D. Cattivelli and S. Gazzola. 2003. Gene transfer of vancomycin and tetracycline resistances among *Enterococcus faecalis* during cheese and sausage fermentations. Int. J. Food Microbiol. 88: 315–323.

Cosenza, G.H., S.K. Williams, D.D. Johnson, C. Sims and C.H. McGowan. 2003. Development and evaluation of a cabrito smoked sausage product. Meat Sci. 64: 119–124.

Das, A.K., A.S.R. Anjaneyulu, R. Thomas and N. Kondaiah. 2009. Effect of different fats on the quality of goat meat patties incorporated with full-fat soy paste. J. Muscle Foods 20: 37–53.

Demeyer, D. and F. Toldrá. 2004. Fermentation. pp. 467–474. *In*: W.K. Jensen, C. Devine and M. Dikeman (eds). Encyclopedia of meat sciences. Elsevier Academic Press, Amsterdam.

Devatkal, S.K., and B.M. Naveena. 2010. Effect of salt, kinnow and pomegranate fruit by-product powders on color and oxidative stability of raw ground meat during refrigerated storage. Meat Sci. 85: 306–311.

Doucet-Populaire, F., P. Trieu-Cuot, I. Dosbaa, A. Andremont and P. Courvalin. 1991. Inducible transfer of conjugative transposon Tn1545 from *Enterococcus faecalis* to *Listeria monocytogenes* in the digestive tracts of gnotobiotic mice. Antimicrob. Agents Chemother. 35: 185–187.

Erkkilä, S., M.L. Suihko, S. Eerola, E. Petäjä and T. Mattila-Sandholm. 2001. Dry sausage fermented by *Lactobacillus rhamnosus* strains. Int. J. Food Microbiol. 64: 205-210.

Ferreira, V., J. Barbosa, J. Silva, M.T. Felício, C. Mena, T. Hogg, P. Gibbs and P. Teixeira. 2007. Characterisation of *alheiras*, traditional sausages produced in the North of Portugal, with respect to their microbiological safety. Food Control 18: 436-440.

Fratianni, F., A. Sada, P. Orlando and F. Nazzaro. 2008. Micro-electrophoretic study of sarcoplasmic fraction in the dry-cured goat raw ham. Open Food Sci. J. 2: 89–94.

Gazzola, S., C. Fontana, D. Bassi and P.S. Cocconcelli. 2012. Assessment of tetracycline and erythromycin resistance transfer during sausage fermentation by culture dependent and -independent methods. Food Microbiol. 30: 348–354.

Health Protection Agency. 2009. Guidelines for assessing the microbiological safety of Ready-to-Eat foods. Health Protection Agency, London.

Hierro, E., L. de la Hoza, A. Juan and J.A. Ordóñez. 2004. Headspace volatile compounds from salted and occasionally smoked dried meats (cecinas) as affected by animal species. Food Chem. 85: 649–657.

ICMSF (International Commission on Microbiological Specifications for Foods). 1997. Establishment of microbiological safety criteria for foods in international trade. Wld. Hlth. Statist. Quart 50: 119–123.

Jahan, M., G.G. Zhanel, R. Sparling and A.R. Holley. 2015. Horizontal transfer of antibiotic resistance from *Enterococcus faecium* of fermented meat origin to clinical isolates of *E. faecium* and *Enterococcus faecalis*. Int. J. Food Microbiol. 199: 78–85.

Lee, C.H. 1994. Importance of lactic acid bacteria in non-dairy food fermentation. pp. 8–25. *In*: C.H. Lee, J. Adler-Nissen and G. Barwald (eds). Lactic acid fermentation of non-dairy food and beverages. Harn Lim Won, Seoul.

Leite, A., S. Rodrigues, E. Pereira, K. Paulos, A.F. Oliveira, J.L. Lorenzo and A. Teixeira. 2015. Physicochemical properties, fatty acid profile and sensory characteristics of sheep and goat meat sausages manufactured with different pork fat levels. Meat Sci. 105: 114–120.

Lu, S., H. Ji, Q. Wang, B. Li, K. Li, C. Xu and C. Jiang. 2015. The effects of starter cultures and plant extracts on the biogenic amine accumulation in traditional Chinese smoked horsemeat sausages. Food Control 50: 869–875.

Lu, Y., O.A. Young and J.D. Brooks. 2014. Physicochemical and sensory characteristics of fermented sheepmeat sausage. Food Sci. Nutr. 2: 669–675.

Lücke, F.K. 2000. Utilization of microbes to process and preserve meat. Meat Sci. 56: 105–115.

Madruga, M.S. and M.C. Bressan. 2011. Goat meats: Description, rational use, certification, processing and technological developments. Small Rum. Res. 98: 39–45.

Matos, T.J.S., A. Bruno-Soares and A.A. Azevedo. 2013. Microbial spoilage of Portuguese *chouriço* along shelf life period. Braz. J. Microbiol. 44: 105–108.

McMillin, K.W. and A.P. Brock. 2005. Production practices and processing for value added goat meat. J. Anim. Sci. 83: 57–68.

Mukherjee, R.S., B.R. Chowdhury, R. Chakraborty and U.R. Chaudhuri. 2006. Effect of fermentation and drying temperature on the characteristics of goat meat (Black Bengal variety) dry sausage. Afr. J. Biotechnol. 5: 1499–1504.

Nassu, R.T., L.A.G. Gonçalves and F.J. Beserra. 2002a. Effect of fat level in chemical and sensory characteristics of goat meat fermented sausage. Pesq. Agropec. Bras. 37: 1169–1173.

Nassu, R.T., L.A.G. Gonçalves and F.J. Beserra. 2002b. Use of different starter cultures in processing of goat meat fermented sausages. Cienc. Rural Santa Maria 32: 1051–1055.

Nassu, R.T., L.A.G. Gonçalves, M.A.A.P. da Silva and F.J. Beserra. 2003. Oxidative stability of fermented goat meat sausage with different levels of natural antioxidant. Meat Sci. 63: 43–49.

Nawaz, M., J. Wang, A. Zhou, M. Chaofeng, X. Wu, J.E. Moore, B.C. Millar and J. Xu. 2011. Characterization and transfer of antibiotic resistance in lactic acid bacteria from fermented food products. Curr. Microbiol. 62: 1081–1089.

Oliveira, A.F., S. Rodrigues, A. Leite, K. Paulos, E. Pereira and A. Teixeira. 2014. Quality of ewe and goat meat cured product mantas to provide value added to culled animals. Can. J. Anim. Sci. 94: 459–462.

Oyewole, J. 1997. Lactic fermented foods in Africa and their benefits. Food Control 8: 289–297.

Paleari, M.A., C. Bersani, M.M. Vittorio and G. Beretta. 2002. Effect of curing and fermentation on the microflora of meat of various animal species. Food Control 13: 195–197.

Paulos, K., S. Rodrigues, A.F. Oliveira, A. Leite, E. Pereira and A. Teixeira. 2015. Sensory characterization and consumer preference mapping of fresh sausages manufactured with goat and sheep meat. J. Food Sci. 80: S1568–S1573.

Pellegrini, L.F.V., C.C. Pires, N.N. Terra, P.C.B. Campagnol, D.B. Galvani and R.M. Chequim. 2008. Elaboração de embutido fermentado tipo salame utilizando carne de ovelhas de descarte. Ciênc. Tecnol. Aliment. 28: 150–153.

Pokorný, J. 1991. Natural antioxidants for food use. Trends Food Sci. Technol. 2: 223–227.

Polpara, Y., T. Sornprasitt and S. Wattanachant. 2008. Quality characteristics of raw and canned goat meat in water, brine, oil and Thai curry during storage. J. Sci. Technol. 30: 41–50.

Risvik, E. 1994. Sensory properties and preferences. Meat Sci. 36: 67–77.

Rodrigues, S. and A. Teixeira. 2009. Effect of sex and carcass weight on sensory quality of goat meat of Cabrito Transmontano. J. Anim. Sci. 87: 711–715.

Rodrigues, S. and A. Teixeira. 2010. Consumers' preferences for meat of Cabrito Transmontano. Effects of sex and carcass weight. Span. J. Agric. Res. 8: 936–945.

Schillinger, U. and F.K. Lücke. 1990. Lactic acid bacteria as protective cultures in meat products. Fleischwirtschaft 70: 1296–1299.

Stajić, S., N. Stanišić, M. Perunović, D. Živković and M. Žujović. 2011. Possibilities for the use of goat meat in the production of traditional sucuk. Biotechnol. Anim. Husb. 27: 1489–1497.

Stringer, M. 2005. Food safety objectives-role in microbiological food safety Management. Food Control 16: 775–794.

Teixeira, A. and S. Rodrigues. 2014. New sheep and goat products: "Mantas" and sausages. An integrated project in co-promotion. pp. 273–278. In: M. Chentouf, A. López-Francos, M. Bengoumi and D. Gabiña (eds). Technology creation and tranfer in small ruminants: roles of research, development services and farmer associations. Options méditerranéennes, Zaragosa, Spain.

Teixeira, A., E. Pereira and S. Rodrigues. 2011. Goat meat quality. Effects of salting, air drying and ageing processes. Small Rumin. Res. 98: 55–58.

Toldrá, F. 2004. Meat: Fermented meats. pp. 399–415. In: J.S. Smith and Y.H. Hui (eds). Food Processing. Principles and Applications. Blackwell Publishing, Ames, Iowa.

Tolentino, G., L.M. Estevinho, A. Pascoal, S. Rodrigues and A. Teixeira. 2016. Microbiological quality and sensory evaluation of new cured products obtained from sheep and goat meat. Anim. Prod. Sci. http://dx.doi.org/10.1071/AN14995.

Toomey, N., D. Bolton and S. Fanning. 2010. Characterisation and transferability of antibiotic resistance genes from lactic acid bacteria isolated from Irish pork and beef abattoirs. Res. Microbiol. 161: 127–135.

van Schothorst, M. 1998. Principles for the establishment of microbiological food safety objectives and related control measures. Food Control 9: 379–384.

van Schothorst, M., M.H. Zwietering, T. Ross, R.L. Buchanan and M.B. Cole. 2009. Relating microbiological criteria to food safety objectives and performance objectives. Food Control 20: 967–979.

Vogel, R.F., M. Becke-Schmid, P. Entgens, W. Gaier and W.P. Hammes. 1992. Plasmid transfer and segregation in *Lactobacillus curvatus* LTH1432 *in vitro* and during sausage fermentations. Syst. Appl. Microbiol. 15: 129–136.

Vuyst, L.D., G. Falony and F. Leroy. 2008. Probiotics in fermented sausages. Meat Sci. 80: 75–78.

Webb, E.C., N.H. Casey and L. Simela. 2005. Goat meat quality. Small Rumin. Res. 60: 153–166.

Witte, W. 1997. Impact of antibiotic use in animal feeding on resistance of bacterial pathogens in humans. pp. 61–75. *In:* D.J. Chadwick and J. Goode (eds.). Antibiotic Resistance. Wiley and Sons Ltd, Chichester, UK.

Wu, W.H., D.C. Rule, J.R. Busboom, R.A. Field and B. Ray. 1991. Starter culture and time/temperature of storage influences on quality of fermented mutton sausage. J. Food Sci. 56: 916–919.

Chapter 7

Hurdle Technologies in Fermented Meat Production

Josef Kameník

1 INTRODUCTION

The spoilage of foodstuffs occurs as the result of the growth and metabolic activity of bacteria or as the result of chemical changes, primarily oxidation (Shah et al. 2014). The chemical composition of foodstuffs, their structure and, most importantly, the conditions of their storage are decisive in whether bacteria and their enzymes contribute to food spoilage or whether oxidation changes occur.

Fresh meat is a foodstuff that undergoes bacterial spoilage extremely rapidly. Certain species of bacteria multiply easily on fresh meat due to its chemical composition, favourable water activity (a_w) and pH value. The growth of bacteria that induce food spoilage is influenced by a number of factors that can be divided into four groups (Bruckner et al. 2012):

- internal factors which are an expression of the physical and chemical properties of the foodstuffs themselves (e.g. the water activity, the content of nutrients and the structure of the foodstuffs)

For Correspondence: Department of Meat Hygiene and Technology, Faculty of Veterinary Hygiene and Ecology, University of Veterinary and Pharmaceutical Sciences Brno, Palackého tř. 1/3, CZ-61242 Brno, Czech Republic, Tel: +420 604 220851, Email: kamenikj@vfu.cz

- external factors, i.e. storage conditions (e.g. storage temperature, composition of the atmosphere)
- technological factors (physical or chemical treatments during processing of the food, e.g. heat treatment)
- implicit factors that reflect synergetic or antagonistic influences among bacteria (e.g. specific rate of growth).

Internal factors (the initial content of psychrotrophs present on the surface of the meat, the water activity, pH value and content of nutrients) and external factors (storage temperature and availability of oxygen) both play a role in the case of fresh meat.

Spoilage of foodstuffs caused by microorganisms can be prevented or delayed in two ways. The first involves the destruction (most frequently by heat treatment) of as many of the bacteria present in the product as possible and subsequent prevention of the growth and multiplication of any surviving individuals. This is also conditional to the prevention of secondary contamination of products that have already been treated, most frequently in the form of appropriate packaging that is generally hermetically sealed. The level and species composition of the bacterial population surviving depends on the intensity of treatment. The product itself cannot prevent the growth of bacteria. Opening the package renews the pathway to secondary contamination and spoilage.

The second way of preventing spoilage is to create an internal environment within the foodstuff that makes the growth of any bacteria present impossible and that supports a gradual decline in their numbers. In meat processing, the first preservation mechanism is used in the production of cooked products, with the most perfect implementation of this principle being applied in the production of canned foods. The second case is applied in frozen meats and, in particular, in dry-fermented sausages and cured air-dried meat products. While frozen food requires an environment in which the temperature is kept beneath freezing point, dry-fermented meat products can be stored at room temperature.

Various technological steps or elements of them that destroy or reduce the population of bacteria present can be used in the processing of foodstuffs. It is up to the producer to decide on the degree of intensity of the production process, as this determines the degree of microbial devitalisation on one hand, and nutritional value and sensory properties on the other. There are manufacturing procedures that involve the progressive application of factors known as hurdles and that determine the microbial stability and safety of foodstuffs and their sensory properties and nutritional value (Leistner 2000). These "hurdles" are a popular analogy used in the description of the concept in which a minimal degree of food processing causes sub-lethal stress to bacterial

Hurdle Technologies in Fermented Meat Production

cells that the bacteria must overcome in order to survive and continue to exist in the food environment (Hill et al. 2002).

Hurdles are applied in the production of fermented sausages that either act in the sausage batter itself (internal) or consist of influences on the external environment (external). Internal hurdles include substances with a preservative action (in particular sodium chloride and nitrites), the redox potential, competitive microflora, a low pH value and a low water activity (a_w) value (Table 1). The principal external factors include the environmental temperature and the use of smoke. Hurdles are applied to a different extent in the case of whole-muscle products (cured air-dried meat products) – the pH value of the meat used for processing, the environmental temperature, the effect of NaCl and a low a_w value.

Table 1 Internal hurdles applied in the production of dry-fermented sausage

Hurdle	Mode of action	Application force	Growth inhibition
NaCl	$\uparrow$Osmotic pressure	+	Aerobic spoilage bacteria, *Enterobacteriaceae*
NaNO$_2$	NO$^{\bullet}$, ONOO^{-}	+	*Enterobacteriaceae*, *Clostridium botulinum*
Redoxpotential	$\downarrow$O$_2$ content	+	Aerobic spoilage bacteria
Starter cultures (LAB)	Competition, Amensalism	++	Food-borne agents, *Enterobacteriaceae*, *Clostridium*
pH value	$\uparrow$Protons (H$^+$)	++	*Enterobacteriaceae*
a_w value	$\uparrow$Osmotic pressure, $\downarrow$free water content	+++	Bacteria general, in some extent molds and yeasts

2 HURDLES IN THE PRODUCTION OF DRY-FERMENTED SAUSAGES

2.1 Fresh Meat and Its Shelf Life

When the raw material used in the preparation of dry-fermented meat products (i.e. fresh meat) is compared with the final products, a significant difference can be found between them in terms of their shelf life. Meat spoils, depending on the surrounding temperature, within the order of hours or, at most, days, while dry-fermented sausages of high quality are not subject to any microbial decay (their spoilage occurs, in practical terms, merely by oxidation with oxygen from the air). How is this possible?

An internal environment is created in dry-fermented meat products during the technological production process that becomes unfavourable to microorganisms and to agents of meat spoilage in particular. This is a traditional and ancient way of preserving foods that man has been using for thousands of years. A single purpose lay behind the development of this process – to extend the shelf life of meat and create products that could be consumed a relatively long period of time after the slaughter or hunting of the animals from which they were made. Such a preservation method also had to be safe – the final product could not pose a threat to the health of the consumer. The technological process was modified over the course of the centuries in dependence on the experience of producers and the development of human knowledge.

Around the year 1500 BC, it was found that the shelf life of meat could be significantly extended if, after mincing finely, it was mixed with salt and aromatic herbs, filled into casings and subsequently dried (Ordóñez and Hoz 2007). Man began using two important technological steps that enabled the creation of a new category of meat products – dry-fermented sausages. These technological steps were fermentation and drying.

Fermented foods generally have a longer shelf life than their original raw materials and their spoilage process has a different character. The antimicrobial effects of fermentation are not confined to spoilage organisms alone and can also affect pathogens that might be present (Adams and Mitchell 2002). In microbiology, the term *fermentation* can be used to describe either microbial processes that produce useful products or a form of anaerobic microbial growth using internally supplied electron acceptors and generating ATP mainly through substrate level phosphorylation (Kim and Gadd 2008). In contrast to respiration, only the incomplete decomposition of substrates occurs during fermentation, for which reason the amount of energy released is also lower. Lactic acid bacteria (LAB) play the predominant role in fermenting the batter made for the preparation of dry-fermented sausages. The LAB, however, make up just part of the total microbial population in meat under natural conditions. In addition, LAB grow in meat in the presence of oxygen from the air far more slowly than certain gram-negative bacteria present, in particular members of the genus *Pseudomonas*.

2.1.1 Meat Spoilage Microbiota

Psychrotrophic species such as *Pseudomonas fragi*, *Ps. lundensis*, *Ps. putida* and *Ps. fluorescens* can be isolated from non-packaged meat showing signs of spoilage. *Ps. fluorescens* occurs most frequently in fresh meat, while *Ps. fragi* becomes dominant during long periods of storage. Bruckner et al. (2012) found *Ps. putida* (around 90% of *Pseudomonas* spp.) to make

Hurdle Technologies in Fermented Meat Production 99

up the largest proportion in fresh pork and poultry meat under aerobic conditions, with *Ps. fluorescens* found less frequently. Depending on the storage temperature, the population of pseudomonas attained values of 9–10 log cfu/g after several days (Table 2).

Table 2   The shelf life of pork and chicken meat at various storage temperatures under aerobic conditions

Temperature (°C)	The bacterial shelf life (in hours) of meat contaminated with Pseudomonas spp. (7.5 log cfu/g)	
	Pork meat (cutlet)	Chicken meat (breast fillet)
2	165.8	126.4
4	122.2	98.6
7	92.9	63.9
10	75.4	41.5
15	45.5	27.1

Source: Bruckner et al. 2012

The population of pseudomonads to the arbitrary level of 7–8 log cfu/g has been attributed to slime and off-odors formation. Both these deviations occur, however, if the pseudomonas have exhausted the glucose and lactic acid in the meat and have begun to metabolise nitrogen compounds, particularly amino acids (Nychas et al. 2008). The characteristic aroma of spoilt meat appears. The proteolytic activity of pseudomonas aids their penetration into the meat. Proteolytic bacteria can take advantage of this capability to obtain a competitive advantage over other bacterial groups or species, as it enables them to access new sources of nutrients that are not available to microbes with weak or non-existent proteolytic properties (Nychas et al. 2008).

It is, for this reason, necessary to create conditions that favour the growth of fermentative bacteria, i.e. LAB, and suppress the growth of aerobic gram-negative microorganisms at the beginning of the production process for dry-fermented sausages. These conditions are absolutely essential conditions of production without which these meat products cannot be prepared, and without which meat is subject to rapid spoilage. What factors are capable of suppressing the growth of aerobic gram-negative bacteria and supporting the growth of microaerophilic and anaerobic gram-positive fermentative bacteria in the initial phase of production of dry-fermented sausages? The addition of salt (NaCl) and reduction of the redox potential. Although the key factors from the viewpoint of the microbial stability of the final products and, in particular, their safety are the pH value and a_w value (Matagaras et al. 2015a), other hurdles to the growth of undesirable bacteria are decisive in the early stage of production, i.e. in the first tens of hours. The addition of salt is, however, unquestionably the most important factor.

100   *Fermented Meat Products: Health Aspects*

2.2   The Effect of Sodium Chloride

Salt has been used to preserve meat and extend its shelf life since time immemorial (Honikel 2007, Sebranek 2009). The principle of its preservative effect lies in increasing the osmotic pressure (Gutierrez et al. 1995). This reduces the water activity in the foodstuff, with a consequent bacteriostatic effect (Duranton et al. 2012).

Evidence of the use of salt to preserve meat dates back to the period around 3,000 BC. Salt had (and still has), of course, an effect on the sensory properties of meat and on a number of technological characteristics important to the further processing of meat in the production of meat products. Sodium chloride has an effect on the taste of sausages, in addition to its preservative effect. Salt reduces the sweet taste of saccharides and the acidic taste of organic acids (Gerhardt 1994).

Salt is, in chemical terms, sodium chloride (NaCl) and contains 39.3% of sodium and 60.7% of chloride ions (Feiner 2006). It is the most important additive used in meat processing. Forgetting the technological advantages of the addition of salt for a moment (breaking down myofibrillar proteins and its effect of the ability of meat to bind water), then sodium chloride plays the following roles from the viewpoint of the shelf life and safety of meat in the production of dry-fermented sausages:

- Salt reduces the water activity. If fresh meat has an a_w value of between 0.990 and 0.980, then the batter for the production of dry-fermented sausages following the addition of NaCl shows a_w values lower by as much as 0.020 to 0.030. Klettner et al. (1999) found a_w values of 0.969, 0.959 and 0.950 following the addition of 2.5, 3.5 and 4.5% of nitrite curing salt to batter for the production of dry-fermented sausages.
- Salt favours the growth of gram-positive bacteria over gram-negative bacteria (Feiner 2006).

The addition of salt to meat increases the osmotic pressure acting on the bacterial cells. The internal osmotic pressure in a bacterial cell is higher than that in the surrounding medium (Gutierrez et al. 1995). The result is pressure acting outwardly against the cell wall, a situation known as *turgor pressure*. The bacterial cell must be capable of maintaining this turgor pressure regardless of differences in the osmotic pressure of the external environment. The response of microorganisms to osmotic stress includes both physiological changes and variations in gene expression (Gandhi and Chikindas 2007). This is known by the term osmoadaptation.

The proportion of NaCl in the batter for dry-fermented sausages is generally beneath 4 percent (Ruiz 2007). A nitrite curing salt (sodium nitrite content 0.4–0.6%) at a proportion of 2.0–3.5% is generally used for salting the batter for the production of dry-fermented sausages (Stahnke and Tjener 2007). The content of salt in the finished product is 3.2–4.5% as the result of the loss of water during drying. The amount of salt added may influence the speed of decline in pH values during ripening. Klettner et al. (1999) found the following pH values when using various doses of nitrite curing salt (4% of glucose; Duploferment 66 starter culture; sausage diameter 60 mm): dose with 2.5% of nitrite curing salt – pH 5.06 on Day 2, dose with 3.5% – pH 5.04 on Day 3, and dose with 4.5% – pH 5.16 only after 4 days had elapsed from batter preparation. After 10 days, all the doses showed the same pH value of 4.95.

2.3 The Effect of Nitrite

The addition of salt has long been known to extend the shelf life of meat. The term meat curing is understood as the treatment of meat and meat products with sodium chlorite with the addition of nitrite and/or nitrate for the purpose of preservation and the stabilisation of colour (Jira 2004, Sebranek 2009). The effect of nitrate on the colouring of meat products was discovered in the nineteenth century. The fact that the pink-red colouring of meat products was caused by nitrite, with the formation of a thermostable pink-red colour, was discovered at the end of the nineteenth century. A few years later – in 1910 and 1914 – Hoagland published his findings, according to which the reactive substance is not the nitrite anion NO_2^-, but nitrous acid HNO_2 or its metabolite nitric oxide NO (Honikel 2007) (Figure 1).

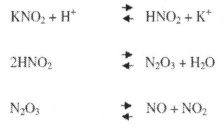

Figure 1 Reaction of nitrite in meat products (Honikel 2007).

It took just a few years for this new knowledge about the action of nitrite in meat to find its way into meat processing practice. Nitrite is, unfortunately, around ten times more toxic than nitrate. The lethal dose of nitrate given for man is a quantity of 80–800 mg/kg body weight, while the lethal dose of nitrite is 33–250 mg/kg. Initially, nitrite was often applied to meat products in excessively high level, as a result of which a number of people were poisoned and subsequently died in, for example, Germany. This problem and the associated risk were resolved in Germany by the nitrite-curing salt law (Nitrit-Pökelsalz-Gesetz) issued in 1934 (Honikel 2007), according to which nitrite could be used in meat processing only when mixed with salt (NaCl) at a concentration of 0.5% of nitrite, later increasing to 0.6%. This decree then spread rapidly to other European countries.

Dissociation occurs following the addition of nitrite to the batter due to the high solubility of nitrite in an aqueous medium at a pH of around 5.5 (Honikel 2008). Under these conditions, around 99% of the nitrite exists as NO_2^- anions. A small amount of undissociated nitrous acid is in a state of balance with N_2O_3 which is itself in balance with NO and NO_2 (see Figure 1). The NO molecule is oxidised to form NO_2 in the presence of oxygen. Nitrite "catches" oxygen in this way and has an antioxidant action in the batter of meat products.

Oxidation of nitrite to form nitrate following its application in the batter explains why nitrate also occurs in products to which no nitrate has been added. Numerous experiments have succeeded in demonstrating that around 10–40% of nitrite is oxidised to form nitrate following its addition to the batter.

Nitrite acts in a number of ways in meat products. It contributes to the colouring of the product and the creation of aroma, and has a preservative and antioxidant effect. No other single substance has yet been found that can substitute for its wide-ranging effects (Lücke 2003). A quantity of nitrite of 30–50 mg/kg of batter is needed to achieve the characteristic colour of meat products, while 20–40 mg/kg has an influence on aroma. The addition of 80–150 mg/kg of batter is enough to provide a preservative effect, while 20–50 mg of nitrite per kg of sausage batter will have an antioxidant effect.

Nitric oxide (NO) reacts with Fe^{2+} ions in bacterial cells and interferes with their energy metabolism whose reactions require the presence of iron ions. Clostridia are sensitive to nitrite, for which reason it can be assumed that Fe^{2+} ions serve as electron carriers in their energy metabolism (Lücke 2008).

Kabisch et al. (2008) tested the effect of nitrite on selected food-borne agents (*Listeria monocytogenes*, Shiga-like toxin-producing/Verotoxin-producing *Escherichia coli*, and *Salmonella* spp.) in the production of various types of fermented sausages and *in vitro* conditions using nutrient media with different a_w and pH values.

Nitrite had an antimicrobial action *in vitro* on selected food-borne agents (*L. monocytogenes*, *Salmonella* spp. and STEC/VTEC) only under certain conditions. Nitrite is not seen to have an inhibitory effect when the pH of the environment is 7.0. At a pH of 5.0, however, nitrite was able to prevent the growth of cells of the tested strains under *in vitro*

Hurdle Technologies in Fermented Meat Production

conditions. In real sausage production, nitrites prevented the growth of *L. monocytogenes*, compared to control batches. The addition of nitrite itself didn't have any effect on *E. coli* numbers in fermented sausages. Finally, the strongest antimicrobial effect *in vivo* was found towards *Salmonella* population (Kabisch et al. 2008).

2.4 Redox Potential

Preparation of the batter for dry-fermented sausages is followed by its filling into casings. Filling of the batter today is performed almost exclusively by vacuum fillers which also suck out the air that gets into the batter during meat comminution and subsequent mixing with additives.

The microflora of the meat used dominates the batter at the beginning of the production process. If natural casings are used, these may also be a source of bacteria (Pisacane et al. 2015). There is a predominance of bacteria of the family *Enterobacteriaceae* and genus *Pseudomonas* during this initial phase (Połka et al. 2015), with LAB, gram-positive catalase-positive cocci and yeast generally present at a population of less than 5 log cfu/g (if starter cultures are not applied). The growth of contaminating bacteria leads to a fall in the level of the oxygen that found its way into the batter during the mixing stage and remained in small quantities following filling into casings. This fall in the level of oxygen in the batter leads to a fall in the redox potential (E_h). A low E_h value forms another hurdle to the growth of undesirable bacteria in the sausage batter.

The redox potential (or oxidation reduction potential – ORP) is one of the basic parameters of the food environment that influences the bacteria present. It characterises the ability of the system to gain (reduce) or lose (oxidise) electrons (Brasca et al. 2007, Larsen et al. 2015). It is a physicochemical factor that determines the reduction and oxidation properties of the environment. The redox potential value is influenced, first and foremost, by temperature, the amount of dissolved oxygen and the pH value (Ignatova et al. 2010). Rödel and Scheurer (1999) found a fall in the value of the redox potential of more than 40 mV in cooked Lyoner sausage during an increase in pH from 5.8 to 6.2.

A positive redox potential value indicates an environment that supports oxidation reactions, while a negative value suggests an environment in which reduction reactions take place. The redox potential value measured in biological materials is converted (for the purposes of comparison) to a pH value of 7.0 (Rödel and Scheuer 1998a). This value is then known as E_h' and the following applies to its calculation:

$$E_h' = E_h + E_N \, (pH_x - 7)$$

E_h is the potential of a standard hydrogen electrode (V), E_N is the Nernst factor for which $2.303 \times RT/F$ applies; 2.303 = conversion factor, R is the universal gas constant (8.314 J K^{-1} mol^{-1}), T is the temperature (Kelvin), F is the Faraday constant (96.5 kJ V^{-1} mol^{-1}), pH$_x$ is the pH value of the measured product.

Dissolved oxygen is the principal factor that determines the value of E_h. Rödel et al. (1992) found an E_h value of -8 mV at a partial oxygen pressure (pO$_2$) of 14 mm Hg and 110 mV at pO$_2$ 100 mm Hg in batter for the production of comminuted meat products at a temperature of 21°C and a 60% of vacuum during cutting.

2.4.1 The Redox Potential of Meat and Meat Products

The redox potential found in fresh meat was between -200 and -300 mV, with a median value of around -200 mV for pork meat and -250 mV for fresh beef (Rödel and Scheuer 1998b). Values of E_h or $E_h{'}$ for meat products are generally positive values, as the technology (comminution and mixing) and additives used change the redox potential. Fermented meat products showed ORP values of 100 to 200 mV, thermally processed meat products -25 to 100 mV (Rödel and Scheuer 1998b). The median values were around 0 to 60 mV in meat products from pre-cooked meat and around 155 mV in meat jellies, while cooked products showed a median of between 20 and 120 mV.

2.4.2 Redox Potential and the Growth of Bacteria

Other important factors determining the redox potential value include the ratio between lactic acid and pyruvic acid and the content of ascorbic acid, thiol groups and metal ions (Larsen et al. 2015). The addition of sodium ascorbate reduces the redox potential – when added at a dose of 0.04% to Lyoner sausage batter, the value of E_h fell from an initial 97 mV to around -50 mV. In contrast, the addition of sodium lactate in a usual dose of 1–3% had practically no effect on the redox potential value (Rödel and Scheuer 1999).

The redox potential expresses the activity of electrons in the system just as the pH value reflects proton (Liu et al. 2013). In various microbial cultures, the E_h value may range from +300 mV for aerobic bacteria to less than -400 mV for anaerobes (Brasca et al. 2007). During the growth of *L. monocytogenes* in a nutrient medium, Ignatova et al. (2010) found a fall in E_h values from +330 mV to -209 mV, and a longer lag phase was recorded in a reductive environment (-360 ± 20 mV) at a pH of 6.0 in comparison with another two redox potential values (0 mV and 380 mV).

The redox potential is a significant selective factor for microorganisms in a given environment (Caldeo and McSweeney 2012). The effect of the redox potential depends on the individual bacterial species and must be evaluated individually for each species (Ignatova et al. 2010). Brasca et al. (2007) tested 10 species of LAB for their ability to change the redox potential during growth in milk over the course of 24 hr. Among the cocci, *Enterococcus faecalis* was one of the strains with the strongest reduction ability; among the rods, *Lactobacillus paracasei* ssp. *paracasei*.

The E_h value falls if oxygen is removed from the batter by a vacuum pump when the batter is filled into casings. The fermentation of saccharides and the consumption of oxygen by LAB cause another fall in redox potential values and create an anaerobic environment. Only facultative or obligate anaerobic bacteria can grow under these conditions. This inhibits aerobic gram-negative bacteria (particularly the genus *Pseudomonas*) which are capable to grow rapidly in meat in the presence of oxygen from the air and cause meat spoilage in a relatively short period of time. A fall in the values of the redox potential in the first few hours after the batter is filled in casings contributes towards the creation of an internal environment that supports the growth of LAB.

2.5 Competitive Microbiota

The approaches outlined above, i.e. the addition of salt, the addition of nitrite, and a dramatic reduction to the level of oxygen (thereby creating a microaerophilic environment in the batter) favour the growth of LAB. These represent a competitive microbiota in the batter capable of suppressing other undesirable microorganisms.

Løvdal (2015) divides undesirable bacteria into those that: i) induce food spoilage and are not in themselves necessarily harmful to the consumer, but degrade the product (taste, aroma, colour, consistency) and thereby limit its shelf life, and ii) pathogenic microorganisms that may be present in small quantities and that do not cause any visible changes (taste, aroma, colour, consistency) in the product, though their ingestion may cause alimentary illness resulting, in certain cases, in death. The LAB are capable of suppressing both groups of undesirable microorganisms in the batter for dry-fermented sausages. This bacterial group is responsible for the fermentative processes that take place during the production of dry-fermented sausages. The LAB in the batter come both from the raw materials used (meat) and from the manufacturing plant environment, though they are now also applied artificially in the form of starter cultures.

2.5.1 Lactic Acid Bacteria

Following the addition of starter cultures, LAB can multiply from an initial population of 6.5 log cfu/g to 8.0–9.0 log cfu/g over the first few days (Kameník et al. 2013, Matagaras et al. 2015a, Figure 2). Połka et al. (2015) found a LAB population from < 2 log cfu/g to 4.52 log cfu/g in samples of Piacentino sausages (northern Italy) without starter cultures immediately after their filling into casings. The addition of starter cultures caused an increase in LAB from the beginning of fermentation to values of between 5.63 and 6.82 log cfu/g. After three weeks of ripening, the LAB population reached values of between 8.15 and 8.89 log cfu/g; at this time there was no evident difference in the number of LAB seen between sausages in which starter cultures were used and sausages in which they were not used. During tests on minisalami filled into cellulose casings of a diameter of 25 mm, Gareis et al. (2010b) recorded an increase in the LAB population (with the addition of starter cultures) of two logarithmic orders during the first three days. The number of LAB did not fall beneath 8 log cfu/g at any time during the course of the experiment (42 days).

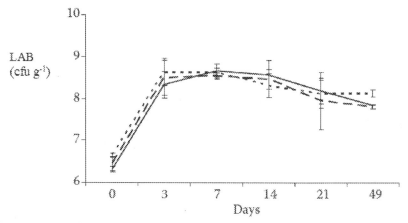

Figure 2 Population of lactic acid bacteria (LAB) in three batches (the same recipe with starter cultures, diameter 75 mm) of dry-fermented sausage (Kameník et al. 2013).

The principal advantage of LAB in the production of dry-fermented sausages lies in the fermentation of saccharides, i.e. the creation of lactic acid and the subsequent fall in pH. This is not, however, the only positive effect of this group of bacteria. It is also important in releasing other active aromatic (taste) substances and producing antimicrobial compounds.

Lactobacilli of predominant importance to dry-fermented sausages, i.e. *Lb. sakei*, *Lb. curvatus* and perhaps *Lb. plantarum*, make DL-lactic acid (Schillinger and Lücke 1987). Study of the dynamics of lactic acid creation have shown that approximately 50% of lactic acid is released during the first four days of fermentation, and another 33% in the period between Days 4 and 14 (List and Klettner 1978). Kameník et al. (2013) found in the Czech salami (diameter of 75 mm) the maximum levels of lactic acid after 3 days of fermentation. The importance of lactic acid to dry-fermented sausages lies in reducing the pH value in the batter and its effect on the aroma of the product.

The pH value attained in the very first days of the production process in dependence on the type of product (generally 4.8–5.0) influences the microbial stability of the batter (see section 2.8 of this Chapter) and the consistency and colour of the product. A lower pH in the batter (product) speeds the drying of the dry-fermented sausages and, thereby, their production as a whole. Stiebing and Rödel (1989) found that improved water binding capacity of meat proteins at higher pH values results in less diffusion of water from the centre of the product to its surface and, thereby, to less water loss by drying. The LAB act as a significant barrier to the growth of undesirable microorganisms, particularly in the first days of the production process. This antagonistic effect is based on a number of mechanisms, and involves competition for nutrients and living space in general. The optimal environment for the development of LAB is created in dry-fermented sausages from the very beginning, enabling them to thrive better than their competitors. They grow more quickly and use nutrients more quickly meaning they are no longer available to other, more slowly developing, microbes. During their lifetime, LAB excrete substances into the external environment, many of which have a direct antibacterial effect. These may be hydrogen peroxide or organic acids. Bacteriocins have a more complicated structure (de Souza Barbosa et al. 2015).

Lactic acid bacteria are not, however, the only microbial group present in the batter for dry-fermented sausages and, later, in the final product. The batter prepared is a mix of raw meat and applied additives that influence its microflora along with the production plant environment. Starter cultures are added to ensure standardisation of the fermentation process. Mutual interactions between individual groups and strains of microbes are then created in the mixed bacterial population.

2.6 Fermentation and Ripening Temperature

The ripening of dry-fermented sausages is comprised of a first fermentation step, generally lasting between two and five days, followed by a drying stage which may last several weeks (Berardo et al. 2015). The typical sensory properties of the final products are formed and their microbial stability and safety assured during this period which is known by the term ripening.

2.6.1 The Two Types of Dry-Fermented Sausage in Europe

There are generally thought to be two basic types of dry-fermented sausage in Europe – the northern type and the southern type. The northern type is characterised by a rapid fall in pH to beneath 5.0 due to the activity of LAB which convert added saccharides into lactic acid.

Such pronounced acidification is not seen in the southern type of dry-fermented sausage due to the lower quantity of saccharides and the lower fermentation temperatures used. The result is a pH > 5.0 which may even rise during ripening due to the metabolic activity of surface molds (Leroy et al. 2013).

The two types of dry-fermented sausage are also associated with different environmental temperatures under which fermentation and ripening take place. The air temperature set assures the development of lactic acid bacteria whose enzymes induce the fermentation process. The temperature selected in the chamber at the beginning of fermentation depends on the type of product, the type of starter culture used and the desired speed at which pH values are to fall. Higher temperatures (generally 24–25°C, with a maximum of 26°C) are set for rapid fermentation and a rapid fall in pH. Temperatures higher than this are not recommended, as effective hurdles to undesirable bacteria (e.g. salmonellas) have not yet been formed in the batter during this stage. Only nitrite is effective in practice.

The southern type of dry-fermented sausage has been made in the Mediterranean region of Europe for centuries. The climatic conditions and higher air temperatures here guaranteed the (more or less) optimal course of drying at a time at which air-conditioned chambers were not yet part of the technology used by meat producers. Outdoor air temperatures are, in contrast, lower on average in more northerly countries such as Germany and Scandinavia. In the past, air had to be heated in natural drying rooms to dry sausages, and an open fire was lit in these rooms for this purpose. In addition to heat, the burning of wood (usually beech wood) also released smoke, for which reason dry-fermented sausages of the northern type are historically smoked. The southern type of product is characterised by the growth of surface molds – the local conditions did not force manufacturers to use fire (and thereby smoke), and the surface of unsmoked products was naturally colonised by molds whose white or grey-white mycelium created the appearance typical of products of the southern type.

Temperature influences the speed of fermentation of saccharides into lactic acid. Stahnke and Tjener (2007) state that an increase in temperature of 5°C approximately doubles the rate of acid creation. 30°C is generally considered the average temperature for the growth of

Hurdle Technologies in Fermented Meat Production

LAB. Traditional processes, however, generally begin at temperatures of around 25°C, gradually falling to 18°C and less after 5–7 days of the start of production.

2.7 Smoke

Smoke is a mixture of air and gases (gaseous phase) in which solid particles of various sizes are dispersed. The composition of smoke is influenced by the type of wood used, the water content in the wood, the temperature, the air supply and the method of smoke generation (Sielaff and Schleusener 2008).

Smoke is created by a biphasic reaction (Jira 2004). First, thermal decomposition products are formed which are subsequently oxidised in the presence of oxygen. Smoke contains more than a thousand various compounds (Stiebing 2008). More than 300 volatile compounds are now known, primarily phenols, organic acids and carbonyl compounds. The non-volatile fraction is dominated by tar, resin and soot, including undesirable compounds such as polycyclic aromatic hydrocarbons (Jira 2004).

The penetration of gases into the product depends on the prevailing air pressure in the smoking room. The higher the air pressure, the more rapidly and more deeply gas penetrates into the smoked product (Sielaff and Schleusener 2008). The desirable effects smoke has on meat products are surface colouring, aromatisation, a preservative effect and an antioxidant effect. Contributing to its preservative effect are aldehydes (e.g. formaldehyde) and phenols (phenol, methylguaiacol), and also numerous acids (e.g. formic, acetic and benzoic acids) which have an antagonistic effect on bacteria, yeasts and molds. This effect is seen on the surface and in the surface layer beneath the casing. In view of the relatively low content of these substances, however, the preservative effect of smoke is relatively small.

Formaldehyde does have a significant effect in strengthening the structure of natural casings during the smoking process. This effect is caused by the action of smoke on proteins. The elimination of water molecules leads to bonding of formaldehyde molecules with two NH groups, thereby bonding smaller molecules to larger molecules (Figure 3).

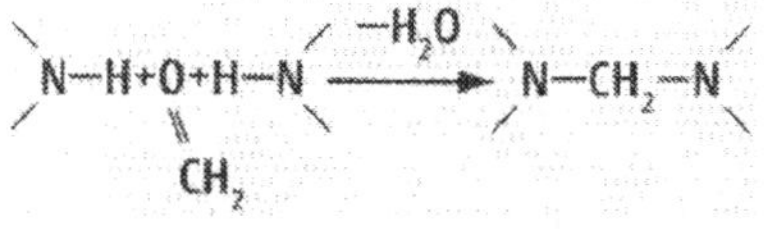

$$N-H+O+H-N \xrightarrow{-H_2O} N-CH_2-N$$
$$\underset{CH_2}{|}$$

Figure 3 The reaction between formaldehyde and protein (Sielaff and Schleusener 2008).

110   *Fermented Meat Products: Health Aspects*

Numerous other bonds cause irreversible cross-linkage which has a great effect on solubility. Native collagen is hardened and becomes insoluble in water. In this way, natural intestine casings obtain the stability against the action of heat required for these products and may then be treated at higher temperatures without splitting (Sielaff and Schleusener 2008).

Smoking dry-fermented sausages is an effective barrier to the growth of surface molds. The cause of unwanted mould growth is generally high air humidity in chambers, inadequate air flow, a low intensity of smoking or an excessive number of sausages on smoking trolleys meaning that the air cannot flow adequately between them and remove water from their surface. Mold almost always appears at the point where two or more sausages touch or where a sausage touches the construction of the smoking trolley.

The use of pimaricin (natamycin), which is a natural product released by the species *Streptomyces natalensis*, is recommended to prevent the growth of mold on the surface of sausages. Pimaricin inhibits the growth of molds on the surface of casings, though its effectiveness falls rapidly. It is effective in the first two weeks. Over a longer period of four weeks or more it is hard to prevent the growth of mold as the drying of the casing surface places considerable limitations of the effectiveness of pimaricin (Labadie 2007). A potassium sorbate solution (20%) can also be used to prevent the growth of moulds on the surface of dry-fermented sausages. According to Feiner (2006), however, natamycin is several times more effective than potassium sorbate.

2.8 Low pH Values

Bacteria maintain a relatively constant pH within their cells even though the external pH may change. Gram-negative bacteria, for example, maintain a pH of between 7.6 and 7.8 (Álvarez-Ordóñez et al. 2012). Organic acids enter microbial cells in undissociated form, and their disassociation then takes place in the intracellular space. This acidifies the cytoplasm and the proton motive force collapses, as a result of which the transport of nutrients to the cell is inhibited and this state leads finally to metabolic exhaustion in the cell interior (Leistner 2000). The low pH of the cytoplasm causes changes to proteins, disruption to normal biological processes and damage to cell structures. It finally results in cell death (Hong et al. 2012).

2.8.1 Bacterial pH Homeostasis

Microorganisms maintain their internal pH (cytoplasm) by means of a pH homeostasis mechanism based on the transport of protons

Hurdle Technologies in Fermented Meat Production　　　111

across the cell membrane. Aerobic bacteria have active H^+ transport associated with electron transfer in the respiratory chain. In contrast, anaerobic organisms perform H^+ transport through molecules of H^+-ATPase with the use of energy gained by ATP hydrolysis (Gandhi and Chikindas 2007). Organic acids can, in this way, contribute towards inhibiting bacteria in the environment as the result of an increase to their energy needs to maintain acid-base balance. Microbes with a fermentative metabolism show a greater range of internal pH values than aerobic bacteria, for which reason they are more tolerant of low pH values, a fact that can be documented by the example of LAB (McDonald et al. 1990).

As LAB do not have a functioning respiratory system, they have to obtain energy by the phosphorylation of suitable substrates. Two basic fermentation pathways are involved in the case of hexoses. The homofermentative pathway is based on glycolysis (known as the Embden-Meyerhof-Parnas pathway) and produces only lactic acid. The heterofermentative pathway also produces a significant amount of carbon dioxide and ethanol or acetic acid, in addition to lactic acid (Wright and Axelsson 2012). The fermentation process in dry-fermented sausages is based on the homofermentation of the saccharides present in the batter.

2.8.2 *Saccharides and pH*

The glucose and glycogen content in the meat of livestock animals after slaughter is practically negligible. The total content of saccharides in the skeletal muscle immediately after slaughter falls within a range of up to one percent, of which around 75% is lactic acid, 13% glucose-6-phosphate, around 8% glycogen, and the remainder glucose and other metabolites (Kauffman 2012). Fermentable saccharides are added to the batter for this reason with the aim of guaranteeing the formation of a quantity of lactic acid optimal from the viewpoint of ensuring the proper course of ripening (Lücke 1985). The composition and quantity of saccharides added to the batter influence the extent of lactic acid formation and, thereby, the final pH value attained. The pH of the batter falls during fermentation from an initial 5.7–6.0 to its lowest level, which fluctuates from 5.5 (dry sausages with a long ripening period with no addition of saccharides) to around 4.6, depending on the type of sausage. The time period is influenced by the temperature at which fermentation takes place, ranging from twelve hours to a number of days. The type of starter culture added also plays a role.

In practice, advantage is taken of the effect of temperature on the course of fermentation and, thereby, on a fall in pH values in the batter to speed up the production cycle for dry-fermented sausages.

Suppliers of additives (starter cultures, combined mixes of seasonings and saccharides, and other additives) offer manufacturers the chance of shortening the ripening period which is, of course, also reflected in a reduction to production costs. The temperature at which fermentation usually began in the past was around 24°C. Today, there are starter cultures on the market for which an initial temperature of 26 to 28°C is recommended. As effective hurdles to undesirable bacteria have yet to be created in the first hours after the filling of dry-fermented sausages in intestine casings, however, it is not a very good idea to set air temperatures in excess of 25°C.

The addition of 0.3% of glucose is the optimal dose for sausages with a medium ripening period (four weeks or more), while 0.5–0.7% is suitable for products with a shorter ripening period (up to three weeks). The recommended doses of saccharides allow a fall in pH to 4.8–5.0 which corresponds to a content of around 25 g of lactic acid per kg of dry matter in the sausages. The initial value of the pH in the sausage batter is also important to the fall in pH values. The doses given above apply if the initial pH is around 5.7. If the initial pH is higher than 6.0, however, the dose of saccharides must be increased proportionately to as much as one percent (Keim and Franke 2007). Klettner et al. (1999) measured the final pH values (ten days after preparation of the batter) following the addition of various levels of dextrose to the batter for dry-fermented sausages (diameter 60 mm, with the addition of a starter culture): 4 g/kg dextrose: 4.94; 7 g/kg: 4.71; 10 g/kg: 4.51.

2.8.3 Acid Tolerance Response

Tolerance to low pH values is important to food-borne agents not merely in view of their survival in fermented foodstuffs, but also from the viewpoint of their virulence. The gastric juices are the first line of defence that these bacteria have to overcome in the human organism (Berk et al. 2005). Although the pH of the gastric juices is around 1.0–3.0, many bacteria overcome this hurdle and are capable of causing illness (Hong et al. 2012). Different bacteria have different strategies for confronting the low pH values in this environment. Many bacteria use a system of decarboxylation of amino acids, such as glutamic acid, lysine or arginine. Amines are formed (γ-aminobutyric acid, cadaverine or agmatine). H^+ protons are consumed during this transformation and the pH of the cytoplasm increases as a result.

The adaptation mechanisms of salmonella are known as Acid Tolerance Response (ATR). The decarboxylation system in *Salmonella enterica* serovar Typhimurium includes lysine (Álvarez-Ordóñez et al. 2012). It is comprised of three constituents – CadC (the operon *cadBA*),

Hurdle Technologies in Fermented Meat Production 113

CadA (the enzyme lysine decarboxylase) and CadB (lysine-cadaverine antiporter). CadC acts as a signal sensor and as a transcription regulator at low pH values in the external environment and in the presence of lysine, resulting in the transcription of operon *cadBA*. This is followed by the decarboxylation of intracellular lysine catalysed by CadA, during which a proton is consumed and the value of the intracellular pH increases. The biogenic amine cadaverine is created by the reaction. Intracellular cadaverine is exchanged for extracellular lysine by the action of antiporter CadB.

Another component of the ATR is comprised of a number of groups of acid shock proteins (ASP). Their role is to prevent macromolecular damage caused by low pH values or to repair this damage (Álvarez-Ordóñez et al. 2012). The defence of bacterial cells against low pH values also includes change to the composition of the fatty acids of the cell membrane in order to preserve its function. In the case of *S. enterica*, serovar Typhimurium, Álvarez-Ordóñez et al. (2012) found a decline in the proportion of unsaturated fatty acids at the expense of saturated fatty acids and, most importantly, an increase in the percentage of cyclical fatty acids as a response to an acidic environment.

2.9 Low a_w Values

The values of pH and a_w are the decisive factors in the reduction of food-borne agents and achieving the microbial stability of dry-fermented sausages (Gareis et al. 2010a, Matagaras et al. 2015b). Fermentation processes themselves do not suffice to assure the long shelf life of fermented meat products. Their importance as hurdles to the growth of undesirable microorganisms is limited by their effectiveness over time. The most important and most stable hurdle begins to emerge during the ripening – a low water activity value or a_w. This is an ancient method of extending the shelf life of meat. The combination of the addition of salt and subsequent drying reduces the content of water available to microorganisms (a_w value). This process has often been connected with smoking the surface of the meat (meat product), with smoke protecting the product with bacteriostatic and mycostatic substances (Honikel 2007).

Gram-negative and gram-positive bacteria respond differently to lower a_w values (Table 3, Medić, Chapter 3 in this book). Gram-negative bacteria show no growth at a_w values < 0.94. In contrast, *Listeria monocytogenes* has been capable to grow at a_w values of as little as 0.89, and *Staphylococcus aureus* down to a value of 0.84 (Gareis et al. 2010a).

Table 3 Water activity and the growth of microorganisms in foodstuffs

Range of a_w	Microorganisms generally inhibited at the lowest value within this range
1.00–0.95	*Pseudomonas, Escherichia, Proteus, Klebsiella, Bacillus, Cl. perfringens*, certain yeasts
0.95–0.91	*Salmonella, Vibrio parahaemolyticus, Cl. botulinum, Serratia, Lactobacillus, Pediococcus*, certain molds, yeasts (*Rhodotorula, Pichia*)
0.91–0.87	*Micrococcus*, many yeasts (*Candida, Torulopsis, Hansenula*)
0.87–0.80	*St. aureus*, the majority of molds (mycotoxigenic *Penicillium* spp.), the majority of *Saccharomyces* spp., *Debaryomyces*
0.80–0.75	The majority of halophilic bacteria, mycotoxigenic *Aspergillus* spp.
0.75–0.65	Xerophilic molds (*Aspergillus chevalieri, A. candidus, Wallemia sebi*), *Saccharomyces bisporus*
0.65–0.60	Osmophilic yeasts (*Saccharomyces rouxii*), certain molds (*A. echinulatus, Monascus bisporus*)
0.60–0.20	No microbial growth

Source: Fontana and Campbell 2004

2.9.1 Influences on a_w in Dry-Fermented Sausages

The a_w value as an important indicator of shelf life is influenced by the addition of NaCl (nitrite curing salt), fermentation and, most importantly, the drying process. The speed of the drying process is determined by the external and internal diffusion of water in the product. This diffusion is dependent on the water content gradient between the product and its surroundings (the relative air humidity and water activity), the type of casing used, the quality of the product surface, the speed of airflow and the air temperature. The composition of the product (the ratio of meat to fat), the pH, the degree of comminution of the batter (the size of the fat particles), the mixing of the batter and the quality of the comminution process all, however, also have an effect on the internal diffusion of water. What reduces internal diffusion of water and, thereby, its transport to the surface of the product? This might be a high fat content, the comminution of the batter into extremely fine particles, or a film of fat covering pieces of meat formed during comminution. The basic precondition to the drying of dry-fermented sausages is the difference in humidity between the surrounding air and the product on one hand, and the core of the product and its surface layer on the other (Feiner 2006).

Water occurs in foodstuffs as free water or is bound in various ways to various components or formations within the food. *Free water* is the essential reaction environment for the great majority of the chemical and microbiological processes that change the properties of foods. The high content of free water in meat is one of the principal reasons for its short shelf life. Foods are dried with the aim of both removing free water (which is the living environment for microbes) from the product and increasing the osmotic pressure in the product. Both these factors worsen living conditions for microorganisms, diminishing their growth and metabolic activity and, in certain cases, leading to the death of vegetative forms of microbes.

Certain principles must be observed during the drying of dry-fermented sausages, as the aim is to obtain a standardised product of high quality. On one hand, operational economics dictate that drying is performed as quickly as possible and achieved with the lowest possible operating costs. On the other hand, the drying process must also take product properties into consideration. A gradual course of drying is absolutely essential in the case of dry-fermented sausages. The uniform movement of water from the centre of the product to its surface, where molecules of water are evaporated into the surrounding air, must be assured.

A distinction is sometimes made between the process of fermentation and the drying process. It is, however, impossible to state unambiguously when the fermentation process ends and the drying process begins. Sausages lose water from the very beginning of fermentation.

Water evaporates from the surface layer of sausages because the relative air humidity in the environment in which drying takes place is lower than a_w of the sausages. This leads to an increase in the salt level in this outer layer. The difference in water content between the centre of the product and its outer layer must be equalised, for which reason water diffuses from the centre of the product to the surface. The external surface layer of the sausages always has a lower water content than the centre of the product. Dry-fermented sausages can be said to dry from the inside out. The speed of evaporation of water from the product surface must be adapted to the speed of water diffusion from the centre to the outside. The formation of case hardening (crusty surface) will occur if moisture is removed from the sausage surface more rapidly than diffusion inside the product can keep up with (Feiner 2006).

The drying process is influenced by the following parameters:

- the size of the particles of meat and fat in the batter
- the diameter of the casing
- the fat content in the batter
- the airflow speed
- the relative air humidity
- the air temperature

A high air temperature, high airflow speed and low relative air humidity increase the intensity of water evaporation from the sausage surface. A reduction in temperature, low airflow speed and increased relative air humidity, in contrast, slow down the drying process. These parameters have to be adjusted in order to dry the product as quickly as possible without the occurrence of case hardening (Feiner 2006).

Drying is also influenced by the degree of comminution of the batter. Molecules of water that migrate from the centre of the product to its surface come up against particles of fat and meat. If the particles of the batter are too small, their number is much higher and the flow of migrating water molecules changes direction many times as a result and its pathway to the surface is lengthened.

Fat in meat products also slows drying, though it also protects the surface against excessive drying. The higher the fat content in the product, the lower is the intensity of drying. The degree of comminution of the batter must also be taken into consideration. If the batter is overloaded, the fat will spread and internal barriers to the migration of water will be formed. Water is then unable to diffuse sufficiently quickly from the centre of the product to the external layer, the casing tends to harden quickly, and the surface may sink and wrinkles form. The movement of water from the batter to the sausage surface during drying may be supported by the addition of fibre (up to 3%). Roth and Sieg (2003) state that wheat fibre creates a three-dimensional network during mixing of the batter. This acts as a drainage system allowing water to get more easily from the centre of the product to the surface where it evaporates into the surrounding air. This can limit or entirely prevent the danger of formation of a soft sausage centre and the formation of case hardening. It may also shorten the period of time necessary for product drying.

Fluctuating relative air humidity is sometimes applied in the ripening and drying of dry-fermented sausages. The drying process speeds up when the relative air humidity falls, as a result of which water cannot migrate quickly enough from the centre of the product to its surface to increase its humidity (and the humidity of casing also). The relative air humidity must, for this reason, be adjusted to ensure the uniformity of water diffusion in the product and avoid errors during the drying process. The speed of airflow is another external control quantity in the process of drying dry-fermented sausages, in addition to the relative air humidity. The lower the relative air humidity, the larger the gradient between the surface a_w value and the humidity of the surrounding air and the faster product drying progresses. Rödel (1985) recommended that the difference between one hundred times the water activity and the relative air humidity be no higher than 5 ($a_w \times 100$ – relative air humidity ≤ 5). According to other authors (Keim and Franke 2007), this difference should be only 3. Hermle et al. (2003) also recommended a difference between the partial pressure of water vapour on the surface of the sausages and the surrounding air of 3–5.

Hurdle Technologies in Fermented Meat Production

The environmental temperature is also important to the drying process. An increase in temperature increases internal water diffusion in the product. Water migrating to the surface more quickly means that either the hardening of the surface of the casing occurs later or the process is not so intensive. The reason for this is the fact that surface water is continually replaced during evaporation by water diffusing from the centre of the product which keeps the surface continually wet (Andrés et al. 2007). On the other hand, the higher the temperature, the greater is the risk of multiplication of undesirable bacteria. There is also a risk of the release of fat which may create hydrophobic barriers in the product with a consequent problem of excessive hardening of the sausage surface. Hard pork back fat with a higher melting point must be used for this reason.

Sausages of a smaller diameter have higher drying losses (and thereby a lower water content) than products of a large diameter. Sausages with a casing of a larger diameter have a smaller surface in comparison with sausages of a smaller diameter. Dry-fermented sausages of small diameter have the largest total surface in relation to their weight, for which reason the speed of drying is highest in these products.

Today, the ripening of dry-fermented sausages takes place in air-conditioned chambers. The external parameters (air temperature, relative air humidity, speed of airflow) in these chambers must be set so as to avoid excessively rapid drying and the formation of case hardening on one hand, while assuring that drying does not take place so slowly as to allow the growth of undesirable moulds, yeasts and bacteria on the product surface on the other hand.

2.9.2 Drying in QDS Technology

QDS (Quick – Dry – Slice) is a new method for preparing sliced dry meat products that can undoubtedly be called 21st century technology (Arnau et al. 2007). The primary aim of QDS technology is to shorten the production period for sliced dry meat products (Stollewerk et al. 2013). Additional effects are also obtained in the form of a "just-in-time" production process, the faster development of new products, and savings in space and energy over the traditional method of production. What is the essence of QDS technology? The traditional production of dry-fermented sausages takes in two basic steps. The first step is preparation of the batter and its filling into casings. The second phase is fermentation and ripening, during which a number of physical, physicochemical and biochemical processes take place that change the sausage batter into final dry products that are highly stable in microbial terms and can be stored at room temperature. The stage of fermentation and subsequent ripening lasts, depending on the type of product, a number of weeks or

118   *Fermented Meat Products: Health Aspects*

even months. Today, modern sales methods on the retail market demand a large proportion of sliced dry sausages generally packed in a modified atmosphere and designed for self-service sales.

The QDS system is designed for the preparation of exactly this kind of packed sausage. Both production process phases (the preparation and filling of the batter, fermentation and ripening) are retained. Fermentation and ripening (drying) are, however, modified in such a way that the entire production cycle takes three days in practice. The batter being filled into casings is followed by two-day fermentation during which the primary texture of the batter is created thanks to lactic acid and a fall in pH. This phase is followed by freezing of the sausages and their slicing. Freezing takes place at temperatures of around –10°C and is necessary in view of the fact that after 48-hr fermentation the product has not developed the kind of texture that would allow for industrial slicing. Product drying follows in a current of air with precisely defined parameters of temperature, relative humidity and flow speed and can generally dry the slices of sausage within 30-45 min. While the traditional production process focuses on the sausage as the basic unit of production, the QDS® process focuses on the individual slice. One fundamental difference (forgetting the time aspect) can be found in the production diagram. While drying comes before slicing in the traditional technology for the production of sliced fermented sausages, slicing comes before drying in the QDS process.

The advantages of the new system are self-evident – a shortening of the production period and assurance of the standardisation of final products. The precisely defined drying process applied in the QDS environment is not possible in traditional drying chambers. It is not possible to obtain an entirely balanced airflow in a large space, as there are a number of variables influencing the process as a whole, in particular how full the chamber is, the size of the sausages, the casings used, etc. These factors are entirely absent during drying in the QDS process. The drying of individual slices takes place more quickly and the evaporation of water vapour cools the slices of sausage, so a higher temperature can be used during QDS drying without impairing the quality of the final products (Table 4).

Table 4 Comparison of the traditional method of preparing dry-fermented sausages with QDS: microclimatic drying conditions

Traditional procedure		*QDS process*®	
Air temperature	10–18°C	Air temperature	25–35°C
Relative air humidity	70–80%	Relative air humidity	20–40%
Temperature of product	10–18°C	Temperature of product	16–22°C

Source: Comaposada et al. 2013b

Hurdle Technologies in Fermented Meat Production   119

A number of tests have been conducted in Spain in recent years to demonstrate the reliability of the QDS system and to compare it with traditional processes (Comaposada et al. 2013a, Garriga et al. 2013, Comaposada et al. 2013b). Extremely similar results were found when comparing the value of TBARS during the drying of a product of the US Pepperoni type and sausages of the German type with the use of conventional technology and the QDS process; the differences found were not statistically significant (Comaposada et al. 2013a). Similarly, only statistically insignificant differences in parameters of instrumental colour analysis (L*, a*, b*) and sensory assessment were recorded.

A consortium of companies (Metalquimia, IRTA and Casademont) has contributed to the development of products in which NaCl is completely replaced by potassium chloride (KCl), potassium lactate and saccharose (Stollewerk et al. 2013). This new procedure is patent protected and has been successfully tested both on sausages (Chorizo) and in the production of dried hams. HPP technology (high pressure processing) was used in testing the safety of these products.

A number of analyses have also been performed to assess the safety of final products prepared with QDS technology in comparison with the traditional procedure, and the results have confirmed the reliability of the modern process (Stollewerk et al. 2011).

3   HURDLES IN THE PRODUCTION OF CURED AIR-DRIED MEAT PRODUCTS

Meat for these products must be kept at the lowest possible level of microbial contamination. The values recommended are log 2–3 cfu/g or cm^2 and pH-value (ultimate) 5.5–5.8. The properties of meat are important from the viewpoint of microbial stability. The absence of comminution means that the rapid onset of the effect of additives (e.g. NaCl, starter cultures) that can stabilise the batter of fermented sausages in the first days of ripening is not possible.

Salt is indispensable to dried whole-muscle products (as it is to dry-fermented sausages). Nitrite and nitrate play a certain role in the production of products with a rapid course of ripening, though only sodium chloride is used for long-ripening hams of high quality. Salting and subsequent drying lead to a fall in a_w which is the decisive factor in the microbial stability of products, for which reason a sufficient quantity of salt must be added to dried hams.

Meat for dried hams should have a pH of 5.5–5.8 which is the first hurdle to the growth of bacteria. It is absolutely essential that the legs (and other types of meat) are chilled rapidly to a temperature of 5°C after slaughter. This factor is even more important to product stability than the pH of the raw meat. The low temperature of the meat must be maintained until such time as the content of salt attains a level decisive to blocking undesirable microbial activity.

Table 5 Parameters limiting the growth of selected food-borne agents

Gram-negative bacteria	Temperature (°C)	Water activity (a_w)	pH
Salmonella enterica	5.2–46	0.94	4.2–9.5
Campylobacter jejuni	30–47	0.98	4.9–9.0
Escherichia coli O157:H7	8–46	0.95	4.2–9.5
Shigella sonnei	6–47	0.95	4.8–9.0
Yersinia enterocolitica	-1.3–42	0.94	4.2–10.0
Gram-positive bacteria			
Listeria monocytogenes	0–43	0.89	4.5–9.0
Staphylococcus aureus	8–45	0.84	4.5–9.3
Clostridium botulinum, proteolytic types A, B, F	10–48	0.94	4.6–9.0
Clostridium botulinum, non-proteolytic types B, E, F	3.3–45	0.97	5.0–9.0
Bacillus cereus	5–50	0.91	4.4–9.3

Source: Gareis et al. 2010a

The dose of salt used is generally 32–35 g/kg of raw meat (Feiner 2006). The product is stable in microbial terms in respect of bacteria of the *Enterobacteriaceae* family following the fall in weight during salting and a balance in the salt content in the meat being attained when the a_w value falls to 0.95 and less. At a_w value of 0.95, the level of NaCl in the product is between 4.3 and 4.5 percent. The hams may then be relocated to areas with a higher temperature at which ripening associated with further drying takes place. At this stage, the characteristic aroma, to which endogenous meat enzymes contribute, develops without the participation of microorganisms (Leistner 1985).

Nitrite is added for product colouring and aromatisation and as a further hurdle to undesirable bacteria at the stage of salting and the balancing of the salt concentration. Certain traditional products (e.g. Parma Ham) are, however, prepared with only the addition of NaCl (sea salt). Ascorbate or erythorbate, generally applied at 0.6–1.0 g/kg of meat, is added to support colouring processes in the meat. Seasonings have an effect on taste and aroma. Garlic, pepper and coriander are used. Protective cultures which influence the stability and fullness of colour may occasionally be applied.

Hurdle Technologies in Fermented Meat Production   121

The product may be moved from refrigeration temperatures (2 to 5°C) to an environment with temperatures of 22 to 24°C to get enzymatic activity going after an a_w value of 0.95 is attained. The product will then remain in this environment for 24–48 hours. The temperature is then reduced to 16–18°C at relative air humidity of 76–80% (2–3 days). If the product is smoked, the first application of smoke occurs after the product has been taken out of refrigeration temperatures; the smoke temperature ranges between 20 and 25°C. The amount of smoke depends on the type of product. Smoking is generally performed two or three times a day for a period of 1–2 hours, with the procedure being repeated as required. After another few days, the temperature is reduced to 12–15°C, the relative air humidity falls to 72–75%, and the drying phase begins. The airflow speed at this stage amounts to 0.3 to 0.4 m/s, and is further reduced to as little as 0.1 m/s at the end of drying (Feiner 2006).

The speed of drying is determined by factors such as the volume of filling of the air-conditioned drying chamber and the size of the individual pieces of dried meat. An increased airflow speed, an increased air temperature and a reduced relative air humidity generally speed up drying. Opposite parameters, i.e. a weak airflow, low temperature and high relative air humidity, slow drying.

Changes occur to products during the course of drying – the weight of meat falls and the product's taste, aroma and tenderness develop. The fall in weight influences the toughness of hams and increases their sliceability. The a_w value falls beneath 0.89 with ongoing drying. The product becomes durable and need not be kept at refrigeration temperatures.

Further preparations are performed before the final products are released onto the market – some products are packed whole (often on the bone), while others are sliced or cut up into smaller parts. Vacuum-packed hams with an a_w value < 0.89 are extremely stable products even when not kept at refrigeration temperatures. Attention must, however, be paid to the possibility of the condensation of water during the packing which may lead to an increase in the value of a_w on the product surface to as much as 1.00, in which case bacteria have enough water for their growth (Feiner 2006).

Surface condensation of water is not generally a problem for products packed whole or sliced into large parts or thick slices (2–5 cm). It is not necessary to chill these products before such preparation. Condensation of moisture in the air on the product surface does not occur if a dried product is taken from the drying chamber (generally at 12 to 15°C) and relocated for slicing and/or packing. The reason for this is the fact that the temperature of the environment in which slicing and packing occur is lower than the temperature of the products. If dried hams are sliced into thin slices, however, the product is often exposed to temperatures beneath 0°C before the slicing operation to achieve improved sliceability. The air temperature in the slicing environment is, however, generally higher. In view of the condensation of water, the a_w can no longer be considered a safe hurdle to undesirable microbial growth and the packed sliced products are stored at refrigeration temperatures. Sliced dried hams are packed in a modified atmosphere (MAP with a mixture of gases: 30–40% CO_2 and 60–70% N_2).

4 CONCLUSION

There are manufacturing procedures that involve the progressive application of factors known as hurdles and that determine the microbial stability and safety of fermented meats and their sensory properties and nutritional value.

Hurdles are applied in the production of fermented sausages that either act in the sausage batter itself (internal) or consist of influences on the external environment (external). Internal hurdles include substances with a preservative action (in particular sodium chloride and nitrites), the redox potential, competitive microflora, a low pH value and a low water activity (a_w) value. The addition of salt and reduction of the redox potential are decisive in the early stage of production, i.e. in the first tens of hours. The key factors from the viewpoint of the microbial stability of the final products and, in particular, their safety are the pH value and a_w value. The principal external factors include the environmental temperature and the use of smoke.

Hurdles are applied to a different extent during the manufacturing of whole-muscle products (cured air-dried meat products) – the pH value of the meat used for processing, the environmental temperature, the effect of NaCl and a low a_w value.

Key words: *NaCl, nitrite, redoxpotential, pH, lactic acid bacteria, water activity, food-borne agents*

REFERENCES

Adams, M. and R. Mitchell. 2002. Fermentation and pathogen control: a risk assessment approach. Int. J. Food Microbiol. 79: 75–83.

Álvarez-Ordóñez, A., M. Prieto, A. Bernardo, C. Hill and M. López. 2012. The acid tolerance response of *Salmonella* spp.: An adaptive strategy to survive in stressful environments prevailing in foods and the host. Food Res. Int. 45: 482–483.

Andrés, A., J.M. Barat, R. Grau and P. Fito. 2007. Principles of drying and smoking. pp. 37–48. *In:* Toldrá (ed.) Handbook of fermented meat and poultry. Blackwell Publishing. Ames. Iowa.

Arnau, J., X. Serra, J. Comaposada, P. Gou and M. Garriga. 2007. Technologies to shorten the drying period of dry-cured meat products. Meat Sci. 77: 81–89.

Berardo, A., E. Claeys, E. Vossen, F. Leroy and S. De Smet. 2015. Protein oxidation affects proteolysis in a meat model system. Meat Sci. 106: 78–84.

Berk, P.A., R. de Jonge, M. H. Zwietering, T. Abee and J. Kieboom. 2005. Acid resistance variability among isolates of *Salmonella enterica* serovar Typhimurium DT104. J. Appl. Microbiol. 99: 859–866.

Brasca, M., S. Morandi, R. Lodi and A. Tamburini. 2007. Redox potential to discriminate among species of lactic acid bacteria. J. Appl. Microbiol. 103: 1516–1524.

Bruckner, S., A. Albrecht, B. Petersen and J. Kreyenschmidt. 2012. Characterization and comparison of spoilage processes in fresh pork and poultry. J. Food Quality 35: 372–382.

Caldeo, V. and P.L.H. McSweeney. 2012. Changes in oxidation-reduction potential during the simulated manufacture of different cheese varieties. Int. Dairy J. 25: 16–20.

Comaposada, J., J. Arnau, J.M. Monfort, D. Sanz, L. Freixanet and J. Lagares. 2013a. Shelf life stability of fermented sausages produced with the QDS process® technology: US Pepperoni and German Salami case studies. pp. 231–243. *In:* Metalquimia. Technological Articles. Metalquimia, S. A. Girona, Spain.

Comaposada, J., J. Arnau, M. Garriga, P. Gou, J.M. Monfort, M. Xargayó, L. Freixanet, J. Lagares, J. Bernardo and M. Corominas. 2013b. Fast drying of dry-cured meat products Quick – Dry – Slice (QDS process®) technology. pp. 253–259. *In:* Metalquimia. Technological Articles. Metalquimia, S. A. Girona, Spain.

De Souza Barbosa, M., S.D. Todorov, I. Ivanova, J.M. Chobert, T. Haertlé and B.D.G. de Melo Franco. 2015. Improving safety of salami by application of bacteriocins produced by an autochthonous *Lactobacillus curvatus* isolate. Food Microbiol. 46: 254–262.

Duranton, F., S. Guillou, H. Simonin, R. Chéret and M.de Lamballerie. 2012. Combined use of high pressure and salt or sodium nitrite to control the growth of endogenous microflora in raw pork meat. Innov. Food Sci. Emerg. 16: 373–380.

Feiner, G. 2006. Meat products handbook. Practical science and technology. CRC Press, Boca Raton, Florida, USA.

Fontana, A. J. and C.S. Campbell. 2004. Water activity. pp. 39–54. *In:* L.M.L. Nollet (ed.). Handbook of food analysis, volume 1: Physical characterization and nutrient analysis. Second Edition, Marcel Dekker, Inc., New York.

Gandhi, M. and M.L. Chikindas. 2007. *Listeria*: A foodborne pathogen that knows how to survive. Int. J. Food Microbiol. 113: 1–15.

Gareis, M., J. Kabisch, R. Pichner and H. Hechelmann. 2010a. Vorkommen von Lebensmittelinfektionserregern in Minisalamis (Occurence of food borne pathogens in minisalamis). Fleischwirtschaft 90, 4: 179–183.

124   *Fermented Meat Products: Health Aspects*

Gareis, M., J. Kabisch, R. Pichner and H. Hechelmann. 2010b. Absterbekinetik von *Salmonella* spp. in Minisalamis (Behaviour and survival of *Salmonella* spp. in minisalami). Fleischwirtschaft 90, 5: 98–106.

Gerhardt, U. 1994. Zutaten und Zusatzstoffe für die Herstellung von schnittfester und streichfähiger Rohwurst. pp. 99–114. *In*: H.J. Buckenhüskes (ed.). Beiträge vom 1. Stuttgarter Rohwurstforum. Gewürzmüller GmbH, Stuttgart.

Garriga, M., G. Ferrini, T. Aymerich, J. Comaposada, D. Sanz, M. Xargayó, L. Freixanet and J. Lagares. 2013. Validation of the QDS Process® of Pepperoni sausages through a challenge test with *E. coli* O157:H7 surrogate. pp. 245–251. *In*: Metalquimia. Technological Articles. Metalquimia, S. A. Girona, Spain.

Gutierrez, C., T. Abee and R. Booth. 1995. Physiology of the osmotic stress response in microorganisms. Int. J. Food Microbiol. 28: 233–244.

Hermle, M., T. Jesinger, P. Gschwind, U. Leutz, V. Kottke and A. Fischer. 2003. Strömungs- und Transportvorgänge in Rohwurst-Reifungsanlagen. Fleischwirtschaft 83, 11: 74–81; 12: 32–35.

Hill, C., P.D. Cotter, R.D. Sleator and C.G.M. Gahan. 2002. Bacterial stress response in *Listeria monocytogenes*: jumping the hurdles imposed by minimal processing. Int. Dairy J. 12: 273–283.

Hong, W., Y. E. Wu, X. Fu and Z. Chang. 2012. Chaperone-dependent mechanisms for acid resistence in enteric bacteria. Trends Microbiol. 20: 328–334.

Honikel, K. O. 2007. Principles of curing. pp. 17–30. *In:* Toldrá (ed.) Handbook of fermented meat and poultry. Blackwell Publishing. Ames. Iowa.

Honikel, K. O. 2008. The use and control of nitrate and nitrite for the processing of meat products. Meat Sci. 78: 68–76.

Ignatova, M., H. Prévost, I. Leguerinel and S. Guillou. 2010. Growth and reducing capacity of *Listeria monocytogenes* under different initial redox potential. J. Appl. Microbiol. 108: 256–265.

Jira, W. 2004. Chemische Vorgänge beim Pökeln und Räuchern. Fleischwirtschaft 84, 5: 235–239; 6: 107–111.

Kabisch, J., R. Scheuer, W. Rödel and M. Gareis. 2008. Untersuchungen zur mikrobiologischen Wirksamkeit von Natriumnitrit bei Rohwursterzeugnissen (Influence on the microbial effect of sodium nitrite in raw fermented sausage). Forschungsprojekt 04OE003/F, Bundesprogramm Ökologischer Landbau, Bonn, BÖL-Bericht – D 14568.

Kameník, J., P. Steinhauserová, A. Saláková, Z. Pavlík, G. Bořilová, L. Steinhauser and J. Ruprich. 2013. Influence of various pork fat types on the ripening and characteristics of dry fermented sausage. Czech J. Food Sci. 31: 419–431.

Kauffman, R.G. 2012. Meat composition. pp. 45–61. *In*: Y. H. Hui (ed.). Handbook of meat and meat processing. 2nd edition. CRC Press, Boca Raton, Florida.

Keim, H. and R. Franke. 2007. Fachwissen Fleischtechnologie. Deutscher Fachverlag, Frankfurt am Main.

Kim, B.H. and G. M. Gadd. 2008. Bacterial physiology and metabolism. Cambridge university press. New York.

Klettner, P.G., W. Rödel and K. Hofmann. 1999. Bedeutung der minimalen Wasserbindung bei schnittfester Rohwurst. Fleischwirtschaft 79, 9: 112–117.

Labadie, J. 2007. Spoilage microorganismus: risks and control. pp. 421–426. *In:* Toldrá (ed.) Handbook of fermented meat and poultry. Blackwell Publishing. Ames. Iowa.

Hurdle Technologies in Fermented Meat Production

Larsen, N., B.B. Werner, F.K. Vogensen and L. Jespersen. 2015. Effect of dissolved oxygen on redox potential and milk acidification by lactic acid bacteria isolated from a DL-starter culture. J. Dairy Sci. 98: 1640–1651.

Leistner, L. 1985. Allgemeines über Rohwurst und Rohschinken. pp. 1–29. *In:* L. Leistner (ed.) Mikrobiologie und Qualität von Rohwurst und Rohschinken. Bundesanstalt für Fleischforschung Kulmbach.

Leistner, L. 2000. Basic aspects of food preservation by hurdle technology. Int. J. Food Microbiol. 55: 181–186.

Leroy, F., A. Geyzen, M. Janssens, L. de Vuyst and P. Scholliers. 2013. Meat fermentation at the crossroads of innovation and tradition: A historical outlook. Trends Food Sci. Tech. 31: 130–137.

List, D. and P.G. Klettner. 1978. Die Milchsäurebildung im Verlauf der Rohwurstreifung bei Starterkulturzusatz. Fleischwirtschaft 58: 136–139.

Liu, C.G., C. Xue, Y.H. Lin and F.W. Bai. 2013. Redox potential control and applications in microaerobic and anaerobic fermentations. Biotechnol. Adv. 31: 257–265.

Løvdal, T. 2015. The microbiology of cold smoked salmon. Food Control 54: 360–373.

Lücke, F.K. 1985. Mikrobiologische Vorgänge bei der Herstellung von Rohwurst und Rohschinken. pp. 85–102. *In:* L. Leistner (ed.) Mikrobiologie und Qualität von Rohwurst und Rohschinken. Bundesanstalt für Fleischforschung Kulmbach.

Lücke, F.K. 2003. Einsatz von Nitrit und Nitrat in der ökologischen Fleischverarbeitung: Vor- und Nachteile. Fleischwirtschaft 83, 11: 138–142.

Lücke, F.K. 2008. Nitrit und die Haltbarkeit und Sicherheit erhitzter Fleischerzeugnisse. Fleischwirtschaft 88, 12: 91–94.

Mataragas, M., F. Rovetto, A. Bellio, V. Alessandria, K. Rantsiou and L. Decastelli. 2015a. Differential gene expression profiling of *Listeria monocytogenes* in Cacciatore and Felino salami to reveal Potential stress resistance biomarkers. Food Microbiol. 46: 408–417.

Matagaras, M., A. Bellio, F. Rovetto, S. Astegiano, L. Decastelli and L. Cocolin. 2015b. Risk-based control of food-borne pathogens *Listeria monocytogenes* and *Salmonella enterica* in the Italian fermented sausages Cacciatore and Felino. Meat Sci. 103: 39–45.

McDonald, L.C., H.P. Fleming and H.M. Hassan. 1990. Acid tolerance of *Leuconostoc mesenteroides* and *Lactobacillus plantarum*. Appl. Environ. Microbiol. 56: 2120–2124.

Nychas, G.-J.E., P.N. Skandamis, C.C. Tassou and K.P. Koutsoumanis. 2008. Meat spoilage during distribution. Meat Sci. 78: 77–89.

Ordóñez, J.A. and L. de la Hoz. 2007. Mediterranean products. pp. 333–347. *In:* Toldrá (ed.) Handbook of fermented meat and poultry. Blackwell Publishing. Ames. Iowa.

Pisacane, V., M.L. Callegari, E. Puglisi, G. Dallolio and A. Rebecchi. 2015. Microbial analyses of traditional Italian salami reveal microorganisms transfer from the natural casing to the meat matrix. Int. J. Food Microbiol. 207: 57–65.

Połka, J., A. Rebecchi, V. Pisacane, L. Morelli and E. Puglisi. 2015. Bacterial diversity in typical Italian salami at different ripening stages as revealed by high-throughput sequencing of 16S rRNA amplicons. Food Microbiol. 46: 342–356.

Rödel, W. 1985. Rohwurstreifung. Klima und andere Einflußgrößen. pp. 60–84. *In:* L. Leistner (ed.) Mikrobiologie und Qualität von Rohwurst und Rohschinken. Bundesanstalt für Fleischforschung Kulmbach.

Rödel, W., R. Scheuer, A. Stiebing and P.G. Klettner. 1992. Messung des Sauerstoffgehaltes in Fleischerzeugnissen. Fleischwirtschaft 72, 7: 966–970.

Rödel, W. and R. Scheuer. 1998a. Das Redoxpotential bei Fleisch und Fleischerzeugnissen. 1. Bedeutung, Messung und Berechnung des Redoxpotentialwertes. Fleischwirtschaft 78, 9: 975–981.

Rödel, W. and R. Scheuer. 1998b. Das Redoxpotential bei Fleisch und Fleischerzeugnissen. 2. Typische Redoxpotentiale bei Fleisch und Fleischerzeugnissen – ein Überblick. Fleischwirtschaft 78, 12: 1286–1289.

Rödel, W. and R. Scheuer. 1999. Redoxpotential bei Fleisch und Fleischerzeugnissen. 3. Steugerung des Redoxpotentialwertes bei der Produktherstellung und Einfluß von Säuregrad, Natriumnitrit, Natriumascorbat, Natriumlaktat und Sauerstoffpartialdruck. Fleischwirtschaft 79, 7: 78–81.

Roth, D. and J. Sieg. 2003. Trocknungsverlust minimieren. Verwendung von Vitacel® Weizenfaser in schnittfester Rohwurst. Fleischwirtschaft 83, 7: 51–54.

Ruiz, J. 2007. Ingredients. pp. 59–76. *In:* Toldrá (ed.) Handbook of fermented meat and poultry. Blackwell Publishing. Ames. Iowa.

Schillinger, U. and F.K. Lücke. 1987. Identification of lactobacilli from meat and meat products. Food Microbiol. 4: 199–208.

Sebranek, J. G. 2009. Basic curing ingredients. *In:* Tarté, R. (ed.): Ingredients in meat products. Springer Science + Business Media, LLC. New York.

Shah, M.A., S.J.D. Bosco and S.A. Mir. 2014. Plant extracts as natural antioxidants in meat and meat products. Meat Sci. 98: 21–33.

Sielaff, H. and H. Schleusener. 2008. Emissionen weiter vermindern. Zusammensetzung und Wirkung des Räucherrauches und Umweltschutz beim Räuchern. Fleischwirtschaft 88, 4: 57–62.

Stahnke, L.H. and K. Tjener. 2007. Influence of processing parameters on cultures performance. pp. 187–194. *In:* Toldrá (ed.) Handbook of fermented meat and poultry. Blackwell Publishing. Ames. Iowa.

Stiebing, A. 2008. EU-Gesetzgeber will Raucharomen privilegieren. Verwendung von Rauchkondensaten zur Herstellung von Fleischerzeugnissen. Fleischwirtschaft 88, 8: 64–70.

Siebing, A. and W. Rödel. 1989. Einfluß des pH-Wertes auf das Trocknungsverhalten von Rohwurst. Fleischwirtschaft 69: 1530–1538.

Stollewerk, K., A. Jofré, J. Comaposada, G. Ferrini and M. Garriga. 2011. Ensuring food safetyby an innovative fermented sausage manufacturing system. Food Control 22: 1984–1991.

Stollewerk, K., A. Jofré, J. Comaposada, J. Arnau and M. Garriga. 2013. Die Lebensmittelsicherheit des Schelltrocknungsverfahrens für gepökelte Fleischprodukte. Implementierung der NaCl-freien Herstellung und der Hochdruckbehandlung. Fleischwirtschaft 93, 11: 109–113.

Wright von, A. and L. Axelsson. 2012. Lactic acid bacteria: an introduction. pp. 1–37. *In:* S. Lantinen, A. C. Ouwehand, S. Salminen and A. von Wright (eds.). Lactic acid bacteria. Microbiological and functional aspects. 4th edition. CRC Press. Boca Raton. Florida.

Chapter 8

Microbial Ecology of Fermented Sausages and Dry-Cured Meats

Bojana Danilović* and Dragiša Savić

1 INTRODUCTION

Spontaneous fermentation has been used as a method for preservation and keeping of the qualities of the highly perishable food for hundreds of years. Meat fermentation probably dates from the Roman Empire, although there are some early records about fermented meat products in China (Zeuthern 2007). Production relies on the lactic acid fermentation in combination with salting and drying (Adams and Moss 2008).

The investigation of the microorganisms involved in meat fermentation started in 1940's (Hutkins 2006). Microbiota which develops during the meat fermentation usually consists of lactic acid bacteria (LAB), Gram-positive catalase-positive cocci (GCC+), yeasts and molds. Microorganisms can originate from meat, spices or can get into the product during the production process. Different factors may affect the development of desirable microbiota and suppression of the hazardous microorganisms. These factors include temperature, pH drop, decrease in the water activity and oxygen availability and the accumulation of certain metabolites (Leistner 1992).

---

*For Correspondence: Faculty of Technology, Bulevar Oslobodjenja 124,16000 Leskovac, Serbia, Tel: +38116247203, Email: danilovic@tf.ni.ac.rs

The total number of mesophilic bacteria in the meat batter varies in the range 4 to 6 log cfu/g (Toldrà 2002), while the initial number of LAB and GCC+ is 3 to 5 log cfu/g (Talon et al. 2007, Spaziani et al. 2009, Ferreira et al. 2009). Beside this, the meat batter can contain Gram-negative bacteria (e.g. *Enterobacteriaceae, Pseudomonas, Achromobacter*) in the range 3 to 5 log cfu/g (Toldrà 2002, Talon et al. 2007). During the production process the number of LAB increases to 8 to 9 log cfu/g (Hammes et al. 2008). The number of GCC+ is usually lower because of the reduction caused by pH drop and eventual presence of nitrites. Higher number of bacteria can be seen in the outer layers of the sausages due to the higher partial pressure of oxygen (Hammes et al. 2008).

Yeasts are usually present in fresh meat in the range 2–4 log cfu/g, while during the production they can reach 6 log cfu/g in fermented sausages, and up to 7 log cfu/g in dry-cured meat products (Selgas and Garcia 2003, 2007). The mold count on the surface of fermented sausages is in the range 7.2–8.4 log cfu/cm^2 (Bernáldez et al. 2013, Canel et al. 2013). The main source of the molds is the air in the ripening chamber (Comi and Iacumin 2013). Also, spices like milled black pepper, garlic powder, nutmeg and crushed caraway can be a great source of mold spores (Mižáková et al. 2002). Additionally, the occurrence of molds in fermented raw meat products is influenced by the season and it is higher in summer months (Mižáková et al. 2002).

2 LACTIC ACID BACTERIA

The presence of LAB in raw meat is usually low but in the environment which favours their growth, they rapidly become dominant (Rantsiou and Cocolin 2006). The accumulation of the products of LAB metabolism, mainly lactic and acetic acid, leads to a significant pH drop. The decrease of pH value is an important issue in the production process since it contributes to the reduction of undesirable microbiota, development of red colour, formation of the characteristic taste and decrease of the water binding capacity of the proteins which ensures the drying process (Baka et al. 2011). The pH value of the product depends on the buffer capacity of meat, metabolic activities of the fermentation microbiota and the amount of the fermentable carbohydrates. Fermented sausages produced in Northern Europe and USA have pH value in the final product in the range 4.6–5.2 which corresponds with the concentration of lactate of 200 mmol/kg dry matter (Hammes et al. 2008). In the Mediterranean products with long curing period pH value is usually in the range 5.4–5.8. Products fermented in the presence of molds have pH value around 6 due to the use of lactate and formation of ammonia (Hammes et al. 2008).

Microbial Ecology of Fermented Sausages and Dry-Cured Meats 129

Acidification to the isoelectric point of the meat proteins and an increase in the ionic strength induces transformation of the protein into gel and therefore important structural changes. Additionally, fast acidification and drying are very important for the inhibition and inactivation of pathogenic microorganisms during the production process (Leistner 1995). Besides the reduction of pH value and water activity (a_w) other specific metabolites like dyacetil, bacteriocins and short chain fatty acids can be responsible for the antimicrobial effect of LAB.

During the fermentation, meat proteins decompose under the influence of endogenic meat enzymes. Since most of the LAB present in fermented products have low proteolytic activity they do not affect significantly protein degradation in the initial stages of the process. On the other hand, LAB influences protein degradation through the reduction of pH value which induces the increase of the muscle proteases activity (Kato et al. 1994). Despite the fact of low proteolytic activity of LAB, some strains of *Lactobacillus sakei, Lb. curvatus* and *Lb. plantarum* can have peptidase (Fadda et al. 1999a, b). The metabolism of those LAB strains can contribute to the generation of free amino acids which are involved in aroma formation (Papamanoli et al. 2003). Proteolysis in the fermented sausages depends on the product formulation, starter cultures and process conditions (Hughes et al. 2002).

One of very important LAB characteristic which affects the safety of the product is the production of bacteriocins. The protective role of bacteriocins in food relies on their characteristics and mainly on inactivity and nontoxicity in eukaryotic cells, decomposition in the presence of digestive enzymes, low influence to intestinal microbial population, pH and temperature tolerance and relatively wide spectrum of antimicrobial activity against food borne pathogens and spoilage bacteria (Vesković-Moračanin 2010).

Among LAB, lactobacilli are the most dominant genera in the production of fermented sausages (Kozačinski et al. 2008, Danilović 2012, Rantsiou et al. 2005, Pragalaki et al. 2013). Lactobacilli present in meat industry are facultative heterofermentative microorganisms that ferment glucose and hexose-phosphate through the Embden-Mayerhof-Parnas pathway with lactic acid being the main product of fermentation (Toldrà 2002). Lactobacilli influence the consistency, aroma and safety of the product through acidification, catalase activity, amino-acid metabolism, antioxidative activity and bacteriocin production (Cocconcelli and Fontana 2008).

Lactobacillus sakei is usually dominant species during the production of different traditional fermented sausages (Table 1). It can represent up to 90–100% of LAB microbiota in the Bologna sausages and Belgian salami (Janssens et al. 2012). Also, it is dominant in "sudžuk" (Kesmen et al. 2012), "botillo" (Garcià-Fontàn et al. 2007), "salchichón" and

Table 1 LAB species isolated from different fermented sausages

Country	Traditional product	LAB isolated from the product	Frequency of isolation (%)	Reference
Serbia	Petrovac sausages	Lb. sakei	48.1	Danilović 2012
		Ln. mesenteroides	36.2	
		Pd. pentosaceus	10.0	
		Other species:	<5.0	
		Lb. curvatus, E. casseliflavus, E. durans		
	Sremska sausages	Lb. fermentum	24.0	Kozačinski et al. 2008
		Lb. curvatus	7.3	
		Lb. brevis	9.3	
		Lb. plantarum	6.0	
		Lb. delbrueckii ssp. delbrueckii	9.3	
		Lc. lactis ssp. lactis	6.6	
		Ln. mesenteroides ssp. mesenteroides	12.6	
		E. faecalis	6.6	
		Other species:	<5.0	
		Lb. delbrueckii ssp. bulgaricus, Lb. cellobiosus, Lb. collinoides, Lb. acidophilus, Lb. paracasei ssp. paracasei ssp. paracasei ssp. mesenteroides ssp. cremoris, E. faecium		
Croatia	dry-fermented sausages	Lb. plantarum	34.0	Kozačinski et al. 2008
		Lb. curvatus	18.0	
		Lb. pentosus	6.7	
		Lb. plantarum	5.3	
		Lb. brevis	20.7	
		Ln. mesenteroides ssp. mesenteroides	5.9	
		Other species :	<5.0	
		Lb. fermentum, Lc. lactis ssp. lactis, Pd. pentosaceus		

Hungary	dry-fermented sausages	*Lb. curvatus*	7.3	Rantsiou et al. 2005
		Lb. sakei	70.8	
		W. paramesenteroides/hellenica[a]	5.8	
		W. viridescens	7.3	
		Other species:	<5.0	
		Lb. paraplantarum/plantarum[a], *Lb. plantarum*, *Ln. citreum*, *Ln. mesenteroides*		
	dry-fermented sausages	*Lb. sakei*	28.7	Kozačinski et al. 2008
		Ln. mesenteroides ssp. *mesenteroides*	16.7	
		Unidentified	17.4	
		Other species :	<5.0	
		Lb. plantarum, *Lb. curvatus*, *Lb. delbrueckii*, *Lb. alimentarius*, *Lb. amylophilus*, *Lb. bavaricus*, *Lb. salivarius*, *Lb. acidophilus*, *Lb. maltoromicus*, *Lb. yamanashiensis*, *Ln. mesenteroides dextranicum*, *Lb. sanfrancisco*, *W. wiridescens*, *Lb. halotolerans*, *Lb. fructivorans*, *Ln. citreum*, *Ln. eonos*		

Table 1 (Contd.)

Table 1 LAB species isolated from different fermented sausages (Contd.)

Greece	dry-fermented sausages	Lb. casei/paracasei[a]	5.3	Rantsiou et al. 2005
		Lb. curvatus	48.2	
		Lb. paraplantarum	5.3	
		Lb. plantarum	14.9	
		Lb. sakei	19.2	
		Other species:	<5.0	
		E. faecium/durans[a], Lb.alimentarius, Lb. paraplantarum/plantarum[a]		
	dry-fermented sausages	Lb. curvatus ssp. curvatus	16.3	Papamanoli et al. 2003
		Lb. buchneri	15.6	
		Lb. paracasei ssp. paracasei	15.0	
		Lb. sakei	33.3	
		Other species:	<5.0	
		Lb. plantarum, Lb. paracasei ssp. tolerans, Lb. casei, Lb. coryniformis ssp. coryniformis, Lb. paraplantarum, Lb. rhamnosus, W. viridescens, Ln. pseudomesenteroides, Leuconostoc spp., Pediococcus spp., E. faecalis, E. faecium		
	dry-fermented sausages	Lb. plantarum	43.3	Kozačinski et al. 2008
		Lb. curvatus	10.7	
		Lb. pentosus	10.7	
		Lb. brevis	8.7	
		Ln. mesenteroides ssp. mesenteroides	6.0	
		Lc. lactis ssp. lactis	6.7	
		Other species:	<5.0	
		Lb. rhamnosus, Lb. sakei, Lb. paracasei ssp. paracasei, Lb. salivarius, Ln. lactis, E. faecium		

	dry-fermented sausages	*Lb. sakei* *Lb. plantarum* Other species: *Lb. brevis, Lb. curvatus, Lb. rhamnosus*	43.3 45.6 <5.0	Drosinos et al. 2007
	dry-fermented sausages	*Lb. sakei* *Lb. curvatus* *Lb. plantarum* *Lb. paracasei*	35.0 20.0 15.0 30.0	Pragalaki et al. 2013
Italy	dry-fermented sausages	*Lb. curvatus* *Lb. plantarum* *Lb. sakei* Other species : *Ln. citreum, Ln. mesenteroides, W. paramesenteroides/ hellenica[a], E. pseudoavium, Lc. lactis* ssp. *lactis, Lb. brevis, Lb. paraplantarum, Lb. paraplantarum/pentosus[a]*	36.0 6.0 42.7 <5.0	Comi et al. 2005
	dry-fermented sausages	*Lb. curvatus* *Lb. paraplantarum* *Lb. plantarum* *Lb. sakei* Other species : *E. pseudoavium, Lc. lactis* ssp. *lactis, Lb. paraplantarum/pentosus[a], Ln. citreum, W. paramesenteroides/hellenica[a]*	29.8 5.8 6.6 48.8 <5.0	Rantsiou et al. 2005
	"filzetta"	*Lb. sakei* *Lb. fermentum* *Lb. brevis* *Ln. mesenteroides* Unidentified	51.1 15.5 2.2 10.0 21.1	Conter et al. 2005

Table 1 (Contd.)

Table 1 LAB species isolated from different fermented sausages (Contd.)

"ciauscolo"	*Lb. plantarum*	78.3	Aquilanti et al. 2007
	Lb. curvatus	21.7	
"salame bergamasco"	*Lb. sakei*	34.4	Cocolin et al. 2009
	Lb. curvatus	58.9	
	Other species	<5.0	
	Lb. paraplantarum, Ln. mesenteroides		
"salame cremonese"	*Lb. sakei*	68.8	Cocolin et al. 2009
	Lb. curvatus	30.7	
	W. hellenica	0.4	
"salame mantovano"	*Lb. sakei*	61.2	Cocolin et al. 2009
	Lb. curvatus	32.0	
	Other species:	<5.0	
	Lb. plantarum, Ln. mesenteroides, Ln. citreum		
"varzi"	*Lb. sakei*	20.0	Di Cagno et al. 2008
	Lb. curvatus	80.0	
"brianza"	*Lb. sakei*	73.3	Di Cagno et al. 2008
	Lb. curvatus	13.3	
	Pd. pentosaceus	13.3	
"piacentino"	*Lb. sakei*	73.3	Di Cagno et al. 2008
	Lb. curvatus	13.3	
	Lb. coryniformis	13.3	
"piedmontese fermented sausage"	*Lb. sakei*	83.3	Greppi et al. 2015
	Ln. mesenteroides	6.6	
	Other species:	<5.0	
	Lb. curvatus, Lb. pentosus, Ln. carnosum, E. gilvus, Cb. divergens		

Argentina	dry-fermented sausage	*Lb. sakei*	55.0	Fontana et al. 2005
		Lb. plantarum	40.0	
		Lb. curvatus	5.0	
	"alheira"	*Lb. plantarum*	26.5	Albano et al. 2009
		Enterococcus spp.	11.3	
		E. faecalis	30.7	
		E. faecium	13.4	
		Other species:	<5.0	
		Lb. paraplantarum, Lb. sakei, Weissella spp., *W. cibaria, W. viridescens, Lb. zeae, Lb. paracasei, Lb. ramnosus, Ln. mesenteroides, Lb. brevis, Pd. pentosaceus, Pd. acidilactici*		
Bosnia and Hercegovina	"sudžuk"	*Lb. plantarum*	40.7	Kozačinski et al. 2008
		Lb. pentosus	18.0	
		Lb. curvatus	16.7	
		Lb. sakei	8.7	
		Lb. brevis	7.3	
		Other species:	<5.0	
		Pd. pentosaceus, Lc. lactis, Lb. salivarius		
Turkey	"sudžuk"	*Lb. sakei*	37.5	Kesmen et al. 2012
		Lb. plantarum	16.7	
		Lb. curvatus	13.5	
		Lb. brevis	10.4	
		Lb. farciminis	5.2	
		Ln. mesenteroides	5.2	
		W. viridescens	6.2	
		Unidentified	2.1	
		Other species:	<5.0	
		Lb. alimentarius, Ln. citreum		

Table 1 (Contd.)

Table 1 LAB species isolated from different fermented sausages (Contd.)

Spain	"botillo"	Lb. sakei	23.3	Garcià-Fontàn et al. 2007
		Lb. alimentarius	17.3	
		Lb. curvatus	15.3	
		Lb. plantarum	12.0	
		Lb. farciminis	10.0	
	"salchichón" and "chorizo"	Lb. sakei	34.0	Casquete et al. 2012
		Lb. curvatus	11.5	
		Lb. lactis	7.0	
		Lb. plantarum	10.0	
		Pd. acidilactici	15.0	
		Pd. pentosaceus	22.5	
	"salchichón"	Pd. acidilactici	59.2	Benito et al. 2008
		Lb. curvatus	2.0	
		Lb. plantarum	38.8	
	"chorizo"	Pd. acidilactici	44.7	Benito et al. 2008
		Lb. plantarum	10.6	
		Lb. brevis	40.4	
		Other species:	<5.0	
		Lb. sakei, Lb. curvatus		
Belgium	Bologna sausage	Lb. sakei	95.6	Janssens et al. 2012
		Other species:	<5.0	
		Ln. carnosum, Lb. paralimentarus/mindensis/crustorum[a], Ln. mesenteroides,		
	Belgian salami	Lb. sakei	100	Janssens et al. 2012

Microbial Ecology of Fermented Sausages and Dry-Cured Meats

Himalay	"lang kargyong"	Lb. sakei	12.5	Rai et al. 2010
		Lb. divergens	8.3	
		Lb. carnis	20.8	
		Lb. sanfrancisco	25.0	
		E. faecium	12.5	
		Ln. mesenteroides	16.7	
		Lb. curvatus	4.2	
	"faak kargyong "	Lb. brevis	5.6	Rai et al. 2010
		Lb. plantarum	11.1	
		Lb. carnis	16.7	
		Ln. mesenteroides	44.4	
		E. faecium	22.2	
	"yak kargyong"	Lb. sakei	16.7	Rai et al. 2010
		Lb. plantarum	16.7	
		Lb. casei	11.1	
		Lb. curvatus	5.6	
		Lb. carnis	11.1	
		Lb. divergens	5.6	
		Lb. sanfrancisco	11.1	
		Ln. mesenteroides	11.1	
		E. faecium	11.1	
	"lang satchu"	Lb. casei	5.3	Rai et al. 2010
		Lb. carnis	5.3	
		Pd. pentosaceus	26.3	
		E. faecium	63.1	
	"yaak satchu"	Pd. pentosaceus	100	Rai et al. 2010
	"suka ko masu"	Lb. plantarum	20.0	Rai et al. 2010
		Lb. carnis	13.3	
		E. faecium	66.7	

Table 1 (Contd.)

Table 1 LAB species isolated from different fermented sausages (Contd.)

Bali	" urutan"	Lb. plantarum	52.1	Antara et al. 2002
		Lb. farciminis	21.1	
		Lb. hildardii	15.5	
		Pd. pentosaceus	7.0	
		Other species:	<5.0	
		Lb. fermentum, Pd. acidilactici		
Thailand	"mum"	Lb. plantarum	24.8	Wanangkarn et al. 2014
		Lb. sakei	20.0	
		Lb. fermentum	4.8	
		Lb. brevis	11.9	
		Pd. pentosaceus	7.6	
		Ln. mesenteroides	13.3	
		Lc. lactis	7.6	
Vietnam	"nem chua"	Lb. pentosus	21.2	Nguyen et al. 2013
		Lb. plantarum	29.7	
		Lb. farciminis	22.7	
		Ln. citreum	9.5	
		Other species:	<5.0	
		Lb. brevis, Lb. paracasei, Lb. fermentum, Lb. acidipiscis, Lb. rosiae, Lb. namurensis, Lb. fuchuensis, Lc. lactis ssp. lactis, Ln. falax, W. paramesenteroides, W. cibaria, Pd. pentosaceus, Pd. acidilactici, Pd. stilesii		

n.d.-not defined; [a]- identification could not be performed

Lb.=Lactobacillus, Ln.=Leuconostoc, Lc.=Lactococcus, Cb.=Carnobacterium, W.=Weissella, Pd.=Pediococcus, E.=Enterococcus

Source: adapted from Milićević et al. 2014

Microbial Ecology of Fermented Sausages and Dry-Cured Meats   139

"chorizo" (Casquete et al. 2012), "filzetta" (Conter et al. 2005) "salame cremonese" (Cocolin et al. 2009), "salame mantovano" (Cocolin et al. 2009), "brianza" (Di Cagno et al. 2008), "piacentino" (Di Cagno et al. 2008), piedmontese fermented sausage (Greppi et al. 2015), Petrovac sausages (Danilović 2012) and traditional sausages produced in Hungary (Rantsiou et al. 2005, Kozačinski et al. 2008), Greece (Papamanoli et al. 2003, Pragalaki et al. 2013), Argentina (Fontana et al. 2005) and Italy (Comi et al. 2005, Rantsiou et al. 2005). Despite such a high presence of *Lb. sakei* in some traditional sausages, there are examples that this species was not detected in Sremska sausage (Kozačinski et al. 2008), "ciauscolo" (Aquilanti et al. 2007), "salchichón" (Benito et al. 2008), "faak kargyong", "lang satchu, "yaak satchu", "suka ko masu" (Rai et al. 2010) "urutan" (Antara et al. 2002), "nem chua" (Nguyen et al. 2013) and traditional sausages from Croatia (Kozačinski et al. 2008).

Lactobacillus curvatus is usually not dominant but it is frequently isolated from different types of fermented sausages (Rantsiou et al. 2005, Kozačinski et al. 2008, Comi et al. 2005, Cocolin et al. 2009). Another species most frequently isolated is *Lb. plantarum*. It can be dominant in "ciauscolo" (Aquilanti et al. 2007), "sudžuk" (Kozačinski et al. 2008), "urutan" (Antara et al. 2002), "mum" (Wanangkarn et al. 2014), "nem chua" (Nguyen et al. 2013) and traditional sausages produced in Croatia (Kozačinski et al. 2008) and Greece (Kozačinski et al. 2008, Drosinos et al. 2007). In a few cases a dominance of other representatives of lactobacilli was stated. Kozačinski et al (2008) reported a dominance of *Lb. fermentum* in Sremska sausages, while Rai et al. (2010) showed that dominant species during the fermentation of "lang kargyong" was *Lb. carnis*.

Pediococci can often be found in starter cultures for the production of fermented sausages in USA because of their good acidifying properties (Hutkins 2006). Beside this, pediococci can produce bacteriocins affective against unwanted microbial population like *Listeria, Pseudomonas* and *Enterobacteriaceae* during the fermentation of meat (Anastasiadou et al. 2008, Semjonovs and Zikmanis 2008). Although pediococci belong to homofermentative LAB, under some conditions (reduced sugar concentra-tion) heterofermentative products may be formed: ethanol, carbon dioxide and acetic acid. Also, serious problem may occur due to the ability of some species to produce peroxide. Hydrogen peroxide can lead to the formation of green colour and can cause lipid oxidation which results in a serious disturbance of aroma (Huttkins 2006). Pediococci isolated from fermented sausages belong to the species *Pediococcus pentosaceus, Pd. acidilactici* and *Pd. stilesii* (Table 1). In some cases they represent the majority of the LAB population, so the microbial population of the Himalayan sausage Yaek satchu was constituted only from *Pd. pentosaceus* (Rai et al. 2010). Also, *Pd. acidilactici* was the dominant species during the production of Spanish sausages "salchicon" and "chorizo" (Benito et al. 2008).

The number of enterococci in fermented sausages is very variable and they can persist during the fermentation because of their wide range of growth temperatures and resistance to salt (Hugas et al. 2003). They are poor acidifiers in meat, but they can produce bacteriocins enterocins which are affective against *Listeria* and other pathogens (Aymerich et al. 2000). The main concern about the use of enterococci in meat fermentation is related to their resistance to antibiotics and the production of biogenic amines. There are literature data (Danilović 2012, Barbosa et al. 2009) that enterococci are resistant to wide range of antibiotics such as tetracycline, chloramphenicol, rifamycine, ciprofloxacin and nitrofurantoin (Zdolec et al., Chapter 14 in this book). Also, Bover-Cid and Holzapfel (1999) found *Enterococcus* spp. as the most intensive tyrosine decarboxylase species among 177 strains of LAB investigated. However, some strains of enterococci can be used as probiotics (Nueno-Palop and Narbad 2011, Bhardwaj et al. 2010, Weiss et al. 2010) while some of them are involved in fermentation of cheeses and dairy products (Sarantinopoulos et al. 2001). Additionally, they can become a dominant microbial population in packed thermally treated meat (Javadi and Farshbaf 2011). Enterococci isolated during the fermentation of dry sausages (Table 1) usually belong to the species *Enterococcus faecalis* and *E. faecium* (Papamanoli et al. 2003, Kozačinski et al. 2008, Rai et al. 2010) but other species like *E. gilvus* (Greppi et al. 2015), *E. pseudoavium* (Comi et al. 2005, Rantsiou et al. 2005), *E. casseliflavus* and *E. durans* (Danilović 2012) can also be found. In some cases enterococci can represent the dominant LAB population as it was stated for *E. faecalis* in "alheira" (Albano et al. 2009) and *E. faecium* in "lang satchu" and "suka ko masu" (Rai et al. 2010).

Heterofermentative metabolism of leuconostocs makes them unsuitable for the use in fermented sausages due to the production of carbon dioxide which leads to the formation of holes. On the other hand, they can contribute to the product quality through the production volatile compounds from free amino acids which originate from the proteolysis process (Aro Aro et al. 2010). The representatives of leuconostoc in fermented sausages are *Leuconostoc mesenteroides, Ln. citreum, Ln. carnosum, Ln. pseudomesenteroides* and *Ln. falax* (Table 1). They are usually present in smaller amounts, but in the case of Himalayan "faak kargyong" *Ln. mesenteroides* represent a dominant species with the frequency of isolation of 44% (Rai et al. 2010).

The total number of aerobic mesophilic bacteria in the interior of the fresh meat pieces for the production of dry-cured products, like ham or lacōn, is approximately 3 log cfu/g, (Huerta et al. 1988, Lorenzo et al. 2010). The same bacteria are present on the surface of the fresh pieces in the range 5 to 7 log cfu/g (Huerta et al. 1988, Lorenzo et al. 2010). Total viable count increases during the first stages of the production, while in the final product this value is in the range 3–4 log cfu/g (Lorenzo et al. 2010, De Jesús et al. 2014). The number of LAB on the surface of the fresh meat is 3.2–4.6 log cfu/g, while in the interior this value is 0.8–1.5 (Huerta et al. 1988, Lorenzo et al. 2010). During the production process the total number of LAB inside the dry-cured meat products usually does not exceed 1.5 log cfu/g and in most cases can be lower than 1 log cfu/g (Lorenzo et al. 2010, De Jesús et al. 2014, Sánchez-Molinero and Arnau 2008). This indicates that LAB from the interior of the product does not play the most important role during the production of dry-cured meats as they are in the fermented sausages production. The LAB species isolated from the dry-cured meat are similar to those present in the sausages: *Lb. curvatus*, *Lb. sakei*, *Lb. plantarum*, *Lb. alimentarius*, *E. durans* and *E. faecalis* (Lorenzo et al. 2010).

3 GRAM-POSITIVE CATALASE-POSITIVE COCCI

Beside LAB, Gram-positive catalase-positive cocci (GCC+) are important bacterial group involved in the fermentation of the sausages. They are mainly represented by the *Staphylococcus* and *Kocuria* genera and their number in fermented sausages is usually up to 7 log cfu/g (Mauriello et al. 2004, Coppola et al. 2000, Drosinos et al. 2007). The dominant species in fermented dry sausages usually are *Staphylococcus xylosus* and *St. saprophyticus* (Table 2). *Staphylococcus xylosus* can represent up to 84% of isolated staphylococci, like in "Soppressata Molisana" (Coppola et al. 1997). Also, it was dominant in Italian sausages (Iacumin et al. 2006, Kozačinski et al. 2008, Bonomo et al. 2009), Naples type salami (Mauriello et al. 2004, Coppola et al. 2000) and "chorizo" and "salchichón" (Martìn et al. 2007). *Staphylococcus saprophyticus* was the dominant species during the fermentation of traditional sausages produced in Greece (Kozačinski et al. 2008, Drosinos et al. 2007, Papamanolli et al. 2002). Besides these two species, Mauriello et al. (2004) stated a significant presence of *St. equorum* in "soppressata ricigliano" and "soppressata gia". Also, Spanish sausages are characterised with relatively high presence of *St. aureus* which can cause serious health issues due to the production of enterotoxins, so it is unwanted in the fermented meats (Martìn et al. 2007). Other staphylococci like *St. capitis*, *St. hominis*, *St. simulans*, *St. warneri*, *St. cohnii*, *St. epidermidis*, *St. succinus*, *St. lentus*, *St. haemolyticus*, *St. caprae*, *St. galinarium*, *St. intermedicus*, *St. novobiosepticus*, *St. vitulinus*, *St. pasteuri*, *St. chromogenes*, *St. fleurettii*, *St. lugdunensis* and *St. sciuri* can, also, be found in fermented sausages (Table 2). *Kocuria* species can also be isolated from fermented sausages (Coppola et al. 2000, Mauriello et al. 2004, Martìn et al. 2007) and dry-cured meat (Lorenzo et al. 2012) and they are mainly represented by *Kocuria varians* and *K. rosea* (Pisacane et al. 2015, Papamanolli et al. 2002, Lorenzo et al. 2012).

Table 2 GCC+ isolated from fermented sausages and dry-cured meats

Origin	Traditional product	GCC+ isolated from the product	Frequency of isolation (%)	Reference
Serbia	Sremska Sausages	St. saprophyticus St. auricularis St. xylosus St. capitis St. simulans St. warneri St. aureus Other species: St. hominis, St. cohnii	21.1 12.2 21.0 12.2 14.4 6.7 6.7 <5.0	Kozačinski et al. 2008
Croatia	dry-fermented sausages	St. xylosus St. capitis St. carnosus St. saprophyticus	29.2 25.0 25.0 20.8	Kozačinski et al. 2008
Hungary	dry-fermented sausages	St. xylosus Micrococcus spp. St. hominis St. lentus St. warneri Other species: St. xylosus, St. capitis, St. epidermidis, St. haemoliticus, St. auricularis, St. saprophyticus, St. cohnii	43.0 16.0 16.0 10.0 6.0 <5.0	Kozačinski et al. 2008

Greece	dry-fermented sausages	St. epidermidis	5.1	Papamanolli et al. 2002
		St. hominis	5.1	
		St. saprophyticus subsp. saprophyticus	27.2	
		St. cohnii subsp. cohnii	6.8	
		St. cohnii subsp. urealyticus	6.8	
		St. xylosus	5.1	
		St. carnosus subsp. carnosus	25.5	
		K. varians	5.1	
		Other species:	<5.0	
		St. capitis, St. warneri, A. agilis, D. nishinomiyaensis, St. auricularis, St. hyicus, St. sciuri		
	dry-fermented sausages	St. epidermidis	7.2	Papamanolli et al. 2002
		St. haemolyticus	12.0	
		St. saprophyticus subsp. saprophyticus	14.4	
		St. xylosus	16.8	
		St. carnosus subsp. carnosus	12.0	
		St. simulans	9.6	
		Other species:	<5.0	
		St. lentus, St. auricularis, St. cohnii subsp. cohnii, St. cohnii subsp. urealyticus, K. varians, A. agilis, St. hyicus, D. nishinomiyaensis		

Table 2 (Contd.)

Table 2 GCC+ isolated from fermented sausages and dry-cured meats (Contd.)

	dry-fermented sausages	*St. saprophyticus*	34.7	Kozačinski et al. 2008
		St. xylosus	14.7	
		St. simulans	11.3	
		St. haemoliticus	11.3	
		St. caprae	8.0	
		St. capitis	6.0	
		St. aureus/intermedius[a]	5.3	
		Other species:	<5.0	
		St. sciuri, St. hominis, St. auricularis, St. warneri, St. cohnii, St. epidermidis,		
	dry-fermented sausages	*St. cohnii*	6.0	Drosinos et al. 2007
		St. saprophyticus	50.3	
		St. simulans	15.3	
		St. xylosus	12.6	
		Other species:	<5.0	
		St. capitis, St. equorum, St. gallinarum, St. hominis novobiosepticus, St. lentus, St. sciuri, St. vitulinus		
Italy	dry-fermented sausages	*St. xylosus*	40.3	Iacumin et al. 2006
		St. epidermidis	5.3	
		St. warneri	14.7	
		St. equorum	10.5	
		St. pasteuri	13.0	
		M. caseolyticus	9.4	
		Other species:	<5.0	
		St. simulans, St. intermedius, St. carnosus, St. saprophyticus, St. cohnii		

Microbial Ecology of Fermented Sausages and Dry-Cured Meats 145

Sample	Species	%	Reference
dry-fermented sausages	*St. hominis*	6.7	Kozačinski et al. 2008
	St. warneri	7.3	
	St. xylosus	74.0	
	Other species:	<5.0	
	St. saprophyticus, St. lentus, St. saprophyticus/simulans[a], *St.hominis/warneri*[a], *St. epidermidis, St. simulans*		
dry-fermented sausages	*St. xylosus*	45.9	Bonomo et al. 2009
	St. equorum	18.9	
	St. pulvereri/vitulus[a]	10.8	
	St. succinus	5.4	
	St. pasteuri	5.4	
	St. saprophyticus	5.4	
	St. warneri	5.4	
	St. caseoliticus	2.7	
"soppressata molisana"	*St. xylosus*	82.2	Coppola et al. 1997
	St. equorum	8.2	
	St. simulans	8.2	
	St. kloosii	1.4	
"Naples type salami "	*St. xylosus*	13.2	Coppola et al. 2000
	Staphylococcus spp.	48.5	
	Unidentified	25.7	
	Other species:	<5.0	
	St. aureus, St. chromogenes, St. epidermidis, St. hominis, St. lugdunensis, St. saprophyticus, St. warneri, Kocuria varians/rosea		

Table 2 (Contd.)

Table 2 GCC+ isolated from fermented sausages and dry-cured meats (Contd.)

"Naples type salami "	*St. xylosus*	44.8	Mauriello et al. 2004
	St. saprophyticus	17.2	
	St. warneri	13.8	
	St. lentus	17.2	
	Other species:	<5.0	
	St. succinus, Kocuria spp.		
"soppressata ricigliano"	*St. xylosus*	29.5	Mauriello et al. 2004
	St. saprophyticus	32.0	
	St. equorum	40.0	
	St. succinus	14.1	
	St. vitulus	5.1	
	Other species:	<5.0	
	St. warneri, St. pasteuri, St. haemolyticus		
"soppressata gia"	*St. xylosus*	12.8	Mauriello et al. 2004
	St. saprophyticus	20.0	
	St. equorum	20.9	
	St. succinus	14.7	
	Other species:	<5.0	
	St. warneri, St. lentus, St. epidermidis, St. haemolyticus		
"piedmontese fermented sausage"	*St. succinuss*	42	Greppi et al. 2015
	St. xylosus	46.7	
	St. equorum	6.0	
	Other species:	<5.0	
	St. saprophyticus, St. cohnii		

Microbial Ecology of Fermented Sausages and Dry-Cured Meats 147

	"salame mantovano"	St. xylosus	38.2	Pisacane et al. 2015
		St. saprophyticus	36.9	
		St. pasteuri	7.2	
		St. equorum	8.0	
		Other species:	<5.0	
		St. simulans, St. epidermidis, K. varians, K. salsicia		
Spain	"chorizo"	St. xylosus	59.4	Martín et al. 2007
		St. aureus	18.9	
		St. lugdunensis	5.4	
		St. sciuri	5.4	
		Kocuria spp.	8.1	
		St. chromogenes	2.7	
	"salchichón"	St. xylosus	60.9	Martín et al. 2007
		St. aureus	26.1	
		Other species:	<5.0	
		St. lugdunensis, Kocuria spp. St. chromogens,		
		St. saprophyticus, St. capitis		
Switzerland	dry-fermented sausages	St. saprophyticus	20.3	Marty et al. 2012
		St. warneri	17.6	
		St. epidermidis	10.8	
		St. xylosus	11.5	
		St. equorum	28.4	
		Other species:	<5.0	
		St. aureus, St. carnosus, St. vitulinus, St. succinus, St. sciuri,		
		St. gallinarum, St. fleurettii		

Table 2 (Contd.)

Table 2 GCC+ isolated from fermented sausages and dry-cured meats (Contd.)

Maroko	dry-cured goat meat	*St. xylosus* *St. equorum* *St. epidermidis* *St. saprophyticus* Other staphylococci	52.4 34.3 7.0 0.8 5.3	Cherroud et al. 2014
Spain	dry-cured ham	*St. equorum* *St. vitulinus* Other species: *St. capitis, St. aureus, St. caprae, St. epidermidis, St. hominis, St. warneri*	75.9 9.2 <5.0	Landeta et al. 2011
	dry-cured ham	*St. equorum* *St. nepalensis* *St. xylosus*	64.7 23.5 11.8	Fulladosa et al. 2010
	dry-cured lacón	*Staphylococcus xylosus* Other species: *St. equorum, St. simulans, St. intermedius, St. capitis,* *St. cohnni, St. gallinarum, K. varians, K. rosea*	85.3 <5.0	Lorenzo et al. 2012

a- identification could not be performed

St.=Staphylococcus, K.=Kocuria, A.=Arthrobacter, D.=Dermococcus

Source: adapted from Milićević et al. (2014)

Microbial Ecology of Fermented Sausages and Dry-Cured Meats   149

According to the available data GCC+ isolated from dry-cured ham belong to the species *St. nepalensis*, *St. cohnni*, *St. xylosus*, *St. equorum*, *St. epidermidis*, *St. vitulinus*, *St. capitis*, *St. aureus*, *St. caprae*, *St. gallinarum*, *St. hominis*, *St. warneri*, *St. saprophyticus*, *St. simulans*, *St. intermedius*, *K. varians* and *K. rosea* (Cherroud et al. 2014, Landeta et al. 2011, Lorenzo et al. 2012, Fulladosa et al. 2010). Among them, the dominant species are *St. xylosus* and *St. equorum*. The dominance of *St. xylosus* was observed in dry-cured goat meat (Cherroud et al. 2014) and dry-cured lacón (Lorenzo et al. 2012) with the frequency of isolation of 52% and 85%, respectively. Additionally, *St. xylosus* was the dominant species in the salt from Iberian and white ham (Cordero and Zumalacárregui 2000). *Staphylococcus equorum* was dominant in the Spanish dry-cured ham with the frequency of isolation of 65–76% (Landeta et al. 2011, Fulladosa et al. 2010).

The role of GCC+ during the production process is in enhancing of the colour stability, prevention of rancidity, decreasing of the processing time and enchasing the flavour (Drosinos et al. 2007). Proteolytic and lipolytic activity of GCC+ lead to the formation of texture and flavour of the product due to the formation of peptides, amino acids, aldehydes, amines and free fatty acids (Mauriello et al. 2004). Although the number of GCC+ is reduced during the process, extracellular enzymes remain in the meat matrix and exhibit their activity for longer period of time. Catalase activity of the GCC+ reduces the concentration of hydrogen and organic peroxides which can be present as the products of microbial metabolism. In such a way, GCC+ suppresses the development of rancidity and off-flavours in the product (Landeta et al. 2013, Mauriello et al. 2004).

The degradation of the amino acids by *St. xylosus* and *St. carnosus* leads to the formation of a great number of volatile compounds (Stahnke 1999a). The formation of the aroma compounds depends on the concentration of salt, nitrates, glucose and oxygen (Stahnke 1999b). Regardless the poor growth in fermented sausages some strains can have impact to the flavour. *Staphylococcus sciuri* and *St. succinus* can increase the synthesis of the aroma compounds as 2-methyl-1-butanal, 1-octen-2-ol and 2-methylpentanoic acid (Berdagué et al. 1993). On the other hand, strains must be present in certain number in order to survive the manufacturing process. Development of flavour by GCC+ is under great influence of the acidification during the production process, while variations in the recipe of the sausage batter does˙ not have a great influence on the bacterial growth and flavour formation (Ravyts et al. 2010). The constitution of GCC+ microbiota in the fermented sausages is affected by the process and the involvement of smoking or molding. During the fermentation and ripening of the smoked artisan-type sausages the domination of *St. saprophyticus* was stated. On the other hand the GCC+ microbiota of the molded sausages prepared from the same batter was constituted mainly by *St. saprophyticus* in the

150   *Fermented Meat Products: Health Aspects*

fermentation phase and *St. equorum* in the ripening phase. This can be explained by the increase of the pH value caused by the molds (Janssens et al. 2013).

4 YEASTS

The role of yeasts during the production of fermented sausages is the reduction of lactic acid content and the increase of ammonia (Flores et al. 2004). The contribution of yeasts to the final aroma of the product is based on their lipolytic and proteolytic activity (Flores et al. 2015). The representative genera in the fermented sausages are *Debaryomyces, Rhodotorula, Candida, Torulopsis, Cryptococcus, Pichia* and *Yarowia* (Table 3). *Debaryomyces hansenii* is the most frequently found in fermented sausages (Table 3). In some cases it can represent a dominant species of yeasts like in South Africa's products (Osei Abunyewa et al. 2000, Wolter et al. 2000), Naples type salami (Coppola et al. 2000), ilama fermented sausages (Mendoza et al. 2014) and traditional Italian sausages (Cocolin et al. 2006). Beside this *Candida zeylanoides* and *Yarrowia lipolitica* can often be found in fermented sausages (Mendoza et al. 2014, Wolter et al. 2000, Osei Abunyewa et al. 2000). The constitution of the yeast population of the fermented sausages differs on the surface and in the inner part of the sausages. In the production of Spanish dry-cured sausage „salchichon" different species have been isolated from the surface and from the centre of the sausage. *Rhodotorula mucilaginosa* was isolated only from the central part of the sausages, while *Candida parapsilosis* was isolated only from the surface. *Yarrowia lipolitica* and *Debaryomyces* spp. were isolated from both parts of the sausages (Mendonça et al. 2013). *Debaryomyces hansenii* can have the potential to improve the flavour of the sausages due to their ability to produce aroma compounds (Cano-García et al. 2014a) by release of volatile compound (alcohols and aldehydes) and at the end acid production (Durá et al. 2004). Also it can contribute to the stabilization of the red colour of the sausages (Olesen and Stahnke 2000). The yeasts growth during the fermentation is under the influence of different factors as sausage diameter or the type of the production process. Due to higher availability of oxygen sausages with smaller diameter have higher counts of these microorganisms (Selgas and Garcia 2007). Fermented sausages which include smoking in the production process have lower number of yeasts (Encinas et al. 2000). Also, addition of some spices like garlic can have inhibitory effect on the yeast population (Olesen and Stahnke 2000). The yeasts population is not affected by the reduction of pH value caused by LAB since they can growth at pH 4 as at pH 6. The reason is in their ability to maintain neutral intracellular pH even in acidic environment thanks to relatively impermeable plasmatic membrane or

Table 3 Yeasts isolated from fermented sausages and dry-cured meat

Origin	Traditional product	Yeasts isolated from the product	Frequency of isolation (%)	Reference
Italy	"salsiccia sotto sugna"	C. famata D. hansenii R. mucilaginosa Y. lipolytica	45.7 6.4 16.0 8.5	Gardini et al. 2001
	dry-fermented sausages	C. zeylanoides D. hansenii Pichia triangularis	8.3 80.0 6.1	Cocolin et al. 2006
	Naples type salami	C. incommunis C. albidus D. hansenii Trychophyton terrestre Ttychosporon pullulans	7.6 21.5 39.2 17.7 13.9	Coppola et al. 2000
	"ciauscolo"	D. hansenii C. psychrophila Kazachstania burnettii	n.d.	Silvestri et al. 2007
Spain	"chorizo"	D. hansenii Trychosporon ovoides C. intermedia/curvata Y. lipolytica C. parapsilosis C. zeylanoides C. matritensis	14.3 28.6 17.1 14.3 14.3 8.6 2.3	Encinas et al. 2000
Argentina	llama fermented sausages	D. hansenii C. zeylanoides	54.0 28.0	Mendoza et al. 2014

Table 3 (Contd.)

Table 3 Yeasts isolated from fermented sausages and dry-cured meat (Contd.)

South Africa	fermented sausages	C. zeylanoides	5.5	Osei Abunyewa et al. 2000
		C.albidus	10.2	
		D. hansenii	20.4	
		D. occidentalis	5.5	
		R. mucilaginosa	14.8	
		Trychosporon beigelii	9.3	
		Y. lipolytica	13.9	
	Salami	C.zeylanoides	7.1	Wolter et al. 2000
		C.laurentii	7.1	
		D.hansenii	57.1	
		R.mucilaginosa	7.1	
		S. cerevisiae	7.1	
		S. roseus	7.1	
		T. beigelii	7.1	
	dry sausage	C. hungaricus	6.7	Wolter et al. 2000
		C. laurentii	13.3	
		D. hansenii	26.7	
		D. vanriji	6.7	
		R. mucilaginosa	13.3	
		S. roseus	6.7	
		Torulaspora delbrueckii	13.3	
		T. beigelii	13.3	
	cabanossi	C. zeylanoides	4.8	Wolter et al. 2000
		C. laurentii	14.3	
		D. hansenii	42.9	
		D. vanriji	19.0	
		T. beigelii	19.0	

Spain	Iberian ham	*D. hansenii*	95.0	Andrade et al. 2010
		C. zeylanoides	5.0	
	Lacón	*D. hansenii*	77.5	Purriños et al. 2013
		C. zeylanoides	7.9	
		C. deformans	5.3	
Portugal	Ham	*D. hansenii*	60.0	Saldanha-da-Gama et al. 1997
		D. polymorphus	10.0	
		C. laurenti	10.0	
		C. humicolus	20.0	
South Africa	"biltong"	*C.laurentii*	8.3	Wolter et al. 2000
		D.hansenii	33.3	
		R. mucilaginosa	8.3	
		S. cerevisiae	8.3	
		S. roseus	8.3	
		Torulaspora delbrueckii	8.3	
		T. beigelii	8.3	
		Y. lipolytica	16.7	
Norway	dry-cured ham	*D. hansenii*	54.0	Asefa et al. 2009a
		C. zeylanoides	18.5	
Turkey	"pastrima"	*C. zeylanoides*	58.0	Ozturk 2015
		C. deformans	12.0	
		C. galli	11.0	
		C. alimentaria	10.0	
		Y. lipolytica	5.0	

Table 3 (Contd.)

Table 3 Yeasts isolated from fermented sausages and dry-cured meat (Contd.)

Italy	dry-cured ham	C. zeylanoides	23.0	Simoncini et al. 2007
		C. famata	6.0	
		D. hansenii	30.0	
		D. maramus	21.0	
		Not identified	8.0	

C. = Candida,
D. = Debaryomyces,
Y. = Yarrowia,
S. = Sacharomyces,
R. = Rhodotorula

Microbial Ecology of Fermented Sausages and Dry-Cured Meats   155

proton transport mechanism (Selgas and Garcia 2007). The investigation of the aroma formation in different in vitro model systems resulted in opportunistic conclusions. According to Olesen and Stahnke (2000) the inoculation with *D. hansenii* had little impact on the production of volatile compounds. On the other hand, results of Cano-García et al. (2014b) indicated that this species can affect the flavour development. As the same authors concluded the influence of *D. hansenii* to the flavour of the fermented sausages is strain depended and influenced by the fermentation parameters. The production of fermented sausages with different amounts of *Debaryomyces* spp. together with the starter culture consisted of LAB and GCC+, had positive effect on flavour and sensory quality of the sausages which manifested through the inhibition of rancidity and generation of ethyl esters. However, large amount of *Debaryomyces* spp. resulted in high generation of acids which can mask the positive effect (Flores et al. 2004).

The yeasts in the production of ham usually belong to the genera *Cryptococcus, Debaryomyces, Rhodotorula, Saccharomyces, Sporobolomyces, Torulaspora, Trichosporon, Yarrowia, Candida* and *Rhodosporidium* (Table 3). Among them the domination of *D. hansenii* is observed in most cases. For example, this species can make up to 95% of the microbiota of the Iberian ham. Although it was dominant during the whole production process the domination was greater during drying-cellar stage (Andrade et al. 2010). During the investigation of the yeast microbiota in the interior and on the surface of Spanish dry-cured meat product lacón the domination of *D. hansenii* was observed in interior and exterior regardless of the salting level (Purriños et al. 2013). The domination of this species was observed above *C. zeylanoides*, which was often isolated from dry-cured meat products (Andrade et al. 2010, Purriños et al. 2013, Asefa et al. 2009a). *D. hansenii* can be characterized by high tolerance to salt and nitrate and optimal temperature for growth and lipolysis of 10°C and above (Purriños et al. 2013).

The inoculation of ham with *Penicillium chrysogenum* and *D. hansenii* can lead to higher content of long chain aliphatic and branched hydrocarbons, furanones, long chain carboxylic acids and their esters (Martìn et al. 2006).

5 MOLDS

The role of the molds in the production of fermented sausages is of technological importance (López-Dìaz et al. 2001, Sunensen and Stahnke 2003) because they:

- have an antioxidative effect and improve the maintenance of the colour regarding their catalase activity and oxygen consumption,

Table 4 Molds isolated from fermented sausages and dry-cured meat

Origin	Traditional product	Molds isolated from the product	Frequency of isolation (%)	Reference
Spain	"chorizo de Cantimpalos"	P. chrysogenum	3.7	López-Dìaz et al. 2001
		P. commune	33.3	
		P. olsonii	24.1	
		Unidentified	18.5	
		Mucorales	29.6	
Argentina	dry-fermented sausages	P. nalgiovense	59.8	Canel et al. 2013
		Aspergillus ochraceus	16.8	
		Mucor racemosus	7.1	
Greece	dry-fermented sausages	P. commune	15.7	Papagianni et al. 2007
		P. echinulatum	5.2	
		P. nalgiovense	18.8	
		P. oxalicum	6.3	
		P. olsonii	8.3	
		P. solitum	26.1	
Slovenia	dry sausage	P. nalgiovense	51.1	Sonjak et al. 2011
		P. nordicum	43.5	
Spain	dry-cured ham	P. olsonii, P. verrucosum, P. camemberti, P. chrysogenum, P. viridicatum, P. polonicum/aurantiogriseum, P. commune, P. griseofulvum, P. solitum, P. echinulatum, P. canescens, P. crustosum, P. brevicompactum, P. griseoroseum, P. citrinum P. jensenii (nalgiovense), Penicillium spp., Eupenicillium spp.	n.d.	Acosta et al. 2009
	Iberian ham	P. commune	27.4	Núñez et al. 1996
		P. chrysogenum	9.2	
		P. expansum	11.9	
		Erotium herbariorum	11.6	
		Eurotium repens	25.2	

Norway	dry-cured ham	*P. atramentosum*	6.1	Asefa et al. 2009b
		P. chrysogenum	6.4	
		P. commune	9.8	
		P. crustosum	5.3	
		P. nalgiovense	37.5	
		P. solitum	12.9	
Croatia	Istrian ham	*P. frequentans*	7.8	Comi et al. 2004
		P. verrocosum	5.5	
		P. lanoso-coeruleum	7.8	
		P. lanoso-griseum	9.4	
		P. chrysogenum	5.5	
		P. commune	5.5	
		Aspergillus repens	7.8	
		Erotium repens	22.8	
Italy	San Daniele ham	*Alternaria alternata*	5.3	Comi and Iacumin 2013
		A. fumigatus	20.0	
		A. niger	6.0	
		A. sydowii	5.3	
		Eurotium amstelodami	10.0	
		P. chrysogenum	12.0	
Slovenia	dry-cured pork neck	*Eurotium* spp.	36.7	Sonjak et al. 2011
		P. "milanense"	23.3	
		P. nordicum	41.7	
	dry-cured ham	*A. versicolor*	20.9	Sonjak et al. 2011
		Cladosporium spp.	7.5	
		Eurotium spp.	31.3	
		P. "milanense"	10.4	
		P. nordicum	37.3	

P.=Penicillium, A.= Aspergillus

- develop a microclimate at the surface and prevent the creation of sticky or slimy surface,
- are involved in the development of the flavour and taste by lactate oxidation, proteolysis, degradation of amino acids, lipolysis, b-oxidation,
- can have a role in the protection against unwanted molds, yeasts and bacteria,
- give the sausages typical white or greyish appearance and
- contribute to the slower evaporation and water loss.

But, on the other hand, the presence of molds can be undesirable since some strains can lead to the defect in the quality and appearance. Also, some strains can produce mycotoxins which exert toxic effects on consumer's health (Iacumin et al. 2009). One of the possible solutions is the use of the mold starter culture which will prevent the growth of the undesirable wild mold population. The use of the commercial protective culture of *Penicillium nalgiovense* in the production of dry-fermented sausage "salchichón" suppresses the growth of the toxigenic molds and therefore the production of mycotoxins (Bernáldez et al. 2013).

The most frequently isolated genera are *Penicillium*, *Aspergillus* and *Erotium* (Table 4). The prevalence of one genus on the surface of the industrial hams is influenced by the temperature, moisture of the meat and relative humidity in climatized ripening chambers (Comi et al. 2004). Investigation of Canel et al. (2013) indicated that the increase of the temperature in the ripening room in the summer can lead to higher frequency of isolation of *Aspergillus ochraceus* which is responsible for the yellowish color of the casing and can be the source of contamination of the sausages with ochratoxin A.

The *Penicillium* species are the most dominant in the mycopopulation of dry-cured meat products and fermented sausages (Table 4). Besides the temperature and the a_w, one of the important factors which affect the *Penicillium* spp. is the content of sodium chloride (López-Díaz et al. 2002). The dominance of *P. nalgiovense*, *P. solitum* and *P. commune* in some meat products and cheeses can be explained by their tolerance to high level of salt. On the other hand smoking can affect the constitution of mycobiota since the increase of the presence of *P. nalgiovense* and the reduction of *P. solitum* and *P. commune* was observed in smoked product (Asefa et al. 2009b)

6 CONCLUSION

The diversity in the ingredients and manufacturing process of fermented sausages and dry-cured meats causes the differences in the composition

Microbial Ecology of Fermented Sausages and Dry-Cured Meats

of microbiota. Generally, LAB and Gram-positive catalase-positive cocci play the significant role in the production of fermented sausages. On the other hand, the production of dry-cured meat usually relies on the presence of Gram-positive catalase-positive cocci, molds and yeasts. The diversity of species among the products may significantly vary due to the differences in the production practices.

Acknowledgments: This work was supported by the Ministry of Education, Science and Technological Development of the Republic of Serbia, grant No.: 31032

Key words: fermented sausages, dry-cured meat, lactic acid bacteria, Gram-positive catalase-positive cocci, yeasts, molds

REFERENCES

Acosta, R., A. Rodríguez-Martín, A. Martín, F. Núñez and M.A. Asensio. 2009. Selection of antifungal protein-producing molds from dry-cured meat products. Int. J. Food Microbiol. 135: 39–46.

Adams, M.R. and M.O. Moss. 2008. Food microbiology, RSC Publishing, United Kingdom.

Albano, H., C.A. van Reenen, S.D. Todorov, D. Cruz, L. Fraga, T. Hogg, L. Dicks and P. Teixeira. 2009. Phenotypic and genetic heterogeneity of lactic acid bacteria isolated from "Alheira", a traditional fermented sausage produced in Portugal. Meat Sci. 82: 389–398.

Anastasiadou, S., M. Papagianni, G. Filiousis, I. Ambrosiadis and P. Koidis. 2008. Growth and metabolism of a meat isolated strain of *Pediococcus pentosaceus* in submerged fermentation: Purification, characterization and properties of the produced pediocin SM-1, Enzyme Microbial Technol. 43: 448–454.

Andrade, M.J., M. Rodríguez, E. Casado and J.J. Córdoba. 2010. Efficiency of mitochondrial DNA restriction analysis and RAPD-PCR to characterize yeasts growing on dry-cured Iberian ham at the different geographic areas of ripening. Meat Sci. 84: 377–383.

Antara, N.S., I.N. Sujaya, A. Yokota, W.R. Aryanta and F. Tomita. 2002. Identification and succession of lactic acid bacteria during fermentation of "urutan", a Balinese indigenous fermented sausage, World. J. Microb. Biot. 18: 255–262.

Aro Aro, J.M., P. Nyam-Osor, K. Tsuji, K. Shimada, M. Fukushima and M. Sekikawa. 2010. The effect of starter cultures on proteolytic changes and amino acid content in fermented sausages. Food Chem. 119: 279–285.

Asefa, D.T., T. Møretrø, R.O. Gjerde, S. Langsrud, C.F. Kure, M.S. Sidhu, T. Nesbakken and I. Skaar. 2009a. Yeast diversity and dynamics in the production processes of Norwegian dry-cured meat products. Int. J. Food Microbiol. 133: 135–140.

160   *Fermented Meat Products: Health Aspects*

Asefa, D.T., R.O. Gjerde, M.S. Sidhu, S. Langsrud, C.F. Kure, T. Nesbakke and I. Skaar. 2009b. Moulds contaminants on Norwegian dry-cured meat products. Int. J. Food Microbiol. 128: 435–439.

Aquilanti, L., S. Santarelli, G. Silvestri, A. Osimani, A. Petruzzelli and F. Clementi. 2007. The microbial ecology of a typical Italian salami during its natural fermentation, Int. J. Food Microbiol. 120: 136–145.

Aymerich, T., M. Garriga, J. Ylla, J. Vallier, J.M. Monfort and M. Hugas, 2000 Application of enterocins as biopreservatives against *Listeria innocua* in meat products. J. Food Protect. 63: 721–726.

Baka, A.C., E.J. Papavergou, T. Pragalaki, J.G. Bloukas and P. Kotzekidou. 2011. Effect of selected autochtonous starter cultures on processing and quality characteristics of Greek fermented sausages. LWT-Food Sci. Tech. 44: 54–61.

Barbosa, J., V. Ferreira and P. Teixeira. 2009. Antibiotic susceptibility of enterococci isolated from traditional fermented meat products. Food Microbiol. 26: 527–532

Benito, M., M.J. Serradilla, S. Ruiz-Moyano, A. Martín, F. Pérez-Nevado and G. Córdoba. 2008. Rapid differentiation of lactic acid bacteria from autochthonous fermentation of Iberian dry-fermented sausages. Meat Sci. 80: 656–661.

Berdagué, J.L., P. Monteil, M.C. Montel and R. Talon. 1993. Effects of starter cultures on the formation of flavor compounds in dry sausage. Meat Sci. 35: 275–287.

Bernáldez, V., J.J. Córdoba, M. Rodríguez, M. Cordero, L. Polo and A. Rodríguez. 2013. Effect of *Penicillium nalgiovense* as protective culture in processing of dry-fermented sausage "salchichón". Food Control 32: 69–76.

Bhardwaj, A., H. Gupta, S. Kapila, G. Kaur, S. Vij and R.K. Malik. 2010. Safety assessment and evaluation of probiotic potential of bacteriocinogenic *Enterococcus faecium* KH 24 strain under *in vitro* and *in vivo* conditions, Int. J. Food Microbiol. 141: 156–164.

Bonomo, M.G., A. Ricciardi, T. Zotta, M.A. Sico and G. Salzano. 2009. Technological and safety characterization of coagulase-negative staphylococci from traditionally fermented sausages of Basilicata region (Southern Italy). Meat Sci. 83: 15–23.

Bover-Cid, S. and W.H. Holzapfel. 1999. Improved screening procedure for biogenic amine production by lactic acid bacteria. Int. J. Food Microbiol. 53: 33–41.

Canel, R.S., J.R. Wagner, S.A. Stenglein and V. Ludemann. 2013. Indigenous filamentous fungi on the surface of Argentinean dry fermented sausages produced in Colonia Caroya (Córdoba). Int. J. Food Microbiol. 164: 81–86.

Cano-García, L., C. Belloch and M. Flores. 2014a. Impact of *Debaryomyces hansenii* strains inoculation on the quality of slow dry-cured fermented sausages. Meat Sci. 96: 1469–1477.

Cano-García, L., S. Rivera-Jiménez, C. Belloch and M. Flores. 2014b. Generation of aroma compounds in a fermented sausage meat model system by *Debaryomyces hansenii* strains. Food Chem. 151: 364–373.

Casquete, R., M. Benito, A. Martín, S. Ruiz-Moyano, E. Aranda and M. Córdoba. 2012. Microbiological quality of salchichón and chorizo, traditional Iberian dry-fermented sausages from two different industries, inoculated with autochthonous starter cultures. Food Control 24: 191–198.

Microbial Ecology of Fermented Sausages and Dry-Cured Meats   161

Cherroud, S., A. Cachaldora, S. Fonseca, A. Laglaoui, J. Carballo and I. Franco. 2014. Microbiological and physicochemical characterization of dry-cured Halal goat meat. Effect of salting time and addition of olive oil and paprika covering. Meat Sci. 98: 129–134.

Cocconcelli, P.S. and C. Fontana. 2008. Characteristics and application of microbial starters in meat fermentation. pp. 129–149. *In:* F. Toldrà (ed.). Meat biotechnology. Springer, New York.

Cocolin, L., R. Urso, K. Rantsiou, C. Cantoni and G. Comi. 2006. Dynamics and characterization of yeasts during natural fermentation of Italian sausages. FEMS Yeast Res. 6: 692–701.

Cocolin, L., P. Dolci, K. Rantsiou, R. Urso, C. Cantoni and G. Comi. 2009. Lactic acid bacteria ecology of three traditional fermented sausages produced in the North of Italy as determined by molecular methods. Meat Sci. 82: 125–132.

Comi, G., S. Orlic, S. Redzepovic, R. Urso and L. Iacumin. 2004. Moulds isolated from Istrian dried ham at the pre-ripening and ripening level. Int. J. Food Microbiol. 96: 29–34.

Comi, G., R. Urso, L. Iacumin, K. Rantsiou, P. Cattaneo, C. Cantoni and L. Cocolin. 2005. Characterisation of naturally fermented sausages produced in the North East of Italy. Meat Sci. 69: 381–392.

Comi, G. and L. Iacumin. 2013. Ecology of moulds during the pre-ripening and ripening of San Daniele dry cured ham. Food Res. Int. 54: 1113–1119.

Conter, M., T. Muscariello, E. Zanardi, S. Ghidini, A. Vergara, G. Campanini and A. Ianieri. 2005. Characterization of lactic acid bacteria isolated from an Italian dry fermented sausage. Annali Della Facolta *Di M*edicina Veterina Del Studi *Di Parma* 25: 167–174.

Coppola, R., M. Iorizzo, R. Saotta, E. Sorrentino and L. Grazia. 1997. Characterization of micrococci and staphylococci isolated from soppressata molisana, a Southern Italy fermented sausage. Food Microbiol. 14: 47–53.

Coppola, S., G. Mauriello, M. Aponte, G. Moschetti and F. Villani. 2000. Microbial succession during ripening of Naples-type salami, a southern Italian fermented sausage. Meat Sci. 56: 321–329.

Cordero, M.R. and J.M. Zumalacárregui. 2000. Characterization of *Micrococcaceae* isolated from salt used for Spanish dry-cured ham. Lett. Appl. Microbiol. 32: 303–306.

Danilović, B. 2012. Changes in the population of lactic acid bacteria during the production of petrovac sausage (Petrovská klobása). PhD thesis, University of Nis, Leskovac, Serbia.

De Jesús, C., G. Hernández-Coronado, J. Girón, J. M. Barat, M. J. Pagan, M. Alcañiz, R. Masot and R. Grau. 2014. Classification of unaltered and altered dry-cured ham by impedance spectroscopy: A preliminary study. Meat Sci. 98: 695–700.

Di Cagno, R., C.C. López, R. Tofalo, G. Gallo, M. De Angelis, A. Paparella, W. Hammes and M. Gobbetti. 2008. Comparison of the compositional, microbiological, biochemical and volatile profile characteristics of three Italian PDO fermented sausages. Meat Sci.79: 224–235.

Drosinos, E., S. Paramithiotis, G. Kolovos, I. Tsikouras and I. Metaxopoulos. 2007. Phenotypic and technological diversity of lactic acid bacteria and staphylococci isolated from traditionally fermented sausages in Southern Greece. Food Microbiol. 24: 260–270.

Durá, M.A., M. Flores and F. Toldrá. 2004. Effect of growth phase and dry-cured sausage processing conditions on *Debaryomyces* spp. generation of volatile compounds from branched-chain amino acids. Food Chem. 86: 391–399.

Encinas, J.P., T.M. López-Díaz, M.L. García-López, A. Otero and B. Moreno. 2000. Yeast populations on Spanish fermented sausages. Meat Sci. 54: 203–208.

Fadda, S., Y. Sanz, G. Vignolo, M. Aristoy, G. Oliver and F. Toldrà. 1999a. Characterisation of muscle sarcoplasmic and myofibrillar protein hydrolysis caused by *Lactobacillus plantarum*. Appl. Environ. Microb. 65: 3540–3546.

Fadda, S., Y. Sanz, G. Vignolo, M. Aristoy, G. Oliver and F. Toldrà. 1999b. Hydrolysis of pork muscle sarcoplasmic proteins by *Lactobacillus curvatus* and *Lactobacillus sakei*. Appl. Environ. Microb. 65: 578–584.

Ferreira, V., J. Barbosa, J. Silva, P. Gibbs, T. Hogg and P. Teixeira. 2009. Microbiological profile of *Salpicãode Vinhais* and *Chouriça de Vinhais* from raw materials to final products: Traditional dry sausages produced in the North of Portugal. Innov. Food Sci. Emerg. Technol. 10: 279–283.

Flores, M., M.A. Durá, A. Marco and F. Toldrá. 2004. Effect of *Debaryomyces* spp. on aroma formation and sensory quality of dry-fermented sausages. Meat Sci. 68: 439–446.

Flores, M., S. Corral, L. Cano-García, A. Salvador and C. Belloch. 2015. Yeast strains as potential aroma enhancers in dry fermented sausages. Int. J. Food Microbiol. 212: 16–24.

Fontana, C., P.S. Cocconcelli and G. Vignolo. 2005. Monitoring the bacterial population dynamics during fermentation of artisanal Argentinean sausages. Int. J. Food Microbiol. 103: 131–142.

Fulladosa, E., M. Garriga, B. Martín, M.D. Guàrdia, J.A. García-Regueiro and J. Arnau. 2010. Volatile profile and microbiological characterization of hollow defect in dry-cured ham. Meat Sci. 86: 801–807.

Gardini, F., G. Suzzi, A. Lombardi, F. Galgano, M.A. Crudele, C. Andrighetto and M.S.R. Tofalo. 2001. A survey of yeasts in traditional sausages of southern Italy. FEMS Yeast Res. 1: 161–167.

Garcià-Fontàn, M., J.M. Lorenzo, S. Martınez, I. Franco and J. Carballo. 2007. Microbiological characteristics of Botillo, a Spanish traditional pork sausage. LWT-Food Sci. Technol. 40: 1610–1622.

Greppi, A., I. Ferrocino, A. La Storia, K. Rantsiou, D. Ercolini and L. Cocolin. 2015. Monitoring of the microbiota of fermented sausages by culture independent rRNA-based approaches. Int. J. Food Microbiol. 212: 67–75.

Hammes, W., D. Haller and M. Gänzle. 2008. Fermented meat. pp. 291–321. *In:* E. Farnworth (ed.). Handbook of fermented functional foods. CRC press. Boca Raton.

Huerta, T., J. Hernández, B. Guamis and E. Hernández. 1988. Microbiological and physico-chemical aspects in dry-salted Spanish ham. Zentralbl. Mikrobiol. 143: 475–482.

Hugas, M., M. Garriga and M.T. Aymerich. 2003. Functionalty of enterococci in meat products. Int. J. Food Microbiol. 88: 223–233.

Hughes, M. C., J.P. Kerry, E.K. Arendt, P.M. Kenneally, P.L.H. McSweeney and E.E.O'Neil. 2002. Characterization of proteolysis during the ripening of semidry fermented sausages. Meat Sci. 62: 205–216.

Hutkins, R. 2008. Microbiology and technology of fermented foods. Blackwell Publishing. Ames, Iowa.

Iacumin, L., G. Comi, C. Cantoni and L. Cocolin. 2006. Ecology and dynamics of coagulase-negative cocci isolated from naturally fermented Italian sausages. Sys. Appl. Microbiol. 29: 480–486.

Iacumin, L., L. Chiesa, D. Boscolo, M. Manzano, C. Cantoni, S. Orlic and G. Comi. 2009. Moulds and ochratoxin A on surfaces of artisanal and industrial dry sausages. Food Microbiol. 26: 65–70.

Javadi, A. and V. Farshbaf. 2011. Contamination level variation of fecal Streptococci (Enterococci) in poultry slaughterhouse premises with hazard analysis and critical control point (HACCP) method. Afr. J. Microbiol. Res. 5: 3801–3804.

Janssens, M., N. Myter, L. De Vuyst and F. Leroy. 2012. Species diversity and metabolic impact of the microbiota are low in spontaneously acidified Belgian sausages with an added starter culture of *Staphylococcus carnosus*. Food Microbiol. 29: 167–177.

Janssens, M., N. Myter, L. De Vuyst and F. Leroy. 2013. Community dynamics of coagulase-negative staphylococci during spontaneous artisan-type meat fermentations differ between smoking and moulding treatments. Int. J. Food Microbiol. 166: 168–175.

Kato, T., T. Matsuda, T. Tahara, M. Sugimoto, Y. Sato and R. Nakamura. 1994. Effects of meat conditioning and lactic fermentation on pork muscle protein degradation. Biosci. Biotech. Bioch. 58: 408–410.

Kesmen, Z., A.E. Yetiman, A. Gulluce, N. Kacmaz, O. Sagdic, B. Cetin, A. Adiguzel, F. Sahin and H. Yetim. 2012. Combination of culture-dependent and culture-independent molecular methods for the determination of lactic microbiota in sucuk. Int. J. Food Microbiol. 153: 428–435.

Kozačinski, L., E. Drosinos, F. Čaklovica, L. Cocolin, J. Gasparik-Reichardt and S. Vesković. 2008. Investigation of microbial association of traditionally fermented sausages. Food Technol. Biotech. 46: 93–106.

Landeta, G., I. Reverón, A.V. Carrascosa, B. de las Rivas and R. Muñoz. 2011. Use of recA gene sequence analysis for the identification of *Staphylococcus equorum* strains predominant on dry-cured hams. Food Microbiol. 28: 1205–1210.

Landeta, G., J.A. Curiel, A.V. Carrascosa, R. Muñoz and B. de las Rivas. 2013. Characterization of coagulase-negative staphylococci isolated from Spanish dry cured meat products. Meat Sci. 93: 387–396.

Leistner, L. 1992. The essentials of producing stable and safe raw fermented sausages. pp. 1–19. *In:* J.M. Smulders, F. Toldra, J. Flores and M. Prieto (eds.). New technologies for meat and meat products. Audet, Nijmegen.

Leistner, L. 1995. Stable and safe fermented sausages worldwide. pp. 160–175. *In:* G. Campbell-Platt and P.E. Cook (eds.). Fermented meats. Blackie Academic & Professional, Glasgow.

López-Dìaz, T.M., J.A. Santos, M.L. García-López and A. Otero. 2001. Surface mycoflora of a Spanish fermented meat sausage and toxigenicity of *Penicillium* isolates. Int. J. Food Microbiol. 68: 69–74.

López-Díaz, T.M., C.J. González, B. Moreno and A. Otero. 2002. Effect of temperature, water activity, pH and some antimicrobials on the growth of *Penicillium olsonii* isolated from the surface of Spanish fermented meat sausage. Food Microbiol. 19: 1–7.

Lorenzo, J.M., M.C. García Fontán, A.Cachaldora, I. Franco and J. Carballo. 2010. Study of the lactic acid bacteria throughout the manufacture of dry-cured

164 *Fermented Meat Products: Health Aspects*

lacón (a Spanish traditional meat product). Effect of some additives. Food Microbiol. 27: 229–235.

Lorenzo, J.M., M.C. García Fontán, M. Gómez, S. Fonseca, I. Franco and J. Carballo. 2012. Study of the *Micrococcaceae* and *Staphylococcaceae* throughout the manufacture of dry-cured Lacón (a Spanish traditional meat product) made without or with additives. J. Food Res. 1: 200–211.

Martín, A., J.J. Córdoba, E. Aranda, M.G. Córdoba and M.A. Asensio. 2006. Contribution of a selected fungal population to the volatile compounds on dry-cured ham. Int. J. Food Microbiol. 110: 8–18.

Martín, A., B. Colín, E. Aranda, M.J. Benito and M.G. Córdoba. 2007. Characterization of Micrococcaceae isolated from Iberian dry-cured sausages. Meat Sci. 75: 696–708.

Marty, E., J. Buchs, E. Eugster-Meier, C. Lacroix and L. Meile. 2012. Identification of staphylococci and dominant lactic acid bacteria in spontaneously fermented Swiss meat products using PCR-RFLP. Food Microbiol. 29: 157–166.

Mauriello, G., A. Casaburi, G. Blaiotta and F. Villani. 2004. Isolation and technological properties of coagulase negative staphylococci from fermented sausages of Southern Italy. Meat Sci. 67: 149–158.

Mendoza, L.M., B. Padilla, C. Belloch and G. Vignolo. 2014. Diversity and enzymatic profile of yeasts isolated from traditional llama meat sausages from north-western Andean region of Argentina. Food Res. Int. 62: 572–579.

Mendonça, R.C.S., D.M. Gouvêa, H.M. Hungaro, A. de F. Sodré and A. Querol-Simon. 2013. Dynamics of the yeast flora in artisanal country style and industrial dry cured sausage (yeast in fermented sausage). Food Control 29: 143–148.

Milićević, B., B. Danilović, N. Zdolec, L. Kozačinski, V. Dobranić and D. Savić. 2014. Microbiota of the fermented sausages: influence to product quality and safety. Bulg. J. Agric. Sci. 20: 1061–1078.

Mižáková A., M. Pipová and P. Turek. 2002. The occurrence of moulds in fermented raw meat products. Czech J. Food Sci. 20: 89–94.

Nguyen, D.H.L., K. Van Hoorde, M. Cnockaert, E. De Brandt, K. De Bruyne, B. Thanh Le and P. Vandamme. 2013. A culture-dependent and -independent approach for the identification of lactic acid bacteria associated with the production of nem chua, a Vietnamese fermented meat product. Food Res. Int. 50: 232–240.

Nueno-Palop, C. and A. Narbad. 2011. Probiotic assessment of *Enterococcus faecalis* CP58 isolated from human gut. Int. J. Food Microbiol. 145: 390–394.

Núñez, F., M.M. Rodríguez, M.E. Bermúdez, J.J. Córdoba and M.A. Asensio. 1996. Composition and toxigenic potential of the mould population on dry-cured Iberian ham. Int. J. Food Microbiol. 32: 185–197.

Olesen, P.T. and L.H. Stahnke. 2000. The influence of *Debaryomyces hansenii* and *Candida utilis* on the aroma formation in garlic spiced fermented sausages and model minces. Meat Sci. 56: 357–368.

Osei Abunyewa, A. A., E. Laing, A. Hugo and B.C. Viljoen. 2000. The population change of yeasts in commercial salami. Food Microbiol. 17: 429–438.

Ozturk, I. 2015. Presence, changes and technological properties of yeast species during processing of pastirma, a Turkish dry-cured meat product. Food Control 50: 76–84.

Microbial Ecology of Fermented Sausages and Dry-Cured Meats

Papagianni, M., I. Ambrosiadis and G. Filiousis. 2007. Mould growth on traditional greek sausages and penicillin production by *Penicillium* isolates. Meat Sci. 76: 653–657.

Papamanoli, E., P. Kotzekidou, N. Tzanetakis and E. Litopoulou-Tzanetaki. 2002. Characterization of *Micrococcaceae* isolated from dry fermented sausage. Food Microbiol. 19: 441–449.

Papamanoli, E., N. Tzanetakis, E. Litopoulou-Tzanetaki and P. Kotzekidou. 2003. Characterisation of lactic acid bacteria isolated from a Greek dry/fermented sausage in respect of their technological and probiotic properties. Meat Sci. 65: 859–867.

Pisacane, V., M.L. Callegari, E. Puglisi, G. Dallolio and A. Rebecchi. 2015. Microbial analyses of traditional Italian salami reveal microorganisms transfer from the natural casing to the meat matrix. Int. J. Food Microbiol. 207: 57–65.

Pragalaki, T., J.G. Bloukas, P. Kotzekidou. 2013. Inhibition of *Listeria monocytogenes* and *Escherichia coli* O157:H7 in liquid broth medium and during processing of fermented sausage using autochthonous starter cultures. Meat Sci. 95: 458–464.

Purriños, L., M.C. García Fontán, J. Carballo and J.M. Lorenzo. 2013. Study of the counts, species and characteristics of the yeast population during the manufacture of dry-cured "lacón". Effect of salt level. Food Microbiol. 34: 12–18.

Rai, A. K., J.P. Tamang and U. Palni. 2010. Microbiological studies of ethnic meat products of the Eastern Himalayas. Meat Sci. 85: 560–567.

Rantsiou, K. and L. Cocolin. 2006. New developments in the study of the microbiota of naturally fermented sausages as determined by molecular methods: A review. Int. J. Food Microbiol. 108: 255–267.

Rantsiou, K., M. Gialitaki, R. Urso, J. Krommer, J. Gasparik-Reichardt, S. Tóth, I. Metaxopoulos, G. Comi and L. Cocolin. 2005. Molecular characterization of *Lactobacillus* species isolated from naturally fermented sausages produced in Greece, Hungary and Italy. Int. J. Food Microbiol. 103: 131–142.

Ravyts, F., L. Steen, O. Goemaere, H. Paelinck, L. De Vuyst and F. Leroy. 2010. The application of staphylococci with flavour-generating potential is affected by acidification in fermented dry sausages. Food Microbiol. 27: 945–954.

Saldanha-da-Gama, A., M. Malfeito-Ferreira and V. Loureiro. 1997. Characterization of yeasts associated with Portuguese pork-based products. Int. J. Food Microbiol. 37: 201–207.

Sánchez-Molinero, F. and J. Arnau. 2008. Effect of the inoculation of a starter culture and vacuum packaging (during resting stage) on the appearance and some microbiological and physicochemical parameters of dry-cured ham. Meat Sci. 79: 29–38.

Sarantinopoulos, P., C. Andrighetto, M.D. Georgalaki, M.C. Rea, A. Lombardi, T.M. Cogan, G. Kalantzopoulos and E. Tsakalidou. 2001. Biochemical properties of enterococci relevant to their technological performance. Int. Dairy J. 11: 621–647.

Selgas, M.D. and M.L. Garcia. 2007. Starter cultures: yeasts. pp. 159–169. *In:* F. Toldrá (ed.). Handbook of fermented meat and poultry. Blackwell Publishing. Ames, Iowa.

Selgas, M.D., J. Ros and M.L. García. 2003. Effect of selected yeast strains on the sensory properties of dry fermented sausages. Eur. Food. Res. Technol. 217: 475–480.

Semjonovs, P. and P. Zikmanis. 2008. Evaluation of novel lactose-positive and exopolysaccharide producing strain of *Pediococcus pentosaceus* for fermented foods. Eur. Food Res. Tech. 227: 851–856.

Silvestri, G., S. Santarelli, L. Aquilanti, A. Beccaceci A. Osimani, F. Tonucci and F. Clementi. 2007. Investigation of the microbial ecology of Ciauscolo, a traditional Italian salami, by culture-dependent techniques and PCR-DGGE. Meat Sci. 77: 413–423.

Simoncini, N., D. Rotelli, R. Virgili and S. Quintavalla. 2007. Dynamics and characterization of yeasts during ripening of typical Italian dry-cured ham. Food Microbiol. 24: 577–584.

Sonjak, S., M. Ličen, J.C. Frisvad and N. Gunde-Cimerman. 2011. The mycobiota of three dry-cured meat products from Slovenia. Food Microbiol. 28: 373–376.

Spaziani, M., M. Del Torre and M.L. Stecchini. 2009. Changes of physicochemical, microbiological, and textural properties during ripening of Italian low-acid sausages. Proteolysis, sensory and volatile profiles. Meat Sci. 81: 77–85.

Stahnke, L.H. 1999a. Volatiles produced by *Staphylococcus xylosus* and *Staphylococcus carnosus* during growth in sausage minces. Part I. Collection and identification. LWT-Food Sci. Tech. 32: 357–364.

Stahnke, L.H. 1999b. Volatiles produced by *Staphylococcus xylosus* and *Staphylococcus carnosus* during growth in sausage minces. Part II. The influence of growth parameters. LWT-Food Sci. Tech. 32: 365–374.

Sunensen, L.O. and L.H. Stahnke. 2003. Mould starter cultures for dry sausages— selection, application and effects. Meat Sci. 65: 935–948.

Talon, R., S. Leroy and I. Lebert. 2007. Microbial ecosystems of traditional fermented meat products: The importance of indigenous starters. Meat Sci. 77: 55–62.

Toldrà, F. 2002. Dry-cured meat products. Food & Nutrition press, Inc.Trumbull, Connecticut.

Vesković-Moračanin, S. 2010. Lactic acid bacteria bacteriocins as natural food protectors – possibilities of applications in meat industry. Tehnologija mesa 51: 83–94. (in Serbian).

Wanangkarn, A., D.C. Liu, A. Swetwiwathana, A. Jindaprasert, C. Phraephaisarn, W. Chumnqoen and F.J. Tan. 2014. Lactic acid bacterial population dynamics during fermentation and storage of Thai fermented sausage according to restriction fragment length polymorphism analysis. Int. J. Food Microbiol. 186: 61–67.

Weiss, A., K. Domig, W. Kneifel and H. Mayer. 2010. Evaluation of PCR-based typing methods for the identification of probiotic *Enterococcus faecium* strains from animal feeds. Anim. Feed Sci. Tech. 158: 187–196.

Wolter, H., E. Laing and B.C. Viljoen. 2000. Isolation and identification of yeasts associated with intermediate moisture meats. Food Tech. Biotech. 38: 69–75.

Zeuthen, P. 2007. A historical perspective of meat fermentation. pp. 3–9. *In:* F. Toldrà (ed.). Handbook of fermented meat and poultry. Blackwell Publishing. Ames, Iowa.

Chapter 9

Application of Molecular Methods in Fermented Meat Microbiota: Biotechnological and Food Safety Benefits

Maria Grazia Bonomo*, Caterina Cafaro, Giovanni Salzano

1 INTRODUCTION

There is a great variety and diversity of fermented meat products all over the world, as a consequence of different formulations used in their production. The ingredients together with manufacturing practices and fermentation techniques are the main factors that lead to products with specific organoleptic profiles and physico-chemical characteristics, making them unique (Bonomo et al. 2008, Cocolin et al. 2008).

Nowadays, fermented meat product manufacture is a very important part of the meat industry in many countries and the use of starter cultures was become common in the manufacture of several types of fermented products. However, many typical fermented products are still produced with artisanal technologies without selected starters (Fonseca et al. 2013). Artisanal products have greater quality than those from controlled fermentations inoculated with industrial starters and possess

---

*For Correspondence: Dipartimento di Scienze, Università degli Studi della Basilicata, Viale dell'Ateneo Lucano 10, 85100 Potenza, Italy, Email: mariagrazia.bonomo@unibas.it

distinctive qualities, due especially to the specific composition of the indigenous microbiota (Lebert et al. 2007, Martin et al. 2007, Talon et al. 2008). Therefore, commercial starter cultures are not always able to compete well with this house flora so that their use often results in loss of desirable sensory properties. So, appropriate cultures have to be selected from indigenous microorganisms, more competitive, well adapted to the particular product and to the specific production technology, and with high metabolic capacities which can beneficially affect product quality and safety, preserving their typicity (Leroy et al. 2006, Bonomo et al. 2009, Babić et al. 2011).

In the last decade, the study of the ecology of fermented meat products was of primary importance to understand the physical and chemical changes occurring during fermentation and ripening (Comi et al. 2005, Lücke 1985). In recent years, an increasing number of studies was focused on the isolation and identification of autochthonous functional starter cultures, with the aim of developing new functional meat products, including microorganisms that generate aroma compounds, health-promoting compounds, bacteriocins and other antimicrobials, possess probiotic qualities, or lack negative properties such as the production of biogenic amines and toxic compounds (Leroy et al. 2006, Ammor and Mayo 2007, Bernardeau et al. 2006, Noonpakdee et al. 2003, Bernardeau et al. 2008, Babić et al. 2011).

Meat fermented products are the result of a complex microbiological activity that mainly consists of a lactic fermentation and several characteristic biochemical changes in which the indispensable participants are lactic acid bacteria (LAB) and coagulase-negative cocci (CoNS) (Rantsiou and Cocolin 2008, Palavecino Prpich et al. 2015). The LAB play an important technological role in meat preservation and fermentation; they decrease the pH by lactic acid production, produce bacteriocins to suppress the growth of spoilage or pathogenic microorganisms, provide diversity of sensory properties by modification of raw material, contribute to the development of flavour, colour and texture and improve the safety, stability and shelf life of meat products (Fontana et al. 2005a, Frece et al. 2005, Frece et al. 2009, Hugas and Monfort 1997, Lücke 2000). Coagulase-negative cocci contribute to the development of sensory properties (flavour, texture and colour) and affect the quality and the stability of the products by their biochemical-metabolic properties, such as nitrate reduction, proteolytic and lipolytic activity (Mauriello et al. 2004, Olesen et al. 2004, Tjener et al. 2004). The ability of CoNS to produce antimicrobial compounds may improve the safety and shelf-life of sausages (Martín et al. 2007, Simonova et al. 2006, Leroy et al. 2009, 2010).

In the last 20 years, the advancements in molecular biology have revolutionized the study of microbial biodiversity with a new range of

Application of Molecular Methods in Fermented Meat Microbiota.... 169

techniques that can help in the understanding of the microbial complexity in natural ecosystems. As a consequence, the traditional microbiological techniques, based on plating, isolation, and biochemical identification, are now supported by new methods that rely on the analysis of the nucleic acids for detection, identification, and quantification of microorganisms (Cocolin et al. 2011, Cocolin et al. 2008).

Nowadays there is a growing interest in developing novel foods containing probiotic microorganisms, such as bifidobacteria and LAB. Such functional cultures may offer organoleptic, technological and nutritional advantages, but more importantly confer a health benefit on the host, as the probiotics administration has been linked to the treatment of various diseases (Deshpandea et al. 2011). The mechanisms and the efficacy of a probiotic effect often depend on the interactions with the host microbiota or with the immuno-competent cells of the intestinal mucosa (Saad et al. 2013).

A properly designed strategy for the incorporation of probiotic microorganisms into foods (formulation strategies, processing, stability and organoleptic quality issues) is a key factor in the development of functional products. Although encapsulation systems have largely been exploited in the pharmaceutical (e.g. drug and vaccine delivery) and agricultural sector (e.g. fertilizers), the food industry has only recently become aware of the immense benefits that these technologies are able to offer (Champagne et al. 2010). Insertion of beneficial bacteria into a food matrix presents a new challenge, not only because of their interactions with other constituents, but also because of the severe conditions often employed during food processing and storage, which might lead to important losses in viability, as probiotic strains are very often thermally labile (on heating and/or freezing) and sensitive to acidity, oxygen or other food constituents (e.g. salts).

There were only a few attempts at incorporating probiotic cultures in dry-fermented sausages (De Vuyst et al. 2008, Muthukumarasamy and Holley 2006) and the results are still considered preliminary for evaluating the effect of probiotic fermented meats on human health. Most concerns are associated with the survival of the probiotic strains during the manufacturing process and detection in high numbers in the end-product (De Vuyst et al. 2008, Leroy et al. 2006). The major difficulty is the monitoring of the probiotic cultures survival in foods, due to the lack of accurate, reliable, convenient and sensitive methods of identification to distinguish the strains of interest among other closely related microorganisms present in the products.

This chapter will discuss the most recent and advanced molecular methods applied to identify, characterize and profile microbial diversity and dynamics of fermented meat products. Understanding the microbial biodiversity and ecology can allow a better control of the transformation process, resulting in products with high quality and safety and unique sensory characteristics and facilitating the development of autochthonous starters.

2 MOLECULAR APPROACHES IN FERMENTED MEAT MICROBIOTA

Molecular techniques have been used as an effective method to identify and characterize the microbiota of fermented meat products for the last 20 years. Developments in the field of molecular biology, allowed for new methods to become available, which could be applied to better understand dynamics and diversity of the microorganisms in fermented meat products. Today it is possible to detect, identify and quantify microorganisms by targeting their nucleic acids (Cocolin et al. 2011).

Nucleic acids can be analyzed either from isolates obtained from the food matrix by traditional microbiological methods (culture-dependent techniques), or by direct extraction from the food sample (culture-independent techniques). In the last 10 years, several studies have underlined that, often, there are significant differences between the results obtained with culture-independent and -dependent methods (Cocolin et al. 2004, Cocolin et al. 2011, Doyle and Buchanan 2013, Ceuppens et al. 2014).

There is a scientific consensus on the fact that culture-dependent methods are not able to properly describe the diversity of complex ecosystems; populations that are present in low numbers or that are in a stressed or injured state will most probably be unintentionally excluded from consideration if traditional microbiological methods are used. Moreover, the cells in a viable but not culturable (VBNC) state cannot be detected, because of their incapability to form colonies on microbiological media (Cocolin et al. 2008, Cocolin et al. 2004). However, also culture-independent methods possess some limitations due to generally high limits of detection, thereby minor populations are not taken into consideration.

Culture-independent approaches are employed to study the ecology and biodiversity in food fermentations, using DNA and RNA as target molecules, that have different properties and meaning and allow obtaining different kind of results. Studying the DNA of a microbial ecosystem will allow definition of the microbial ecology and diversity as DNA is a very stable molecule and it is long present also after the cell has died, while the RNA analysis will highlight more properly the microbial populations that are metabolically active, thereby contributing to the fermentation process as RNA, and especially messenger RNA (mRNA), can have very short life (Cocolin et al. 2008, Chakraborty et al. 2014).

Application of Molecular Methods in Fermented Meat Microbiota.... 171

Recently the combination of culture-dependent and culture-independent molecular methods have become the preferred approach for determining and analyzing the species composition of different microbial communities (Ercolini et al. 2001a, b, Temmerman et al. 2003, 2004, Silvestri et al. 2007, Kesmen et al. 2012).

2.1 Culture-Independent Methods

The advent of culture-independent methods to characterize the microbiota from fermented food products allow to analyze multiple samples simultaneously without the need for preceding culturing or for prior knowledge of the ecosystems diversity (Cocolin et al. 2001). In this way the investigation of microbial ecology has dramatically changed and this process is in constant evolution (Solieri et al. 2013). For a long period, food-associated microorganisms and their dynamics have been studied through culture based-methods (Doyle and Buchanan 2013). However, these revealed to be often weak to accomplish a complete microbial characterization of many ecosystems, among which foodstuffs (Ceuppens et al. 2014). Problems and shortcomings of culturing methods basically involve the underestimation of microbial diversity, and even the failure of a precise detection of some species or genera. The introduction of molecular technologies permitted to identify food related microorganisms and to evaluate their relative abundance, providing a fast, accurate and economic detection tool increasingly important in food microbiology, complementing or substituting classical methods (Ceuppens et al. 2014, Chakraborty et al. 2014, Galimberti et al. 2015).

2.1.1 Denaturing Gradient Gel Electrophoresis (DGGE)

The culture-independent method that has become more popular to study the diversity of microbial communities and has been more extensively applied to sausage fermentation is the analysis of PCR products by using denaturing gradient gel electrophoresis (DGGE) (Rantsiou and Cocolin 2008, Cocolin et al. 2001).

The technique consists of the electrophoretic separation of PCR-generated double stranded DNA in a polyacrylamide gel containing a gradient of chemical denaturants (urea and formamide); when the DNA molecule meets with an appropriate denaturant concentration, a sequence-dependent, partial denaturation of the double strand occurs and this conformation change of the DNA structure causes a reduced migration rate of the molecule. When the method is used for microbial profiling, PCR is carried out with universal primers, able to prime amplification for all the microbes present in the sample, and then, the complex mixture of the DNA molecules obtained can be differentiated and characterized by separation on denaturing gradient gels.

Every single band that is visible in DGGE gels represents a component of the microbiota; the more bands are visible, the more complex is the ecosystem. By using this method, it is possible not only to profile the microbial populations, but also to follow their dynamics during time. Modern image analysis systems have proven to be of value for the analysis of DGGE bands and their associated patterns. It should be noted that these methods are not quantitative (Rantsiou and Cocolin 2006). Direct PCR amplification of different regions of the 16S rRNA gene and subsequent analysis by DGGE has been used to study the ecology of the microbial processes involved in the production of many Italian fermented sausages (Aquilanti et al. 2007, Cocolin et al. 2009, Cocolin et al. 2001, Rantsiou et al. 2005, Silvestri et al. 2007, Villani et al. 2007), but studies on the fermentation dynamics of Argentinean (Fontana et al. 2005a, Fontana et al. 2005b) and Portuguese (Albano et al. 2008), Greek (Sidira et al. 2014), Vietnamese (Nguyen et al. 2013) and Turkish (Kesmen et al. 2012) sausages are available as well. Moreover, the use of DGGE in food microbiology was reviewed by Ercolini (2004) and recently by other authors (Cocolin et al. 2011, Galimberti et al. 2015) that checked the contribution of molecular methods for a better comprehension of complex food ecosystems as biodiversity and dynamics of meat fermentations.

It was reported that PCR-DGGE was the most suitable technique to investigate the microbial community in different food matrices in which the cultivation of many microorganisms is difficult or is thought to be impossible. However the detection limit of DGGE depends on the target species, the concentration of the other members of the microbial community and nature of the food matrix (Fontana et al. 2005b). Moreover the efficiency of DNA extraction and possible competition among templates during PCR amplification also affect the limit of detection (Ercolini 2004).

The study of Cocolin et al. (2001) was the first work that used the direct DGGE analysis to monitor the bacterial population dynamics during natural fermentation of Italian sausages. The PCR-amplified V1 region of the 16S rRNA gene analysis highlighted how fermentation of sausages is a highly competitive process in which a wide species diversity can be found at the beginning of fermentation and then a rapid evolution of the predominant populations is observed evidencing the presence of the active population responsible for the changes that occurred during ripening of the sausages. This was confirmed also by other studies, for example the general consideration that resulted from the study of Rantsiou et al. (2005) was that the main differences detected in the ecology, between traditional fermented sausages from the same region of Italy but three different plants, were not represented by the microbial species identified by band sequencing, but by their relative

Application of Molecular Methods in Fermented Meat Microbiota.... 173

distribution between fermentations. In general, in different studies, targeting the V1 and V3 regions 16S rRNA, it was possible to identify the main lactic acid bacteria and staphylococci species and to prove the important role that these microbial groups have in the specific fermented sausages production (Fontana et al. 2005b, Villani et al. 2007, Albano et al. 2008, Cocolin et al. 2009). DGGE was used in monitoring the growth kinetics during the ripening process and the effects of these on the safety and quality, allowing tracking of the microbial 'typicity' and collecting of important information for the designing of autochthonous starter cultures (Aquilanti et al. 2007, Albano et al. 2008).

Recently, novel fermented meat products were investigated, always, preserving the combination of culture-dependent and -independent molecular methods to characterize the typical microbiota attributed to the specific conditions prevalent at the location of production for a better understanding of their influence on the typical characteristics of the product (Nguyen et al. 2013, Kesmen et al. 2012). As a result of these studies, it was concluded that polyphasic strategies are necessary for accurate and reliable screening of the microbial composition of fermented products. This application could provide an opportunity to better understand and control the transformation process during fermentation. Moreover the profiling of bacterial populations occurring in these artisanal products can be useful to determine the technologically important strains to be employed as an autochthonous starter culture to obtain high quality and safety properties and the desired sensory profiles in the final product (Nguyen et al. 2013, Kesmen et al. 2012).

Moreover, recently, fermentation has been shown to have not only preservative effects and the ability of aiding the modification of the physicochemical properties of different foods but also the capability to provide significant impact on the nutritional quality and functional performances of the raw material (Kos et al. 2003), so it has been observed a growing interest in developing novel fermented meat products containing probiotic microorganisms, with functional properties that can offer organoleptic, technological and nutritional advantages and particularly confer a health benefit on the host. A new frontier goal for fermented meat is the use of functional starter cultures, i.e. starter cultures able to improve food safety, and to preserve the typical sensory quality of traditional sausages but with an "added function", potential health benefits, as proposed by some authors (Babić et al. 2011, Bevilacqua et al. 2015).

A recent study of Sidira et al. (2014) proved a novel and interesting picture of fermented meat products in which the use of immobilized *Lactobacillus casei* on wheat as probiotic starter culture in dry-fermented sausages was assessed. Insertion of beneficial bacteria into a food matrix presents a new challenge, not only because of their interactions with other constituents, but also because of the severe conditions often employed during food processing and storage, which might lead to important losses in viability; therefore, to overcome such deficiencies, immobilization techniques are usually applied in order to maintain cell viability, activity and functionality, in order to allow the formulation of new types of foods fortified with immobilized health-promoting bacteria that are only released upon reaching the human gut (Champagne et al. 2010, Bosnea et al. 2009, Charalampopoulos et al. 2003).

A few attempts at incorporating probiotic cultures in dry-fermented sausages were performed (De Vuyst et al. 2008, Muthukumarasamy and Holley 2006), and the results are still considered preliminary for evaluating the effect of probiotic fermented meats on human health; moreover, the monitoring of the probiotic cultures survival in foods is hampered by the lack of accurate, reliable, convenient and sensitive methods of identification to distinguish the strains of interest among other closely related microorganisms present in the products. For this reason, Sidira et al. (2014) investigated the microbial interactions and dynamics in a novel fermented meat product by V1 region of the 16S rRNA gene DGGE analysis during manufacture and ripening. This analysis allowed documenting the survival of *Lb. casei* cells at the required levels for providing the health benefits at the time of consumption (during ripening and after mild heat treatment), repression of spoilage and pathogenic bacteria, improvement of quality characteristics and extension of products' shelf-life.

The application of PCR–DGGE in the field of fermented sausages offers a better understanding of the biodiversity and dynamics of the populations involved in the transformation. However it should be mentioned that pitfalls, associated with sampling, DNA extraction, DNA purity, PCR conditions, formation of heteroduplex and chimeric molecules, may still exist, thereby the results obtained need to be verified and validated (Ercolini 2004). One important aspect that has to be taken into consideration when applying DGGE in food fermentation is the sensitivity limit. It has been demonstrated that populations that are below 10^3–10^4 colony forming units (cfu)/g will not be detected (Cocolin et al. 2001). This is especially valid when in the same ecosystem two populations, one at high and the other at low counts, exist, as it usually happens in sausage fermentations. Moreover, due to the extensive application of sequencing, databases have seen tremendous growth of the sequences deposited. Often, these entries are classified as 'unculturable microorganism', since they have been detected only by culture-independent methods and no significant similarity to available sequences was obtained. This aspect introduces potential difficulties in the understanding of the ecology of fermented foods and at the same time underlines the need to improve the traditional cultivation methods (Cocolin et al. 2011).

2.1.2 Next Generation Sequencing (NGS)

Since advances in technology have always driven discoveries and changes in microorganism taxonomy, taxonomic identification is an issue of primary importance when approaching the study of food microbiota. In this scenario, genomics now underlies a renaissance in food microbiology therefore accelerating food safety monitoring and food production processes (Ceuppens et al. 2014). Microbial taxonomy directly influences a number of basic scientific and applied fields where microorganisms are involved (Tautz et al. 2003) including food production, conservation and probiotic activity. Depending on the level of investigation required, the taxonomic resolution of microorganisms can vary. Aiming to differentiate microorganisms at the species level, methods based on DNA sequencing are currently the most adopted. In many cases, when a fast and accurate response is needed, a 'DNA barcoding-like' approach is the most reliable (Chakraborty et al. 2014). Many scientists used 16S rRNA gene as a universal marker for species-level typing of microorganisms (Bokulich et al. 2012, Claesson et al. 2010, Janda and Abbot 2007). This genomic region is considered a 'bacterial barcode' due to its peculiar properties (Patel 2001): it is present in all the bacterial species, it contains sufficient information (1500 bp long) to differentiate species and, in some cases, strains (Muñoz-Quezada et al. 2013) and finally, the 16S rRNA relies upon an impressive archive of reference sequences such as Greengenes (De Santis et al. 2006) and SILVA (Pruesse et al. 2007). Amplicons belonging to whole genomic extraction conducted on food products matrices are sequenced and the reads are compared to reference databases to identify the Operational Taxonomic Units (OTUs).

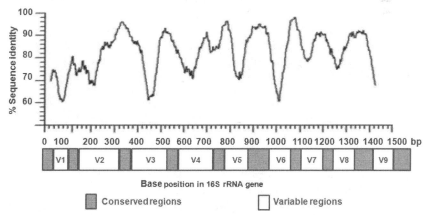

Figure 1 Schematic overview of the 16S rRNA gene. The Sequence identity of the 16S rRNA gene of more than 6,000 bacteria is shown. V1–V9 are the hypervariable regions while the others are the conserved regions (modified from Wahl et al. 2015).

176 *Fermented Meat Products: Health Aspects*

Figure 1 shows the nine hypervariable regions (V1–V9) contained in the approximately 1500 bp long 16S rRNA which can be used for identification of bacteria; a part of the gene is amplified by PCR with primers binding in highly conserved regions, subsequently, the resulting PCR product spanning one or more hypervariable regions is sequenced and used for taxonomic classification.

Progresses in sequencing technologies and bioinformatics analysis of data, led nowadays to a more complex scenario of food microbial communities, offering a panel of analytical tools able to screen the whole microbial community of food matrices. The use of universal markers produces several DNA barcode fragments, corresponding to the each bacterial species present in a food sample. With the ultimate goal of characterizing the complete spectrum of microorganisms, several novel approaches, referred to as 'Next Generation Sequencing' (NGS) and, more recently, 'High Throughput Sequencing' (HTS), have been developed (Ercolini 2013, Mayo et al. 2014, Solieri et al. 2013, Galimberti et al. 2015).

Food microbiology deals with the study of microorganisms that have both beneficial and deleterious effects on the quality and safety of food products. The fast and low-cost NGS approaches have revolutionized microbial taxonomy and classification and have changed the landscape of genome sequencing projects for food-associated microbial species (Coenye et al. 2005). The NGS-driven advances have been exploited mainly to re-sequence strains and individuals for which reference genome sequences are available in order to sample genomic diversity within microbial species. The NGS approaches have greatly increased the ability of researchers to profile food microbial communities, as well as to elucidate the molecular mechanisms of interesting functionalities in food ecosystems. These applications enable the culture-independent sequencing of collective sets of DNA or RNA molecules obtained from mixed microbial communities to determine their content (Solieri et al. 2013). Figure 2 presents a general flow chart of Next-Generation Sequencing (NGS) applications in food microbiology.

NGS techniques have promoted the emergence of new, high-throughput technologies, such as genomics, metagenomics, transcriptomics and metatranscriptomics, etc. As compared to previous culture-independent methods, the number of nucleic acid sequences analyzed by NGS techniques is exceedingly higher, allowing a deeper description of the microbial constituents of the ecosystems. These technologies can be used in two substantially different ways: sequencing the total microbial nucleic acids (shotgun sequencing) and gene-specific sequencing (targeted sequencing). For the latter, segments of highly conserved DNA or cDNA sequences are first amplified by PCR using universal or group-specific primers. Targeted techniques provide a snapshot of the diversity and phylogeny of the different elements making up microbial populations. The term phylobiome has been introduced recently to refer to the phylogenetic information gathered using this approach (van Hijum et al. 2013). In addition, shotgun techniques inform on the genetics and functional capabilities of the microbial constituents of food ecosystems, providing insights into the number and potential function of genes within the community (Wilmes et al. 2009, Solieri et al. 2013). Both shotgun and targeted techniques have already been used to study the microbiology of a series of foods and food fermentations, and pertinent reviews have recently been compiled (Solieri et al. 2013, Liu 2011, Bokulich and Mills 2012, Ercolini 2013). However, research in this area is so active that findings must be continually reviewed, and the current and potential applications of these constantly updated.

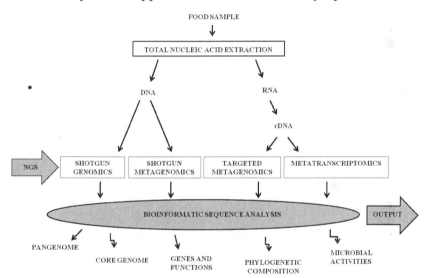

Figure 2 General flow chart of Next-Generation Sequencing (NGS) applications in food microbiology.

Interesting review articles on various aspects of the impact of NGS technologies on food microbial genomics were drafted, and provided complete information about the most common NGS systems and platforms and then addressed how NGS techniques have been employed in the study of food microbiota and food fermentations, discussing their limits and perspectives. The most important findings are reviewed, including those made in the study of the microbiota of milk, fermented dairy products, and plant-, meat- and fish derived fermented foods (Ercolini 2013, Solieri et al. 2013, Mayo et al. 2014).

NGS platforms involve many different technologies (Glenn 2011) all of which generate large, genome-scale datasets. However, they differ substantially in terms of their engineering, sequencing chemistry, output (length of reads, number of sequences), accuracy and cost. Current commercial platforms include the 454 (Roche), Illumina (Illumina), SOLiD and Ion Torrent (Life Technologies), and PacBio (Pacific Biosciences) systems. Comparisons of their advantages and disadvantages have recently been published (Liu et al. 2012, Quail et al. 2012). Many other 'third-generation' techniques, such as DNA nanoball sequencing, heliscope single molecule sequencing, nanopore DNA sequencing, tunneling current DNA sequencing, sequencing with mass spectrometry, and microscopy-based techniques, are currently under development (Schadt et al. 2010, Mayo et al. 2014, Galimberti et al. 2015).

In general, all of the NGS analysis platforms require three principal steps, each of which must be optimized, that are library preparation, template amplification and sequencing, and are shown in Figure 3 and described in details in Loman et al. (2012) and in Mayo et al. (2014).

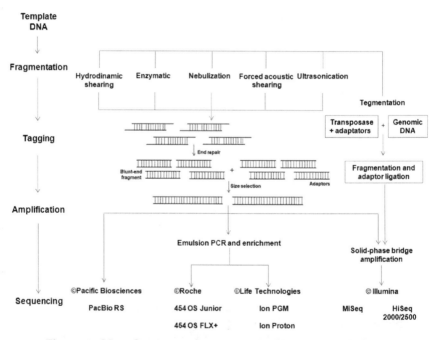

Figure 3 Next-Generation Sequencing (NGS) analysis platforms (modified from Loman et al. 2012).

Recently, various studies used NGS approaches to study the microbial ecosystem (in terms of diversity and dynamics) of different fermented foods (Ercolini et al. 2012, Masoud et al. 2011, Bokulich and Mills 2012, Bokulich et al. 2012, Jung et al. 2012, Kiyohara et al. 2011, Koyanagi et al. 2011, Roh et al. 2010, Nam et al. 2012a, Nam et al. 2012b, Polka et al. 2015), and, in most cases, the obtained results could be of great impact on the food supply chain to improve industrial biotransformation processes, enhance quality of final products, extend the shelf-life and valuating local productions.

Currently, some reports on meat and meat products have been published; their basic goal was to characterize microbial diversity and community dynamics in order to understand the relationships between microorganisms and their impact on food sensorial properties and safety. The application of NGS approach can contribute to describe microorganisms and their activities so as to work out their roles in changes in the quality of fresh foods. Some studies revealed an high bacterial diversity in beef, minced and other meat sort, showing how microbial diversity, growth dynamics, and metabolite production can change according to different storage packaging conditions and how the association between microbial development and chemical changes occurring during the storage of meat is a potential means of revealing indicators of meat quality or freshness (De Filippis et al. 2013, Ercolini et al. 2011, Stoops et al. 2015, Kiermeier et al. 2013, Xiao et al. 2013).

Similarly, Nieminen et al. (2012) compared microbial communities in marinated and unmarinated broiler meat by metagenomic analysis, evidencing a number of bacterial taxa that have not been associated with late shelf-life meat before. In this study, the sequencing depth allowed observation of functional differences between metagenomes that reflect the differences in phylogenetic composition of the bacterial communities rather than functions that could be specifically related to bacterial growth in the investigated meat preparations. A more informative comparison of spoilage-related functions in different meat products could be achieved by studying the actively expressed genes using metatranscriptomics.

In the study of Chaillou et al. (2014), bacterial communities associated to spoilage of four meat products were explored by V1–V3 region of the 16S rRNA gene pyrosequencing to describe microbial diversity at the species level as spoilage could be species-dependent. They recognized reservoirs of spoilage bacteria and deduced the impact of different storage conditions and food properties on these communities, with novel species that could be involved in spoilage and unpredicted assemblages of dominant species.

Investigation of the bacterial communities involved in fermented meat products can be carried out by microbiological and molecular methods. The latter approaches are now being revolutionized by the introduction of NGS, most specifically Illumina platforms which provide million of reads up to 300×2 bp length, with further advancements expected in the near future (Bokulich and Mills 2012, Shokralla et al. 2012). We are rapidly moving to the postgenomic era, when a complete assessment of the genes present in a certain sample will be obtained,

180   *Fermented Meat Products: Health Aspects*

and their expression and activity assessed by metatrascriptomic and metaproteomics. An important step in this direction is represented by the quantitative assessment of bacterial taxa, which can be achieved by sequencing of 16S rRNA amplicons: the identification on the basis of 16S NGS data of the species present in a food sample can greatly support metagenomic analyses to be conducted on the whole microbial DNA. In the work of Polka et al. (2015), Illumina MiSeq approach was applied to characterize the bacterial community of Salame Piacentino PDO (Protected Designation of Origin), a fermented sausage from different factories and at different ripening stages.

Culture-based microbiological analyses and PCR-DGGE were carried out in order to be compared with NGS results obtained by Illumina MiSeq analyses of the two bacterial 16S hypervariable regions V3 and V4. The initial approach was the same as PCR-DGGE, i.e., the amplification of hypervariable regions in a gene marker of interest such as the 16S rRNA, but rather than being separated on denaturing gel and eventually sequenced after cloning or cutting of bands of interests, amplicons were directly analyzed on an Illumina MySeq platform employing a 250 × 2 bp paired reads chemistry. As, previously (Vasileiadis et al. 2012), it has been demonstrated that analyses of single hypervariable regions can support a good taxonomical assessment of bacterial diversity, Polka et al. (2015) showed that a manual curation of 16S rRNA databases such as the GreenGenes database can greatly improve the taxonomical assignment of sequences covering the V3 and V4 regions of bacteria; a manual amendment of the database by adding reference sequences of genera that were found in salami samples, lead to a taxonomical assignment up to the species level for 99.4% of the 722.196 bacterial sequences retrieved from the salami. These results proved that the microbial community composition of dry-fermented sausages is largely composed of known species whose 16S sequences are already available in public databases; at methodological level, it also demonstrated that manual curation of reference database is very important to improve the taxonomical classification of NGS data, and that the sequencing of two hypervariable 16S regions permitted by Illumina technology is sufficient for diversity assessment of these fermented foods. The comparison between NGS and PCR-DGGE approaches showed that NGS confirms, with a better resolution and quantitative assessments, the samples patterns identified by PCR-DGGE. The greater performance of NGS was fully appreciated when taxonomical analyses were carried out; NGS also allowed identifying a diversity of species that is not appreciated by PCR-DGGE. PCR-DGGE, in fact, did not allow reliable quantification of population abundances, and the sequencing of excised bands identified only the main species, while the whole data set obtained by NGS analyses presented 13 main families and 98 rare ones, 27 of

Application of Molecular Methods in Fermented Meat Microbiota....   181

which were present in at least 10% of the samples, even if, most probably, many of these sequences belong to environmental contaminants that do not play an ecological role in the sausage fermentation and with casings that were confirmed to be the major sources of rare species; since in local DOP products the use of casings from pig intestine is mandatory, this result confirm the importance of such practice to guarantee and preserve the bacterial diversity in fermented raw meats (Polka et al. 2015). This highlights the great potential of the NGS application to microbial ecology of fermented meat products to gain a complete and in-depth picture of the bacterial species and identify species that cannot be detected with classical microbiological and molecular methods.

Moreover, RNA-based analysis can significantly increase the ability to identify the impact of the microbial population on organoleptic characteristics of typical food products; when RNA is analyzed, the microbial populations that are metabolically active can be potentially detected and identified, and these are the populations that contribute the most to the fermentation process (Cocolin et al. 2013). In a recent study (Greppi et al. 2015) the diversity of metabolically active microbiota occurring during the natural fermentation of a traditional Piedmontese sausage was evaluated by using RT-PCR-DGGE coupled with RNA-based pyrosequencing of 16S rRNA gene. As to the study of Polka et al. (2015) on typical Italian dry-fermented sausages, that reported the presence of 32 different *Staphylococcaceae* and 33 *Lactobacillaceae* identified by using DNA-based NGS with DGGE techniques, Greppi et al. (2015) showed that only a few species belonging to those genera may be metabolically active and can really contribute to determine the final characteristics of the products.

2.2   Culture-Dependent Methods

Since the nineties, molecular techniques for the identification and characterization of microorganisms isolated from fermented meat products, started to be used side by side or to substitute morphological, phenotypical and biochemical tests. Although the different approaches, in most of the cases, arrived at similar results, molecular techniques immediately showed a higher level of reproducibility, automatism and rapidity (Cocolin et al. 2008, 2011).

There is an important distinction between culture-dependent and -independent methods, as already mentioned; the culture-dependent methods are commonly used to molecularly identify and characterize microbial isolates, while the culture-independent ones are used to profile directly the microbial populations, dynamics and changes in the microbial ecology during fermentation and ripening processes (Cocolin et al. 2008). In recent decades, the study of fermented meat products microflora was carried out by different studies, most of which reported the molecular identification and characterization of isolated LAB and CNC strains considered the technologically relevant microbial groups in the fermented meat products (Ben Amor et al. 2007, Olive and Bean 1999, Temmerman et al. 2004, Kesmen et al. 2012). The 16S rRNA is the gene more often targeted for the identification of isolates, as it is common to all bacteria, it possesses variable and conserved regions used for the differentiation, it is as an evolutionary clock that provides phylogenetic information and, also, a large database of sequences is already available (Singh et al. 2009).

The increasing availability of the sequences of the 16S rRNA gene and the intergenic region between 16S rRNA and 23S rRNA genes allowed the development of different methods for the identification of microbial species of interest in the field of sausage fermentation (Rantsiou and Cocolin, 2008). Ribosomal RNA probes (Nissen and Dainty 1995, Hertel et al. 1991), species-specific PCR (Berthier and Ehrlich 1998, Yost and Nattress 2000, Blaiotta et al. 2003b, Corbiere Morot-Bizot et al. 2003, Rossi et al. 2001), randomly amplified polymorphic DNA (RAPD)-PCR (Berthier and Ehrlich 1999, Andrighetto et al. 2001, Comi et al. 2005), restriction analysis of amplified 16S rDNA gene (ARDRA)-PCR (Aymerich et al. 2006, Rantsiou and Cocolin 2006, Bonomo et al. 2008), restriction fragment length polymorphism (RFLP) analysis of the 16S rDNA gene (Sanz et al. 1998, Lee et al. 2004, Belgacem et al. 2009), multiplex PCR (Corbiere Morot-Bizot et al. 2004, Fonseca et al. 2013), PCR amplification of repetitive bacterial DNA elements (rep-PCR) (Gevers et al. 2001, Danilović et al. 2011), pulsed field gel electrophoresis (PFGE) (Di Maria et al. 2002) and DGGE (Cocolin et al. 2001b, Ercolini et al. 2001, Blaiotta et al. 2003a) have been applied for the identification and characterization of LAB and CNC isolated from fermented sausages.

The application of these molecular strategies has become a routine step for identification and characterization purposes in the last 20 years and a considerable literature, reporting in detail a review of the molecular methods applied in different studies, is offered (Rantsiou and Cocolin, 2008, Cocolin et al. 2008, Cocolin et al. 2011). All studies presented the predominance of *Lb. sakei*, *Lb. curvatus* and *Lb. plantarum*, considered the dominant flora responsible for the transformation process, as well as staphylococci such as *Staphylococcus xylosus*, *St. saprophyticus*, *St. carnosus* and *St. equorum* (Ammor et al. 2005, Aymerich et al. 2003, Cocolin et al. 2001, Comi et al. 2005, Papamanoli et al. 2003, Rantsiou et al. 2004, Rantsiou et al. 2005, Aymerich et al. 2003, Hugas et al. 1993, Papamanoli et al. 2003, Rebecchi et al. 1998, García Fontán et al. 2007a,b, Mauriello et al. 2004, Talon et al. 2007).

Some examples of the most recent analyses of microbiota in fermented meat products are discussed below. All of them highlighted

Application of Molecular Methods in Fermented Meat Microbiota.... 183

as, recently, the combination of culture-dependent and culture-independent molecular methods have become the preferred approach for determining and analyzing the species composition of targeted microbial communities (Kesmen et al. 2012, Fonseca et al. 2013, Nguyen et al. 2013, Federici et al. 2014, Wanangkarn et al. 2014, Kesmen et al. 2014). In all, the applied techniques are those established in the years for the identification and characterization but the novelty is that the analyzed microbiota was isolated from the not yet investigated products and also the new application purpose.

Kesmen et al. (2012) used different molecular strategies, based on both culture-dependent and culture-independent methods, to determine the total microbial diversity of the sucuk from different regions of Turkey by PCR-DGGE analysis of the PCR amplified V1 and V3 region of the 16S rRNA gene (rDNA), to identify and characterize LAB strains isolated from the sucuk samples and cultivated on opportune culture media by rep-PCR fingerprinting technique and then by the sequencing of the 16S rDNA and 16S–23S rRNA intergenic spacer regions. Similarly, the study of Nguyen et al. (2013) gave a more extensive and detailed description of the LAB diversity of nem chua, a Vietnamese fermented meat product, using a polyphasic approach, consisting of $(GTG)_5$-PCR fingerprinting and sequence analysis of the phenylalanyl-tRNA synthase (*pheS*) and/or RNA polymerase α subunit (*rpoA*) genes, used in combination with PCR-DGGE in order to obtain an even more complete picture of the nem chua LAB community.

The study of Wanangkarn et al. (2014) applied restriction fragment length polymorphism (RFLP) analysis to identify the LAB isolated from "mum" Thai fermented sausages during fermentation and storage, in order to facilitate the development of LAB starter cultures that enable controlled processing during mum manufacturing, and provide a more consistent and higher quality product. The authors aimed to identify the dominant LAB species and evaluate the variations in LAB community composition in Thai fermented (mum) sausages manufactured using the conventional processing method and to evaluate the feasibilities and efficiencies of restriction enzymes used for the differentiation of the LAB in RFLP analysis. Their results showed that RFLP analysis is capable of rapidly and easily differentiating and identifying the LAB isolated from mum sausages during fermentation and storage, indicating *Lb. sakei* and *Lb. plantarum* as predominant during fermentation, and marked increases of *Leuconostoc mesenteroides* during the storage. These findings also illustrated that chilled storage combined with vacuum packaging validly influenced the microbial ecology of the mum sausage.

Moreover, the dynamics of the bacterial population throughout the ripening of Galician chorizo, a traditional dry-fermented sausage produced in the north-west of Spain, were investigated by using classical

184   *Fermented Meat Products: Health Aspects*

and molecular approaches (Fonseca et al. 2013). The combination of the results obtained from microbial counts, species and genus-specific PCR as well as real-time quantitative PCR (qPCR) allowed the identification for the dominant bacterial species and the study of the variation in the community composition over the ripening period. In this study, the real-time PCR was shown to be an efficient tool for the study of the complex associations developed in meat fermentations and for the characterization of dominant populations.

At last, the viable microbial community present in Ciauscolo salami of Central Italy was profiled by ARDRA and RAPD-PCR (Federici et al. 2014). The molecular identification highlighted the presence of a high variety of LAB species confirming previous data reporting that *Lb. sakei*, *Pd. pentosaceus* and *Lb. plantarum* are the most common lactobacilli in Italian (Comi et al. 2005, Greco et al. 2005) and European (Aymerich et al. 2003) fermented meat sausages. Ciauscolo productions showed a high genotypic heterogeneity of LAB population based on both the high number of different species identified and intra-specific genotypic variation by RAPD analyses. The isolates were also examined for their potential use as probiotics, since probiotic properties are very strain-dependent. In this study, in vitro methods were selected to investigate the diversity of biological properties of LAB of food origin in comparison with LAB culture collection set. In addition, we showed that food LAB strains have probiotic abilities. In fact, isolation and characterization of novel LAB strains from not investigated niches could have the twofold advantage of revealing taxonomic characteristics and obtaining strains with interesting functional traits.

3 CONCLUSION

Understanding the dynamics, diversity and behavior of microbiota during meat products fermentation is a very interesting and challenging task. As underlined above, microorganisms present in fermented meat products possesses specific characteristics that lead to the development of typical textures and flavour, and putative health-conferring properties; so, the comprehension of the ecology of the fermentation process can help the producers to reach high quality of their products. The surprising advancements in the field of molecular biology, in the last years, revolutionized the way microorganisms are detected and characterized and allowed development of new approaches to study the microbiota without any cultivation. The information collected so far, by applying molecular methods in this field, confirm the findings that were produced by the first studies on the ecology of fermented sausages, but with a deeper level of comprehension.

Application of Molecular Methods in Fermented Meat Microbiota.... 185

DGGE approach allows to compare different products and understand how similar they are, based on the profile of the populations that are detected, moreover the molecular characterization of isolates during the transformation process permits to understand species and strains biodiversity.Recently, the knowledge achieved for the study of the microbiota of fermented meat products is substantial, with a great step forward regard to the investigation of strain dynamics and successions.

The outcomes from the first studies on this specific subject, underline the complexity of the ecology at strain level, during sausage fermentation; biotypes of the same species are coming from the raw materials and the processing plants and their ability to grow and dominate depends on a lot of parameters, as microbial ecosystem is characterized by a species-site-dominance, and this dominance is closely related to the environmental parameters. In fact, traditional fermented sausages show a distinctive organoleptic profile that can be explained by admitting that the isolated strains, although they belong to the species commonly regarded as responsible for sausages fermentation, they possess specific physiological and technological characteristics that make these traditional products unique. Therefore, microbial populations affect significantly the organoleptic and sensory characteristics of the final product by specific and important metabolic activities that contribute to improve quality and safety of fermented sausages.

The possibility to follow specific strains nowadays is possible through the application of molecular methods for the characterization, even if cultivation and isolation are still necessary. However, protocols for direct characterization, using total DNA extracted from the sample, are now available; the application of NGS techniques contribute to describe microorganisms, their activities and dynamics and their role in changes of the quality of these products that is strictly connected to the populations able to develop and to carry out the transformation process, and more specifically to certain biotypes within a species. These techniques able to profile complex microbial ecosystems; if regions of the DNA able to differentiate between strains of the same species are properly selected, their diversity in terms of abundance of a specific sequence can be evidenced, while targeting the RNA, important insights of specific activities of the different biotypes can be obtained. This aspect is extremely relevant, when the differences at the sensory profile, between sausages produced with the same ingredients but in plants of different geographical areas, are taken into consideration, and also because, in recent times, microbial species used in fermented meat products are gaining increasing attention in the area of probiotics. For this reason, the isolation and characterization of microbial species from un-investigated niches of novel fermented meat products are spreading

186   *Fermented Meat Products: Health Aspects*

as they could have the two-fold advantage of revealing taxonomic characteristics and obtaining strains with interesting functional traits that can be used in technological and/or probiotic applications as live microbial feed supplements that beneficially affect the health.

Key words: *culture-dependent and culture-independent molecular methods, Denaturing Gradient Gel Electrophoresis (DGGE), Next Generation Sequencing (NGS), identification and characterization, dynamics and changes of microbial ecology.*

REFERENCES

Albano, H., I. Henriques, A. Correia, T. Hogg and P. Teixeira. 2008. Characterization of microbial population of 'alheira' (a traditional portuguese fermented sausage) by PCR–DGGE and traditional cultural microbiological methods. J. Appl. Microbiol. 105: 2187–2194.

Aymerich, T., B. Martin, M. Garriga and M. Hugas. 2003. Microbial quality and direct PCR identification of lactic acid bacteria and nonpathogenic staphylococci from artisanal low-acid sausages. Appl. Environ. Microbiol. 69: 4583–4594.

Aymerich, T., B. Martin, M. Garriga, M.C. Vidal-Carou, S. Bover-Cid and M. Hugas. 2006. Safety properties and molecular strain typing of lactic acid bacteria from slightly fermented sausages. J. Appl. Microbiol. 100: 40–49.

Ammor, S., E. Dufour, M. Zagorec, S. Chaillou and I. Chevallier. 2005. Characterization and selection of Lactobacillus sakei strains isolated from traditional dry sausage for their potential use as starter cultures. Food Microbiol. 22: 529–538.

Ammor, M.S. and B. Mayo. 2007. Selection criteria for lactic acid bacteria to be used as functional starter cultures in dry sausage production: an update. Meat Sci. 76: 138–146.

Andrighetto, C., L. Zampese and A. Lombardi. 2001. RAPD-PCR characterization of lactobacilli isolated from artisanal meat plants and traditional fermented sausages of Veneto region (Italy). Lett. Appl. Microbiol. 33: 26–30.

Aquilanti, L., S. Santarelli, G. Silvestri, A. Osimani, A. Petruzzelli and F. Clementi. 2007. The microbial ecology of a typical italian salami during its natural fermentation. Int. J. Food Microbiol. 120: 136–145.

Babić, I., K. Markov, D. Kovačević, A. Trontel, A. Slavica, J. Đugum, D. Čvek, I.K. Svetec, S. Posavec and J. Frece. 2011. Identification and characterization of potential autochthonous starter cultures from a Croatian "brand" product "Slavonski kulen". Meat Sci. 88: 517–524.

Belgacem, Z.B., X. Dousset, H. Prévost and M. Manai. 2009. Polyphasic taxonomic studies of lactic acid bacteria associated with Tunisian fermented meat based on the heterogeneity of the 16S–23S rRNA gene intergenic spacer region. Arch. Microbiol. 191: 711–720.

Ben Amor, K., E.E. Vaughan and W.M. De Vos. 2007. Advanced molecular tools for the identification of lactic acid bacteria. The J. Nutr. 137: 741–747.

Application of Molecular Methods in Fermented Meat Microbiota.... 187

Berthier, F. and S.D. Ehrlich. 1998. Rapid species identification within two groups of closely related lactobacilli using PCR primers that target the 16S/23S rRNA spacer region. FEMS Microbiol. Lett. 161: 97–106.

Berthier, F. and S.D. Ehrlich. 1999. Genetic diversity within *Lactobacillus sakei* and *Lactobacillus curvatus* and design of PCR primers for its detection using randomly amplified polymorphic DNA. Int. J. Syst. Bacteriol. 49: 997–1007.

Bernardeau, M., M. Guguen and J.P. Vernoux. 2006. Beneficial lactobacilli in food and feed: long-termuse, biodiversity and proposals for specific and realistic safety assessments. FEMS Microbiol. Rev. 30: 487–513.

Bernardeau, M., J.P. Vernoux, S. Henri-Dubernet and M. Guéguen. 2008. Safety assessment of dairy microorganisms: the *Lactobacillus* genus. Int. J. Food Microbiol. 126: 278–285.

Bevilacqua, A., M.R. Corbo, B. Speranza, B. Di Maggio, M. Gallo and M. Sinigaglia. 2015. Functional starter cultures for meat: a case study on technological and probiotic characterization. Food Nutr. Sci. 6: 511–522.

Blaiotta, G., C. Pannacchia, D. Ercolini, G. Moschetti and F. Villani. 2003a. Combining denaturing gradient gel electrophoresis of 16S rDNA V3 region and 16S–23S rDNA spacer region polymorphism analyses for the identification of staphylococci from Italian fermented sausages. Syst. Appl. Microbiol. 26: 423–433.

Blaiotta, G., Pennacchia C., Parente E. and F. Villani. 2003b. Design and evaluation of specific PCR primers for rapid and reliable identification of *Staphylococcus xylosus* strains isolated from dry fermented sausages. Syst. Appl. Microbiol. 26: 601–610.

Bokulich, N.A., C.L. Joseph, G. Allen, A.K. Benson and D.A. Mills. 2012. Next-generation sequencing reveals significant bacterial diversity of botrytized wine. PLoS ONE 7:e36357.

Bokulich, N.A. and D.A. Mills. 2012. Next-generation approaches to the microbial ecology of food fermentations. BMB Reports 45: 377–389.

Bonomo, M.G., A. Ricciardi, T. Zotta, E. Parente and G. Salzano. 2008. Molecular and technological characterization of lactic acid bacteria from traditional fermented sausages of Basilicata region (Southern Italy). Meat Sci. 80: 1238–1248.

Bonomo, M.G., A. Ricciardi, T. Zotta, M.A. Sico and G. Salzano. 2009. Technological and safety characterization of coagulase-negative staphylococci from traditionally fermented sausages of Basilicata region (Southern Italy). Meat Sci. 83: 15–23.

Bosnea, L., Y. Kourkoutas, N. Albantaki,C. Tzia,A.A. Koutinas and M. Kanellaki. 2009. Functionality of freeze-dried *L. casei* cells immobilized on wheat grains. LWT- Food Sci. Technol. 42: 1696–1702.

Ceuppens, S., D. Li, M. Uyttendaele, P. Renault, P. Ross and M.V. Ranst. 2014. Molecular methods in food safety microbiology: Interpretation and implications of nucleic acid detection. Compr. Rev. Food Sci. F. 13: 551–577.

Chakraborty, C., C.G.P. Doss, B.C. Patra and S. Bandyopadhyay. 2014. DNA barcoding to map the microbial communities: Current advances and future directions. Appl. Microbiol. Biotechnol. 98: 3425–3436.

Chaillou, S., A. Chaulot-Talmon, H. Caekebeke, M. Cardinal, S. Christieans, C. Denis, M.H. Desmonts, X. Dousset, C. Feurer, E. Hamon, J.J. Joffraud, S. La Carbona, F. Leroi, S. Leroy, S. Lorre, S. Mace, M.F. Pilet, H. Prevost,

188 *Fermented Meat Products: Health Aspects*

M. Rivollier, D. Roux, R. Talon, M. Zagorec and M.C. Champomier-Verges. 2014. Origin and ecological selection of core and food-specific bacterial communities associated with meat and seafood spoilage. The ISME J. 1–14.

Charalampopoulos, D., S.S. Pandiella and C. Webb. 2003. Evaluation of the effect of malt, wheat and barley extracts on the viability of potentially probiotic lactic acid bacteria under acidic conditions. Int. J. Food Microbiol. 82: 133–141.

Champagne, C.P., B.H. Lee and L. Saucier. 2010. Immobilization of cells and enzymes for fermented dairy or meat products. pp. 345–365. *In*: N.J. Zuidam, and V.A. Nedović (eds.). Encapsulation technologies for active food ingredients and food processing. Springer, London.

Claesson, M.J., Q. Wang, O. O'Sullivan, R. Greene-Diniz, J.R. Cole and R.P. Ross. 2010. Comparison of two next-generation sequencing technologies for resolving highly complex microbiota composition using tandem variable 16S rRNA gene regions. Nucleic Acids Res. 38:e200.

Cocolin, L., M. Manzano, C. Cantoni and G. Comi. 2001. Denaturing gradient gel electrophoresis analysis of the 16S rRNA gene V1 region to monitor dynamic changes in the bacterial population during fermentation of Italian sausages. Appl. Environ. Microbiol. 67: 5113–5121.

Cocolin, L., K. Rantsiou, L. Iacumin, R. Urso, C. Cantoni and G. Comi. 2004. Study of the ecology of fresh sausages and characterization of populations of lactic acid bacteria by molecular methods. Appl. Environ. Microbiol. 70: 1883–1894.

Cocolin L., P. Dolci and K. Rantsiou. 2008. Molecular methods for identification of microorganisms in traditional meat products. pp. 91–127. *In*: F. Toldrá (ed.). Meat biotechnology. Springer, New York.

Cocolin, L., P. Dolci, K. Rantsiou, R. Urso, C. Cantoni and G. Comi. 2009. Lactic acid bacteria ecology of three traditional fermented sausages produced in the north of Italy as determined by molecular methods. Meat Sci. 82: 125–132.

Cocolin L., P. Dolci and K. Rantsiou. 2011. Biodiversity and dynamics of meat fermentations: The contribution of molecular methods for a better comprehension of a complex ecosystem. Meat Sci. 89: 296–302.

Cocolin, L., V. Alessandria, P. Dolci, R. Gorra and K. Rantsiou. 2013. Culture independent methods to assess the diversity and dynamics of microbiota during food fermentation. Int. J. Food Microbiol. 167: 29–43.

Coenye, T., D. Gevers, Y. Van de Peer, P. Vandamme and J. Swings. 2005. Towards a prokaryotic genomic taxonomy. FEMS Microbiol. Rev. 29: 147–167.

Comi, G., R. Urso, L. Iacumin, K. Rantsiou, P. Cattaneo, C. Cantoni and L. Cocolin. 2005. Characterisation of naturally fermented sausages produced in the North East of Italy. Meat Sci. 69: 381–392.

Corbiere Morot-Bizot, S., R. Talon and S. Leroy. 2003. Development of specific primers for a rapid and accurate identification of *Staphylococcus xylosus*, a species used in food fermentation. J. Microbiol. Meth. 55: 279–286.

Corbiere Morot-Bizot, S., R. Talon and S. Leroy. 2004. Development of a multiplex PCR for the identification of *Staphylococcus* genus and four staphylococcal species isolated from food. J. Appl. Microbiol. 97: 1087–1094.

Danilović, B., N. Joković, L. Petrović, K. Veljović, M. Tolinački and D. Savić. 2011. The characterization of lactic acid bacteria during the fermentation of an artisan Serbian sausage. Meat Sci. 88: 668–674.

Application of Molecular Methods in Fermented Meat Microbiota.... 189

De Filippis, F., A. La storia, F. Villani and D. Ercolini. 2013. Exploring the source of bacterial spoilers in beefsteaks by culture-independent high-throughput sequencing. PLoS ONE 8: e70222.

De Santis, T.Z., P. Hugenholtz, N. Larsen, M. Rojas, E.L. Brodie and K. Keller. 2006. Greengenes, a chimera-checked 16S rRNA gene database and workbench compatible with ARB. Appl. Environ. Microbiol. 72: 5069–5072.

Deshpandea, G., S. Rao and S. Patolea. 2011. Progress in the field of probiotics: Year 2011. Curr. Opin. Gastroenterol. 27: 13–18.

De Vuyst, L., G. Falony and F. Leroy. 2008. Probiotics in fermented sausages. Meat Sci. 80: 75–88.

Di Maria, S., A.L. Basso, E. Santoro, L. Grazia and R. Coppola. 2002. Monitoring of *Staphylococcus xylosus* DSM 20266 added as starter during fermentation and ripening of soppressata molisana, a typical Italian sausage. J. Appl. Microbiol. 92: 158–164.

Doyle, M.P. and R.L. Buchanan. 2013. Food microbiology: Fundamentals and frontiers (4th ed.). ASM Press, Washington, DC.

Ercolini, D., G. Moschetti, G. Blaiotta and S. Coppola. 2001a. The potential of a polyphasic PCR–DGGE approach in evaluating microbial diversity of natural whey cultures for water-buffalo Mozzarella cheese production: bias of "culture dependent" and "culture dependent" approaches. Syst. Appl. Microbiol. 24: 610–617.

Ercolini, D., G. Moschetti, G. Blaiotta and S. Coppola. 2001b. Behavior of variable V3 region from 16S rDNA of lactic acid bacteria in denaturing gradient gel electrophoresis. Curr. Microbiol. 42: 199–202.

Ercolini, D. 2004. PCR–DGGE fingerprinting: Novel strategies for detection of microbes in food. J. Microbiol. Meth. 56: 297–314.

Ercolini, D., I. Ferrocino, A. Nasi, M. Ndagijimana, P. Vernocchi, A. La Storia, L. Laghi, G. Mauriello, M.E. Guerzoni and F. Villani. 2011. Monitoring of microbial metabolites and bacterial diversity in beef stored under different packaging conditions. Appl. Environ. Microbiol. 77: 7372–7381.

Ercolini, D., F. De Filippis, A. La Storia and M. Iacono. 2012. "Remake" by high-throughput sequencing of the microbiota involved in the production of water buffalo mozzarella cheese. Appl. Environ. Microbiol. 78: 8142–8145.

Ercolini, D. 2013. High-throughput sequencing and metagenomics: Moving forward in the culture-independent analysis of food microbial ecology. Appl. Environ. Microbiol. 79: 3148–3155.

Federici, S., F. Ciarrocchi, R. Campana, E. Ciandrini, G. Blasi and W. Baffone. 2014. Identification and functional traits of lactic acid bacteria isolated from Ciauscolo salami produced in Central Italy. Meat Sci. 98: 575–584.

Fonseca, S., A. Cachaldora, M. Gómez, I. Franco and J. Carballo. 2013. Monitoring the bacterial population dynamics during the ripening of Galician chorizo, a traditional dry fermented Spanish sausage. Food Microbiol. 33: 77–84.

Fontana, C., P.S. Cocconcelli and G. Vignolo. 2005a. Monitoring the bacterial population dynamics during fermentation of artisanal Argentinean sausages. Int. J. Food Microbiol. 103: 131–142.

Fontana, C., G. Vignolo and P.S. Cocconcelli. 2005b. PCR–DGGE analysis for the identification of microbial populations from Argentinean dry fermented sausages. J. Microbiol. Meth. 63: 254–263.

Frece, J., B. Kos, J. Beganović, S. Vuković and J. Šušković. 2005. In vivo testing of functional properties of three selected probiotic strains. World J. Microbiol. Biotechnol. 21: 1401–1408.

Frece, J., B. Kos, I.K. Svetec, Z. Zgaga, J. Beganović and A. Leboš. 2009. Synbiotic effect of Lactobacillus helveticus M92 and prebiotics on the intestinal microflora and immune system of mice. J. Dairy Res. 76: 98–104.

Galimberti, A., A. Bruno, V. Mezzasalma, F. De Mattia, I. Bruni and M. Labra. 2015. Emerging DNA-based technologies to characterize food ecosystems. Food Res. Int. 69: 424–433.

García Fontán, M.C., J.M. Lorenzo, S. Martínez, I. Franco and J. Carballo. 2007a. Microbiological characteristics of Botillo, a Spanish traditional pork sausage. LWT – Food Sci. Technol. 40: 1610–1622.

García Fontán, M.C., J.M. Lorenzo, A. Parada, I. Franco and J. Carballo. 2007b. Microbiological characteristics of "androlla", a Spanish traditional pork sausage. Food Microbiol. 24: 52–58.

Gevers, D., G. Huys and J. Swings. 2001. Applicability of rep-PCR fingerprinting for identification of Lactobacillus species. FEMS Microbiol. Lett. 205: 31–36.

Glenn, T.C. 2011. Field Guide to Next Generation DNA Sequencers. Mol. Ecol. Resources 11: 759–769.

Greppi, A., I. Ferrocino, A. La Storia, K. Rantsiou, D. Ercolini and L. Cocolin. 2015. Monitoring of the microbiota of fermented sausages by culture independent rRNA-based approaches. Int. J. Food Microbiol. 212: 67–75.

Hertel, C., W. Ludwig, M. Obst, R.F. Vogel, W.P. Hammes and K.H. Schleifer. 1991. 23S rRNA-targeted oligonucleotide probes for the rapid identification of meat lactobacilli. Syst. Appl. Microbiol. 14: 173–177.

Hugas, M., and J.M. Monfort. 1997. Bacterial starter cultures for meat fermentation. Food Chem. 59: 547–554.

Janda, J.M. and S.L. Abbott. 2007. 16S rRNA gene sequencing for bacterial identification in the diagnostic laboratory: Pluses, perils, and pitfalls. J. Clin. Microbiol. 45: 2761–2764.

Jung, M.J., Y.D. Nam, S.W. Roh and J.W. Bae. 2012. Unexpected convergence of fungal and bacterial communities during fermentation of traditional Korean alcoholic beverages inoculated with various natural starters. Food Microbiol. 30: 112–123.

Kesmen, Z., B. Yarimcam, H. Aslan, E. Ozbekar and H. Yetim. 2014. Application of different molecular techniques for characterization of catalase-positive cocci isolated from sucuk. J. Food Sci. 79: 222–229.

Kesmen, Z., A.E. Yetiman, A. Gulluce, N. Kacmaz, O. Sagdic, B. Cetin, A. Adiguzel, F. Sahin and H. Yeti. 2012. Combination of culture-dependent and culture-independent molecular methods for the determination of lactic microbiota in sucuk. Int. J. Food Microbiol. 153: 428–435.

Kiermeier, A., M. Tamplin, D. May, G. Holds, M. Williams and A. Dann. 2013. Microbial growth, communities and sensory characteristics of vacuum and modified atmosphere packaged lamb shoulders. Food Microbiol. 36: 305–315.

Kiyohara, M., T. Koyanagi, H. Matsui, K. Yamamoto, H. Take, Y. Katsuyama, A. Tsuji, H. Miyamae, T. Kondo and S. Nakamura. 2011. Changes in microbiota population during fermentation of narezushi as revealed by pyrosequencing analysis. Biosci. Biotechnol. Biochem. 76: 48–52.

Application of Molecular Methods in Fermented Meat Microbiota.... 191

Kos, B., J. Šušković, S. Vuković, M. Šimpraga, J. Frece and S. Matošić. 2003. Adhesion and aggregation ability of probiotic strain *Lactobacillus acidophilus* M92. J. Appl. Microbiol. 94: 981–987.

Koyanagi, T., M. Kiyohara, H. Matsui, K. Yamamoto, T. Kondo, T. Katayama and H. Kumagai. 2011. Pyrosequencing survey of the microbial diversity of 'narezushi', an archetype of modern Japanese sushi. Lett. Appl. Microbiol. 53: 635–640.

Lebert, I., S. Leroy, P. Giammarinaro, A. Lebert, J.P. Chacornac and S. Bover-Cid. 2007. Diversity of microorganisms in the environment and dry fermented sausages of small traditional French processing units. Meat Sci. 76: 112–122.

Lee, J., J. Jang, B. Kim, J. Kim, G. Jeong and H. Han. 2004. Identification of *Lactobacillus sakei* and *Lactobacillus curvatus* by multiplex PCR-based restriction enzyme analysis. J. Microbiol. Meth. 59: 1–6.

Leroy, F., J. Verluyten and L. De Vuyst. 2006. Functional meat starter cultures for improved sausage fermentation. Int. J. Food Microbiol. 106: 270–285.

Leroy, S., I. Lebert, J.P. Chacornac, P. Chavant, T. Bernardi and R. Talon. 2009. Genetic diversity and biofilm formation of *Staphylococcus equorum* isolated from naturally fermented sausages and their manufacturing environment. Int. J. Food Microbiol. 134: 46–51.

Leroy, S., P. Giammarinaro, J.P. Chacornac, I. Lebert and R. Talon. 2010. Biodiversity of indigenous staphylococci of natural fermented dry sausages and manufacturing environments of small-scale processing units. Food Microbiol. 27: 294–301.

Liu, G.E. 2011. Recent applications of DNA sequencing technologies in food, nutrition and agriculture. Recent Pat. Food Nutr. Agric. 3: 187–191.

Liu, L., Y. Li, S. Li, N. Hu, Y. He, R. Pong, D. Lin, L. Lu and M. Law. 2012. Comparison of Next-Generation Sequencing Systems. J. Biomed. Biotechnol. 251364.

Loman, N.J., C. Constantinidou, J.Z.M. Chan, M. Halachev, M. Sergeant, C.W. Penn, E.R. Robinson and M.J. Pallen. 2012. High-throughput bacterial genome sequencing: an embarrassment of choice, a world of opportunity. Nat. Rev. Microbiol. 10: 598–606.

Lücke, F.K. 1985. Fermented sausages. *In*: B.J.B. Wood (ed.). Microbiology of fermented foods. Elsevier Applied Science, New York, USA.

Lücke, F.K. 2000. Utilization of microbes to process and preserve meat. Meat Sci. 56: 105–115.

Martin, A., B. Colin, E. Aranda, M.J. Benito and M.G. Cordoba. 2007. Characterization of Micrococcaceae isolated from Iberian dry-cured sausages. Meat Sci. 75: 696–708.

Masoud, W., M. Takamiya, F.K. Vogensen, S. Lillevang, W.A. Al-Soud, S.J. Sørensen and M. Jakobsen. 2011. Characterization of bacterial populations in Danish raw milk cheeses made with different starter cultures by denaturating gradient gel electrophoresis and pyrosequencing. Int. Dairy J. 21: 142–148.

Mauriello, G, A. Casaburi, G. Blaiotta and F. Villani. 2004. Isolation and technological properties of coagulase negative staphylococci from fermented sausages of Southern Italy. Meat Sci. 67: 149–158.

Mayo, B., C. Rachid, Á. Alegría, A. Leite, R. Peixoto and S. Delgado. 2014. Impact of next generation sequencing techniques in food microbiology. Curr. Genomics 15: 293–309.

Masoud, W., M. Takamiya, F.K. Vogensen, S. Lillevang, W.A. Al-Soud, S.J. Sørensen and M. Jakobsen. 2011. Characterization of bacterial populations in Danish raw milk cheeses made with different starter cultures by denaturating gradient gel electrophoresis and pyrosequencing. Int. Dairy J. 21: 142–148.

Muñoz-Quezada, S., E. Chenoll, J. María Vieites, S. Genovés, J. Maldonado and M. Bermúdez-Brito. 2013. Isolation, identification and characterisation of three novel probiotic strains (*Lactobacillus paracasei* CNCM I-4034, *Bifidobacterium breve* CNCM I-4035 and *Lactobacillus rhamnosus* CNCM I-4036) from the faeces of exclusively breast-fed infants. Br. J. Nutr. 109: S51–S62.

Muthukumarasamy, P. and R.A. Holley. 2006. Microbiological and sensory quality of dry fermented sausages containing alginate-microencapsulated *Lactobacillus reuteri*. Int. J. Food Microbiol. 111: 164–169.

Nam, Y.D., S.Y. Lee and S.I. Lim. 2012a. Microbial community analysis of Korean soybean pastes by next-generation sequencing. Int. J. Food Microbiol. 155: 36–42.

Nam, Y.D., S.H. Yi and S.I. Lim. 2012b. Bacterial diversity of cheonggukjang, a traditional Korean fermented food, analyzed by barcoded pyrosequencing. Food Control 28: 135–142.

Nguyen, D.T.L., K. Van Hoorde, M. Cnockaert, E. De Brandt, K. De Bruyne, B.T. Le and P. Vandamme. 2013. A culture-dependent and -independent approach for the identification of lactic acid bacteria associated with the production of nem chua, a Vietnamese fermented meat product. Food Res. Int. 50: 232–240.

Nieminen, T.T., K. Koskinen, P. Laine, J. Hultman, E. Sade, L. Paulin, A. Paloranta, P. Johansson, J. Bjorkroth and P. Auvinen. 2012. Comparison of microbial communities in marinated and unmarinated broiler meat by metagenomics. Int. J. Food Microbiol. 157: 142–149.

Nissen, H. and R. Dainty. 1995. Comparison of the use of rRNA probes and conventional methods in identifying strains of *Lactobacillus sake* and *L. curvatus* isolated from meat. Int. J. Food Microbiol. 25: 311–315.

Noonpakdee, W., C. Santivarangkna, P. Jumriangrit and S. Panyim. 2003. Isolation of nisin-producing *Lactococcus lactis* WNC 20 strain from nham, a tradition Thai fermented sausage. Int. J. Food Microbiol. 81: 137–145.

Olesen, P.T., A.S. Meyer and L.H. Stahnke. 2004. Generation of flavour compounds in fermented sausages—the influence of curing ingredients, *Staphylococcus* starter culture and ripening time. Meat Sci. 66: 675–687.

Olive, D.M. and P. Bean. 1999. Principles and applications of methods for DNA-based typing of microbial organisms. J. Clin. Microbiol. 37: 1661–1669.

Palavecino Prpich, N.Z., M.P. Castro, M.E. Cayré, O.A. Garro and G.M. Vignolo. 2015. Indigenous starter cultures to improve quality of artisanal dry fermented sausages from Chaco (Argentina). Int. J. Food Sci. 1–9.

Papamanoli, E., N. Tzanetakis, E. Litopoulou-Tzanetaki and P. Kotzekidou. 2003. Characterization of lactic acid bacteria isolated from a Greek dry-fermented sausage in respect of their technological and probiotic properties. Meat Sci. 65: 859–867.

Patel, J.B. 2001. 16S rRNA gene sequencing for bacterial pathogen identification in the clinical laboratory. Mol. Diagnosis 6: 313–321.

Application of Molecular Methods in Fermented Meat Microbiota.... 193

Połka, J., A. Rebecchi, V. Pisacane, L. Morelli and E. Puglisi. 2015. Bacterial diversity in typical Italian salami at different ripening stages as revealed by high-throughput sequencing of 16S rRNA amplicons. Food Microbiol. 46: 342–356.

Pruesse, E., C. Quast, K. Knittel, B.M. Fuchs, W. Ludwig and J. Peplies. 2007. SILVA: a comprehensive online resource for quality checked and aligned ribosomal RNA sequence data compatible with ARB. Nucleic Acids Res. 35: 7188–7196.

Quail, M.A., M. Smith, P. Coupland, T.D. Otto, S.R. Harris, T.R. Connor, A. Bertoni, H.P. Swerdlow and Y. Gu. 2012. A tale of three next generation sequencing platforms: comparison of Ion torrent, pacific biosciences and illumina MiSeq sequencers. BMC Genomics 13: 341.

Rantsiou, K. and L. Cocolin. 2006. New developments in the study of the microbiota of naturally fermented sausages as determined by molecular methods: A review. Int. J. Food Microbiol. 108: 255–267.

Rantsiou, K., G. Comi and L. Cocolin. 2004. The rpoB gene as a target for PCR-DGGE analysis to follow lactic acid bacterial population dynamics during food fermentations. Food Microbiol. 21: 481–487.

Rantsiou, K., R. Urso, L. Iacumin, C. Cantoni, P. Cattaneo, G. Comi and L. Cocolin. 2005. Culture dependent and independent methods to investigate the microbial ecology of Italian fermented sausages. Appl. Environ. Microbiol. 71: 1977–1986.

Rantsiou, K. and L. Cocolin. 2008. Fermented meat products. pp. 91–118. *In*: L. Cocolin and D. Ercolini (eds.). Molecular methods and microbial ecology of fermented foods. Springer, New York.

Rebecchi, A., S. Crivori, P.G. Sarra and P.S. Cocconcelli. 1998. Physiological and molecular techniques for the study of bacterial community development in sausage fermentation. J. Appl. Microbiol. 84: 1043–1049.

Roh, S.W., K.H. Kim, Y.D. Nam, H.W. Chang, E.J. Park and J.W. Bae. 2010. Investigation of archaeal and bacterial diversity in fermented seafood using barcoded pyrosequencing. ISME J. 4: 1–16.

Rossi, F., R. Tofalo, S. Torriani and G. Suzzi. 2001. Identification by 16S–23S rDNA intergenic region amplification, genotypic and phenotypic clustering of *Staphylococcus xylosus* strains from dry sausages. J. Appl. Microbiol. 90: 365–371.

Saad, N., C. Delattre, M. Urdaci, J.M. Schmitter and P. Bressollier. 2013. An overview of the last advances in probiotic and prebiotic field. LWT- Food Sci. Technol. 50: 1–16.

Sanz, Y., M. Hernandez, M.A. Ferrus and J. Hernandez. 1998. Characterization of Lactobacillus sake isolates from dry-cured sausage by restriction fragment length polymorphism analysis of the 16S rRNA gene. J. Appl. Microbiol. 84: 600–606.

Schadt, E.E., S. Turner and A. Kasarskis. 2010. A window into third generation sequencing. Hum. Mol. Gen. 19: R227–240.

Shokralla, S., J.L. Spall, J.F. Gibson and M. Hajibabaei. 2012. Next-generation sequencing technologies for environmental DNA research. Mol. Ecol. 21: 1794–1805.

194

Sidira, M., A. Karapetsas, A. Galanis, M. Kanellaki and Y. Kourkoutas. 2014. Effective survival of immobilized *Lactobacillus casei* during ripening and heat treatment of probiotic dry-fermented sausages and investigation of the microbial dynamics. Meat Sci. 96: 948–955.

Silvestri, G., S. Santarelli, L. Aquilanti, A. Beccaceci, A. Osimani and F. Tonucci. 2007. Investigation of the microbial ecology of Ciauscolo, a traditional Italian salami, by culture-dependent techniques and PCR-DGGE. Meat Sci. 77: 413–423.

Simonova, M., V. Strompfova, M. Marcinakova, A. Laukova, S. Vesterlund and M.L. Moratalla. 2006. Characterization of *Staphylococcus xylosus* and *Staphylococcus carnosus* isolated from Slovak meat products. Meat Sci. 73: 559–564.

Singh, S., P. Goswami, R. Singh and K.J. Heller. 2009. Application of molecular identification tools for *Lactobacillus*, with a focus on discrimination between closely related species: a review. LWT- Food Sci. Technol. 42: 448–457.

Solieri, L., T.C. Dakal and P. Giudici. 2013. Next-generation sequencing and its potential impact on food microbial genomics. Ann. Microbiol. 63: 21–37.

Stoops, J., S. Ruyters, P. Busschaert, R. Spaepen, C. Verreth, J. Claes, B. Lievens and L. Van Campenhout. 2015. Bacterial community dynamics during cold storage of minced meat packaged under modified atmosphere and supplemented with different preservatives. Food Microbiol. 48: 192–199.

Talon, R., S. Leroy and I. Lebert. 2007. Microbial ecosystems of traditional fermented meat products: the importance of indigenous starters. Meat Sci. 77: 55–62.

Talon, R., S. Leroy, I. Lebert, P. Giammarinaro, J.P. Chacornac and M. Latorre-Moratalla. 2008. Safety improvement and preservation of typical sensory qualities of traditional dry fermented sausages using autochthonous starter cultures. Int. J. Food Microbiol. 126: 227–234.

Tautz, D., P. Arctander, A. Minelli, R.H. Thomas and A.P. Vogler. 2003. A plea for DNA taxonomy. Trends Ecol. Evol. 18: 70–74.

Temmerman, R., I. Scheirlinck, G. Huys and J. Swings. 2003. Culture- independent analysis of probiotic products by denaturing gradient gel electrophoresis. Appl. Environ. Microbiol. 69: 220–226.

Temmerman, R., G. Huys and J. Swing 2004. Identification of lactic acid bacteria: culture dependent and culture-independent methods. Trends Food Sci. Technol. 15: 348–359.

Tjener, K., L.H. Stahnke, L. Andersen and J. Martinussen. 2004. Growth and production of volatiles by *Staphylococcus carnosus* in dry sausages: influence of inoculation level and ripening time. Meat Sci. 67: 447–452.

van Hijum, S.A.F.T., E.E. Vaughan and R.F. Vogel. 2013. Application of state-of-art sequencing technologies to indigenous food fermentations. Curr. Opin. Biotechnol. 24: 178–186.

Vasileiadis, S., E. Puglisi, M. Arena, F. Cappa, P.S. Cocconcelli and M. Trevisan. 2012. Soil bacterial diversity screening using single 16S rRNA gene V regions coupled with multi-million read generating sequencing technologies. PLoS ONE 7: e42671.

Villani, F., A. Casaburi, C. Pennacchia, L. Filosa, F. Russo and D. Ercolini. 2007. The microbial ecology of the Soppressata of Vallo di Diano, a traditional dry

Application of Molecular Methods in Fermented Meat Microbiota....   195

fermented sausage from southern Italy, and in vitro and in situ selection of autochthonous starter cultures. Appl. Environ. Microbiol. 73: 5453–5463.

Wahl, B., F. Ernst, Y. Kumar, B. Müller, K. Stangier and T. Paprotka. 2015. Characterisation of bacteria in food samples by Next Generation Sequencing. www.gatc-biotech.com

Wanangkarn, A., D.C. Liu, A. Swetwiwathana, A. Jindaprasert, C. Phraephaisarn, W. Chumnqoen and F.J. Tan. 2014. Lactic acid bacterial population dynamics during fermentation and storage of Thai fermented sausage according to restriction fragment length polymorphism analysis Int. J. Food Microbiol. 186: 61–67.

Wilmes, P., S.L. Simmons, V.J. Denef and J.F. Banfield. 2009. The dynamic genetic repertoire of microbial communities. FEMS Microbiol. Rev. 33: 109–132.

Xiao, X., Y. Dong, Y. Zhu and H. Cui. 2013. Bacterial diversity of analysis of Zhenjiang yoa meat during refrigerated and vacuum-packed storage by 454 pyrosequencing. Curr. Microbiol. 66: 398–405.

Yost, C.K. and F.M. Nattress. 2000. The use of multiplex PCR reactions to characterize populations of lactic acid bacteria associated with meat spoilage. Lett. Appl. Microbiol. 31: 129–133.

Chapter 10

Foodborne Pathogens of Fermented Meat Products

Spiros Paramithiotis* and Eleftherios H. Drosinos

1 INTRODUCTION

Fermented meat products constitute a perfect example of safety assurance through the hurdle concept. Antagonism for the available nutrients that takes place at low temperatures, fast acidification, low water activity (a_w) values as well as the inhibitory action of sodium chloride, curing agents, spices and smoke, create a hostile environment for most pathogenic bacteria. However, inadequacy of the process that results from improper design, poor quality of the raw materials as well as the ability of the pathogens to adapt and survive may lead to the manufacture of products that may contain significant populations of pathogens that concomitantly pose a public health risk.

The most important foodborne pathogens regarding fermented meat products are *Listeria monocytogenes*, *Salmonella* spp., *Escherichia coli* O157:H7 and *Staphylococcus aureus*. The ubiquitous nature of the microorganisms classified to the *Listeria* species has been already established in the literature; their ecological niches are very diverse and extend in nearly all agricultural and food associated environments where

---

*For Correspondence: Laboratory of Food Quality Control and Hygiene, Department of Food Science and Human Nutrition, Agricultural University of Athens, Iera Odos 75, GR-11855 Athens, Greece, Tel: +302105294705 Fax: +302105294683, Email: sdp@aua.gr

Foodborne Pathogens of Fermented Meat Products

they survive as saprophytes. *Listeria* species consists of ten members, namely *L. innocua, L. monocytogenes, L. welshimeri, L. seeligeri, L. ivanovii, L. grayi, L. rocourtiae, L. marthii, L. weihenstephanensis* and *L. fleischmannii.* Listeriosis is a very important disease due to high morbidity and mortality, especially in vulnerable populations. Due to the widespread nature of *L. monocytogenes* contamination of either raw materials or the final product itself in populations higher that the limit of 100 cfu (colony forming units)/g is likely to occur (De Cesare et al. 2007, Thevenot et al. 2005). The growth potential of the pathogen during fermentation is very low but depending upon several factors such as fermentation conditions, post-fermentation treatment and the physiological status of the microorganism, it may survive even during storage and distribution. Prevalence of *L. monocytogenes* in fermented meat products has been reported to vary between 0 and 40% (FAO/WHO 2004, De Cesare et al. 2007, Martin et al. 2011). Despite this notable prevalence, no confirmed outbreaks have been assigned to that pathogen, at least so far.

Salmonella is another microorganism that is characterized by thorough distribution in natural environments as well as persistence in a variety of food production lines. Generally, human foodborne salmonellosis is an infection of serious public health concern; it is increasing worldwide and appears to be the primary cause of confirmed foodborne outbreaks. Approximately 1.0 million salmonellosis cases are reported annually in the U.S; it is the second most frequent causative agent of foodborne illness with Norovirus being the most frequent. The non-typhoidal serotypes Typhimurium, Enteritidis, Newport, Heidelberg and Javiana were the 5 most frequently reported serotypes isolated from human sources from 1999 to 2009; the typhoidal serotypes Paratyphi B var. L(+) tartrate+ and Typhi were in the 20 most frequently reported serotypes isolated from human sources during the same decade (CDC 2012). *Salmonella* in food is mainly associated with meat and meat products; however presence on fruits and vegetables is increasingly reported as well as the respective outbreaks. In the case of fermented meat products, an average of 5–10% prevalence is usually reported (Siriken et al. 2006) with extreme values, such as the 88% of the chorizos made in Guadalajara, being occasionally reported (Escartin et al. 1999). As far as the minimum infective dose was concerned, it has been reported to exceed 10^5 cfu/g (Kothary and Babu 2001).

Production of verotoxins by some *E. coli* is a very important characteristic that distinguishes them and designate them as Shiga-toxin producing or Vero-toxin producing (STEC or VTEC). A subset of VTEC is capable of causing haemolytic uremic syndrome (HUS) and therefore is termed enterohaemorrhagic (EHEC). The most common EHEC that causes disease, irrespective the food vehicle, is *E. coli* O157:H7. Prevalence in fermented meat products is usually very low (Siriken et al. 2006).

Table 1 Outbreaks associated with consumption of fermented meat products

Year	Location	Causative agent	Cases/Deaths	Reference
1987-88	England	S. Typhimurium DT 124	101/0	Cowden et al. 1989
1992	Australia	E. coli O111:H–	21	Paton et al. 1996
1994	Washington & California	E. coli O157:H7	17/0	Tilden et al. 1996
1995	Northern Italy	S. Typhimurium PT 193	83/0	Pontello et al. 1998
1998	Southern Ontario, Canada	E. coli O157:H7	39/0	Williams et al. 2000
1999	British Columbia, Canada	E. coli O157:H7	143/0	McDonald et al. 2004
2001	Alaska	Cl. botulinum	14	Anonymous 2001
2002	Sweden	E. coli O157:H7	38/0	Sartz et al. 2008
2004	Italy	S. Typhimurium DT104A	63/0	Luzzi et al. 2007
2005	Sweden	S. Typhimurium NST	15/0	Hjertqvist et al. 2006
2006	Norway	S. Kedougou	54/1	Emberland et al. 2006
2007	Italy	E. coli O157	2/0	Conedera et al. 2007
2009-2010	USA-multistate	S. Montevideo	272/0	Gieraltowski et al. 2013
2010	Denmark	S. Typhimurium	20/0	Kuhn et al. 2011

NST: not specifically typeable

Staphylococcus aureus is of significant importance in fermented meat products. In the case of incompetent fermentation and/or storage, the microorganism may grow and produce heat stable enterotoxins that may cause poisoning outbreaks (Metaxopoulos et al. 1981a, b, Nychas and Arkoudelos 1990). Prevalence in fermented meat products is usually reported at ca. 10% with extreme values occasionally reported (Skandamis and Nychas 2007).

Regarding all these pathogenic microorganisms, their growth potential in fermented meat products is rather low, provided that the formulation is well designed and fermentation well balanced. The parameters that allow efficient growth control are the use of raw materials of good quality, fermentation below 22°C with a final pH value of less than 5.3 and ageing at temperature less than 15°C with a target a_w value of less than 0.93 (Lücke 2000).

In Table 1 the outbreaks associated with the consumption of fermented meat products are presented. In nearly all of them, *E. coli* and *Salmonella* spp. were implicated as the causative agent; only one outbreak in a southwest Alaska village was caused by the ingestion of type E toxin by *Clostridium botulinum*. Regarding the latter, the product implicated was a homemade fermented beaver product that is a typical traditional meal of that territory. As far as the rest of the cases were concerned, when further investigation was possible, the meat used was held responsible for the contamination of the product in the majority of the cases and in one case the contamination occurred through red and black pepper (Gieraltowski et al. 2013).

2   DETECTION OF FOODBORNE PATHOGENS

Detection of foodborne pathogens may take place either by classical microbiological techniques or by the more recent culture-independent approach. The former approach is currently officially accepted through the application of the respective ISO methods that are regularly amended according to the latest respective literature. Detection of foodborne pathogens through the application of culture-independent methods has been the subject of intensive study. Several techniques based on specific nucleic acid sequences as well as specific proteins that may be used as markers of the pathogen's presence have been described. However application of such techniques is, in most of the cases, still far from being officially accepted for the reasons that will be explained below.

DNA-based detection is based on the amplification through PCR of sequences specific to the microbial target at the desired level of detection (i.e., genus/species/serotype/strain). Theoretically, this approach may offer improved selectivity, sensitivity, specificity, and reliability compared to the classical microbiological techniques. However, there are several issues, such as the ability to extract DNA of sufficient quality and quantity, the efficiency of primers to identify the microbial target at the desired level, that need to be resolved before this improved detection takes place. Because of this, the results obtained by the application of this approach, may only serve as an indication of pathogen existence that should be verified by the respective standardized protocols.

The low population of the pathogens in the presence of a dominant microbial population as well as the high protein and fat content of sausages severely restrict the application and greatly influence the result of this approach. In the following paragraphs, the most effective interventions employed to provide with a solution to the above mentioned challenges are discussed.

It has been generally accepted that the limit of reproducible detection of microorganisms through PCR-based techniques is 10^3 cfu/g. Given the fact that pathogenic microorganisms are in the majority of the cases present in much lower populations and are accompanied by the dominant lactic acid consortium, there is a need for either selective enrichment or concentration of the target cells. The former is the intervention that is employed in most of the cases; it may be achieved by incubation in selective growth media, mostly liquid, that allow proliferation of the target microorganism and at the same time suppression of the growth of the antagonistic microbiota. This is a field of intensive study. The optimal enrichment conditions for all foodborne pathogens have been described and improvements are constantly presented (Gnanou Besse et al. 2005, Arrausi-Subiza et al. 2014, Abulreesh et al. 2014, Yoshitomi et al. 2015, Stromberg et al. 2015, Dailey et al. 2014, 2015). Moreover, several enrichment strategies have been described that allow simultaneous enrichment of more than one pathogen (Kawasaki et al. 2005, Li et al. 2005, Murphy et al. 2007, Kim and Bhunia 2008).

Alternatively, techniques such as immunomagnetic separation and buoyant density gradient centrifugation may be used for the concentration of the target pathogen. The former is based on the use of antibodies immobilized on the surface of magnetic beads and their specific interaction with antigens present on the surface of the target cells. Then, the bead-target complex is separated from the food matrix resulting in both concentration and separation from the food sample. This approach has been successfully applied in the detection of S. Typhimurium in raw meats (Moreira et al. 2008), *Campylobacter jejuni* in spiked chicken wash samples (Waller and Ogata 2000), *E. coli* O157:H7, *Salmonella*, and *Shigella* present in ground beef (Wang et al. 2007), salmonellae in poultry (Fluit et al. 1993), and ground beef (Mercanoglu and Griffiths 2005), as well as *L. monocytogenes* in turkey meat (Bilir Ormanci et al. 2008).

Buoyant density gradient centrifugation is based on the separation of the sample constituents when a solution of density gradient is applied. This approach has been successfully applied for the detection of *Sh. flexneri* (Lindqvist et al. 1997), *E. coli* O157 (Lindqvist 1997), *Yersinia enterocolitica* (Lambertz et al. 2000, Lantz et al. 1998, Wolffs et al. 2004), and *C. jejuni* (Wolffs et al. 2005) from various meat products. Regarding effective DNA extraction, all factors already discussed apply in this case as well.

Several genomic targets have been employed for the detection of pathogens through PCR. The multiplex format is preferred due to the simultaneous detection of more than one target. In Table 2, an example of target genes, primers used along with the amplicon sizes that have been effectively used in pathogen detection through multiple PCR, DNA-based macroarray and multiple qPCR in meat and meat products are presented. The former is merely a detection step that follows selective enrichment or concentration of the target and therefore quantification of the population of the pathogen is not possible. The latter is possible by application of qPCR schemes.

Fluorescence in Situ Hybridization (FISH) is another promising approach regarding the culture-independent detection of pathogens. It is based on the cytometric observation of cells in which fluorescently labeled probes have been hybridized to the target nucleic acid sequence, depending on the desired taxonomic level. The main field of application of this approach is medicine (Jehan et al. 2012); the effect of the food matrix as well as the low population of the target microorganisms is currently limiting widespread application. Apart from them, the limitations of this approach quell from the specificity offered by the molecule on which hybridization is designed to occur. In order to alleviate these limitations, several interventions have been proposed (Rohde et al. 2015). Despite the improvements, this technique is still characterized by a large number of false negative and false positive results.

This technique has been applied for *in situ* detection of pathogens in various meat types such as *E. coli* in raw ground beef (Regnault et al. 2000) and campylobacters in chicken products (Moreno et al. 2001, Schmid et al. 2005) but, to the best of our knowledge, not yet successfully in fermented meat products.

3 FATE OF FOODBORNE PATHOGENS DURING FERMENTATION

The fate of several foodborne pathogens during fermentation, ripening and storage of fermented meat products has been extensively studied.

Table 2 Primer sequences, target genes and amplicon sizes used in multiplex format for pathogen detection through PCR, DNA-based macroarray and qPCR in meat and meat products

Target pathogen	Target gene	Name and Sequence (5′-3′)	Amplicon size	Reference
Multiplex PCR				
Salmonella spp.	*invA*	139F: GTGAAATTATCGCCACGTTCGGGCAA 141R: TCATCGCACCGTCAAAGGAACC	284	Jofre et al. 2005
L. monocytogenes	*prfA*	Lip1F: GATACAGAAACATCGGTTGGC Lip2R: GTGTAACTTGATGCCATCAGG Lip3R: TGACCGCAAATAGAGCCAAG	274 (Lip1–Lip2) 215 (Lip1–Lip3)	
S. Enteritidis	*fimI*	SF: CCTTTCTCCATCGTCCTGAA SR: TGGTGTTATCTGCCTGACC	85	Wang et al. 2004
L. monocytogenes	*hly*	LF: TCCGCAAAAGATGAAGTTC LR: ACTCCTGGTGTTTCTCGATT	98	
S. enterica	*invA*	SAL-F: AATTATCGCCACGTTCGGGCAA SAL-R: TCGCACCGTCAAAGGAACC	278	Germini et al. 2009
L. monocytogenes	*prfA*	LIS-F: TCATCGACGGCAACCTCGG LIS-R: TGAGCAACGTATCCTCCAGAGT	217	
E. coli O157:H7	*eaeA*	ESC-F: GGCGGATAAGACTTCGGCTA ESC-R: CGTTTTGGCACTATTTGCCC	151	
Salmonella spp.	NA	TS-11: GTCACGGAAGAAGAGAAATCCGTACG TS-5: GGGAGTCCAGGTTGACGGAAAATTT	375	Kawasaki et al. 2005
L. monocytogenes	NA	LM1: CGGAGGTTCCGCAAAAGATG LM2: CCTCCAGAGTGATCGATGTT	234	
E. coli O157:H7	NA	VS8: GGCGGATTAGACTTCGGCTA VS9: CGTTTTGGCACTATTTGCCC	120	

Organism	Gene	Primer sequence	Size (bp)	Reference
C. jejuni		C1: CAAATAAAGTTAGAGGTAGAATGT C4: GGATAAGCACTAGCTAGCTGAT	159	Wang and Slavik 2005
E. coli O157:H7		UidAa: GCGAAAACTGTGGAATTGGG UidAb: TGATGCTCCATAACTTCCTG	252	
S. Typhimurium		S29: CAGTATCAGGGCAAAAACGGC S30: TTCAAAGTTCTGCGCTTTGTT	360	
L. monocytogenes		FP: AGCTCTTAGCTCCATGAGTT RP: ACATTGTAGCTAAGGCGACT	450	
E. coli O157:H7	uidA	PT-2: GCGAAAACTGTGGAATTGGG PT-3: TGATGCTCCATCACTTCCTG	252	Li et al. 2005
S. Typhimurium	NA	ST-11: AGCCAACCATTGCTAAATTGGCGCA ST-15: GGTAGAAATTCCCAGCGGGTACTG	429	
Sh. flexneri	ipaH	ipaH-1: GTTCCTTGACCGCCTTTCCGATACCGTC ipaH-2: GCCGGTCAGCCACCCTCTGAGAGTAC	620	
St. aureus	16S rDNA	16S-F: GTGCACATCTTGACGGTACC 16S-B: CGAAGGGGAAGGCTCTATC	565	Chen et al. 2012
L. monocytogenes	hly	hlyA-F: CAAGTCCTAAGACGCCAATC hlyA-B: ATAAAGTGTAGTGCCCCAGA	1412	
E. coli O157:H7	eaeA	eaeA-F: AGGTCGTCGTGTCTGCTA eaeA-B: CCGTGGTTGCTTGCGTTTG	255	
S. Enteritidis	invA	invA-F: TCCCTTTGCGAATAACATCC invA-B: ATTACTTGTGCCGAAGAGCC	786	
Sh. flexneri	ipaH	ipaH-F: GGGAGAACCAGTCCGTAAA ipaH-B: CGCATCTCTGAAACATCTTGA	1088	

Table 2 (Contd.)

Table 2 Primer sequences, target genes and amplicon sizes used in multiplex format for pathogen detection through PCR, DNA-based macroarray and qPCR in meat and meat products (*Contd.*)

		DNA-based macroarray		
L. monocytogenes	htpG-like	Lm1: GAAGAAATTAATAGTTGGAC Lm2: GAAATAGTCTTGGAAATGCG Probe: TTTTTTTTTTTTTGAAATAGTCTTGGAAATGCG	120	Chiang et al. 2012
Salmonella spp.	random	Sspp1S: TTTGCGTTGCGTCTGTCC Sspp1A: GCTTATCGTCTGCGGCTC Probe: TTTTTTTTTTTTTTTAGGCGATAGTACGGCCATTT	237	
St. aureus	hsp	SAU3: ACAAATAATAAAGGTGGC SAU4: TATCGCCAGTTTGTACTT Probe: AACAATTGTAGGGGAGACAGATTTTTTTTTTTTTTTT	406	
Eb. sakazakii	hsp	Esa1: TTGAAGACCACGATCGCGA Esa2: ATTCGTACGTTCAGTTTGTTCG Probe: TTTTTTTTTTTTTTTAGAAGCTGGACGATCAGGAAGT	490	
Str. agalactiae	hsp	SAG1: TAGATGGCGAATTCACTGAGA SAG2: ATTGAGCAATCCCTATCACG Probe: TTTTTTTTTTTTTTTATTGAGCAATCCCTATCACG	112	
E. coli O157:H7	random	HAF1: AATAAACAGGTACGGCT HAR1: TATCACCTCATCAACCAAAATC Probe: TTTTTTTTTTTTTTTAATTGAATCCATTCGTCTACTT	316	
V. parahaemolyticus	pR72H	VP33: TGCGAATTCGATAGGGTGTTAACC VP32: CGAATCCTGAACATACGCAGC Probe: TTTTTTTTTTTTTTTTAAGCCGTTCTAAGGTGC	387	
P. fluorescens	aprX	PFF: ACCGAGAACACCAGCTTGTC PFR: CTCACGGTCAATGGCAAAC Probe: CTCACGGTCAATGGCAAACTTTTTTTTTTTTTTTTT	778	

Multiplex qPCR				
E. coli	*uidA*	*E. coli*-1: TTGACCCACACTTTGCCGTAA *E. coli*-2: GCGAAAACTGTGGAATTGGG *E. coli-p*: VIC-TGACCGCATCGAAACGCAGCTTAMRA		Wang et al. 2007
Salmonella spp.	*NA*	Sal-1: GCTATTTTCGTCCGGCATGA Sal-2: GCGACTATCAGGTTACCGTGGA *Salmonella-p*: FAMTAGCCAGCGAGGTGAAAACGACAAAGGTAMRA		
Shigella spp.	*ipaH*	Shig-1: CTTGACCGCCTTTCCGATA Shig-2: AGCGAAAGACTGCTGTCGAAG *Shigella-p*: TET-AACAGGTCGCTGCATGGCTGGAATAMRA		
S. Typhimurium	*fliC*	SfC-F: TGCAGAAAATTGATGCTGCT SfC–R: TTGCCCAGGTTGGTAATAGC ST-JOE: JOE-ACCTGGGTGCGGTACAGAACCGTBHQ1a		Lee et al. 2009
S. Enteritidis	*sefA*	SsA-F: GGTAAAGGGGCTTCGGTATC SsA-R: TATTGGCTCCCTGAATACGC SE-Cy5: Cy5-TGGTGGTGTAGCCACTGTCCCGTBHQ1a		
L. monocytogenes	*hly*	Lm515: GCACCTTTGTAGTATTGTAAATTC Lm625: TAACCAATGGGATCCACAAG LmLNA543: Rox-TGGTCCCGTTCTCACTG-BHQ2		Nitecki et al. 2015
E. coli O157:H7	*stx*	Ec967: TAAGCATGAAGAAGATGTTTATGG Ec1076: ATCCTCATTATACTTGGAAAACTC EcLNA1042: Hex-TTAGCACAATCCGCCG-BHQ1		
S. Enteritidis	*safA*	F: GGTTGCTAACACGACACTG R: TGGGGCATTGGTATCAAAG Probe: FAM-CTCCTCCCATTCCACATTTGCG-BHQ1		Maurischat et al. 2015

Table 2 (*Contd.*)

Table 2 Primer sequences, target genes and amplicon sizes used in multiplex format for pathogen detection through PCR, DNA-based macroarray and qPCR in meat and meat products (*Contd.*)

S. Typhimurium	*fliA-IS200*	F: CATTACACCTTCAGCGGTAT		
		R: CTGGTAAGAGAGCCTTATAGG		
		Probe: YY-CGGCATGATTATCCGTTTCTACAGAGG-BHQ1		
Biphasic STM	*fljB-hin*	F: TGGTGCTGTTAGCAGAC		
		R: TCAACACTAACAGTCTGTCG		
		Probe: TR-AACCGCCAGTTCACGCAC-BHQ2		
Biphasic STM	*hin-iroB*	F: GTGTGGCATAAATAAACCGA		
		R: AGGCTTACCTGTGTCATCCA		
		Probe: Cy5-TAACGCGCTCACGATAAGGC-BHQ3		
IAC	*pUC18/1*	F: GTCGGGAAACCTGTCG		
		R: GCTCACATGTTCTTTCCTGC		
		Probe: LC705-CGGGGAGAGGCGGTT-BHQ3		

Eb.=Enterobacter (Chronobacter), Sh.=Shigella

From a microbiological perspective, the raw materials and the processing environment constitute the sources of contamination. In spontaneous fermentations, the outcome of the fermentation, i.e. the quality of the final products depends solely on this naturally occurring microbiota.

The level, extent and type of microbiological contamination are determined by the conditions the animals are reared and slaughtered. The sources of contamination include the abiotic environment in contact with the animal and the processing equipment; specific characteristics of the animal, the season of the year as well as the geographical origin may also affect the level of contamination. The initial microbiota of fresh carcasses is dominated by Gram-negative rods (mainly pseudomonads) and micrococci. Other Gram-negative bacteria such as *Acinetobacter* spp., *Alcaligenes* spp., *Moraxella* spp. and *Enterobacteriaceae*, Gram-positive species including spore-forming bacteria, lactic acid-producing bacteria and *Brochothrix thermosphacta*, as well as yeasts and molds, may also form a secondary microbiota (Paramithiotis et al. 2009). All these microorganisms may contaminate the processing environment of the facilities and as a result of insufficient cleaning and disinfection procedures may re-contaminate the products at various production stages. This was verified by studies addressing the microbiological quality of processing environment and relative equipment (Chevallier et al. 2006, Lebert et al. 2007, Gounadaki et al. 2008). Furthermore, these studies highlight the importance of successful fermentation, in terms of properly reduced a_w and pH value, in the stability and safety of the final product.

Casings and additives are sources of contamination that should be taken into consideration. Casings are usually preserved by salting, curing and/or drying; when adequately salted they are considered microbiologically acceptable. Houben (2005) reported on the presence of *Salmonella* spp., *L. monocytogenes* and sulphite reducing clostridia. A total of 214 samples were examined for the former two pathogens and 138 for the latter. In the first case absence of the pathogens was verified. However, in the case of clostridia 53 samples were found positive with average populations ranging from 5.6 ± 3.75 to 187.1 ± 558.19 cfu/g. The microbiological quality of 13 spices, 38 spice mixtures and 15 additives was examined by Paramithiotis and Drosinos (2010). The additives examined were of excellent microbiological quality since all microbial categories assessed were below detection limit and this was attributed to the nature of those products. More accurately, it was reported that the level of treatment that they have undergone along with good hygiene practices during distribution and storage has resulted in a high microbiological quality and safety, even if some of them could possibly support microbial growth. Regarding the spices examined, no foodborne pathogens were detected; on the contrary presence of *Salmonella* spp., *Bacillus cereus*, *Cl. perfringens*, *St. aureus* and *E. coli* has been reported in a wide range of herbs and spices (Schwab et al. 1982, De Boer et al. 1985, McKee 1995, Garcia et al. 2001, Banerjee and Sarkar 2003, Sagoo et al. 2009). Garlic, leek and mustard seeds were reported to be of satisfactory microbiological quality according to Recommendation 2004/24/EC (Anonymous 2004). The remaining were characterized as unacceptable due to the presence of high *Enterobacteriaceae* populations that were assigned to result either from the absence of sufficient decontamination or deviations from good hygiene practices applied during production, distribution and storage in the meat industry. Moreover analyses for aflatoxin B1 content revealed the absence or, at least, presence of this contaminant below the detection limit of 1.0 ppb in all samples.

Addition of sodium chloride and nitrite as well as the effect of fermentation, namely microbial antagonism, decrease of the redox potential, low pH and a_w values, negatively affect development of spoilage and pathogenic microbiota. A large number of different sausage types currently exist differing in formulation, size and production procedure. All these differences should be taken into consideration when attempting to validate safety of specific products. The effect of the manufacturing conditions and the ingredients used in sausage formulation on survival of foodborne pathogens, especially regarding *L. monocytogenes*, *Salmonella* serovars and *E. coli*, has been thoroughly studied. In Tables 3, 4 and 5 representative studies assessing the survival of *L. monocytogenes*, *Salmonella* serovars and *E. coli* during sausage fermentation are presented, respectively. Generally, fermentation is a very efficient step in reducing the numbers of these pathogens and preventing their growth but not always regarding their survival.

In the case of *St. aureus* and synthesis of toxin A, use of proper starter culture and formulation that supply with adequate amount of carbon sources for rapid lactic acid bacteria (LAB) development and decrease of the pH value are the decisive factors for growth inhibition (Gonzalez-Fandos et al. 1999). Additionally, fermentation at lower temperatures (i.e. 20°C) may ensure suppression of the pathogen's growth. During ripening, temperature and pH value are the most important parameters that affect survival of the pathogen. The lower the ripening temperature (i.e. below 10°C) and the pH value, the lower the surviving population (Pereira et al. 1982, Incze 1987). As far as the individual ingredients were concerned, the nitrite levels used in sausage production may only marginally inhibit *St. aureus* growth (Collins-Thompson et al. 1984). However, in the context of the hurdle concept, i.e. in combination with pH, a_w and temperature, the antimicrobial effect is improved. Regarding synthesis of toxin A, the lowest level of *St. aureus* required lies between 10^6 and 10^7 cfu/g (Gonzalez-Fandos et al. 1999). Therefore, production procedure should be adjusted so that the pathogen does not reach that level. Metaxopoulos et al. (1981a, b) in a classical study proved the role

Table 3 Representative studies assessing the survival of *L. monocytogenes* during sausage fermentation

Product	Comment	Reference
Chourico de Vinho	Reduction of the initial population (ca. 8 log cfu/g) for ca. 2 log cfu/g after 15 d of drying (15°C at 85% relative humidity environment). Undetectable after 15 d of drying when starter culture (*Lb. sakei*) and fermentable carbohydrates were added. Undetectable after 30 d of drying.	Garcia Diez and Patarata 2013
Salami	The addition of starter cultures prevented growth, but not always the survival of the pathogen.	Campanini et al. 1993
Swedish salami	Inactivation rates increased with temperature; at 8°C storage temperature > 16 d were required for 1-log cfu/g inactivation. At 20°C 4-7 d were required.	Lindqvist and Lindblad 2009
Cacciatore	Inactivation was ca. 0.38±0.23 log cfu/g of the initial 5-6 log cfu/g inoculum	Mataragas et al. 2015
Felino	Inactivation was ca. 0.39±0.2 log cfu/g of the initial 5-6 log cfu/g inoculum	Mataragas et al. 2015
Salami	In heavily inoculated (ca. 5.5 log cfu/g) products, population decrease was dependent upon the starter culture applied. In medium inoculum level (ca. 3.1 log cfu/g) no significant differences were observed. Regarding the latter the pathogen was eliminated after 49 d.	Lahti et al. 2001
Soudjouk	Fermentation and drying reduced initial numbers (ca. 9.0 log cfu/g) by 0.07 and 0.74 log cfu/g for sausages fermented to pH 5.3 and 4.8, respectively. An additional reduction by 0.08-1.80 log cfu/g was observed during storage at 4, 10 or 21°C. During storage of surface inoculated commercially manufactured products the D-values generated were 10.1 d (4°C); 6.4 d (10°C); 1.4 d (21°C) and 0.9 d (30°C).	Porto-Fett et al. 2008
Norwegian salami	A 5 log cfu/g increase of the initial population (3-4 log cfu/g) during fermentation was observed and after 5.5 months storage at 4°C or 20°C it was not detectable. When inoculum level was 5-7 log cfu/g a population decrease during fermentation was reported. After 5.5 months storage at 4°C only 90% reduction of the initial numbers was reported but it was below detection limit at 20°C.	Nissen and Holck 1998

Table 3 (*Contd.*)

Table 3 Representative studies assessing the survival of *L. monocytogenes* during sausage fermentation (*Contd.*)

Low fat salami	Population decrease by 0.5 log cfu/g of the initial population (5 log cfu/g). During the shelf life under moderate thermal abuse (8-12°C) the portioned and vacuum packed product was not able to support growth of the pathogen.	Dalzini et al. 2015
Sremska	The production procedure resulted in a 0.8 log cfu/g reduction of the initial 6 log cfu/g population. An additional 0.9 log cfu/g reduction was noted after 45 d of storage at 4°C. Post-processing pasteurization failed to eliminate the pathogen.	Ducic et al. 2016
Sudzuk	The production procedure resulted in a 0.5 log cfu/g reduction of the initial 6 log cfu/g population. No additional was noted after 45 d of storage at 4°C. Post-processing pasteurization failed to eliminate the pathogen.	Ducic et al. 2016
Soudjouk-style	A reduction of 0.1 to 0.5 log cfu/g of the initial inoculum (6.5 log cfu/g) was reported during fermentation to final pH of 5.2 to 4.6. Additional reduction from 0 to 0.4 log cfu/g was observed after drying to a_w of 0.92 to 0.86. The reduction rate during storage was higher in sausages with lower pH and a_w values that stored at higher temperatures.	Hwang et al. 2009

Table 4 Representative studies assessing the survival of *Salmonella* serovars during sausage fermentation

Product	Comment	Reference
S. Typhimurium		
Lebanon bologna	Fermentation (12 h at 26.7°C; then at 37.8°C until pH 5.2 or 4.7) reduced initial population (ca. 8 log cfu/g) by <2 log cfu/g. Combination of fermentation and heating (43.3°C for 20 h; 46.1°C for 10 h; 48.9°C for 3h) by >7 log cfu/g.	Ellajosyula et al. 1998
Soudjouk	Fermentation and drying reduced initial numbers (ca. 9.0 log cfu/g) by 1.52 and 3.15 log cfu/g for sausages fermented to pH 5.3 and 4.8, respectively. An additional reduction by 0.88-3.74 log cfu/g was observed during storage at 4, 10 or 21°C. During storage of surface inoculated commercially manufactured products the D-values generated were 7.6 d (4°C); 4.3 d (10°C); 0.9 d (21°C) and 1.4 d (30°C).	Porto-Fett et al. 2008

Low fat salami Sremska	Population decrease by 1.65 log cfu/g of the initial population (5 log cfu/g). The production procedure resulted in a 1.4 log cfu/g reduction of the initial 6 log cfu/g population. An additional 1.9 log cfu/g reduction was noted after 45 d of storage at 4°C. Post-processing pasteurization eliminated the pathogen and maintained acceptable sensorial properties.	Dalzini et al. 2015 Ducic et al. 2016
Soudjouk-style	A reduction of 0 to 2.2 log cfu/g of the initial inoculum (6.5 log cfu/g) was reported during fermentation to final pH of 5.2 to 4.6. Additional reduction from 0.3 to 2.4 log cfu/g was observed after drying to a_w of 0.92 to 0.86. The reduction rate during storage was higher in sausages with lower pH and a_w values that stored at higher temperatures.	Hwang et al. 2009
Salmonella spp.		
Chourico de Vinho	Reduction of the initial population (ca. 8 log cfu/g) for ca. 2 log cfu/g after 15 d of drying (15°C at 85% relative humidity environment). Undetectable after 30 d of drying.	Garcia Diez and Patarata 2013
S. enterica		
Cacciatore	Inactivation was ca. 1.10±0.24 log cfu/g of the initial 5-6 log cfu/g inoculum	Mataragas et al. 2015
Felino	Inactivation was ca. 1.62±0.38 log cfu/g of the initial 5-6 log cfu/g inoculum	Mataragas et al. 2015
S. Kentucky		
Norwegian salami	A decrease of the initial population (3-4 log cfu/g) during fermentation below detection limit (150 cfu/sample) was observed. When inoculum level was 5-7 log cfu/g a population decrease during fermentation was also reported. After 5.5 months storage at 4°C and 20°C it was below detection limit.	Nissen and Holck 1998

Table 5 Representative studies assessing the survival of *E. coli* during sausage fermentation

Product	Comment	Reference
E. coli O157:H7		
Lebanon bologna	Fermentation (12 h at 26.7°C; then at 37.8°C until pH 5.2 or 4.7) reduced initial population (ca. 8 log cfu/g) by <2 log cfu/g. Heating (43.3°C for 20 h; 46.1°C for 10 h; 48.9°C for 3 h) by <3 log cfu/g. Combination of fermentation and heating by >7 log cfu/g.	Ellajosyula et al. 1998
Salami	Numbers were reduced from the original ca. 7 log cfu/g by 2.1, 1.6 or 1.1 log g^{-1} when batter was tempered frozen and thawed, frozen and thawed or refrigerated, respectively. Regardless of storage temperature or atmosphere, after 7 days salami prepared by tempered frozen and thawed, frozen and thawed or refrigerated batter exhibited population reduction by 2.7-4.9, 2.3-4.8 or 1.6-3.1 log cfu/g.	Faith et al. 1998
Salami	Population decrease was dependent upon the starter culture applied. However, the pathogen was detectable after fermentation.	Lahti et al. 2001
Salami	A reduction by 1.0 and 0.7 log cfu/g of the initial population (7.4 log cfu/g) at the end of fermentation and drying, respectively, was reported. Unencapsulated *Lb. reuteri* with or without *B. longum* reduced *E. coli* O157:H7 by 3.0 log cfu/g and *B. longum* caused a 1.9 log cfu/g reduction. Microencapsulation decreased the inhibitory action of the LAB.	Muthukumarasamy and Holley 2007
Soudjouk	Fermentation and drying reduced initial numbers (ca. 9.0 log cfu/g) by 0.03 and 1.11 log cfu/g for sausages fermented to pH 5.3 and 4.8, respectively. An additional reduction by 0.68-3.17 log cfu/g was observed during storage at 4, 10 or 21°C. During storage of surface inoculated commercially manufactured products the D-values generated were 5.9 d (4°C); 2.9 d (10°C); 1.6 d (21°C) and 0.25 d (30°C).	Porto-Fett et al. 2008
Norwegian salami	A 3 log cfu/g increase of the initial population (3-4 log cfu/g) during fermentation was observed and after 5.5 months storage at 4°C or 20°C it was not detectable. When inoculum level was 5-7 log cfu/g a population decrease during fermentation was reported. After 5.5 months storage at 4°C it survived at low populations (500 cfu/sample) but it was below detection limit at 20°C.	Nissen and Holck 1998

Soudjouk-style	A reduction of 0 to 0.9 log cfu/g of the initial inoculum (6.5 log cfu/g) was reported during fermentation to final pH of 5.2 to 4.6. Additional reduction from 0 to 3.5 log cfu/g was observed after drying to a_w of 0.92 to 0.86. The reduction rate during storage was higher in sausages with lower pH and a_w values that stored at higher temperatures.	Hwang et al. 2009
E. coli VTEC		
Morr & Salami	High levels of NaCl, NaNO$_2$, glucose (low pH) and fermentation temperature gave enhanced VTEC reduction, while high fat and large casing diameter (a_w) gave the opposite effect. Interaction effects were small; that no single variable had a dominant effect on VTEC reductions.	Heir et al. 2010
Low fat salami	Population decrease by 2.5 log cfu/g of the initial population (5 log cfu/g).	Dalzini et al. 2015
Sudzuk	The production procedure resulted in a 1.3 log cfu/g reduction of the initial 6 log cfu/g population. An additional 1.3 log cfu/g reduction was noted after 45 d of storage at 4°C. Post-processing pasteurization eliminated the pathogen and maintained acceptable sensorial properties.	Ducic et al. 2016
E. coli generic		
Swedish salami	Inactivation rates increased with temperature; at 8°C storage temperature 21 d were required for 1-log cfu/g inactivation. At 20°C 7-11 d were required.	Lindqvist and Lindblad 2009

214 Fermented Meat Products: Health Aspects

of competition between LAB and *St. aureus* in combination of pH value of the mix; the lower the initial pH and the higher the initial LAB, the greater the inhibition of *St. aureus* in fermented meat products.

The inactivation of *Y. enterocolitica* in fermented sausages has also been assessed. Addition of more than 80 mg/kg sodium nitrite was required to eliminate 5 log of *Y. enterocolitica* O:3 after 28 days whereas the pathogen was detected throughout the test period of 35 days upon addition of 50 mg/kg or less (Asplund et al. 1993). Similarly, the utilization of starter cultures and the reduction of the pH value to 4.7 resulted in the elimination of the same population of the pathogen during sucuk fermentation. On the contrary, spontaneous fermentation resulted in a final pH value of 5.6 and the pathogen was still detectable after 4 days of fermentation and 12 days of drying (Ceylan and Fung 2000). Inactivation of an acid tolerance induced *Y. enterocolitica* strain during storage was also studied by Lindqvist and Lindblad (2009). It was reported that the storage time required for one log reduction of the initial 6 log inoculum was 18 days at 8°C and between 1 and 4 days at 22°C with Baranyi and Roberts model (Baranyi and Roberts 1994) being the best fitting one in nearly all temperatures studied.

4 BIOCONTROL OF FOODBORNE PATHOGENS

Among the several approaches that have been employed for effective biocontrol of pathogens, the use of bacteriocins and bacteriocinogenic strains has been the most effective. However, it should be stressed that use of bacteriocins should be considered only in the context of the hurdle concept and not as a sole preventing measure.

Bacteriocins are ribosomally synthesized peptides the encoding genes of which are frequently associated with mobilisable elements on the chromosome or on plasmids (Deegan et al. 2006, Cleveland et al. 2001). Their production is related to the microbial antagonistic interactions within a certain ecosystem; the producer strain may gain a competitive advantage since it is immune to its action. Bacteriocins produced by LAB have been intensively studied over the last decades. The bacteriocinogenic species isolated from various meat products include *Lactococcus lactis* (Noonpakdee et al. 2003), *Lactobacillus sakei* (Aymerich et al. 2000b), *Lb. curvatus* (Mataragas et al. 2003), *Lb. plantarum* (Messi et al. 2001), *Pediococcus acidilactici* (Albano et al. 2007a), *Pd. parvulus* (Schneider et al. 2006), *Pd. pentosaceus* (Albano et al. 2007b), *Carnobacterium divergens* (Holck et al. 1996), *Cb. piscicola* (Jack et al. 1996), *Enterococcus faecium* (Cintas et al. 1998), *Leuconostoc mesenteroides* (Drosinos et al. 2006a), *Ln. carnosum* (Osmanagaoglu 2007), *Ln. gelidum* (Harding and Shaw 1990), *St. warneri* (Prema et al. 2006) and *St. xylosus* (Villani et al. 1997).

Foodborne Pathogens of Fermented Meat Products 215

The factors affecting the *in vitro* bacteriocin production include the microbial strain and its physiological status, the composition and pH value of the growth medium as well as growth temperature. A huge amount of literature currently exists regarding optimization and modeling of bacteriocin production.

In situ bacteriocin application requires thorough study of the effects that sausage formulation may have on the production and/or activity of bacteriocins. The effect of various ingredients on growth of the sakacin K producer *Lb. sakei* CTC494 *in vitro* and in model fermented sausages as well as on the antilisterial activity of the bacteriocin was studied by Hugas et al. (2002). The influence of sausage ingredients on sakacin K production was reported to be variable; sodium chloride mixed with nitrate and nitrite decreased total bacteriocin production, whereas sodium chloride and pepper had no effect despite the stimulating action on the growth of the strain. Sodium chloride seemed to protect *L. monocytogenes* from the action of the bacteriocin, but this was surpassed both *in vitro* and *in situ* by black pepper and sodium nitrite that enhanced sakacin K activity. The same effect was also observed when manganese was used as a sausage ingredient instead of pepper, suggesting that one of the active factors in pepper was manganese.

The effect of sodium chloride, sodium nitrite and different spices on curvacin A, a listericidal bacteriocin produced by *Lb. curvatus* LTH 1174 was studied by Verluyten et al. (2003, 2004a, b). The strain was highly sensitive to nitrite; growth was inhibited even at concentration as low as 10 ppm. Moreover, changes in sodium chloride concentration affected both growth and bacteriocin activity. It clearly slowed down the growth of *Lb. curvatus* LTH 1174, but more importantly, it had a detrimental effect on specific curvacin A production and hence on overall bacteriocin activity. Even at salt concentration of 2% wt/vol a decrease in bacteriocin production was observed without effect on growth. Regarding the effect of spices, pepper, nutmeg, rosemary, mace, and garlic decreased the maximum specific growth rate; on the contrary paprika was reported to increase it. Garlic was the only spice enhancing specific bacteriocin production, resulting in a higher bacteriocin activity in the cell-free culture supernatant. Pepper addition enhanced lactic acid production but this was not assigned to the presence of manganese. Significant effects of sodium chloride, nitrate and pepper on Enterocin A and B production by *E. faecium* strain CTC492 during sausage fermentation was also reported by Aymerich et al. (2000a). Drosinos et al (2006b) found that the addition of bacteriocin-producing *Lb. sakei* strains reduced the time required for a 4-log reduction of *L. monocytogenes* (t_{4D}) in fermented sausages studied. *In situ* application of bacteriocins has been the field of intensive study. However, several limitations occurring due to a variety of reasons including relatively narrow spectrum of action, the possible inactivation in food systems by proteases or as a result of their interactions with food constituents should be addressed. Furthermore, emergence of resistant mutants might also lead to a loss of antimicrobial activity (Rekhif et al. 1994).

Currently, there are two approaches for such an application; it is either inclusion of the bacteriocinogenic strain in the starter culture or addition of the bacteriocin itself either as such or immobilized on a proper agent. The former is the preferred one mainly due to legal and economic implications.

The effectiveness of *Pd. acidilactici* strains PAC 1.0, H and PO2 against *L. monocytogenes* as well as *Pd. acidilactici* PA-2 against *E. coli* O157:H7 and *L. monocytogenes* during the manufacture of dry sausage has been studied by Foegeding et al. (1992), Luchansky et al. (1992) and Lahti et al. (2001). In the first case, the contribution of pediocin to the inactivation of *L. monocytogenes* strains was confirmed; in the second case, the incorporation of the bacteriocinogenic strain of *Pd. acidilactici* (PA-2) as a starter culture decreased the population of *L. monocytogenes* but only when it was present in high numbers (5.05–5.41 log cfu/g) but had no effect on *E. coli* O157:H7, regardless of the inoculum level.

Villani et al. (1997) reported on the effect of the bacteriocinogenic strain *St. xylosus* 1E on *L. monocytogenes*. The viable counts of the pathogen in Naples-type sausages after 21 days of maturation. Moreover, no *L. monocytogenes* were recovered after 75 days in sausages inoculated with *St. xylosus* 1E, while the pathogen was still present at this time in control sausages in which *L. monocytogenes* was challenged. Dicks et al. (2004) used the plantaricin 423 producer *Lb. plantarum* strain 423 and the curvacin DF126 producer *Lb. curvatus* strain DF126 as starter cultures for the production of Ostrich meat salami. Both bacteriocins succeded in inhibiting growth of *L. monocytogenes*. However, after 15 h of fermentation, the viable counts of *L. monocytogenes* LM1 increased, most probably due to a decrease in activity of the bacteriocins and/or the development of resistant bacterial cells. The anti-staphylococcal activity of bacteriocin AS-48 that was added either as a semi-purified form or by incorporation of the producer strain *E. faecalis* A-48-32 in the starter culture was studied by Ananou et al. (2005). The producer strain managed to develop well and the best result was achieved with a bacteriocinogenic strain inoculum of 10^7 cfu/g. Benkerroum et al. (2005) combined lyophilized bacteriocinogenic strains (*Lc. lactis* subsp. *lactis* LMG21206 and *Lb. curvatus* LBPE) with commercial starter culture in order to study their *in situ* effectiveness to control *L. monocytogenes* in dry-fermented sausages. The meat batter was contaminated with a mixture of four different *L. monocytogenes* strains. It was reported that the pathogen failed to grow in any of the contaminated batches. The effectiveness of enterocin CCM 4231 in controlling *L. monocytogenes* contamination in dry-fermented

Hornad salami was examined by Laukova et al. (1999). Immediately upon addition of the enterocin *L. monocytogenes* counts reduced from the initial 8 log cfu/g by 1.67 log cfu/g. However, during the second day of fermentation *L. monocytogenes* population increased by 3.38 log cfu/g. After 1 week of ripening the population of the pathogen in the presence of the enterocin was 3 log cfu/g lower than the control sample, i.e. without enterocin addition. However, bacteriocin activity could not be detected analytically. Similar population decrease was caused by the addition of semipurified bacteriocins to the batter of artificially contaminated by 4-5 log cfu/g *L. monocytogenes*. The counts of the pathogen were reduced by 2 and 1.5 log cfu/g after 10 and 20 days, respectively (Barbosa et al. 2015a). A new insight was offered by Barbosa et al. (2015b) by encapsulating a bacteriocinogenic *Lb. curvatus* strain in calcium alginate and using it as a bioprotective culture during fermentation of a batter artificially inoculated with 5 log cfu/g *L. monocytogenes*. Although entrapped cells survived better, no effect on bacteriocin production and concomitantly on *L. monocytogenes* inhibition was observed.

5 CONCLUSION

Safety of fermented meat products is obtained as the outcome of a very fine balance between minimization of pathogenic microorganisms' presence through the use of high quality raw materials, the antagonistic activities of the lactic acid microbiota and the inhibitory action of the additives. Survival and growth of foodborne pathogens in fermented meat products has been extensively studied and the actual control measures that may ensure their absence or at least restrict their growth potential have been determined. Then the question that arises is why outbreaks keep occurring. With the exception of the very early ones that actually triggered research and pointed at certain direction, the rest of the cases, at least the ones that further research has taken place, have occurred due to deviations from the above mentioned balance.

Given the advances in molecular biology and the research needs of this specific field, in the near future an increase of studies referring to the exploitation at the molecular basis of the survival and/or proliferation of the pathogens during manufacturing of fermented meat products. Such types of studies will improve the understanding of the physiology of the pathogens and their response to the variety of environmental stimuli that may occur during production procedure and may assist in the development of novel strategies and interventions to combat pathogens.

Key words: *fermented meat products, L. monocytogenes, Salmonella* spp., *St. aureus,*
E. coli O157:H7.

REFERENCES

Abulreesh, H.H., T.A. Paget and R. Goulder. 2014. A pre-enrichment step is essential for detection of *Campylobacter* sp. in turbid pond water. Trop. Biomed. 31: 320–326.

Albano, H., S.D. Todorov, C.A. van Reenen, T. Hogg, L.M.T. Dicks and P. Teixeira. 2007a. Characterization of two bacteriocins produced by *Pediococcus acidilactici* isolated from 'Alheira', a fermented sausage traditionally produced in Portugal. Int. J. Food Microbiol. 116: 239–247.

Albano, H., M. Oliveira, R. Aroso, N. Cubero, T. Hogg and P. Teixeira. 2007b. Antilisterial activity of lactic acid bacteria isolated from 'Alheiras' (traditional Portuguese fermented sausages): *In situ* assays. Meat Sci. 76: 796–800.

Ananou, S., M. Maqueda, M. Martinez-Bueno, A. Galvez and E. Valdivia. 2005. Control of *Staphylococcus aureus* in sausages by enterocin AS-48. Meat Sci. 71: 549–556.

Anonymous. 2001. Botulism outbreak associated with eating fermented food – Alaska, 2001. Morbidity and Mortality Weekly Report 50: 680–682.

Anonymous. 2004. European Commission (EC). Commission Recommendation of 19 December 2003 concerning a coordinated programme for the official control of food stuffs for 2004 (2004/24/EC). Official Journal of European Union, L6, 29–37.

Arrausi-Subiza, M., J.C. Ibabe, R. Atxaerandio, R.A. Juste and M. Barral. 2014. Evaluation of different enrichment methods for pathogenic *Yersinia* species detection by real time PCR. BMC Vet. Res. 10: 192.

Asplund, K., E. Nurmi, J. Hirn, T. Hirvi and P. Hill. 1993. Survival of *Yersinia enterocolitica* in fermented sausages manufactured with different levels of nitrite and different starter cultures. J. Food Protect. 56: 710–712.

Aymerich, T., M.G. Artigas, M. Garriga, J.M. Monfort and M. Hugas. 2000a. Effect of sausage ingredients and additives on the production of enterocins A and B by *Enterococcus faecium* CTC492. Optimization of *in vitro* production and anti-listerial effect in dry fermented sausages. J. Appl. Microbiol. 88: 686–694.

Aymerich, M.T., M. Garriga, J.M. Monford, I. Nes and M. Hugas. 2000b. Bacteriocin producing lactobacilli in Spanish-style fermented sausages: characterization of bacteriocins. Food Microbiol. 17: 33–45.

Banerjee, M. and P.K. Sarkar. 2003. Microbiological quality of some retail spices in India. Food Res. Int. 36: 469–474.

Baranyi, J. and T.A. Roberts. 1994. A dynamic approach to predicting bacterial growth in food. Int. J. Food Microbiol. 23: 277–294.

Barbosa, M.S., S.D. Todorov, I. Ivanova, J.M. Chobert, T. Haertle and B.D.G.M. Franco. 2015a. Improving safety of salami by application of bacteriocins produced by an autochthonous *Lactobacillus curvatus* isolate. Food Microbiol. 46: 254–262.

Barbosa, M.S., S.D. Todorov, C.H. Jurkiewicz and B.D.G.M. Franco. 2015b. Bacteriocin production by *Lactobacillus curvatus* MBSa2 entrapped in calcium alginate during ripening of salami for control of *Listeria monocytogenes*. Food Control 47: 147–153.

Benkerroum, N., A. Daoudi, T. Hamraoui, H. Ghalfi, C. Thiry, M. Duroy, P. Evrart, D. Roblain and P. Thonart. 2005. Lyophilized preparations of bacteriocinogenic *Lactobacillus curvatus* and *Lactococcus lactis* subsp. *lactis* as potential protective adjuncts to control *Listeria monocytogenes* in dry-fermented sausages. J. Appl. Microbiol. 98: 56–63.

Bilir Ormanci, F.S., I. Erol, N.D. Ayaz, O. Iseri and D. Sariguzel. 2008. Immunomagnetic separation and PCR detection of *Listeria monocytogenes* in turkey meat and antibiotic resistance of the isolates. Brit. Poultry Sci. 49: 560–565.

Campanini, M., I. Pedrazzoni, S. Barbuti and P. Baldini. 1993. Behaviour of *Listeria monocytogenes* during the maturation of naturally and artificially contaminated salami: effect of lactic acid bacteria starter cultures. Int. J. Food Microbiol. 20: 169–175.

CDC. 2012. Pathogens causing US foodborne illnesses, hospitalizations, and deaths, 2000–2008. Available in http://www.cdc.gov/salmonella/ Accessed September 2014.

Ceylan, E. and D.Y.C. Fung. 2000. Destruction of *Yersinia enterocolitica* by *Lactobacillus sake* and *Pediococcus acidilactici* during low-temperature fermentation of Turkish dry sausage (sucuk). J. Food Sci. 65: 876–879.

Chen, J., J. Tang, J. Liu, Z. Cai and X. Bai. 2012. Development and evaluation of a multiplex PCR for simultaneous detection of five foodborne pathogens. J. Appl. Microbiol. 112: 823–830.

Chevallier, I., S. Ammor, A. Laguet, S. Labayle, V. Castanet, E. Dufour and R. Talon. 2006. Microbial ecology of a small-scale facility producing traditional dry sausage. Food Control 17: 446–453.

Chiang, Y.C., H.Y. Tsen, H.Y. Chen, Y.H. Chang, C.K. Lin, C.Y. Chen and W.Y. Pai. 2012. Multiplex PCR and a chromogenic DNA macroarray for the detection of *Listeria monocytogens, Staphylococcus aureus, Streptococcus agalactiae, Enterobacter sakazakii, Escherichia coli* O157:H7, *Vibrio parahaemolyticus, Salmonella* spp. and *Pseudomonas fluorescens* in milk and meat samples. J. Microbiol. Meth. 88: 110–116.

Cintas, L.M., P. Casaus, H. Holo, P.E. Hernandez, I.F. Nes and L.S. Havarstein. 1998. Enterocins L50A and L50B, two novel bacteriocins from *Enterococcus faecium* L50, and related to staphylococcal hemolysins. J. Bacteriol. 180: 1988–1994.

Cleveland, J., T.J. Montville, I.F Nes and M.L. Chikindas 2001. Bacteriocins: Safe, natural antimicrobials for food preservation. Int. J. Food Microbiol. 71: 1–20.

Collins-Thompson, D.L., B. Krusky, W.R. Usborne and A.H.W. Hauschild 1984. The effect of nitrite on the growth of pathogens during manufacture of dry and semi-dry sausage. Can. I. Food Sc. Tech. J. 17: 102–106.

Conedera, G., E. Mattiazzi, F. Russo, E. Chiesa, I. Scorzato, S. Grandesso, A. Bessegato A. Fioravanti and A. Caprioli. 2007. A family outbreak of *Escherichia coli* O157 haemorrhagic colitis caused by pork meat salami. Epidemiol. Infect. 135: 311–314.

Cowden, J.M., M. O'Mahony, C.L.R. Bartlett, B. Rana, B. Smyth, D. Lynch, H. Tillett, L. Ward, D. Roberts, R.J. Gilbert, A.C. Baird-Parker and D.C. Kilsby. 1989. A national outbreak of *Salmonella* Typhimurium DT 124 caused by contaminated salami sticks. Epidemiol. Infect. 103: 219–225.

Dailey, R.C., K.G. Martin and R.D. Smiley. 2014. The effects of competition from non-pathogenic foodborne bacteria during the selective enrichment of *Listeria monocytogenes* using buffered Listeria enrichment broth. Food Microbiol. 44: 173–179

Dailey, R.C., L.J. Welch, A.D. Hitchins and R.D. Smiley. 2015. Effect of *Listeria seeligeri* or *Listeria welshimeri* on *Listeria monocytogenes* detection in and recovery from buffered Listeria enrichment broth. Food Microbiol. 46: 528–534.

Dalzini, E., E. Cosciani-Cunico, V. Bernini, B. Bertasi, M.N. Losio, P. Daminelli and G. Varisco. 2015. Behaviour of *Escherichia coli* O157 (VTEC), *Salmonella* Typhimurium and *Listeria monocytogenes* during the manufacture, ripening and shelf life of low fat salami. Food Control 47: 306–311

De Boer, E., W.M. Spiegelenberg and F.W. Janssen. 1985. Microbiology of spices and herbs. Anton. Leeuw. Int. J. G. 51: 435–438.

De Cesare, A., R. Mioni and G. Manfreda. 2007. Prevalence of *Listeria monocytogenes* in fresh and fermented Italian sausages and ribotyping of contaminating strains. Int. J. Food Microbiol. 120: 124–130.

Deegan, L.H., P.D. Cotter, C. Hill and P. Ross. 2006. Bacteriocins: Biological tools for biopreservation and shelf-life extension. Int. Dairy J. 16: 1058–1071.

Dicks, L.M.T., F.D. Mellett and L.C. Hoffman. 2004. Use of bacteriocin-producing starter cultures of *Lactobacillus plantarum* and *Lactobacillus curvatus* in production of ostrich meat salami. Meat Sci. 66: 703–708.

Drosinos, E.H., M. Mataragas and J. Metaxopoulos. 2006a. Modeling of growth and bacteriocin production by *Leuconostoc mesenteroides* E131. Meat Sci. 74: 690–696.

Drosinos E.H., M. Mataragas, S. Veskovic-Moracanin, J. Gasparik-Reichardt, M. Hadziosmanovic and D. Alagic. 2006b. Quantifying nonthermal inactivation of *Listeria monocytogenes* in European fermented sausages using bacteriocinogenic lactic acid bacteria or their bacteriocins: a case study for risk assessment. J. Food Protect. 69: 2648–2663.

Ducic, M., N. Klisara, S. Markov, B. Blagojevic, A. Vidakovic and S. Buncic. 2016. The fate and pasteurization-based inactivation of *Escherichia coli* O157, *Salmonella* Typhimurium and *Listeria monocytogenes* in dry, fermented sausages. Food Control 59: 400–406.

Ellajosyula, K.R., S. Doores, E.W. Mills, R.A. Wllson, R.C. Anantheswaran and S.J. Knabev. 1998. Destruction of *Escherichia coli* 0157: H7 and *Salmonella* Typhimurium in Lebanon Bologna by interaction of fermentation pH, heating temperature, and time. J. Food Protect. 61: 152–157.

Emberland, K.E., K. Nygard, B.T. Heier, P. Aavitsland, J. Lassen, T.L. Stavnes and B. Gondrosen. 2006. Outbreak of *Salmonella* Kedougou in Norway associated with salami, April-June 2006. EuroSurveillance 11(27). Available online: http://www.eurosurveillance.org/ViewArticle.aspx? ArticleId=2995

Escartin, E.F., A. Castillo, A. Hinojosa-Puga and J.S. Lozano. 1999. Prevalence of *Salmonella* in chorizo and its survival under different storage temperatures. Food Microbiol. 16: 479–486.

Faith, N.G., N. Parniere, T. Larson, T.D. Lorang, C.W. Kaspar and J.B. Luchansky. 1998. Viability of *Escherichia coli* 0157:H7 in salami following conditioning of batter, fermentation and drying of sticks, and storage of slices. J. Food Protect. 61: 377–382.

Foodborne Pathogens of Fermented Meat Products

FAO/WHO. 2004. Risk Assessment; Appendix 4—Prevalence and incidence of *L. monocytogenes* in fermented meat products (QMRA).

Fluit, A.C., M.N. Widjojoatmodjo, A.T.A. Box, R. Torensma and J. Verhoef. 1993. Rapid detection of salmonellae in poultry with the magnetic immunomagnetic–polymerase chain reaction assay. Appl. Environ. Microbiol. 59: 1342–1346.

Foegeding, P.M., A.B. Thomas, D.H. Pilkington and T.R. Klaenhammer. 1992. Enhanced control of *Listeria monocytogenes* by *in situ*-produced pediocin during dry fermented sausage production. Appl. Environ. Microbiol. 58: 884–890.

Garcia Diez, J. and L. Patarata. 2013. Behavior of *Salmonella* spp., *Listeria monocytogenes*, and *Staphylococcus aureus* in Chourico de Vinho, a dry fermented sausage made from wine-marinated meat. J. Food Protect. 76: 588–594.

Garcia, S., F. Iracheta, F. Galvan and N. Heredia. 2001. Microbiological survey of retail herbs and spices from Mexican markets. J. Food Protect. 64: 99–103.

Germini, A., A. Masola, P. Carnevali and R. Marchelli. 2009. Simultaneous detection of *Escherichia coli* O175:H7, *Salmonella* spp., and *Listeria monocytogenes* by multiplex PCR. Food Control 20: 733–738.

Gieraltowski, L., E. Julian, J. Pringle, K. Macdonald, D. Quilliam, N. Marsden-Haug, L. Saathoff-Huber, D. Von Stein, B. Kissler, M. Parish, D. Elder, V. Howard-King, J. Besser, S. Sodha, A. Loharikar, S. Dalton, I. Williams and C.B. Behravesh. 2013. Nationwide outbreak of *Salmonella* Montevideo infections associated with contaminated imported black and red pepper: warehouse membership cards provide critical clues to identify the source. Epidemiol. Infect. 141: 1244–1252.

Gnanou Besse, N., N. Audinet, A. Kerouanton, P. Colin and M. Kalmokoff. 2005. Evolution of *Listeria* populations in food samples undergoing enrichment culturing. Int. J. Food Microbiol. 104: 123–134.

Gonzalez-Fandos, M.E., M. Sierra, M.L. Garcia-Lopez, M.C. Garcia-Fernandez and A. Otero. 1999. The influence of manufacturing and drying conditions on the survival and toxinogenesis of *Staphylococcus aureus* in two Spanish dry sausages (chorizo and salchichon). Meat Sci. 52: 411–419.

Gounadaki, A.S., P.N. Skandamis, E.H. Drosinos and G.J.E. Nychas. 2008. Microbial ecology of food contact surfaces and products of small-scale facilities producing traditional sausages Food Microbiol. 25: 313–323.

Harding, C.D. and B.G. Shaw. 1990. Antimicrobial activity of *Leuconostoc gelidum* against closely related species and *Listeria monocytogenes*. J. Appl. Bacteriol. 69: 648–654.

Heir, E., A.L. Holck, M.K. Omer, O. Alvseike, M. Hoy, I. Mage and L. Axelsson. 2010. Reduction of verotoxigenic *Escherichia coli* by process and recipe optimisation in dry-fermented sausages. Int. J. Food Microbiol. 141: 195–202.

Hjertqvist, M., I. Luzzi, S. Loefdahl, A. Olsson, J. Radal and Y. Andersson. 2006. Unusual phage pattern of *Salmonella* Typhimurium isolated from Swedish patients and Italian salami. EuroSurveillance 11(6). Available online: http://www.eurosurveillance.org/ViewArticle. aspx?ArticleId=2896

Holck, A., L. Axelsson and U. Schillinger. 1996. Divergicin 750, a novel bacteriocin produced by *Carnobacterium divergens* 750. FEMS Microbiol. Lett. 136: 163–168.

Houben, J.H. 2005. A survey of dry-salted natural casings for the presence of *Salmonella* spp., *Listeria monocytogenes* and sulphite-reducing *Clostridium* spores. Food Microbiol. 22: 221–225.

Hugas, M., M. Garriga, M. Pascual, M.T. Aymerich and J.M. Monfort. 2002. Enhancement of sakacin K activity against *Listeria monocytogenes* in fermented sausages with pepper or manganese as ingredients. Food Microbiol. 19: 519–528.

Hwang, C.A., A.C.S. Porto-Fett, V.K. Juneja, S.C. Ingham, B.H. Ingham and J.B. Luchansky. 2009. Modeling the survival of *Escherichia coli* O157:H7, *Listeria monocytogenes*, and *Salmonella* Typhimurium during fermentation, drying, and storage of soudjouk-style fermented sausage. Int. J. Food Microbiol. 129: 244–252.

Incze, K. 1987. The technology and microbiology of Hungarian salami. Tradition and current status. Fleischwirtschaft 67: 445–447.

Jack, R.W., J. Wan, J. Gordon, K. Harmark, B.E. Davidson, A.J. Hillier, R.E.H. Wettenhall, M.W. Hickey and M.J. Coventry. 1996. Characterization of the chemical and antimicrobial properties of piscicolin 126, a bacteriocin produced by *Carnobacterium piscicola* JG126. Appl. Environ. Microbiol. 62: 2897–2903.

Jehan, Z., S. Uddin and K.S. Al-Kuraya. 2012. *In-situ* hybridization as a molecular tool in cancer diagnosis and treatment. Curr. Med. Chem. 19: 3730–3738.

Jofre, A., B. Martin, M. Garriga, M. Hugas, M. Pla, D. Rodriguez-Lazaro and T. Aymerich. 2005. Simultaneous detection of *Listeria monocytogenes* and *Salmonella* by multiplex PCR in cooked ham. Food Microbiol. 22: 109–115.

Kawasaki, S., N. Horikoshi, Y. Okada, K. Takeshita, T. Sameshima and S. Kawamoto. 2005. Multiplex PCR for simultaneous detection of *Salmonella* spp., *Listeria monocytogenes*, and *Escherichia coli* O157:H7 in meat samples. J. Food Protect. 68: 551–556.

Kim, H. and A.K. Bhunia. 2008. SEL, a selective enrichment broth for simultaneous growth of *Salmonella enterica*, *Escherichia coli* O157:H7, and *Listeria monocytogenes*. Appl. Environ. Microbiol. 74: 4853–4866.

Kothary, M.H. and U.S. Babu. 2001. Infective dose of foodborne pathogens in volunteers: A review. J. Food Safety 21: 49–73.

Kuhn, K.G., M. Torpdahl, C. Frank, K. Sigsgaard and S. Ethelberg. 2011. An outbreak of *Salmonella* Typhimurium traced back to salami, Denmark, April to June 2010. EuroSurveillance 16(19) Available online: http://www. eurosurveillance.org/ViewArticle. aspx?ArticleId=19863

Lahti, E., T. Johansson, T. Honkanen-Buzalski, P. Hill and E. Nurmi. 2001. Survival and detection of *Escherichia coli* O157:H7 and *Listeria monocytogenes* during the manufacture of dry sausage using two different starter cultures. Food Microbiol. 18: 75–85.

Lambertz, S.T., R. Lindqvist, A. Ballagi-Pordany and M.-L. Danielsson-Tham. 2000. A combined culture and PCR method for detection of pathogenic *Yersinia enterocolitica* in food. Int. J. Food Microbiol. 57: 63–73.

Lantz, P.-G., R. Knutsson, Y. Blixt, W.A. Al-Soud, E. Borch and P. Radstrom. 1998. Detection of pathogenic *Yersinia enterocolitica* in enrichment media and pork by a multiplex PCR: A study of sample preparation and PCR-inhibitory components. Int. J. Food Microbiol. 45: 93–105.

Laukova, A., S. Czikkova, S. Laczkova and P. Turek. 1999. Use of enterocin CCM 4231 to control *Listeria monocytogenes* in experimentally contaminated dry fermented Hornad salami. Int. J. Food Microbiol. 52: 115–119.

Lebert, I., S. Leroy, P. Giammarinaro, A. Lebert, J.P. Chacornac, S. Bover-Cid, M.C. Vidal-Carou and R. Talon. 2007. Diversity of microorganisms in the environment and dry fermented sausages of small traditional French processing units. Meat Sci. 76: 112–122.

Lee, S.H., B.Y. Jung, N. Rayamahji, H.S. Lee, W.J. Jeon, K.S. Choi, C.H. Kweon and H.S. Yoo. 2009. A multiplex real-time PCR for differential detection and quantification of *Salmonella* spp., *Salmonella enterica* serovar Typhimurium and Enteritidis in meats. J. Vet. Sci. 10: 43–51.

Li, Y., S. Zhuang and A. Mustapha. 2005. Application of a multiplex PCR for the simultaneous detection of *Escherichia coli* O157:H7, *Salmonella* and *Shigella* in raw and ready-to-eat meat products. Meat Sci. 71: 402–406.

Lindqvist, R. 1997. Preparation of PCR samples from food by a rapid and simple centrifugation technique evaluated by detection of *Escherichia coli* O157:H7. Int. J. Food Microbiol. 37: 78–82.

Lindqvist, R. and M. Lindblad. 2009. Inactivation of *Escherichia coli*, *Listeria monocytogenes* and *Yersinia enterocolitica* in fermented sausages during maturation/storage. Int. J. Food Microbiol. 129: 59–67.

Lindqvist, R., B. Norling and S.T. Lambertz. 1997. A rapid sample preparation method for PCR detection of food pathogens based on buoyant density centrifugation. Lett. Appl. Microbiol. 24: 306–310.

Luchansky, J.B., K.A. Glass, K.D. Harsono, A.J. Degnan, N.G. Faith, B. Cauvin, G. Baccus-Taylor, K. Arihara, B. Bater, A.J. Maurer and R.G. Cassens. 1992. Genomic analysis of pediococcus starter cultures used to control *Listeria monocytogenes* in Turkey summer sausage. Appl. Environ. Microbiol. 58: 3053–3059.

Lücke, K.H. 2000. Fermented meats. pp. 420-444. *In*: B.M. Lund, A.C. Baird-Parker and G.W. Gould (eds.). The microbiological safety and quality of food. Aspen Publ., Gaithersburg.

Luzzi, I., P. Galetta, M. Massari, C. Rizzo, A. Dionisi, E. Filetici, A. Cawthorne, A. Tozzi, M. Argentieri, S. Bilei, L. Busani, C. Gnesivo, A. Pendenza, A. Piccoli, P. Napoli, R. Loffredo, M.O. Trinito, E. Santarelli and M.L. Ciofi degli Atti. 2007. An Easter outbreak of *Salmonella* Typhimurium DT 104A associated with traditional pork salami in Italy. EuroSurveillance 12(4) Available online: http://www.eurosurveillance.org/em/v12n04/1204-226.asp

Martin, B., M. Garriga and T. Aymerich. 2011. Prevalence of *Salmonella* spp. and *Listeria monocytogenes* at small-scale Spanish factories producing traditional fermented sausages. J. Food Protect. 74: 812–815.

Mataragas, M., A. Bellio, F. Rovetto, S. Astegiano, L. Decastelli and L. Cocolin. 2015. Risk-based control of food-borne pathogens *Listeria monocytogenes* and *Salmonella enterica* in the Italian fermented sausages Cacciatore and Felino. Meat Sci. 103: 39–45.

Mataragas, M., J. Metaxopoulos, M. Galiotou and E. H. Drosinos. 2003. Influence of pH and temperature on growth and bacteriocin production by *Leuconostoc mesenteroides* L124 and *Lactobacillus curvatus* L442. Meat Sci. 64: 265–271.

Maurischat, S., B. Baumann, A. Martin and B. Malorny. 2015. Rapid detection and specific differentiation of *Salmonella enterica* subsp. *enterica* Enteritidis, Typhimurium and its monophasic variant 4,[5],12:i: – by real-time multiplex PCR. Int. J. Food Microbiol. 193: 8–14.

McDonald, D.M., M. Fyfe, A. Paccagnella, A. Trinidad, K. Louie and D. Patrick. 2004. *Escherichia coli* O157:H7 outbreak linked to salami, British Columbia, Canada, 1999. Epidemiol. Infect. 132: 283–289.

McKee, L.H. 1995. Microbial contamination of spices and herbs: a review. Lebensm. Wiss. Technol. 28: 1–11.

Mercanoglu, B. and M.W. Griffiths. 2005. Combination of immunomagnetic separation with real time PCR for rapid detection of *Salmonella* in milk, ground beef, and alfalfa sprouts. J. Food Protect. 68: 557–561.

Messi, P., M. Bondi, C. Sabia, R. Battini and G. Manicardi. 2001. Detection and preliminary characterization of a bacteriocin (plantaricin 35d) produced by a *Lactobacillus plantarum* strain. Int. J. Food Microbiol. 64: 193–198.

Metaxopoulos, J., C. Genigeorgis, M.J. Fanelli, C. Franti and E. Cosma. 1981a. Production of Italian dry salami. I. Initiation of staphylococcal growth in salami under commercial manufacturing conditions. J. Food Protect. 44: 347–352.

Metaxopoulos, J., C. Genigeorgis, M.J. Fanelli, C. Franti and E. Cosma. 1981b. Production of Italian dry salami: Effect of starter culture and chemical acidulation on staphylococcal growth in salami under commercial manufacturing conditions. Appl. Environ. Microbiol. 42: 863–871.

Moreira, A.N., F.R. Conceicao, R. de Cassia, S. Conceicao, R.J. Ramos, J.B. Carvalhal, O.A. Dellagostin and J.A.G. Aleixo. 2008. Detection of *Salmonella* Typhimurium in raw meats using in-house prepared monoclonal antibody coated magnetic beads and PCR assay of the *fimA* gene. J. Immunoassay Immunochem. 29: 58–69.

Moreno, Y., M. Hernandez, M.A. Ferrus, J.L. Alonso, S. Botella, R. Montes and J. Hernandez. 2001. Direct detection of thermotolerant campylobacters in chicken products by PCR and *in situ* hybridization. Res. Microbiol. 152: 577–582.

Murphy, M., A. Carroll, C. Walsh, P. Whyte, M. O'Mahony, W. Anderson, E. McNamara and S. Fanning. 2007. Development and assessment of a rapid method to detect *Escherichia coli* O26, O111 and O157 in retail minced beef. Int. J. Hyg. Environ. Health 210: 155–161.

Muthukumarasamy, P. and R.A. Holley. 2007. Survival of *Escherichia coli* O157:H7 in dry fermented sausages containing micro-encapsulated probiotic lactic acid bacteria. Food Microbiol. 24: 82–88.

Nissen, H. and A. Holck. 1998. Survival of *Escherichia coli* O157:H7, *Listeria monocytogenes* and *Salmonella* Kentucky in Norwegian fermented, dry sausage. Food Microbiol. 15: 273–279.

Nitecki, S.S., N. Teape, B.F. Carney, J.W. Slater and W.M. Brueck. 2015. A duplex qPCR for the simultaneous detection of *Escherichia coli* O157:H7 and *Listeria monocytogenes* using LNA probes. Lett. Appl. Microbiol. 61: 20–27.

Noonpakdee, W., C. Santivarngkna, P. Jumriangrit, K. Sonomoto and S. Panyim. 2003. Isolation of nisin-producing *Lactococcus lactis* WNC 20 strain from nham, a traditional Thai fermented sausage. Int. J. Food Microbiol. 81: 137–145.

Nychas, G.J.E. and J.S. Arkoudelos. 1990. Staphylococci: Their role in fermented sausages. J. Appl. Bacteriol. 69 (Suppl.): 167S–188S.

Osmanagaoglu, O. 2007. Detection and characterization of Leucocin OZ, a new anti-listerial bacteriocin produced by *Leuconostoc carnosum* with a broad spectrum of activity. Food Control 18: 118–123.

Paramithiotis, S. and E.H. Drosinos. 2010. Microbiological quality and aflatoxin B1 content of some spices and additives used in meat products. Qual. Assur. Saf. Crop. 2: 41–45.

Paramithiotis, S., P.N. Skandamis and G.J.E. Nychas. 2009. Insights into fresh meat spoilage. pp. 55–82. *In*: F. Toldra (ed.). Safety of meat and processed meat. Springer, New York.

Paton, A.W., R.M. Ratcliff, R.M. Doyle, J. Seymour-Murray, D. Davos, J.A. Lanser and J.C. Paton. 1996. Molecular microbiological investigation of an outbreak of hemolytic-uremic syndrome caused by dry fermented sausage contaminated with Shiga-like toxin-producing *Escherichia coli*. J. Clin. Microbiol. 34: 1622–1627.

Pereira, J.L., S.P. Salzberg and M.S. Bergdoll. 1982. Effect of temperature, pH and sodium chloride concentration and production of staphylococcal enterotoxins A and B. J. Food Protect. 45: 1306–1309.

Pontello, M., L. Sodano, A. Nastasi, C. Mammina, M. Astuti, M. Domenichini, G. Belluzzi, E. Soccini, M.G. Silvestri, M. Gatti, E. Gerosa and A.A. Montagna. 1998. Community-based outbreak of *Salmonella enterica* serotype Typhimurium associated with salami consumption in Northern Italy. Epidemiol. Infect. 120: 209–214.

Porto-Fett, A.C.S., C.-A. Hwang, J.E. Call, V.K. Juneja, S.C. Ingham, B.H. Ingham and J.B. Luchansky. 2008. Viability of multi-strain mixtures of *Listeria monocytogenes*, *Salmonella* Typhimurium, or *Escherichia coli* O157:H7 inoculated into the batter or onto the surface of a soudjouk-style fermented semi-dry sausage. Food Microbiol. 25: 793– 801.

Prema, P., S. Bharathy, A. Palavesam, M. Sivasubramanian and G. Immanuel. 2006. Detection, purification and efficacy of warnerin produced by *Staphylococcus warneri*. World J. Microbiol. Biotechnol. 22: 865–872.

Regnault, B., S. Martin-Delautre, M. Lejay-Collin, M. Lefevre and P.A.D. Grimont. 2000. Oligonucleotide probe for the visualization of *Escherichia coli/Escherichia fergusonii* cells by *in situ* hybridization: specificity and potential applications. Res. Microbiol. 151: 521–533.

Rekhif, N., A. Atrih and G. Lefebvre. 1994. Selection and properties of spontaneous mutants of *Listeria monocytogenes* ATTC 15313 resistant to different bacteriocins produced by lactic acid bacteria strains. Curr. Microbiol. 28: 237–242.

Rohde, A., J.A. Hammerl, B. Appel, R. Dieckmann and S. Al Dahouk. 2015. FISHing for bacteria in food - A promising tool for the reliable detection of pathogenic bacteria? Food Microbiol. 46: 395–407.

Sagoo, S.K., C.L. Little, M. Greenwood, V. Mithani, K.A. Grant, J. McLauchlin, E. de Pinna and E.J. Threlfall. 2009. Assessment of the microbiological safety of dried spices and herbs from production and retail premises in the United Kingdom. Food Microbiol. 26: 39–43.

Sartz, L., B. De Jong, M. Hjertqvist, L. Plym-Forshell, R. Asterlund, S. Loefdahl, B. Osterman, A. Stahl, E. Eriksson, H.B. Hansson and D. Karpman. 2008. An outbreak of *Escherichia coli* O157:H7 infection in southern Sweden associated with consumption of fermented sausage; aspects of sausage production that increase the risk of contamination. Epidemiol. Infect. 136: 370–380.

Schmid, M., W.A. Lehner, R. Stephan, K.-H. Schleifer and H. Meier. 2005. Development and application of oligonucleotide probes for *in situ* detection of thermotolerant *Campylobacter* in chicken faecal and liver samples. Int. J. Food Microbiol. 105: 245–255.

Schneider, R., F.J. Fernandez, M.B. Aquilar, I. Guerrero-Legarreta, A. Alpuche-Solis and E. Ponce-Alquicira. 2006. Partial characterization of a class IIa pediocin produced by *Pediococcus parvulus* 133 strain isolated from meat (Mexical 'chorizo'). Food Control 17: 909–915.

Schwab, A.H., A.D. Harpestad, A. Swartzentruber, J.M. Lanier, B.A. Wentz, A.P. Duran, R.J. Barnard and R.B. Jr. Read. 1982. Microbiological quality of some spices and herbs in retail markets. Appl. Environ. Microbiol. 44: 627–630.

Siriken, B., S. Pamuk, C. Ozakin, S. Gedikoglu and M. Eyigor. 2006. A note on the incidences of *Salmonella* spp., *Listeria* spp. and *Escherichia coli* O157:H7 serotypes in Turkish sausage (Soudjouck). Meat Sci. 72: 177–181.

Skandamis, P. and G.J.E. Nychas. 2007. Pathogens: Risks and Control. pp. 427-454. *In*: F. Toldra (ed.). Handbook of fermented meat and poultry. Blackwell Publishing, Oxford, UK.

Stromberg, Z.R., G.L. Lewis, D.B. Marx and R.A. Moxley. 2015. Comparison of enrichment broths for supporting growth of Shiga toxin-producing *Escherichia coli*. Curr. Microbiol. 71: 214–219.

Thevenot, D., M.L. Delignette-Muller, S. Christieans and C. Vernozy-Rozand. 2005. Prevalence of *Listeria monocytogenes* in 13 dried sausage processing plants and their products. Int. J. Food Microbiol. 102: 85–94.

Tilden, J. Jr., W. Young, A.M. McNamara, C. Custer, B. Boesel, M.A. Lambert-Fair, J. Majkowski, D. Vugia, S.B. Wemer, J. Hollingsworth and J.G. Jr. Morris. 1996. A new route of transmission for *Escherichia coli*: infection from dry fermented salami. Am. J. Public Health 86: 1142–1145.

Verluyten, J., F. Leroy and L. De Vuyst. 2004a. Effects of different spices used in production of fermented sausages on growth of and curvacin A production by *Lactobacillus curvatus* LTH 1174. Appl. Environ. Microbiol. 70: 4807–4813.

Verluyten, J., W. Messens and L. De Vuyst. 2003. The curing agent sodium nitrite, used in the production of fermented sausages, is less inhibiting to the bacteriocin-producing meat starter culture *Lactobacillus curvatus* LTH 1174 under anaerobic conditions. Appl. Environ. Microbiol. 69: 3833–3839.

Verluyten, J., W. Messens and L. De Vuyst. 2004b. Sodium chloride reduces production of curvacin A, a bacteriocin produced by *Lactobacillus curvatus* strain LTH 1174, originating from fermented sausage. Appl. Environ. Microbiol. 70: 2271–2278.

Villani, F., L. Sannino, G. Moschetti, G. Mauriello, O. Pepe, R. Amodio-Cocchieri and S. Coppola. 1997. Partial characterization of an antagonistic substance produced by *Staphylococcus xylosus* 1E and determination of the effectiveness of the producer strain to inhibit *Listeria monocytogenes* in Italian sausages. Food Microbiol. 14: 555–566.

Waller, D.F. and S.A. Ogata. 2000. Quantitative immunocapture PCR assay for detection of *Campylobacter jejuni* in foods. Appl. Environ. Microbiol. 66: 4115–4118.

Wang, H. and M.F. Slavik. 2005. A multiplex polymerase chain reaction assay for rapid detection of *Escherichia coli* O157:H7, *Listeria monocytogenes*, *Salmonella Typhimurium* and *Campylobacter jejuni* in artificially contaminated food samples. J. Rapid Meth. Aut. Mic. 13: 213–223.

Wang, L., Y. Li and A. Mustapha. 2007. Rapid and simultaneous quantitation of *Escherichia coli* O157:H7, *Salmonella*, and *Shigella* in ground beef by multiplex real-time PCR and immunomagnetic separation. J. Food Protect. 70: 1366–1372.

Wang, X., N. Jothikumar and M.W. Griffiths. 2004. Enrichment and DNA extraction protocols for the simultaneous detection of *Salmonella* and *Listeria monocytogenes* in raw sausage meat with multiplex real-time PCR. J. Food Protect. 67: 189–192.

Williams, R.C., S. Isaacs, M.L. Decou, E.A. Richardson, M.C. Buffett, R.W Slinger, M.H. Brodsky, B.W. Ciebin, A. Ellis, J. Hockin and the *E. coli* O157:H7 Working Group. 2000. Illness outbreak associated with *Escherichia coli* O157:H7 in Genoa salami. Can. Med. Assoc. J. 162: 1409–1413.

Wolffs, P., R. Knutsson, B. Norling and P. Radstrom. 2004. Rapid quantification of *Yersinia enterocolitica* in pork sample preparation method, flotation, prior to real-time PCR. J.Clin. Microbiol. 42: 1042–1047.

Wolffs, P., B. Norling, J. Hoorfar, M. Griffiths and P. Radstrom. 2005. Quantification of *Campylobacter* spp. in chicken rinse samples by using flotation prior to real-time PCR. Appl. Environ. Microbiol. 71: 5759–5764.

Yoshitomi, K.J., K.C. Jinneman, P.A. Orlandi, S.D. Weagant, R. Zapata and W.M. Fedio. 2015. Evaluation of rapid screening techniques for detection of *Salmonella* spp. from produce samples after pre-enrichment according to FDA BAM and a short secondary enrichment. Lett. Appl. Microbiol. 61: 7–12.

Chapter 11

Protective Cultures and Bacteriocins in Fermented Meats

Maria João Fraqueza*, Luis Patarata and Andrea Lauková

1 INTRODUCTION

Meat fermentation results from a particular manufacturing process that involves natural microbiota well adapted to the meat substrate or intentionally added as starter cultures, contributing to their preservation and sensorial characteristics. Around the world, many meat products are based in a spontaneous fermentation with particularities related to local and ancestral know-how that lead to a great diversity of named traditional fermented meat products according to their origin and country. These meat products won the consumer preferences, not only due to their sensorial characteristics, but also due to the biopreservation used in the production, with a minimum use of chemical additives.

The main fermentative microbiotas are lactic acid bacteria (LAB) including *Lactobacillus*, *Pediococcus*, *Enterococcus*, *Leuconostoc*, *Lactococcus* and *Weissella*. LAB constitutes a diverse group widely distributed throughout nature, being also an important component of indigenous microbiota in healthy humans and animals. LAB diversity in fermented

---

*For Correspondence: CIISA, Faculty of Veterinary Medicine, University of Lisbon, Avenida da Universidade Técnica, Pólo Universitário do Alto da Ajuda, 1300-477 Lisbon, Portugal, Tel: 213 652 884, Email: mjoaofraqueza@fmv.ulisboa.pt

meat products is linked to manufacturing practices, and it has been reported that indigenous starters development is very promising approach since it enables meat products production with high sanitary and sensory characteristics. Another bacterial group with relevance in the manufacture of fermented meat products is the Gram-positive catalase positive cocci (GCC+). They are responsible by color stability, control of spoilage, decreases in the processing time and contribute to flavor development. Those most frequently isolated from fermented meat products are mainly GCC+ such as *Staphylococcus xylosus*, *St. equorum*, *St. succinus*, and *St. saprophyticus* (Morot-Bizot et al. 2006, Drosinos et al. 2005, Fontana et al. 2005, Mauriello et al. 2004, Simonová et al. 2006). Also, yeasts are related to fermented meat products contributing to the flavor and aroma through fermentation of different sugars, metabolization and reduction of lactic acid (producing a sweeter end product), proteolysis and amino acid degradation, which generates volatile compounds and ammonia (Durá et al. 2004).

The well adapted species and strains to the processing units' environmental conditions constitute the so-called "house microbiota" that naturally colonizes meat products during processing, being responsible for the spontaneous fermentation of meat with production of lactic acid and other metabolites involved in the development of the sensory characteristics of sausages (flavor, colour, texture), the enhance of safety, stability and the shelf life of the fermented meat produscts (van Kranenburg et al. 2002, Fadda et al. 2010). LAB´s protective effects are also attributed to bacteriocins which are ribosomally synthesized peptides with antimicrobial activity against more or less related bacteria food pathogens including (Nes et al. 2007).

LAB have a long history of safe use in fermented food production and consumption that supports their GRAS (generally recognized as safe) and QPS (qualified presumption of safety) status provided by FDA (US Food and Drug administration) and EFSA (European Food Safety Authority), respectively. However the detection of antibiotic resistant (AR) strains among LAB has resulted in their recognition as reservoir of AR genes horizontally transmissible to pathogens through the food chain, this being a matter of concern (Marshall et al. 2009). Also there are particular LAB species that could be decarboxylase positive and promote the production of biogenic amines in meat products.

This chapter review describes the diversity and role of LAB in fermented meat products, stressing safety requirements for LAB or other species used as starters. The application of protective starters and their effects in different meat products will be resumed to describe the potential of different species and strains. The role of bacteriocins in meat products protection against pathogens will be demonstrated as well the application and development that has been done in meat products. The

230   *Fermented Meat Products: Health Aspects*

use of protective microbiota on the concept a biopreservation technology added to other emergent technologies will be addressed describing their potential to control meat products safety and shelf life. Finally, some remarks are pointed out considering safety concerns for protective starters and bacteriocins in meat products.

2 BIODIVERSITY AND ROLE OF FERMENTATIVE MICROBIOTA

Bacterial communities coexist in fermented meat products, allowing for microbial diversity. Spontaneous fermentation has been used empirically in fermented meat products as one of the oldest biopreservation technologies. Many factors influence the final microbiota of fermented meat sausages. The presence of ecological determinants influences the establishment of a specific microbial consortium that will determine the rate of colonization. The selective influence of intrinsic (concentration and availability of nutrients, pH, redox potential, buffering capacity, water activity -a_w, meat structure) and extrinsic (temperature, relative humidity and oxygen availability) factors may determine differences in microbial ecosystem of raw meat substrate. This initial microbiota will be influenced by technological particularities used in fermented meat products processing and will adapt to this special niche. The addition of ingredients such as sodium chloride, nitrate/nitrite, sugars, wine, condiments (garlic, pepper), as well as the particular a_w (0.85–0.92), temperature (12–18°C to 24–30°C), oxygen gradient during ripening and smoking application will select a microbiota able to develop in fermented meat sausages.

Among the different genera belonging to the LAB group, those that are frequently found in fermented meat products are *Lactobacillus*, *Enterococcus*, *Pediococcus* and *Leuconostoc* (Table 1). LAB includes a diverse group of Gram-positive non-spore forming cocci, coccobacilli or rods, with common morphological, metabolic and physiological characteristics (Batt 2000). They are facultative anaerobic with variable oxygen tolerance in different species. LAB growth depends on the presence of fermentable carbohydrates. They are classified as homofermentative or heterofermentative based on end products of glucose metabolism. While homofermentative LAB convert glucose mainly to lactic acid using the glycolysis (Embden–Meyerhof–Parnas or Embden–Meyerhof) pathway, the heterofermentative LAB use the phosphoketolase (6-phosphogluconate) pathway and convert glucose to lactic acid, carbon dioxide and ethanol or acetic acid. The diverse metabolic capabilities make LAB easily adaptable to a wide range of conditions, allowing them to thrive in acid foods' fermentations. This heterogeneous group

Protective Cultures and Bacteriocins in Fermented Meats 231

of bacteria comprises about 20 genera within the phylum Firmicutes, class *Bacilli*, and order *Lactobacillales*. *Aerococcaceae, Carnobacteriacea, Enterococcaceae, Lactobacillaceae, Leuconostocaceae,* and *Streptococcaceae* are the different families included in the former order (Ludwig et al. 2009). From a practical point of view the principal LABs are included in the genera *Aerococcus, Carnobacterium, Enterococcus, Lactobacillus, Lactococcus, Leuconostoc, Oenococcus, Pediococcus, Streptococcus, Tetragenococcus, Vagococcus* and *Weissella* (Axelsson 2004, Wright and Axelsson 2012, Vandamme et al. 2014).

LAB are widely found in nature, adapted to different carbohydrate rich environmental niches, as well as to terrestrial and marine animals. LAB are found in dairy products (yoghurt and cheese), in fermented vegetables (olives, sauerkraut), in meat and fermented meats products and in sourdough bread (Coeuret et al. 2003, Medina et al. 2011, Hurtado et al. 2012, Panagou et al. 2013, De Vuyst et al. 2014). LAB play a recognized role in fermented foods preservation and safety, thus promoting final products microbial stability. The preservation ability of LAB is based on competition for nutrients and the production of antimicrobial active metabolites such as organic acids (mainly lactic acid and acetic acid), and other substances, such as ethanol, fatty acids, acetoin, hydrogen peroxide, diacetyl, antifungal compounds (propionate, phenyl-lactate, hydroxyphenyl-lactate, cyclic dipeptides and 3-hydroxy fatty acids), bacteriocins (nisin, reuterin, reutericyclin, pediocin, lacticin, enterocin and others) and bacteriocin-like inhibitory substances (BLIS) that can inhibit growth of spoilage and pathogenic bacteria and act as multiple hurdles (Reis et al. 2012). These metabolites are active during food fermentation and/or subsequent ripening and storage. The production of weak organic acids, such as acetic and lactic acids, inhibits microbial growth through multiple actions, including membrane disruption, inhibition of metabolic reactions, disturbance of pH homeostasis, and accumulation of toxic anions in the cell. The other antimicrobials already referred often acts in synergy.

Several LAB species well-adapted to the fermented meat products environment (Table 1), were isolated from different products produced worldwide and further identified (Adiguzel and Atasever 2009, Danilović et al. 2011, Panagou et al. 2013, Wanangkarn et al. 2014, Federici et al. 2014).

Meat origin, sausage composition, sugar absence, ripening temperature, smoking or contamination by environmental microbiota during home- or small-scale production units can explain the broad species diversity (Marty et al. 2012, Wanangkarn et al. 2014). The predominant species in fermented meat products are *Lactobacillus sakei, Lb. curvatus, Lb. plantarum, Leuconostoc mesenteroides, Pediococcus* spp., *Enterococcus* spp., which growth is modulated and adapted to the existing

Table 1 Overview of lactic acid bacteria (LAB) species diversity of dry- fermented sausages from different geographical origins (from Fraqueza 2015).

Region/ Country	Product	Composition	Main process Conditions	Major known function	LAB Level (log cfu/g)	Identification method	Reported LAB species diversity	References
Italy	Ciauscolo salami	Pork	Short-ripened fermented sausage 30 d 20–25°C, cold smoked 2 d	Micro-organisms inhibition by acidification	nd	Species identification by 16S rRNA gene sequencing	*Lb. sakei/ Pd. pentosaceus/ Lb. plantarum/ Lb. paraplantarum/ E. faecalis/ Lb. paracasei/ Lb. lactis/ Lb. johnsonii/ Lb. brevis/ Lactococcus* spp./ *Carnobacterium* spp./ *W. helénica/ Leuconostoc mesenteroides*	Federici et al. 2014
Northeastern region of Thailand	Thai fermented sausage	Beef , bovine liver spleen, roasted rice powder garlic, salt, spices and seasonings	Fermented sausage 3 d at room temperature	Acidification pH 4.1	nd	Species identification by 16S rRNA gene sequencing	*Lb. sakei/ Lb. plantarum/ Ln. mesenteroides Lb. brevis/ Pd. pentosaceus/ Lc. lactis/ Lb. fermentum*	Wanangkarn et al. 2012, 2014
Spain	Dry-fermented sausage	Pork	Fermented sausage	Acidification, nitrate reduction activity, lipolytic activity	nd	Species identification by 16S rRNA gene sequencing	*Lb. sakei /Lb. plantarum /Lb. paracasei/ Lb. coryniformi/E. faecium*	Landeta et al. 2013

Protective Cultures and Bacteriocins in Fermented Meats

Region	Product	Meat	Process	Function	pH	Method	Microorganisms	Reference
Swiss	Dry-fermented sausages not named	Wildlife meat (deer, wild boar, chamois) or meat from cattle, sheep and pork	Fermentation and ripening conducted at 9–17°C, 15 d, cold smoking on last stage	Microorganisms inhibition by acidification (pH 4.6–6.6)	4.6–9.1	16S rRNA, gene-based PCR – restriction fragment length polymorphism (RFLP)	*Lb. sakei /Lb. curoatus / E. faecalis/E. faecium /Pd. pentosaceus* and *Streptococcus* spp.	Marty et al. 2012
Western Himalayas/India and Nepal	Jamma or Geema/Kargyong	Ethnic chevron (goat) meat	Boiling, 15 min. Smoking, 15–20 d uncontrolled temperature and humidity	Acidification (pH 5.5), flavor	7.8–7.5	16S rRNA and phenylalanyl-tRNA synthase (pheS) genes sequencing	*E. durans/E. faecalis/E. faecium/E. hirae, Ln. citreum, Ln. mesenteroides, Pd.pentosaceus,* and *W. cibaria*	Oki et al. 2011
North Portugal	Alheira	Pork and poultry	Smoking, max. 8 d at temperatures below 37 °C and with uncontrolled humidity	Aroma and taste, extension of shelf life	nd	Species- and genus-specific primers and 16S rRNA gene sequencing	*Lb. plantarum/Lb. paraplantarum/ Lb. brevis/ Lb. rhamnosus/ Lb. sakei/ Lb. zeae/ Lb. paracasei/ Ln. mesenteroides, Pd. pentosaceus, Pd. acidilactici, W. cibaria/W. viridescens and E. faecium/ E. faecalis*	Albano et al. 2009

Table 1 (*Contd.*)

Table 1 Overview of lactic acid bacteria (LAB) species diversity of dry- fermented sausages from different geographical origins (from Fraqueza 2015) (*Contd.*)

Italy	Piacentino salami	Pork meat	Fermentation 8 d (RH = 40% to 90% at 15–25°C) Ripening (45 d) at RH = 70–90% and 12–19°C.	Aroma and taste	3–9	Species identification by 16S rRNA gene sequencing	*Lb. sakei/* *Lb. curvatus/* *Lb. plantarum/* *Lb. brevis/* *Lb. rhamnosus,* *Lb. paracasei and* *Lb. reuteri*	Zonenschain et al. 2009
South-east Europe (Greece, Bosnia and Herzegovina, Croatia, Hungary, Italy and Serbia).	Sudzuk, Sremska, Friuli	Pork and beef meat	Ripening (28 d)	Acidification (pH 4.7 to 5.7), aroma and taste	3–5	Phenotypic identification API 50 CHL	*Lb. sakei/* *Lb. curvatus/* *Lb. plantarum/* *Lb. pentosus/* *Lb. rhamnosus/* *Lb. brevis/* *Lb. paracasei/* *Lb. alimentarius/* *Lb. fermentum/* *Lb. bavaricus;* *Lc. lactis;* *Pd. pentosaceus/* *Pd. acidilactici;* *Ln. mesenteroides;* *E. faecium*	Kozačinski et al. 2008

Protective Cultures and Bacteriocins in Fermented Meats 235

Italy	Fermented sausage	Pork meat, lard, sodium chloride, nitrite and nitrate, and black pepper	Ripening (28–45 d)	Acidification pH=5.62–5.65	8–9	Species-specific PCR and DGGE analysis followed by16S rRNA gene sequencing	*Lb. sakei/ Lb. plantarum/ Lb. curvatus/ Lb. casei/Lb. brevis/ Lb. paraplantarum/ Lc. garvieae/ Lc. lactis/ Ln. carnosum/ Ln. mesenteroides W. helénica/W. paramesenteroides*	Urso et al. 2006
Spain	Fuet, Chorizo and Salchichon	Pork meat	Cold ripening	Acidification (pH 5.3 to 6.2), aroma and taste	6.86–8.99	Species-specific PCR and 16S rRNA gene sequencing	*Lb. sakei/ Lb. plantarum/ Lb. curvatus/ Ln. mesenteroides/ E. faecium*	Aymerich et al. 2003, 2006a
Trás-os-Montes, North Portugal	Chouriço	Pork meat and fat; salt, wine, garlic, bay, eventually paprika	Cold smoked, 1 to 4 wk drying at a low temperature	Acidification (pH=5.5), aroma and taste	8	Phenotypic and species specific PCR	*Lb. sakei/ Lb. plantarum*	Patarata 2002

Table 1 (*Contd.*)

stringent conditions of processing (Wanangkarn et al. 2014, Federici et al. 2014). Some *Lactobacillus* species will dominate because they can develop continuously throughout the fermentation process, since they have a greater acid tolerance and capability to adapt to redox changes than other bacterial species. Of note the genome of *Lb. sakei* 23K, isolated from a French sausage, showed the existence of particularities in its genetic repertoire, absent from other lactobacilli, such as a versatile redox metabolism giving resilience against changing redox and oxygen levels, or energy production pathways not normally associated with a lactic acid bacterium (Chaillou et al. 2005). This suggests biological functions that could have a role in species adaptation to meat products. Genes potentially responsible for biofilm formation and cellular aggregation that may assist the microorganism to colonize meat surfaces were also identified (Chaillou et al. 2005, Muscariello et al. 2013). *Lb. sakei* and *Lb. plantarum* are surprisingly well equipped to cope with changing oxygen conditions (Serrano et al. 2007, Guilbaud et al. 2011) present in dry-fermented meat sausage's processing. *Lb. curvatus* CRL705, a strain isolated from an Argentinean artisanal fermented sausage showed the ability to produce the two-component lactocin 705 bacteriocin and "AL705," a bacteriocin with antilisterial activity (Hebert et al. 2012), which are bacteria tools that contribute to its dominance.

From the group of LAB, enterococci presence in meat products could be highly controversial. Enterococci can be seen as indicators of fecal contamination and lack of good hygiene practices and manufacturing practices, responsible for meat products spoilage, being highly related to the presence of toxic substances such as biogenic amines in meat products (Latorre-Moratalla et al. 2010a, b). Tyramine production is reported in the majority of enterococci. However they can have an important role in flavor development and bioprotection of dry-fermented meat products since are frequently mentioned as producers of bacteriocins (Callewaert et al. 2000, Mareková et al. 2007).

The genetic diversity is not only found at specific but also at intra-specific level, with well adapted strains to the processing units' environmental conditions, constituting the so-called "house microbiota" that naturally colonizes meat sausages during processing, being responsible for the spontaneous fermentation of meat with production of lactic acid and other metabolites involved in the development of the sensory characteristics of sausages (flavor, colour, texture), the enhance of safety, stability and the shelf life of the fermented sausages (van Kranenburg et al. 2002, Fadda et al. 2010). Figure 1 shows different fingerprinting profiles (PCR rep with GTG5) of *Lb. plantarum* isolated from different Portuguese dry-fermented sausages confirming a high diversity but with main clusters associated with processing units' origin. A high biodiversity inside 295 isolates from the predominant

Protective Cultures and Bacteriocins in Fermented Meats 237

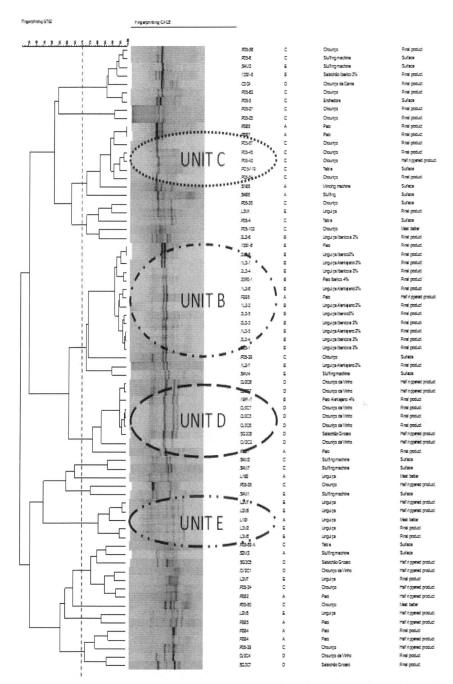

Figure 1 PCR Fingerprinting profile of *Lactobacillus plantarum* isolates ($n = 73$) from different dry-fermented sausages belonging to different processing units.

238   *Fermented Meat Products: Health Aspects*

LAB species in different kinds of sausages produced in Hungary, Italy, and Greece was also found by Rantsiou et al. (2005). The clustering analysis at 70% homology gave five different clusters for the 27 isolates of *Lb. plantarum*, nine clusters for the 100 isolates of *Lb. curvatus*, and 19 clusters for the 168 isolates of *Lb. sakei*. The majority of the clusters were country-specific.

The use of small portions of fermented meat (backslopping) for sausage manufacture purposes became the origin of the starters composed of selected strains intentionally added to meat sausage to better controlling the fermentation process. The idea of starter cultures application in fermented sausages was developed with Jensen and Paddock in 1940 (US Patent 2,225,783), becoming increasingly frequent in order to confer protection and standardization of the dry-fermented sausages process (Vignolo et al. 2015). The competitiveness of bacterial starter cultures is in their ability to compete with the adventitious microbiota of meat, to colonize this environment, and to dominate the microbial community during the fermentation process. The starter culture must compete with the natural microbiota of the raw material and undertake the metabolic activities expected of it, being conditioned by its growth rate and survival in the conditions prevailing in the sausage. Linked to the use of a starter is the concept of biopreservation referring to the enhancement of food safety and stability by using natural or controlled microbiota and/or antimicrobial compounds (Ananou et al. 2007).

Biopreservation can cop with consumers' demands for products with less chemical additives and less salt and fat content and is nowadays achieved by: (a) application of a technological microbiota harboring protective effects; (b) application of an adjunct culture producing antimicrobial metabolites *in situ* or *ex situ* that does not influence food quality; or (c) application of antimicrobial metabolites without the producing strain (e.g. fermented and bacteriocin extracts).

3   SAFETY REQUIREMENTS FOR STARTERS

Fermented meat products, and their LAB microbiota, fit in the concept of "history of safe use", defined in EFSA´s safety assessment guidance (EFSA 2005), due to all evidences of safe production and consumption by genetically diverse human population over the years. In fact, dry-fermented sausages are highly valued and constitute a frequently consumed food item, it might be said that LAB are ingested in high levels along with fermented meat products consumption. This happen from early times, actually records about fermented meat processing are very old. Also, reported cases of disease related with the most frequent LAB species isolated from these products are, as far as we know, rare. According to Bernardeau et al. (2006), *Lactobacillus* infections occur at a very low rate in the generally healthy population - estimated 0.5/1 million per year. So, there is reasonable certainty that no harm will result from dry-fermented sausages consumption. However, the introduction of particular microbial cultures deliberately added as starters to the raw meat for sausages manufacturing purposes, should be seen with caution in order to prevent the introduction of potential hazards and assure dry-fermented meat products safety.

LAB have the GRAS (generally regarded as safe) status provided by the US Food and Drug Administration (FDA) (Chamba and Jamet, 2008). In addition, the European Food Safety Authority (EFSA) has adopted a generic approach to the safety assessment of the microorganisms used in food/feed resulting in the application of the concept 'Qualified Presumption of Safety' (QPS) to a selected group of microorganisms including LAB (EFSA 2005). Thus, as LAB are inherent components of traditional fermented sausages, the QPS concept is applicable to the defined strains of this microbial group intended to be deliberately added as starters to the raw meat for sausages manufacturing purposes.

The safety assessment of the strains used as starter or protective cultures for dry-fermented sausages' production should discard the presence of antibiotic resistance genes in the selected strains to avoid their transmission to commensal or pathogenic bacteria (EFSA 2012, 2013, Fraqueza 2015). Likewise, strains should not produce other hazardous substances such as biogenic amines (EFSA 2011). Starter cultures for meat products manufacture should lack amino-acid decarboxylase activity. Fermented meat products are potentially at risk regarding biogenic amines, since they are rich in protein and use LAB and eventually coagulase negative *Staphylococcus* for their processing. Biogenic amines have been reported in several meat products that have a LAB microbiota, naturally present or inoculated as a starter culture (Latorre-Moratalla et al. 2010b). Usually biogenic amines production required the presence of bacteria that are capable of decarboxylating amino acids. Histamine and tyramine are probably the two most important biogenic amines of bacterial origin in food, due to their toxicological effects (Bover-Cid and Holzapfel 1999). When strains are assessed to be incorporated in fermented meat products, a screening for their (in)ability to mediate de production of biogenic amines should be performed (Latorre-Moratalla et al. 2010a, EFSA 2011, Tabanelli et al. 2014).

Different factors or circumstances could change the *"reasonable certainty of no harm"* and since LAB are present in human gastrointestinal tract, and are also intentionally added to the food use in the diet, concerns arouse about the antimicrobial resistance of these beneficial bacteria. Bacteria used as starter cultures or protective cultures for dry-fermented sausages' production could possibly contain antibiotic resistance genes, which might be transferred to commensal or pathogenic bacteria and this fact it is not accepted by EFSA. Antimicrobial resistance was introduced as a possible safety concern for the assessment and inclusion of bacterial species in the QPS list (EFSA 2012). QPS approach has proved to be a useful tool to harmonize and prioritize safety assessment linked to the taxonomy, familiarity, pathogenicity and end use of the microorganisms (EFSA 2005, 2013). Virulent factors, biogenic amines production (EFSA 2011) and antibiotic resistance determinants of human and veterinary clinical significance are requirements for QPS qualification, being imperative their absence.

According to Bourdichon et al. (2012), to include a microbial specie in a safety list, its presence should be documented, regarding not only the occurrence of a specific microorganism in a fermented food product, but also by providing evidences whether the presence of the microorganism is beneficial, fortuitous or undesired. EFSA (2013) has stated that a total of 35 *Lactobacillus* species can be considered to have QPS-status, being present in this group the most frequent species identified also in dry-fermented sausages such as *Lb. sakei*, *Lb. curvatus*, *Lb. plantarum*, *Lb. fermentum*, *Lb. brevis*, *Lb. rhamnosus* and *Lb. alimentarius*. In addition to *Lactobacillus* species, also other LAB species have been granted QPS-status. They include four Leuconostocs, (*Ln. citreum*, *Ln. lactis*, *Ln. mesenteroides* and *Ln. pseudomesenteroides*), three Pediococci (*Pd. acidilactici*, *Pd. dextrinicus* and *Pd. pentosaceus*), *Lactococcus lactis* and *Streptococcus thermophilus*. *Enterococcus faecium* is in the QPS list indicated as non plus (*Lactobacillus* is indicated as plus) in spite of the recent scientific knowledge allowing a differentiation of pathogenic from non-pathogenic strains. *E. faecium* can be used but it depends on assessment of each strain separately which has to be absent of virulence factor genes and transmissible antibiotic markers (EFSA 2011).

Jahan and Holley (2014) stated that *E. faecalis* strains more frequently carried virulence genes than *E. faecium* strains and that the temperature optimum for biofilm formation was strain rather than species dependent. In a study conducted by Jahan et al. (2015) e. g. the transfer of resistance determinants from *E. faecium* to both clinical *E. faecalis* and *E. faecium* strains via a natural conjugation mechanism was demonstrated and the possible mechanism was suggested to involve an integron. This is the first evidence of a class 1 integron being present in any food isolate, especially from an *E. faecium* strain obtained from dry sausage. Previously, class 1 integrons have only been detected in clinical isolates of *E. faecium* (Xu et al. 2010). These results reinforce the concern expressed elsewhere about the safety of enterococcal strains present in foods, particularly in commercially fermented meat where viable antibiotic resistant enterococci may be present.

Protective Cultures and Bacteriocins in Fermented Meats 241

Some yeast are in QPS list particularly *Debaromyces hansenii* with antimould activity and particular interest to be use on fermented and cured meat products surface. The inclusion of several yeast species in the QPS list is mainly based on the apparent history of safety. However, in recent update (EFSA 2013) some yeasts were removed from the list due to a study came up pointing to an additional factor that could contribute to virulence in opportunistic variants of e.g. *Saccharomyces cerevisiae* (ability to cope with oxidative stress (Llopis et al. 2012). More comparative studies of virulence factors in opportunistic yeasts are needed before a general picture can evolve.

4 CONTRIBUTION OF PROTECTIVE STARTERS TO SAFETY AND SHELF-LIFE IMPROVEMENT IN MEAT PRODUCTS

According to EFSA report (EFSA and ECDC 2015) the most common pathogens causing foodborne disease and associated to meat products are *Campylobacter* (causing campylobacteriosis), *Salmonella* (causing salmonellosis), *Listeria monocytogenes* (causing listeriosis), pathogenic *Escherichia coli*, *Yersinia* as well the toxins of *Staphylococcus aureus*, *Clostridium perfringens* and *Clostridium botulinum*. The control of pathogens on meat products should be thought on proactive measures coupled with technological interventions. In the last years, the protective role of LAB in fermented meat products in relation to pathogens has been highlighted and has been demonstrated by several authors.

In a fermented sausage manufacturing process, even considering the acidification achieved in certain processes, not all biological safety problems are completely controlled. This problem is particularly noticeable in fast-cured products, which are only slightly dried. Despite the control of pseudomonas and enterobacteriaceae achieved by the reduced a_w, preventing the deterioration of the product, the probability of survival, and possible toxinogenesis, by *L. monocytogenes*, *E. coli*, *Salmonella* spp., and *St. aureus* must be considered. As reviewed by Moore (2004) and Leroy et al. (2006) fermented meat products had been involved in punctual cases of foodborne poisoning outbreaks associated to the above-mentioned pathogens. On the other hand, several studies indicate that pathogens are found in ready to eat fermented sausage, which should be regarded as a concern (Talon et al. 2012, Henriques and Fraqueza 2014). In order to protect the health of consumers, EU set limits to the presence of some of those pathogens in foods. It is required the absence of *Salmonella* spp. in 25 g of product. Regarding *L. monocytogenes*, it is not mandatory its absence in 25 g, provided that the food operator ensures that the food does not allow growth of the pathogen during its shelf life and that it is not consumed by risk groups. When these assumptions are not guaranteed, the product must leave the responsibility of the food operator with *L. monocytogenes* absent in 25 g. Given the potential difficulty in managing these pathogens, particularly *L. monocytogenes*, it has been observed at the industry level a trend to pasteurize dry cured/fermented sausages, with potential losses at sensory quality level. The use of selected protective starter cultures may be an important strategy to adequately control pathogens present in dry-cured fermented sausage. It is not possible to individualize the mechanisms involved in that inhibition, once there is a complex interaction between different phenomena to the production of a combined effect. Among the different inhibitory tools presented by bioprotective cultures, which are modulated by genetic and environmental factors, the main are the direct competition in the meat product ecosystem, the organic acids such as lactic and/or acetic acid, hydrogen peroxide, reuterin and bacteriocins (O'Bryan et al. 2015).

4.1 Organic Acids

The acid production is might be a critical process in the manufacture of fermented meat products. It is a common practice to add fermentable carbohydrates to the mixture of raw materials, when the safety of the product relies on the reduced pH, alone or in combination with reduced a_w (Talon et al. 2012). Nonetheless the importance of acidification, to produce high sensory quality products it is not recommended a strong acidification, due to the impact, directly on the taste of the product, and the indirectly through the inhibition of several catabolic phenomena that are involved in the flavor development (Ravyts et al. 2010, Montanari et al. 2016).

When the reduced pH is determinant to ensure safety, two major practices are common. A rapid acidification, achieved by the addition of glucose, that is readily metabolized; or a slower acidification, using disaccharides, such as lactose or sucrose (Petaja-Kanninen and Puolanne 2007). Most LAB ferment glucose to lactic acid in normal substrate and limited access to oxygen, using the homofermentative pathway. However, even for strains that are normally homofermentive, there are several factors influencing this pattern, namely: the availability or type of carbohydrate, the pH of the medium and the presence of oxygen. Any deviation from the homofermentative path might result in the formation of fermentation products of other than lactic acid. That shift from homofermentaive path to a mixed fermentation results in the accumulation of acetic acid, formic acid, ethanol, carbon dioxide and other compounds (Gänzle 2015).

Fermented sausages present a characteristic evolution of pH (Figure 2), with an initial decrease, due to fermentation of carbohydrates, mainly hexoses. This initial decrease is a key factor in the control of unwanted spoilage and pathogen microbiota in this step of the process. After the sharp initial reduction, it is usually observed a slight increase in pH. This is due to the diffusion of acid in the meat and to the production of alkaline compounds during the drying of the sausage, mainly from the NPN fraction (Ikonić et al. 2013). The pH reduction is associated with modification of the water binding capacity of proteins favoring dehydration and the cohesion of the sausage, resulting in a firm and sliceable texture. The presence of acids interferes with the characteristic flavor and aroma of the product, however, where excessive amounts originate abnormal flavor (Cocconcelli and Fontana 2008).

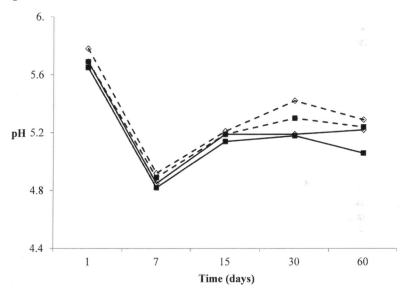

Figure 2 Evolution of pH in Portuguese *chouriço* inoculated with a starter culture of L. sakei (■) or non-inoculated (◇), with (—) or without (- - -) 2% of carbohydrates (glucose+lactose) during the fermentation and drying period (adapted from Diez and Patarata 2013).

The antimicrobial effect of organic acids is due mainly to its lipophilic nature and inherent ability to cross the cell membrane undissociated form and by interference on membrane proteins responsible for the transport of nutrients and reduction of ATP production capacity. In the cytoplasm the acid dissociates, and the cell must export the H+ using an ATP consuming pump. The bacteriostatic effect is achieved by that continuous consumption of energy. If the exporting process is insufficient to maintain the intracellular equilibrium, the bactericidal effect is achieved as a result of structural damage on the cell membrane and macromolecules such as DNA and proteins that compromise the viability of the bacteria (Reis et al. 2012, O'Bryan et al. 2015). The resistance of LAB to the low pH arises from several mechanisms. The first line of defense is the one that is common to all bacteria – the ATP dependent proton pump. Additionally LAB may attenuate the effect of high concentration of H^+ by the production alkaline compounds, mainly from amino acids. Beyond these strategies, the protection might also be achieved by the synthesis of protective proteins and changes in cell wall composition. These mechanisms balance the negative impact of lower intracellular pH, which can lead to loss of activity of the glycolytic enzymes – which are relatively sensitive to pH reduction – which would compromise the capacity to produce ATP and irreparably compromise the primary mechanism of defense to the pH reduction, the ATP-dependent proton pump (Cotter and Hill 2003).

4.2 Bacteriocins

The role of bacteriocins in nature has been wildly debated, but it is accepted that its products benefits the producer organism with an ecological advantage over their most probable competitor (Zacharof and Lovitt 2012). These antimicrobial peptides are usually active against a specific group of microorganisms, usually taxonomically close to the producer, but there are also bacteriocins that have a broad spectrum of antimicrobial activity, that are probably the general mechanism of defense and competition. It is believed that all members of *Eubacteria* and *Archea*, when recently isolated from their ecosystems have the ability to express their characteristic bacteriocins. The most part of the bacteriocins described in the literature are produced by species of *Lactobacillus, Lactococcus, Enterococcus, Pediococcus*, and *Leuconostoc* (Heng et al. 2007, Papagianni and Sergelidis 2013).

Bacteriocins are a heterogeneous group of compounds of peptidic nature. Despite their common characteristic of inhibiting other organisms, they have considerable differences on their genetic, structure, target microorganisms and mode of action. Several attempts have been made to group bacteriocins. Klaenhammer (1993) proposed a classification in four classes that is still generally accepted nowadays, although several modifications and adaptations had been proposed. These four classes are composed by the Class I, designated by lantibiotics that are small membrane active peptides (< 5 kDa) characterized by the presence of lanthionine, β-methyl lanthionine. The Class II is composed by heat stable, non-lanthionine containing peptides (< 10 kDa). That class is subdivided into three subgroups: IIa (anti-listeria peptides), IIb (porotion

Protective Cultures and Bacteriocins in Fermented Meats 245

complexes consisting of two peptides for activity) and IIc (thiol-activated peptides requiring reduced cysteine residues for activity). The Class III includes proteins of higher molecular weight (> 30 kDa) and heat-labile and Class IV more complex molecules that includes carbohydrates or lipids on its composition to achieve the required activity. A fifth class was further proposed by Kemperman et al. (2003) composed by cyclic peptides. Heng et al. (2007) noted that this classification scheme did not have sustainable evidence fulfilling the criteria for Class IV, and it has been considered to withdraw it of the classification of Klaenhamemer (1993). Maqueda et al. (2008) proposed to upgrade to a new class IV for circulins – the circular, posttranslationally modified bacteriocins. From the several attempts to revise the classification of Klaenhamemer (1993), one that has been widely accepted is that proposed by Cotter et al. (2005). These author proposed a simpler classification dividing the bacteriocins only in two major groups: Class I – lantibiotics, corresponding to the lanthionine-containing bacteriocins and Class II – non-lanthionine-containing bacteriocins. The former Class III composed by high molecular weight heat-labile proteins is proposed to be reclassified as bacteriolysins. Nontheless, Heng and Tagg (2006) suggest keeping the Class III of large heat labile proteins. The allocation of Circulins, has been in discussion. Cotter et al. (2005) suggest including them in the Class II while the former authors classify them in a new Class IV that had been proposed to be due to the lack of consolidated evidences of its importance. The majority of the bacteriocins of interest for potential application by the food industry fall into Classes I and II (Deegan et al. 2006).

Lactic acid bacteria bacteriocins exert their antimicrobial effect mainly by damages in the cell membrane. Some bacteriocins, like nisin, have a broad spectrum of activity against most Gram positive bacteria, while other presents only a narrow spectrum of activity, commonly against the species of the producer, as observed for lactocin A (Drider et al. 2006). The mechanism of action involves several steps, first the bacteriocin is attracted to the target cell by electrostatic attraction. The action of the bacteriocin in the target cell involves the pore formation, dissipation of proton motive force and loss of capacity to maintain the pH equilibrium in the cytoplasm, depletion of ATP and amino acids and ions leakage resulting in the lethal effect (Balciunas et al. 2013). The effects of bacteriocins on cell envelop and on pH homeostasis results in an increased global antimicrobial effect in fermented products due to the simultaneous production of bacteriocins and organic acids. Even if the cell succeed repairing the cell envelop damages, the concomitant production of organic acids and the absolute need to betake to ATP-dependent proton-pump to avoid cytoplasm acidification reduces the odds to survive (Deegan et al. 2006, Reis et al. 2012). Concomitantly with the damages in the cell envelope, some bacteriocins also have the ability to interfere with the cell wall synthesis through the binding of the lipid II, the precursor of peptidoglycan. This antimicrobial action based in a dual mechanism is responsible for the high efficiency of those bacteriocins that manifest its activity even at nM concentrations (Cleveland et al. 2001).

The activity of LAB bacteriocins is generally directed against low-GC Gram positive species. Once the target of bacteriocin is the cell membrane, the outer lipopolysaccharide layer of Gram negative bacteria exerts a protective effect against the action of bacteriocins. However, it has been experimentally observed inhibitory action due to the action of bacteriocins. That inhibition might be increased if any strategy is used to compromise the integrity of the outer layer of the Gram negative membrane (Heng et al. 2007). The use of chelating agents has been shown to improve the sensitivity of Gram negative bacteria, due to the chelation of the divalent cations of the lipopolysacharid layer. Among these chelating agents, ethylene-diaminetetraacetate (EDTA) has shown to be very effective in improving the sensibility of Gram-negative bacteria to the action of bacteriocins (Castellano et al. 2011). Other strategies that might be used to affect outer membrane and increase the sensitivity of Gram-negatives to the action of bacteriocins might include heat shock, high pressure or specific enzymes (Heng et al. 2007).

Bacteriocin producers have a mechanism to protect themselves from the action of their bacteriocins, through the production of dedicated immunity proteins. Two types of immunity have been described for lantibiotics, one is based on a specific immunity protein and the other is composed by a multicomponent transporter mechanism. Depending on the producer microorganism and bacteriocin involved, immunity might be achieved by only one mechanism or for both of them. The single component mechanism acts preventing the insertion of the lantibiotic in the cell membrane, while the multicomponent ones probably act by expelling the bacteriocin from the cell membrane. Lantibiotics generally have a one component mechanism based on immunity proteins (Cotter et al. 2005). Several research works have been made on the screening and identification of bacteriocins of LAB with interest in meat products particularly by *Lb. sakei*, *Lb. plantarum*, among others (Table 2).

The works summarized in Table 2 involved different methodological approaches to study the antagonistic activity. Generally in a first phase it is established the ability of LAB to inhibit sensitive bacteria. These sensitive bacteria are usually pathogens, spoilage bacteria or other LAB. The screening for antagonistic activity is based on phenotypic methods involving spot on agar and well diffusion tests based on the procedures described by Tagg et al. (1976) with small variations introduced by each author. To evaluated the production of bacteriocin like inhibitory

Protective Cultures and Bacteriocins in Fermented Meats 247

Table 2 Examples of bacteriocins identified in *Lb. sakei, Lb. plantarum, Lb. curvatus, Lc. lactis* and *Pediococcus* spp. isolated from meat and meat products

Bacteriocin Producer	Sensitive microorganism	Reference
Sakacin *Lb. sakei*	*L. monocytogenes* *St. aureus* *Enterococcus* spp. *B. thermophacta* *Pseudomonas* spp. *Campylobacter* spp. *E. coli* *Klebsiella* spp. Other LAB	Tichaczek et al. 1992, Aasen et al. 2000, De Martinis and Freitas 2003, Alves et al. 2006, Urso et al. 2006, Dortu et al. 2008, Jones et al. 2008, Castro et al. 2011, Vesković-Moračanin et al. 2010, Hartmann et al. 2011, Todorov et al. 2013, Barbosa et al. 2014
Plantaricin *Lb. plantarum*	*L. monocytogenes* *St. aureus* *Cl. perfringens* *Cl. tyrobutyricum* *B. cereus* *Enterococcus* spp. *B. thermophacta* *Pseudomonas* spp *Salmonella* spp. *E. coli* Other LAB	Kanatani and Oshimura 1994, Enan et al. 1996, Suma et al. 1998, Messi et al. 2001, Essid et al. 2009, Todorov 2009, Hartman et al. 2011, Fontana et al. 2015
Curvacin *Lb. curvatus*	*L. monocytogenes* *St. aureus* *B. thermophacta* *Pseudomonas* spp *E. coli* Other LAB	Tchikazec et al. 1992, Verluyten et al. 2004, Cocolin and Rantsiou 2007, Hartman et al. 2011, Barbosa et al. 2015
Nisin *Lc. lactis*	*L. monocytogenes* *St. aureus* *Cl. tyrobutyricum* Other LAB	Dal Bello et al. 2010, Biscola et al. 2013
Pediocins *Pediococcus* spp.	*L. monocytogenes* *Enterococcus* spp. Other LAB	Abrams et al. 2011, Engelhardt et al. 2015

substances (BLIS) the supernatant of the producer is treated to eliminate the effect of acid, neutralized with a base or dialyzing it in a neutral buffer, and the effect of hydrogen peroxide is eliminated by its degradation with catalase before the application of the supernatant in the wells, or introducing the catalase in the agar medium used for the screening. The peptide nature of the compound is evaluated by the study of its resistance to heat and by the use of proteolytic enzymes (Hartman et al. 2011). Among the bacteria tested as sensitive, there is clear interest for *L. monocytogenes*. This pathogen is reputably difficult

Table 3 LAB protective starters applied to meat products

Meat product	Main process conditions	Major function	Mechanism involved	Level of inoculation	Species tested	Strain indicated as protective starter	References
Dry-fermented sausages Salami	Manufacturing process: dried at 20–18°C and 65–60% RH for 3 d and finally ripened at 14°C and 76% RH for 28 d up to moisture content of 27–34%	Inhibition of staphylococci, *Enterobacteriaceae* and *L. monocytogenes*	Numeric competition, bacteriocin activity	10^7 cfu/g of product	*Lb. curvatus* 54M16	*Lb. curvatus* 54M16	Casaburi et al. 2016
Dry-cured meat products	*In vitro*	Antimold activity	Compete with *Penicillium nordicum* and to inhibit ochratoxin A (OTA) accumulation	10^4, 10^5 and 10^6 cells/mL	Ten *Debaromyces hansenii* strains	*D. hansenii*	Andrade et al. 2014
Fermented sausages	Fermentation at 22°C with a RH=90–95% for 2 d, and then ripened at 12°C with a RH = 70–75% for 36 d	Control the spoilage and pathogenic microorganisms,	Numeric competition, pH decrease, bacteriocin activity	10^5–10^7 cfu/g	*Lb. sakei* C2	*Lb. sakei* C2	Gao et al. 2014

Dry-cured meat products	In vivo	Antimold Activity	Competition	10^6 cells/cm^2 on meat sample surface	Two strains of D. hansenii (code: D. hansenii 78 and D. hansenii 147) and one strain of H. burtonii (code: H. burtonii 120)	D. hansenii 147	Simoncini et al. 2014
Frozen ground-beef patties	5°C during 9 d With a chelator Na2EDTA	Reduction of E. coli O157:H7	Bacteriocin activity	10^7 cfu/g of ground-beef	Lb. curvatus CRL705 and Lc. lactis CRL1109	Lb. curvatus CRL705 and Lc. lactis CRL1109 with a chelator Na2EDTA	Castellano et al. 2011
Vacuum-packaged sliced beef meat	Vacuum packaged and stored at 4 ± 1°C during 28 d	Inhibition of spoilage bacteria, improvement of shelf life product	Antimicrobial peptides BLIS production pH decrease	$10^{8.5}$ cells/g	Lb. sakei CECT 4808 Lb. curvatus CECT 904T	Lb. sakei CECT 4808	Katikou et al. 2005
Cooked, sliced, and vacuum-packed meat products	Commercially sliced and vacuum-packed meat products such as ham, salami, cooked loin, and smoked bacon	Inhibition of L. monocytogenes	Production of leucocins A and B –4010	$10^{7.8}$ and $10^{8.5}$ Ln. carnosum /g product	Ln. carnosum Carnobacterium piscicola	Ln. carnosum 4010	Budde et al. 2003

Table 3 (*Contd.*)

Table 3 LAB protective starters applied to meat products (*Contd.*).

Cooked ham	Applied before slicing and vacuum-packaging by spraying bottle	Inhibition of *L. monocytogenes*	Not specified	10^5 cfu/g of product	*Lb. sakei*		*Lb. sakei* TH1	Bredholt 2001
Spanish-style dry-fermented sausage	Fermentation was performed at 22 to 24°C, and 90–95% RH. Sausages were dried at 15°C and and 70–80% RH during 4 wk	Inhibition of *L. monocytogenes*	Competitive during meat fermentation, production of enterocins *Ent* 4231	10^6 cfu/g of meat batter	Strains *E. faecium* RZS C13 and *E. faecium* CCM 4231 and *Lb. sakei* CTC 494		*E. faecium* RZS C13 and *E. faecium* CCM 4231	Callewaert et al. 2000

Protective Cultures and Bacteriocins in Fermented Meats 251

to control in dry meat products, particularly because of its ability to survive and growth in products with a reduced water activity (a_w) that are as reduced as 0.79 to 0.86 to survival and 0.92 to growth (Rocourt and Cossart 1997). Additionally, the psychrotrophic character of this pathogen contributes to the difficulty on its control in the meat products industry. Due to its ubiquitous nature, it should be always considered the possibility of its presence in the raw materials or, eventually, in the industry environment if failures on the cleaning and disinfection procedures occurs (Fraqueza and Barreto 2015). Additionally, once *L. monocytogenes* is a Gram-positive taxonomically close to LAB it responses very well to the effect of bacteriocins (Engelhardt et al. 2015).

4.3 Control of Biological Hazards and Shelf-Life Improvement

Some examples of the direct application of bioprotective cultures in fermented meat products are shown in Table 3. The different mechanisms to achieve pathogens or spoilage microbiota reduction are a part of pH reduction particularly related to the production of bacteriocins active against pathogens. The direct inoculation, usually at 10^7 cell/g, of specific strains of *Lb. sakei*, *Lb. curvatus*, *Leuconostoc*, *Enterococcus* on meat products is frequent as presented on Table 3.

LAB have been found to be effective in inhibiting *L. monocytogenes* in cooked meat products (Hugas 1998, Bredholt et al. 2001, Koo et al. 2012) and in vacuum packaged meats (Juven et al. 1998). The use of protective cultures by the meat industry is increasing because of the promising results obtained particularly in *L. monocytogenes* reduction and shelf-life extension. A strain of *Lb. sakei*, previously isolated from cooked ham was able to inhibit growth of *L. monocytogenes* and *E. coli* O157:H7 in cooked ham (Bredholt et al. 2001). *Lb. sakei* was applied to the cooked products at a concentration of 10^5–10^6 cfu/g immediately before slicing and vacuum-packaging using a hand-operated spraying bottle. The LAB strain inhibited growth of 10^3 cfu/g of a cocktail of three rifampicin resistant mutant *L. monocytogenes* strains both at 8°C and 4°C. The products were still acceptable after storage for 28 days at 4°C and, after opening the packages, for a further 5 days at 4°C.

Moreover, there also have been several studies on the potential use of lactic acid bacteria to prevent the growth of *L. monocytogenes* in fermented meat products: *Lb. curvatus* 54M16 used in salami (Casaburi et al. 2016), *Lb. sakei* C2 used in dry-fermented sausages (Gao et al. 2014), among others examples given in Table 3. Callewaert et al. (2000) used as starter cultures for the production of a Spanish-style dry-fermented sausage the strains *E. faecium* RZS C13 and *E. faecium* CCM 4231. These *Enterococcus* strains were competitive during meat fermentation and inhibited the growth of *Listeria* spp. due to the presence of enterocins. Likewise, Lauková et al. (1999) had reported the effectiveness of *Ent* 4231 (produced by the earlier-mentioned *E. faecium* CCM 4231) in controlling *L. monocytogenes* contamination in dry-fermented Slovak salami Hornad or *Listeria innocua* growth in Púchov salami (Lauková and Turek 2011). Ent M produced by *E. faecium* strain AL41 was succesfully used to decrease *L. innocua* cells in Slovak Gombasek sausage (Lauková et al. 2003). Budde et al. (2003) have been used *Leuconostoc carnosum* 4010 on commercially sliced and vacuum-packed meat products such as ham, salami, cooked loin, and smoked bacon assuring a decrease of *L. monocytogenes* related to the production of leucocins A and B- 4010.

Yeasts are mentioned on meat products particularly related to flavor development or with antimold properties. *Candida famata* and *D. hansenii* are frequently related to flavor development of fermented meat products. *D. hansenii* are active against molds as demonstrated on works of Andrade et al. (2014) and Simoncini et al. (2014).

Beside many studies have been conducted and protective achievement of the strains demonstrated the commercial implementation of the starter strains have not been successfully done in the majority of the cases. While for dry-fermented cured products such salami, chorizo or similar fermented meat products the use of a starter cultures with protective properties should be unproblematic with regard to both application and legislation. These protective cultures applied on other meat products, such as fresh and cooked meat products could be problematic. Starter cultures should not be considered preservatives or as additives, but rather as a hurdle technology system: biopreservation. In not fermented meat products the technical requirement of adding an extra processing step after cooking to use the culture, could be a hindrance for some industries. The application of protective cultures to meat products, such as bacon, smoked filet, emulsion products and other cooked meat products, is not part of general starter culture acceptance. Supervisory authorities could not accept the high initial level of LAB in such meat products. However, there is a strong opinion regarding consumers' acceptability of the use of protective cultures as a natural way to ensure the quality and safety of food products.

Nevertheless, some renamed enterprises have patented starter protective cultures and commercialize than for fermented products. Bactoferm™ F-LC is a patented culture blend presented by the Danish manufacturer "Chr. Hansen", capable of acidification as well as preventing growth of *L. monocytogenes* due to the culture production of pediocin and bavaricin. This blend is composed by *St. xylosus, Pd. acidilactici* and *Lb. curvatus*. Low fermentation temperature (<25°C, 77°F) results in a traditional acidification profile whereas high fermentation temperature (35–45°C, 95–115°F) gives a US style product. Also, Lyocarni

Protective Cultures and Bacteriocins in Fermented Meats 253

VBL–53 from Clerici-Sacco Group consists of *St. xylosus*, *St. carnosus*, *Lb. sakei*, and *Lb. curvatus* which in combination ensure a uniform and controlled production of fast fermented salami with added safety. BiobaK L is another blend starter and maturing cultures for the reduction of *Listeria* sp. during the fermentation process of raw sausages, presented by Wiberg®. DANISCO from ®DuPont Nutrition & Health, also presents antimicrobial blends named NovaGARD and MicroGARD antimicrobial blends (MicroGARD® 730) for cooked and cured meat products (DuPont Nutrition & Health, 2015). The great constraint of these commercial starters is the fact that there is a lack of adaptation to the traditional fermented meat products with unavoidable differences on sensorial quality presented by these products (Urso et al. 2006).

The application of antimicrobial metabolites bacteriocins without the producing strain could be a strategy particularly for Enterococcus strains with all related constraints of virulence factor, mobile antibiotic resistance elements and high amino decarboxylase ability with production of biogenic amines.

The purification of bacteriocins enhanced is capability to inhibit pathogen indicators. Nisin is one of the purified bacteriocins that have been accepted to be use on food as an additive (but not commercially in meat). Also pediocin PA1 is available on the market as food additive. LAB bacteriocins do not confer taste or odor to meat products being considered safe for humans and with GRAS status. Nichie et al. (2012) performed a completed overview about bacteriocins and their potential application. All the procedure to detect, purify and identify novel type of antimicrobial peptides has shown to be tedious, time-consuming and expensive. With the recent technological advances in mass spectrometry, genomics and informatics, a new proteogenomic approach has been developed to enable us performing peptide identification with high sensitivity for the discovery of microorganisms that produce both known and putatively novel bacteriocins based on their molecular mass, antibacterial spectra and selectivity.

5 PROTECTIVE STARTERS AND BACTERIOCINS WITH EMERGENT TECHNOLOGIES TO IMPROVE SAFETY AND MEAT PRODUCTS' SHELF-LIFE

Nowadays, the combination of emergent technologies with conventional methods seems to be an appropriate choice to assure meat products safety and extend their shelf life. The application of protective cultures or their metabolites can be complemented with active packaging, active coatings, encapsulation, or associated with athermic technologies aiming to maintain the singular sensorial characteristics of meat products.

5.1 Active Packaging

Active packaging is a system in which the food, the package, and the environment interact in a positive way to extend shelf life, improve the condition of packaged food, or to achieve food safety and maintenance of sensory properties (Realini and Marcos 2014). According to Vanderroost et al. (2014) this system could be coupled with the next generation of food packaging that will be intelligent, called "smart package" giving important information on the food condition or packaging integrity not only beneficial for the customer, but also enables the detection of calamities and possible abuse through the entire supply chain, from farm to fork. This will require the development of new technologies, e.g. printed electronic systems with integrated sensors which are connected to other, yet to be developed, devices enabling for example the release or absorption of substances or activating certain non-thermal preservation processes through different kinds of techniques, including UV irradiation, gamma irradiation, ultrasound treatment, and application of high-voltage pulsed electric fields (PEFs) (Ortega-Rivas 2012).

For the time being, active packaging technologies include a large variety of possibilities and are gaining relevance in the meat industry. Antimicrobial packaging appears to be one of the most promising applications of active food packaging technology, allowing for spoilage microorganisms and pathogens control (Chen and Brody 2013). Some active packaging solutions use multiple function active systems, such as the combination of oxygen scavengers with carbon dioxide and/or antimicrobial releasing systems, while others have slow releasing behavior (Chen and Brody 2013, Ozdemir and Floros 2008). Sánchez-Ortega et al. (2014) presents a complete review of antimicrobial edible films and coatings for meat and meat products preservation based on different active compounds and coating material. A rapid growth is anticipated for antimicrobial packaging in the near future, especially with LAB and nisin, when cost and performance factors limitations are overcome (Realini and Marcos, 2014). Marcos et al. (2013) was able to reduce *L. monocytogenes* growth during shelf-life of sliced dry-fermented sausages active packaged in polyvinyl alcohol films impregnated with the bacteriocin nisin. In another study using Wiener sausages packed in polymer films containing lactocins, Massani et al. (2014) were able to demonstrate growth control of *L. innocua*. Basch et al. (2013) reported the synergistic effect of nisin and potassium sorbate incorporated in tapioca starch–hydroxypropyl methylcellulose films against *L. innocua* and *Zygosaccharomyces bailii in vitro*. The main drawback of using commercial extracts of nisin is their low content of nisin (2.5% Delvo® Nis, Royal DSM and Nisaplin®, Danisco), therefore big amounts of the extract need to be added to the films to obtain an antimicrobial effect. Many research

Protective Cultures and Bacteriocins in Fermented Meats 255

efforts have been done on the development of antimicrobial films using other semi-purified bacteriocins. Among them, enterocins produced by *E. faecium* have proved to be effective for controlling *Listeria* spp. growth in meat products (Aymerich et al. 2000a, b, Vignolo et al. 2000, Ananou et al. 2005a,b, Aymerich et al. 2006b, Marcos et al. 2007). Pullulan films containing sakacin A (1 mg/cm^2) produced by *Lb. sakei* also proved to be effective against *L. monocytogenes* growth, inducing a 3 log reduction in turkey breast during 3 weeks of refrigerated storage (Trinetta et al. 2010). Pediocin has also been tested on sliced ham, overlapping the slices of ham with the films (control, 25% and 50% of pediocin), with a reduction of 2 log cycles of *L. innocua* after 15 days of storage (Santiago-Silva et al. 2009)

The big challenge for active packaging is to develop active materials able to preserve their original mechanical and barrier properties and also being biodegradable. An active packaging film based on whey protein isolate was developed by incorporating nisin to promote microbial food safety on meat products such as cooked ham (Rossi-Marquez et al. 2009). The addition of purer active substances rather than the use of non-purified extracts or the use of active compounds in the form of nanoparticles would reduce the amount of required active substance and would therefore contribute to preserve the original properties of the material.

Antimicrobial packaging solutions for meat products are in the market being presented by some enterprises but only few presented bacteriocins based systems. SANICO® developed a natamycin based antifungal coating for sausages (Laboratoires STANDA 2015).

5.2 Encapsulation

Encapsulation or entrapment of LAB has been used to improve viability of cells in the intestinal tract and in food products such as yoghurts, cheeses, cream and fermented milk (Rathore et al. 2013), to our knowledge the study of microencapsulated protective starters and its application to meat products have been scarce. This technology has been more studied with probiotic bacteria applied in dairy products (Khan et al. 2011, Burgain et al. 2011).

Microencapsulation has the capacity to maintain the viability, activity and functionality of bacteria. This approach enhances bacterial cell tolerance to alcohols, phenols, antibiotics or quaternary ammonium sanitizers (Trauth et al. 2001, Lacroix et al. 2005) and resistance to adverse processing techniques such as freezing (Sheu et al. 1993) and freeze-drying (Kearney et al. 1990) and hostile environments such as simulated gastric environment (Cui et al. 2000). The most frequently used materials for the microencapsulation include: alginate, starch, gelatin and casein. It is common the encapsulation/entrapment of LAB with calcium alginate, a nontoxic linear heteropolysaccharide extracted from different types of algae (Cook et al. 2012). The alginate recovers the bacterial cells and forms a barrier, protecting them against environmental conditions. The alginate barrier constitutes a semi-permeable spherical fine coat, which nutrients and metabolites readily cross (Kailasapathy 2002, Anal and Singh 2007). Some studies have shown that entrapment of LAB in calcium alginate improves lactic acid production (Idris and Wahidrin 2006, Rao et al. 2007), but little is known on bacteriocin production by entrapped LAB. However, the prodution of calcium alginate beads is able to deliver *Lactobacillus* spp. with preserved viability and antibacterial activity against several multi-resistant strains.

The potential use of microencapsulation to protect the bacteria in meat products was investigated but with a sense of probiotic bacteria protection (Muthukumarasamy and Holley 2006, 2007). The introduction of encapsulated probiotic cells into dry-fermented sausages did not affect sensory properties of the product. The viability of encapsulated probiotic cells was improved compared to free cells. It was demonstrated that probiotics could reduce *E. coli* O157:H7 in number, but the microencapsulation decreased this potential.

5.3 High Hydrostatic Pressure

High hydrostatic pressure (HPP) is an emerging technology with an important potential to be used in ready to eat foods, as fermented sausages. It represents a valuable alternative to thermal pasteurization, assuring the control of specific pathogens without the sensory depreciation caused by the use of heat (Jofré et al. 2009). The diversity of microorganism that compose the microbial ecosystem of a food, particularly in those where there is a positive technological microbiota, the effect of HPP should be correctly evaluated to avoid an unexpected reconfiguration of the microbiota, with potential flaws in the preservation, namely by the overgrowth of barotolerant spoilage microorganism, with consequences in shorten shelf life, or worst, the loss of safety, due to the survival or overgrowth of pathogens (Huang et al. 2014). Considering the knowledge on the composition of the microbiota of fermented meat products, major problem that industrials have to face is the survival of low infectious doses *Salmonella* spp. and *E. coli* O157:H7. As they are gram-negative, it is possible to inactivate them with moderate HPP treatments. Furthermore, experimental evidences has shown that gram-negative bacteria, usually with considerable resistance to antimicrobials, became more sensitive due to the multiple injuries produced by the HPP, namely to the action of bacteriocins (Govers and Aertsen 2015). Another pathogen of high concern in meat products industry is *L. monocytogenes*.

Once it is a gram-positive, it is more difficult to control only by HPP, being necessary to use higher pressure to kill it. However, as discussed earlier, this pathogen is of particular sensitivity to bacteriocins, allowing designing pressure processing that ensures the survival of bacteriocin-producing LAB. With that combination, even though the survival of *L. monocytogenes* to HPP, the additional hurdle that bacteriocins represent will be responsible to its control during the post-processing period. Moreover, it is believed that HPP acts synergistically with bacteriocins, improving its diffusion on the food matrix and inherently increasing the probability of having the bacteriocin in contact with the target pathogen and due to the injuries in cell membrane produced by the HPP that is the main site of action of bacteriocins (Ojha et al. 2015). Several authors had promising results applying HPP simultaneously with purified bacteriocins, particularly nisin and enterocins, in sliced meat products (Jofré et al. 2009, Liu et al. 2012, Alba et al. 2013, Marcos 2013). As bacteriocins are very simple peptides, particularly those from classes I and II, that primary chain is composed by covalent bonds of amino acid sequences is not affected by the HPP (Huang et al. 2014), maintaining the activity after the exposure to high pressure. The use of HPP combined or not with other technologies has been experimented by several research teams, presenting promising results, some of them patented and in industrial application. A very complete revision on these works is presented by Oliveira et al. (2015).

6 CONCLUSION

The use of protective cultures in the manufacturing of fermented meat products is a well-established technology. Many evidences demonstrate that the use of starter cultures with specific proprieties contributes to improve the safety of fermented sausages. The inoculation of foods with microorganisms implies that no doubt should prevail on its safety for human health implications, namely concerning the antibiotic resistance of these microorganisms and absence of virulence factors. Among the different attributes of protective microbiota, the production of bacteriocins active against pathogens, particularly *L. monocytogenes*, have an important contribution to the safety of fermented meat products and meat products. Several starter cultures with protective capabilities are commercialized worldwide and processing using bacteriocins or bacteriocin-producing LAB are well established in the industry. The combination of protective cultures with traditional methods or with emerging technologies has been tested and the results are satisfactory. The use of protective cultures assures safety and improves shelf life of meat products without compromising their sensory attributes and nutritional value.

258 *Fermented Meat Products: Health Aspects*

Acknowledgments: This work was supported by national funds through FCT- Fundação para a Ciência e a Tecnologia (FCT) project "Portuguese traditional meat products: strategies to improve safety and quality" (PTDC/AGR-ALI/119075/2010).

Key words: *protective cultures, bacteriocins, biodiversity, safety requirements, shelf-life, biological hazards, active packaging, encapsulation, high hydrostatic pressure*

REFERENCES

Aasen, I.M., T. Møretrø, T. Katla, L. Axelsson and I. Storrø. 2000. Influence of complex nutrients, temperature and pH on bacteriocin production by *Lactobacillus sakei* CCUG 42687. Appl. Microbiol. Biotechnol. 53: 159–166.

Abrams, D., J. Barbosa, H. Albano, J. Silva, P. Gibbs and P. Teixeira. 2011. Characterization of bacPPK34 a bacteriocin produced by *Pediococcus pentosaceus* strain K34 isolated from "Alheira". Food Control 22: 940–946.

Adiguzel, G.C. and M. Atasever. 2009. Phenotypic and genotypic characterization of lactic acid bacteria isolated from Turkish dry fermented sausage. Rom. Biotech. Lett. 14: 4130–4138.

Alba, M., D. Bravo and M. Medina. 2013. Inactivation of *Escherichia coli* O157:H7 in dry-cured ham by high-pressure treatments combined with biopreservatives. Food Control 31: 508–513.

Albano, H., C. van Reenen, S. Todorov, D. Cruz, L. Fraga, T. Hogg, L. Dicks and P. Teixeira. 2009. Phenotypic and genetic heterogeneity of lactic acid bacteria isolated from "Alheira", a traditional fermented sausage produced in Portugal. Meat Sci. 82: 389–398.

Alves, V.F., R. Martinez, M. Lavrador and E. De Martinis. 2006. Antilisterial activity of lactic acid bacteria inoculated on cooked ham. Meat Sci. 74: 623–627.

Anal, A.K. and H. Singh. 2007. Recent advances in microencapsulation of probiotics for industrial applications and targeted delivery. Trends Food Sci. Tech. 18: 240–251.

Ananou, S., A. Gálvez, M. Martínez-Bueno, M. Maqueda and E. Valdivia. 2005a. Synergistic effect of enterocin AS-48 in combination with outer membrane permeabilizing treatments against *Escherichia coli* O157:H7. J. Appl. Microbiol. 99: 1364–1372.

Ananou, S., M. Garriga, M. Hugas, M. Maqueda, M. Martínez-Bueno, A. Gálvez, and E. Valdivia. 2005b. Control of *Listeria monocytogenes* in model sausages by enterocin AS-48. Int. J. Food Microbiol. 103: 179–190.

Ananou, S., M. Maqueda, M. Martínez-Bueno and E. Valdivia. 2007. Bio-preservation, an ecological approach to improve the safety and shelf-life of foods. pp. 475–486. *In*: A. Mendez Vilas (ed.). Communicating current research and educational topics and trends in applied microbiology. Formatex.

Andrade, M., L. Thorsen, A. Rodríguez, J. Córdoba and L Jespersen. 2014. Inhibition of ochratoxigenic moulds by *Debaryomyces hansenii* strains for biopreservation of dry-cured meat products. Int. J. Food Microbiol. 170: 70–77.

Axelsson, L. 2004. Lactic Acid Bacteria: Classification and Physiology. pp. 1–66. *In*: S. Salminen, A. von Wright and A. Ouwehand (eds.) Lactic Acid Bacteria. Microbiological and functional aspects. Marcel Dekker, New York.

Aymerich, T., M. Garriga, J. Ylla, J. Vallier, J.M. Monfort and M. Hugas. 2000a. Application of enterocins as biopreservatives against *Listeria innocua* in meat products. J. Food Protect. 63: 721–726.

Aymerich, T., M.G. Artigas, M. Garriga, J.M. Monfort and M. Hugas. 2000b. Effect of sausage ingredients and additives on the production of enterocin A and B by *Enterococcus faecium* CTC492. Optimization of in vitro production and anti-listerial effect in dry fermented sausages. J. Appl. Microbiol. 88: 686–694.

Aymerich, T., B. Martín, M. Garriga and M. Hugas. 2003. Microbial quality and direct PCR identification of lactic acid bacteria and nonpathogenic staphylococci from artisanal low-acid sausages. Appl. Environ. Microbiol. 69: 4583–4594.

Aymerich, T., B. Martin, M. Garriga, M.C. Vidal-Carou, S. Bover-Cid and M. Hugas. 2006a. Safety properties and molecular strain typing of lactic acid bacteria from slightly fermented sausages. J. Appl. Microbiol. 100: 40–49.

Aymerich, T., M. Garriga, A. Jofré, B. Martín and J.M. Monfort. 2006b. The use of bacteriocins against meat-borne pathogens. pp. 371–399. *In*: L. M. L. Nollet and F. Toldrà (eds.). Advanced technologies for meat processing. Marcel Dekker, New York.

Balciunas, E.M., F. Martinez, S. Todorov, B. Franco, A. Converti and R. Oliveira. 2013. Novel biotechnological applications of bacteriocins. A review. Food Control 32: 134–142.

Barbosa, M.S., S.D. Todorov, Y. Belguesmia, Y. Choiset, H. Rabesona, I.V. Ivanova, J.M. Chobert, T. Haertlé and B.D. Franco. 2014. Purification and characterization of the bacteriocin produced by *Lactobacillus sakei* isolated from Brazilian salami. J. Appl. Microbiol. 116: 1195–1208.

Barbosa, M., S. Todorov, I. Ivanova, J.-M. Chobert, T. Haertlé and B.G. Franco. 2015. Improving safety of salami by application of bacteriocins produced by an autochthonous *Lactobacillus curvatus* isolate. Food Microbiol. 46: 254–262.

Basch, C., R. Jagus and S. Flores. 2013. Physical and antimicrobial properties of tapioca starch-HPMC edible films incorporated with nisin and/or potassium sorbate. Food Bioprocess. Tech. 6: 2419–2428.

Batt, C.A. 2000. Lactococcus: Introduction. pp. 1164–1166. *In*: R.K. Robinson, C.A. Batt and P.D. Patel (eds). Encyclopedia of food microbiology. Academic Press, London, U.K.

Dal Bello, B., K. Rantsiou, A. Bellio, G. Zeppa, R. Ambrosoli, T. Civera and L. Cocolin. 2010. Microbial ecology of artisanal products from North West of Italy and antimicrobial activity of the autochthonous populations. LWT - Food Sci. Technol. 43: 1151–1159.

Bernardeau, M., M. Gueguen and J.P. Vernoux. 2006. Beneficial lactobacilli in food and feed: long-term use, biodiversity and proposals for specific and realistic safety assessments. FEMS Microbiol. Rev. 30: 487–513.

Biscola, V., S.D. Todorov, V.S.C. Capuano, H. Abriouel, A. Gálvez and B.D. Franco. 2013. Isolation and characterization of a nisin-like bacteriocin produced by a *Lactococcus lactis* strain isolated from charqui, a Brazilian fermented, salted and dried meat product. Meat Sci. 93: 607–613.

Bourdichon, F., S. Casaregola, C. Farrokh, J.C. Frisvad, M.L. Gerds, W.P. Hammes, J. Harnett, G. Huys, S. Laulund, A. Ouwehand, I.B. Powell, J.B. Prajapati, Y. Seto, E.T. Schure, A. Van Boven, V. Vankerckhoven, A. Zgoda, S. Tuijtelaars and E.B. Hansen. 2012. Food fermentations: Microorganisms with technological beneficial use. Int. J. Food Microbiol. 154: 87–97.

Bover-Cid, S. and W.H. Holzapfel. 1999. Improved screening procedure for biogenic amine production by lactic acid bacteria. Int. J. Food Microbiol. 53: 33–41.

Bredholt, S., T. Nesbakken and A. Holck. 2001. Industrial application of an antilisterial strain of *Lactobacillus sakei* as a protective culture and its effect on the sensory acceptability of cooked, sliced, vacuum-packaged meats. Int. J. Food Microbiol. 66: 191–196.

Budde, B.B., T. Hornbæk, T. Jacobsen, V. Barkholt and A.G. Koch. 2003. *Leuconostoc carnosum* 4010 has the potential for use as a protective culture for vacuum-packed meats: culture isolation, bacteriocin identification, and meat application experiments. Int. J. Food Microbiol. 83: 171–184.

Burgain, J., C. Gaiani, M. Linder and J. Scher. 2011. Encapsulation of probiotic living cells: From laboratory scale to industrial applications. J. Food Eng. 104: 467–483.

Callewaert, R., M. Hugas and L. De Vuyst. 2000. Competitiveness and bacteriocin production of Enterococci in the production of Spanish-style dry fermented sausages. Int. J. Food Microbiol. 57: 33–42.

Casaburi, A., V. Di Martino, P. Ferranti, L. Picariello and F. Villani. 2016. Technological properties and bacteriocins production by *Lactobacillus curvatus* 54M16 and its use as starter culture for fermented sausage manufacture. Food Control 59: 31–45.

Castellano, P., C. Belfiore and G. Vignolo. 2011. Combination of bioprotective cultures with EDTA to reduce *Escherichia coli* O157:H7 in frozen ground-beef patties. Food Control 22: 1461–1465.

Castro, M.P., N.Z. Palavecino, C. Herman, O. Garro and C. Campos. 2011. Lactic acid bacteria isolated from artisanal dry sausages: characterization of antibacterial compounds and study of the factors affecting bacteriocin production. Meat Sci. 87: 321–329.

Chaillou, S., M.-C. Champomier-Verge`s, M. Cornet, A.-M, Crutz-Le Coq, A.-M. Dudez, V. Martin, S. Beaufils, E. Darbon-Ronge`re, R. Bossy, V. Loux and M. Zagorec. 2005. The complete genome sequence of the meat-borne lactic acid bacterium *Lactobacillus sakei* 23K. Nat. Biotechnol. 23: 1527–1533.

Chamba, J.F. and E. Jamet. 2008. Contribution to the safety assessment of technological microflora found in fermented dairy products. Int. J. Food Microbiol. 126: 263–266.

Chen, J. and A.L. Brody. 2013. Use of active packaging structures to control the microbial quality of a ready-to-eat meat product. Food Control 30: 306–310.

Cleveland, J., T. Montville, I. Nes and M. Chikindas. 2001. Bacteriocins: safe, natural antimicrobials for food preservation. Int. J. Food Microbiol. 71: 1–20.

Cocconcelli, P.S. and C. Fontana. 2008. Characteristics and applications of microbial starters in meat fermentations. pp. 129–148. *In*: F. Toldrá (ed.). Meat biotechnology. Springer, New York.

Cocolin, L. and K. Rantsiou. 2007. Sequencing and expression analysis of sakacin genes in *Lactobacillus curvatus* strains. Appl. Microbiol. Biotechnol. 76: 1403–1411.

Coeuret, V., S. Dubernet, M. Bernardeau, M. Gueguen and J.P. Vernoux. 2003. Isolation, characterization and identification of lactobacilli focusing mainly on cheeses and other dairy products. Lait 83: 269–306.

Cook, M.T., G. Tzortzis, D. Charalampopoulos and V.V. Khutoryanskiy. 2012. Microencapsulation of probiotics for gastrointestinal delivery. J. Control Release. 162: 56–67.

Cotter P.D., C. Hill and R.P. Ross. 2005. Bacteriocins: developing innate immunity for food. Nat. Rev. Microbiol. 3: 777–788.

Cotter, P.D. and C. Hill. 2003. Surviving the acid test: responses of gram-positive bacteria to low pH. Microbiol. Mol. Biol. Rev. 67: 429–453.

Cui, J.H., J.S. Goh, P.H. Kim, S.H. Choi and B.J. Lee. 2000. Survival and stability of bifidobacteria loaded in alginate poly-l-lysine microparticles. Int. J. Pharm. 210: 51–59.

Danilović, B., N. Joković, L. Petrović, K. Veljović, M. Tolinački and D. Savić. 2011. The characterisation of lactic acid bacteria during the fermentation of an artisan Serbian sausage (Petrovská Klobása). Meat Sci. 88: 668–674.

De Vuyst, L., S. Van Kerrebroeck, H. Harth, G. Huys, H.-M. Daniel and S. Weckx. 2014. Microbial ecology of sourdough fermentations: Diverse or uniform? Food Microbiol. 37: 11–29.

Deegan, L., P.D. Cotter, C. Hill and P. Ross. 2006. Bacteriocins: Biological tools for bio-preservation and shelf-life extension, Int. Dairy J. 16: 1058–1071.

Diez, J. and L. Patarata. 2013. Behavior of *Salmonella* spp., *Listeria monocytogenes*, and *Staphylococcus aureus* in Chouriço de Vinho, a dry fermented sausage made from wine-marinated meat. J. Food Protect. 76: 588–594.

Dortu, C., M. Huch, W.H. Holzapfel, C.M. Franz and P. Thonart. 2008. Antilisterial activity of bacteriocin-producing *Lactobacillus curvatus* CWBI-B28 and *Lactobacillus sakei* CWBI-B1365 on raw beef and poultry meat. Lett. Appl. Microbiol. 47: 581–586.

Drider, D., G. Fimland, Y. Hechard, L. M. McMullen, and H. Prevost. 2006. The continuing story of class IIa bacteriocins. Microbiol. Mol. Biol. Rev. 70: 564–582.

Drosinos, E.H., M. Mataragas, N. Xiraphi, G. Moschonas, F. Gaitis and J. Metaxopoulos. 2005. Characterization of the microbial flora from a traditional Greek fermented sausage. Meat Sci. 69: 307–317.

DuPont Nutrition & Health. 2015. DANISCO http://www.danisco.com/product-range/antimicrobials/novagard-antimicrobial-blends/ (Last accessed: 18/08/2015)

Durá, M.A., M. Flores and F. Toldrà. 2004. Effect of *Debaryomyces* spp. on the proteolysis of dry fermented sausages. Meat Sci. 68: 319–328.

EFSA and ECDC. 2015.(European Food Safety Authority and European Centre for Disease Prevention and Control). 2015. The European Union summary report on trends and sources of zoonoses, zoonotic agents and food-borne outbreaks in 2013. EFSA Journal 2015 13: 3991.

EFSA. 2005. QPS. Qualified Presumption of Safety of micro-organisms in food and feed. EFSA scientific colloquium summary report. European Food Safety Authority – October 2005. Parma, Italy. pp. 1–143.

EFSA. 2011. EFSA Panel on Biological Hazards (BIOHAZ); Scientific Opinion on risk based control of biogenic amine formation in fermented foods. EFSA Journal 2011 9: 2393.

EFSA. 2012. EFSA Panel on additives and products or substances used in animal feed (FEEDAP). Guidance on the assessment of bacterial susceptibility to antimicrobials of human and veterinary importance. EFSA Journal 2012 10: 2740.

EFSA. 2013. EFSA BIOHAZ Panel (EFSA Panel on Biological Hazards), 2013. Scientific Opinion on the maintenance of the list of QPS biological agents intentionally added to food and feed (2013 update). EFSA Journal 2013 11: 3449.

Enan, A., A. Essawy, M. Uyttendaele and J. Debevere. 1996. Antibacterial activity of *Lactobacillus plantarum* UG1 isolated from dry sausage: Characterization, production and bactericidal action of plantaricin UG1. Int. J. Food Microbiol. 30: 189–215.

Engelhardt, T., H. Albano, G. Kiskó, C. Mohácsi-Farkas and P. Teixeira. 2015. Antilisterial activity of bacteriocinogenic *Pediococcus acidilactici* HA6111-2 and *Lactobacillus plantarum* ESB 202 grown under pH and osmotic stress conditions. Food Microbiol. 48: 109–115.

Essid, I., M. Medini and M. Hassouna. 2009. Technological and safety properties of *Lactobacillus plantarum* strains isolated from a Tunisian traditional salted meat. Meat Sci. 81: 203–208.

Fadda, S., C. López and G. Vignolo. 2010. Role of lactic acid bacteria during meat conditioning and fermentation: peptides generated as sensorial and hygienic biomarkers. Meat Sci. 86: 66–79.

Federici, S., F. Ciarrocchi, R. Campana, E. Ciandrini, G. Blasi and W. Baffone. 2014. Identification and functional traits of lactic acid bacteria isolated from Ciauscolo salami produced in Central Italy. Meat Sci. 98: 575–584.

Fontana, C., P.S. Cocconcelli, G. Vignolo and L. Saavedra. 2015. Occurrence of antilisterial structural bacteriocins genes in meat borne lactic acid bacteria. Food Control. 47: 53–59.

Fontana, C., P.S. Cocconcelli and G. Vignolo. 2005. Monitoring the bacterial population dynamics during fermentation of artisanal Argentinean sausages. Int. J. Food Microbiol. 103: 131–142.

Fraqueza, M.J. 2015. Antibiotic resistance of lactic acid bacteria isolated from dry-fermented sausages. Int. J. Food Microbiol. 212: 76–88.

Fraqueza, M.J. and A.S. Barreto. 2015. HACCP. pp. 469–485. *In*: F. Toldrá (ed). Handbook of fermented meat and poultry. Wiley-Blackwell, Chichester.

Gänzle, M.G. 2015. Lactic metabolism revisited: metabolism of lactic acid bacteria in food fermentations and food spoilage. Curr. Opin. Food Sci. 2: 106–117.

Gao, Y., D. Li and X. Liu. 2014. Bacteriocin-producing *Lactobacillus sakei* C2 as starter culture in fermented sausages. Food Control 33: 1–6.

Govers, S. and A. Aertsen. 2015. Impact of high hydrostatic pressure processing on individual cellular resuscitation times and protein aggregates in *Escherichia coli*. Int. J. Food Microbiol. 213: 17–23.

Guilbaud, M., M. Zagorec, S. Chaillou and M.C. Champomier-Vergès. 2011. Intraspecies diversity of *Lactobacillus sakei* response to oxidative stress and variability of strain performance in mixed strains challenges. Food Microbiol. 29: 197–204.

Protective Cultures and Bacteriocins in Fermented Meats 263

Hartmann, H.A., T. Wilke and R. Erdmann. 2011. Efficacy of bacteriocin-containing cell-free culture supernatants from lactic acid bacteria to control *Listeria monocytogenes* in food. Int. J. Food Microbiol. 146: 192–199.

Hebert, E.M., L. Saavedra, M.P. Taranto, F. Mozzi, C. Magni, M.E.F. Nader, G. Font de Valdez, F. Sesma, G. Vignolo and R.R. Raya. 2012. Genome sequence of the bacteriocin-producing *Lactobacillus curvatus* strain CRL705. J. Bacteriol. 194: 538–539.

Heng, N. C. K. and J.R. Tagg 2006. What's in a name? Class distinction for bacteriocins. Nat. Rev. Microbiol. 4: http://www.nature.com/nrmicro/journal/v4/n2/full/nrmicro1273-c1.html. doi:10.1038/nrmicro1273-c1.

Heng, N.K., P. Wescombe, J. Burton, R. Jack and J. Tagg. 2007. The diversity of bacteriocins in gram-positive bacteria. pp. 45–92. *In*: M. Riley and M. Chavan (eds.). Bacteriocins: ecology and evolution. Springer, Berlin.

Huang, H.-W., H.-M. Lung, B.B. Yang and C.-Y. Wang. 2014. Responses of micro-organisms to high hydrostatic pressure processing. Food Control. 40: 250–259.

Hugas, M. 1998. Bacterocinogenic lactic acid bacteria for the biopreservation of meat and meat products. Meat Sci. 49: 139–150.

Hurtado, A., C. Reguant, A. Bordons and N. Rozès. 2012. Lactic acid bacteria from fermented table olives. Food Microbiol. 31: 1–8.

Idris, A. and S. Wahidrin. 2006. Effect of sodium alginate concentration, bead diameter, initial pH and temperature on lactic acid production from pineapple waste using immobilized *Lactobacillus delbrueckii*. Process. Biochem. 41: 1117–1123.

Ikonić, P., T. Tasić, L. Petrović, S. Škaljac, M. Jokanović, A. Mandić, B. Ikonić. 2013. Proteolysis and biogenic amines formation during the ripening of Petrovská klobása, traditional dry-fermented sausage from Northern Serbia. Food Control 30: 69–75.

Jahan, M. and R.A. Holley. 2014. Incidence of virulence factors in enterococci from raw and fermented meat and biofilm forming capacity at 25°C and 37°C. Int. J. Food Microbiol. 170: 65–69.

Jahan, M., G.G. Zhanel, R. Sparling and R.A. Holley. 2015. Horizontal transfer of antibiotic resistance from *Enterococcus faecium* of fermented meat origin to clinical isolates of *E. faecium* and *Enterococcus faecalis*. Int. J. Food Microbiol. 199: 78–85.

Jofré, A., T. Aymerich, N. Grèbol and M. Garriga. 2009. Efficiency of high hydrostatic pressure at 600 MPa against food-borne microorganisms by challenge tests on convenience meat products. LWT - Food Sci. Technol. 42: 924–928.

Jones, R. J., H.M. Hussein, M. Zagorec, G. Brightwell and J.R. Tagg. 2008. Isolation of lactic acid bacteria with inhibitory activity against pathogens and spoilage organisms associated with fresh meat. Food Microbiol. 25: 228–234.

Juven, B., S. Barefoot, M. Pierson, L. McCaskill and B. Smith. 1998. Growth and survival of *Listeria monocytogenes* in vacuum-packaged ground beef inoculated with *Lactobacillus alimentarius* FloraCarn L-2. J. Food Protect. 5: 551–556.

Kailasapathy, K. 2002. Microencapsulation of probiotic bacteria: technology and potential applications. Curr. Issues Intest.Microbiol. 3: 39–48.

Kanatani K. and M. Oshimura. 1994. Plasmid-associated bacteriocin production by *Lactobacillus plantarum* strain. Biosci. Biotechnol. Biochem. 58: 2084–208.

Katikou, P., I. Ambrosiadis, D. Georgantelis. P. Koidis and S.A. Georgakis. 2005. Effect of *Lactobacillus*-protective cultures with bacteriocin-like inhibitory substances' producing ability on microbiological, chemical and sensory changes during storage of refrigerated vacuum-packaged sliced beef. J. Appl. Microbiol. 99: 1303–1313.

Kearney, L., M. Upton and A.M.C. Loughlin. 1990. Enhancing the viability of *Lactobacillus plantarum* inoculum by immobilizing the cells in calcium-alginate beads incorporating cryoprotectants. Appl. Environ. Microbiol. 56: 3112–3116.

Kemperman, R., M. Jonker, A. Nauta, O.P. Kuipers and J. Kok. 2003. Functional analysis of the gene cluster involved in production of the bacteriocin circularin A by *Clostridium beijerinckii* ATCC 25752. Appl. Environ. Microbiol. 69: 5839–5848.

Khan, M.I., M.S. Arshad, F.M. Anjum, A. Sameen and W.T. Gill. 2011. Meat as a functional food with special reference to probiotic sausages. Food Res. Int. 44: 3125–3133.

Klaenhammer, T. R. 1993. Genetics of bacteriocins produced by lactic acid bacteria. FEMS Microbiol. Rev. 12: 39–86

Koo, O.-K., M. Eggleton, C.A. O'Bryan, P.G. Crandall and S.C. Ricke. 2012. Antimicrobial activity of lactic acid bacteria against *Listeria monocytogenes* on frankfurters formulated with and without lactate/diacetate. Meat Sci. 92: 533–537.

Kozačinski, L., E. Drosinos, F. Čaklovica, L. Cocolin, J. Gasparik-Reichardt, and S. Vesković-Moračanin. 2008. Investigation of microbial association of traditionally fermented sausages. Food Technol. Biotechnol. 46: 93–106.

Laboratories STANDA. 2015. SANICO® is our range of antifungal coatings for the agrofood industry. http://www.standa-fr.com/eng/laboratoires-standa/sanico/ (Last accessed: 18/08/2015).

Lacroix, C., F. Grattepanche, Y. Doleyres and D. Bergmaier. 2005. Immobilised cell technologies for the dairy industry. pp. 295–321. *In*: V. Nedovic and R. Willaert (eds.). Applications of cell immobilisation biotechnology. Springer, Berlin.

Landeta, G., J.A. Curiel, A.V. Carrascosa, R. Muñoz and B. de las Rivas. 2013. Technological and safety properties of lactic acid bacteria isolated from Spanish dry-cured sausages. Meat Sci. 95: 272–280.

Latorre-Moratalla, M. L., S. Bover-Cid, R. Talon, M. Garriga, E. Zanardi, A. Ianieri, M.J. Fraqueza, M. Elias, E.H. Drosinos and M.C. Vidal-Carou. 2010a. Strategies to reduce biogenic amine accumulation in traditional sausage manufacturing. LWT - Food Sci. Technol. 43: 20–25.

Latorre-Moratalla, M. L., S. Bover-Cid, R. Talon, T. Aymerich, M. Garriga, E. Zanardi, A. Ianieri, M.J. Fraqueza, M. Elias, E.H. Drosinos, A. Laukova and M.C. Vidal-Carou. 2010b. Distribution of aminogenic activity among potential autochthonous starter cultures for dry fermented sausages. J. Food Protect. 73: 524–528.

Lauková, A. and P., Turek. 2011. Effect of enterocin 4231 in Slovak fermented salami Púchov after its experimental inoculation with *Listeria innocua* Li1. Acta Sci. Pol., Technol. Aliment. 10: 423–431.

Lauková, A., P. Turek, M. Mareková and P. Nagy. 2003. Use of ent M, new variant of ent P to control *Listeria innocua* in experimentally contaminated Gombasek sausage. Arch. Lebensmittelhyg. 54: 46–48.

Laukova, A., S. Czikková, S. Laczková and P. Turek. 1999. Use of Enterocin Ccm 4231 to control *Listeria monocytogenes* in experimentally contaminated dry fermented Hornád salami. Int. J. Food Microbiol. 52: 115–119.

Leroy, F., J. Verluyten and L. Vuyst. 2006. Functional meat starter cultures for improved sausage fermentation. Int. J. Food Microbiol. 106: 270–285.

Liu, G., Y. Wang, M. Gui, H. Zheng, R. Dai and P. Li. 2012. Combined effect of high hydrostatic pressure and enterocin LM-2 on the refrigerated shelf life of ready-to-eat sliced vacuum-packed cooked ham. Food Control 24: 64–71.

Llopis, S., A. Querol, A. Heyken, B. Hube, L. Jespersen, M.T. Fernandez-Espinar and R.P. Torrado. 2012. Transcriptomics in human blood incubation reveals the importance of oxidative stress response in *Saccharomyces cerevisiae* clinical strains. BMC Genomics 13: 1–12.

Ludwig, W., K.H. Schleifer and W.B.Whitman. 2009. Order II. Lactobacillales ord. nov. p. 464. " *In*: De Vos, P., Garrity, G.M., Jones, D., Krieg, N.R., Ludwig, W., Rainey, F.A., Schleifer, K.H., Whitman, W.B. (eds).Bergey's Manual of Systematic Bacteriology, Second edition, vol. 3 (The Firmicutes), Springer, New York.

Maqueda, M, S.Hidalgo, F.Montalbán-López, M. Valdivia and M. Martínez-Bueno. 2008. Genetic features of circular bacteriocins produced by Gram-positive bacteria. FEMS Microbiol Rev. 32: 2–22.

Marcos, B., T. Aymerich, J.M. Monfort and M. Garriga, 2007. Use of antimicrobial biodegradable packaging to control *Listeria monocytogenes* during storage of cooked ham. Int. J. Food Microbiol. 120: 152–158.

Marcos, B., T. Aymerich, M. Garriga and J. Arnau. 2013. Active packaging containing nisin and high pressure processing as post-processing listericidal treatments for convenience fermented sausages. Food Control 30: 325–330.

Mareková, M., A. Lauková, M. Skaugen and I.F. Nes. 2007. Isolation and characterization of a new bacteriocin, termed enterocin M, produced by environmental isolate *Enterococcus faecium* AL41. J. Ind. Microbiol. Biotechnol. 34: 533–537.

Marshall, B.M., D.J. Ochieng and S.B. Levy. 2009. Commensals: underappreciated reservoir of antibiotic resistance. Microbe 4: 231–238.

De Martinis, E.C.P. and F.Z. Freitas. 2003. Screening of lactic acid bacteria from Brazilian meats for bacteriocin formation. Food Control 14: 197–200.

Marty, E., J. Buchs, E. Eugster-Meier, C. Lacroix and L. Meile. 2012. Identification of Staphylococci and dominant Lactic Acid Bacteria in spontaneously fermented Swiss meat products using PCR-RFLP. Food Microbiol. 29: 157–166.

Massani, M., V. Molina, M. Sanchez, V. Renaud, P. Eisenberg and G. Vignolo. 2014. Active polymers containing *Lactobacillus curvatus* CRL705 bacteriocins: Effectiveness assessment in Wieners, Int. J. Food Microbiol. 178: 7–12.

Mauriello, G, A. Casaburi, G. Blaiotta and F. Villani. 2004. Isolation and technological properties of coagulase negative staphylococci from fermented sausages of Southern Italy. Meat Sci. 67: 149–158.

Medina, R.B., R. Oliszewski, M.C.A. Mukdsi, C.P. Van Nieuwenhove and S.N. González. 2011. Sheep and goat's dairy products from South America: Microbiota and its metabolic activity. Small Ruminant Res. 101: 84–91.

Messi, P., M. Bondi, C. Sabia, R. Battini and G. Manicardi. 2001. Detection and preliminary characterization of a bacteriocin (plantaricin 35d) produced by a *Lactobacillus plantarum* strain. Int. J. Food Microbiol. 64: 193–198.

Montanari, C. E. Bargossi, A. Gardini, R. Lanciotti, R. Magnani, F. Gardini and G. Tabanelli. 2016. Correlation between volatile profiles of Italian fermented sausages and their size and starter culture. Food Chem. 192: 736–744.

Moore, J.E. 2004. Gastrointenstinal outbreaks associated with fermented meats. Meat Sci. 67: 565–568.

Morot-Bizot S.C., S. Leroy and R. Talon. 2006. Staphylococcal community of a small unit manufacturing traditional dry fermented sausages. Int. J. Food Microbiol. 108: 210–217.

Muscariello, L., C. Marino, U. Capri, V. Vastano, R. Marasco and M. Sacco. 2013. CcpA and three newly identified proteins are involved in biofilm development in *Lactobacillus plantarum*. J. Basic Microbiol. 53: 62–71.

Muthukumarasamy, P. and R.A. Holley. 2007. Survival of *Escherichia coli* O157:H7 in dry fermented sausages containing micro-encapsulated probiotic lactic acid bacteria. Food Microbiol. 24: 82–88.

Muthukumarasamy, P. and R.P. Holley. 2006. Microbiological and sensory quality of dry fermented sausages containing alginate-microencapsulated *Lactobacillus reuteri*. Int. J. Food Microbiol. 111: 164–169.

Nes, I.F., S.-S. Yoon and D.B. Diep. 2007. Ribosomally Synthesiszed Antimicrobial Peptides (Bacteriocins) in Lactic Acid Bacteria: A Review. Food Sci. Biotechnol. 16: 675–690.

O'Bryan, C.A., P.G. Crandall, S.C. Ricke and J.B. Ndahetuye. 2015. Lactic acid bacteria (LAB) as antimicrobials in food products: Types and mechanisms of action. pp. 117–113. *In*: T.M. Taylor (eds.). Handbook of natural antimicrobials for food safety and quality, Woodhead Publishing, Oxford.

Ojha, K., J.P. Kerry, G. Duffy, T. Beresford and B.K. Tiwari. 2015. Technological advances for enhancing quality and safety of fermented meat products. Trends Food Sci. Technol. 44: 105–116.

Oki, K., A.K. Rai, S. Sato, K. Watanabe and J.P. Tamang. 2011. Lactic acid bacteria isolated from ethnic preserved meat products of the Western Himalayas. Food Microbiol. 28: 1308–1315.

Oliveira, T., A.L.S. Ramos, E.M. Ramos, R.H. Piccoli and M. Cristianini. 2015. Natural antimicrobials as additional hurdles to preservation of foods by high pressure processing, Trends Food Sci. Technol. 45: 60–85.

Ortega-Rivas, E. 2012. Non-thermal food Engineering operations. Springer. New York.

Ozdemir, M. and J.D. Floros. 2008. Optimization of edible whey protein films containing preservatives for mechanical and optical properties, J. Food Eng. 84: 116–123.

Panagou, E.Z., G.-J.E. Nychasand and J.N. Sofos. 2013. Types of traditional Greek foods and their safety. Food Control 29: 32–41.

Papagianni, M. and D. Sergelidis. 2013. Effects of the presence of the curing agent sodium nitrite, used in the production of fermented sausages, on bacteriocin production by *Weissella paramesenteroides* DX grown in meat simulation medium. Enzyme Microb. Technol. 53: 1–5.

Patarata, L.A.2002. Caracterização e Avaliação da Aptidão Tecnológica de Bactérias do Ácido Láctico e *Micrococcaceae* em Produtos de Salsicharia. Efeito da sua utilização em culturas de arranque e de formulação acidificante no fabrico de linguiça tradicional transmontana. PhD thesis. Universidade de Trás-os-Montes e Alto Douro, Vila Real.

Petaja-Kanninen, E. and E. Puolane. 2007. Meat Fermentation Worldwide: History and Principles. Principles of meat fermentation. pp. 31–36. In: F. Toldrá, Y. Hui, I. Astiasarán, W. Nip, J. Sebranek, E. Silveira, L. Stahnke and R. Talon (eds.). Handbook of fermented meat and poultry. Blackwell Publishing, Oxford.

Rantsiou, K., E. Drosinos, M. Gialitaki, R. Urso, J. Krommer, J. Gasparik-Reichardt, S. Tóth, I. Metaxopoulos, G. Comi and L. Cocolin. 2005. Molecular characterization of *Lactobacillus* species isolated from naturally fermented sausages produced in Greece, Hungary and Italy. Food Microbiol. 22: 19–28.

Rao, C.S., R.S. Prakasham, A.B. Rao and J.S. Yadav. 2007. Functionalized alginate as immobilization matrix in enantioselective L (+) lactic acid production by *Lactobacillus delbrucekii*. Appl. Biochem. Biotechnol. 149: 219–228.

Rathore, S., P.M. Desai, C.V. Liew, L.W. Chan and P.W.S. Heng. 2013. Microencapsulation of microbial cells, J. Food Eng. 116: 369–381.

Ravyts, F., L. Steen, O. Goemaere, H. Paelinck, L. De Vuyst and F. Leroy. 2010. The application of staphylococci with flavour-generating potential is affected by acidification in fermented dry sausages. Food Microbiol. 27: 945–954.

Realini, C.E. and B. Marcos. 2014. Active and intelligent packaging systems for a modern society. Meat Sci. 98: 404–419.

Reis, J.A., A.T. Paula, S.N. Casarotti and A.L.B. Penna. 2012. Lactic acid bacteria antimicrobial compounds: Characteristics and applications. Food Eng. Rev. 4: 124–140.

Rocourt, J. and P. Cossart. 1997. *Listeria monocytogenes*. pp. 337-352. In: M.P. Doyle, L.R. Beuchat, and T.J. Montville (eds.). Food Microbiology – fundamentals and frontiers. ASM Press, Washington.

Rossi-Marquez, G., J.H. Han, B. Garcia-Almendarez, E. Castano-Tostado and C. Regalado-Gonzalez. 2009. Effect of temperature, pH and film thickness on nisin release from antimicrobial whey protein isolate edible films. J. Sci. Food Agr. 98: 2492–2497.

Sánchez-Ortega, I., B.E. García-Almendárez, E.M. Santos-López, A. Amaro-Reyes, J.E. Barboza-Corona and C. Regalado. 2014. Antimicrobial edible films and coatings for meat and meat products preservation. The Scientific World Journal, Article ID 248935. 18 pages. http://dx.doi.org/10.1155/2014/248935

Santiago-Silva, P., N.F.F. Soares, J.E. Nobrega, M. Junior, K. Barbosa, A. Volp, E. Zerdas and N. Wurlitzer. 2009. Antimicrobial efficiency of film incorporated with pediocin (ALTA 2351) on preservation of sliced ham. Food Control 20: 85–89.

Serrano, L.M., D. Molenaar, M. Wels, B. Teusink, P.A. Bron, W.M. de Vos and E.J. Smid. 2007. Thioredoxin reductase is a key factor in the oxidative stress response of *Lactobacillus plantarum* WCFS1. Microb. Cell Fact. 28: 6–29.

Sheu, T.Y., R.T. Marshall and H. Heymann. 1993. Improving survival of culture bacteria in frozen desserts by microentrapment. J. Dairy Sci. 76: 1902–1907.

Simoncini, N., R. Virgili, G. Spadola and P. Battilani. 2014. Autochthonous yeasts as potential biocontrol agents in dry-cured meat products. Food Control 46: 160–167.

Simonová, M., V. Strompfová, M. Marciňáková, A. Lauková, S. Vesterlund, M.L. Moratalla, S. Bover-Cid and C. Vidal-Carou. 2006. Characterization of *Staphylococcus xylosus* and *Staphylococcus carnosus* isolated from Slovak meat products. Meat Sci. 73: 559–564.

Vesković-Moračanin, S., D. Obradović, B. Velebit, B. Borović,, M. Škrinjar and L. Turubatović. 2010. Antimicrobial properties of indigenous Lactobacillus sakei strain. Acta Vet. 60: 59–66.

Suma, K., M.C. Misra and M.C. Varadaraj. 1998. Plantaricin LP84, a broad spectrum heat-stable bacteriocin of *Lactobacillus plantarum* NCIM 2084 produced in a simple glucose broth medium. Int. J. Food Microbiol. 40: 17–25.

Tabanelli, G., C. Montanari, E. Bargossi, R. Lanciotti, V. Gatto, G. Felis, S. Torriani and F. Gardini. 2014. Control of tyramine and histamine accumulation by lactic acid bacteria using bacteriocin forming lactococci. Int. J. Food Microbiol. 190: 14–23.

Tagg, J.R., A.S. Dajani and L.W. Wannamaker. 1976. Bacteriocins of Grampositive bacteria. Bacteriol. Rev. 40: 722–756.

Talon, R., I. Lebert, A. Lebert, S. Leroy, M. Garriga, T. Aymerich, E.H. Drosinos, E. Zanardi, A. Ianieri, M.J. Fraqueza, L. Patarata and A. Laukova. 2012. Microbial ecosystem of traditional dry fermented sausages in Mediterranean countries and Slovakia. pp. 115–127. *In*: G.S. Williams (ed). Mediterranean ecosystems: dynamics, management and conservation. Nova Science Publishers, New York.

Tichaczek, P.S., J. Nissen-Meyer, I. Nes, R. Vogel and W. Hammes. 1992. Characterization of the bacteriocins curvacin A from *Lactobacillus curvatus* LTH1174 and sakacin P from *L. sake* LTH673. Syst. Appl. Microbiol. 15: 460–468.

Todorov, S.D., M. Vaz-Velho, B. Franco and W. Holzapfel. 2013. Partial characterization of bacteriocins produced by three strains of *Lactobacillus sakei*, isolated from salpicao, a fermented meat product from North-West of Portugal. Food Control 30: 111–121.

Todorov, S.D. 2009. Bacteriocins from *Lactobacillus plantarum* – production, genetic organization and mode of action. A review. Braz. J. Microbiol. 40: 209–221.

Trauth, E., J.P. Lemaître, C. Rojas, C. Diviès and R. Cachon. 2001. Resistance of immobilized lactic acid bacteria to the inhibitory effect of quaternary ammonium sanitizers. Lebensm. Wiss. Technol. 34: 239–243.

Trinetta, V., J. Floros and C. Cutter. 2010. Sakacin A-containing pullulan film: An active packaging system to control epidemic clones of *Listeria monocytogenes* in ready-to-eat foods. J. Food Safety 30: 366–381.

Urso, R., K. Rantsiou, C. Cantoni, G. Comi and L. Cocolin. 2006. Technological characterization of a bacteriocin-producing *Lactobacillus sakei* and its use in fermented sausages production. Int. J. Food Microbiol. 110: 232–239.

van Kranenburg, R., M. Kleerebezem, J. Vlieg, B. Ursing, J. Boekhorst, B. Smit, E. Ayad, G. Smit and R. Siezen. 2002. Flavour formation from amino acids by lactic acid bacteria: predictions from genome sequence analysis. Int. Dairy J. 12: 111–121.

Vandamme, P., K. Bruyne and B. Pot. 2014. Phylogenetics and systematics. pp.31–39. *In*: W. H Holzapfel and B.J.B. Wood (eds.). Lactic acid bacteria biodiversity and taxonomy. John Wiley & Sons, Oxford.

Vanderroost, M., P. Ragaert, F. Devlieghere and B. Meulenaer. 2014. Intelligent food packaging: The next generation. Trends Food Sci. Tech. 39: 47–62.

Verluyten, J., W. Messens and L. De Vuyst. 2004. Sodium chloride reduces production of curvacin A, a bacteriocin produced by *Lactobacillus curvatus* strain LTH 1174, originating from fermented sausage. Appl. Environ. Microbiol. 70: 2271–2278.

Vignolo, G., J. Palacios, M. Farías, F. Sesma, U. Schillinger, W. Holzapfel and G. Oliver. 2000. Combined effect of bacteriocins on the survival of various *Listeria* species in broth and meat system. Curr. Microbiol. 41: 410–416.

Vignolo, G., P. Castellano and S. Fadda. 2015. Bioprotective cultures. pp. 129–137. *In*: Toldra, F. (ed.). Handbook of fermented meat and poultry. Wiley-Blackwell, Oxford.

Wanangkarn, A., D.-C. Liu, A. Swetwiwathana and F.-J. Tan. 2012. An innovative method for the preparation of mum (Thai fermented sausages) with acceptable technological quality and extended shelf-life. Food Chem. 135: 515–521.

Wanangkarn, A., D.-C. Liu, A. Swetwiwathana, A. Jindaprasert, C. Phraephaisarn, W. Chumnqoen and F.-J. Tan. 2014. Lactic acid bacterial population dynamics during fermentation and storage of Thai fermented sausage according to restriction fragment length polymorphism analysis. Int. J. Food Microbiol. 186: 61–67.

Wright, A. and L. Axelsson. 2012. Lactic Acid Bacteria: *An Introduction*. pp. 2–14. *In*: S. Lahtinen, A.C. Ouwehand, S. Salminen and A. Wright (eds). Lactic acid bacteria, microbiological and functional aspects. CRC Press, Boca Raton.

Xu, Z., L. Li, M. Shirtliff, B. Peters, Y. Peng, M. Alam, S. Yamasaki and L. Shi. 2010. First report of class 2 integron in clinical *Enterococcus faecalis* and class 1 integron in *Enterococcus faecium* in South China. Diagn. Microbiol. Infect. Dis. 68: 315–317.

Zacharof, M.P. and R.W. Lovitt. 2012. Bacteriocins produced by lactic acid bacteria. APCBEE Procedia 2: 50–56.

Zonenschain, D., A. Rebecchi and L. Morelli. 2009. Erythromycin- and tetracycline-resistant lactobacilli in Italian fermented dry sausages. J. Appl. Microbiol. 107: 1559–1568.

Chapter 12

Autochthonous Starter Cultures

Jadranka Frece* and Ksenija Markov

1 INTRODUCTION

Native microbial population of the traditional fermented sausages is a rich source of potential autochthonous starter cultures for traditional or industrial applications. Lactic acid bacteria (LAB) and coagulase-negative staphylococci (CoNS) are the main microbial groups involved in the meat fermentation and routinely implemented in sausage production as starters, in order to upgrade product's safety and quality. The main challenge in selection of members of starter cultures is to improve food safety, and also to preserve the typical sensory quality of traditional sausages (Talon et al. 2008). The most promising microorganisms for starter cultures are those which are selected from autochthonous microbiota since they are well adapted to the meat environment and to the specific manufacturing process and are capable of dominating the microbiota of the product due to their specific metabolic capabilities. Nowadays, increasing number of studies has been focused on the isolation and identification of autochthonous functional starter cultures, with the aim of developing new functional meat products, which will be recognized and labelled as autochthonous due to the influence of climate

---

*For Correspondence: Laboratory for General Microbiology and Food Microbiology Faculty of Food Technology and Biotechnology University of Zagreb Pierottijeva 6, 10000 Zagreb Croatia, Tel: +385 1 4605 284, Fax: + 385 1 4836 424, Email: jfrece@pbf.hr

and vegetation of the region where they are produced (Babić et al. 2011, Frece et al. 2014a, c). Examples include microorganisms that generate aroma compounds, health-promoting compounds, bacteriocins and other antimicrobials, contribute to cured meat colour, possess probiotic qualities, or lack negative properties such as the production of biogenic amines and toxic compounds (Leroy et al. 2006). The strain selection is based on general, technological and functional properties which should be tested by phenotypic and genotypic methods. Thus competitive and protective autochthonous functional starter cultures contribute to the reduction or elimination of microbiological and toxicological risks but also contribute in traditional sensory properties of meat product.

2 Autochthonous Starter Cultures for Fermented Meat Products

Consumers are increasingly interested in traditional food products because of their unique characteristics associated with traditional methods of production and specific sensory features. On the European market is a large selection of traditional fermented products as a result of the diversity of raw materials, recipes and manufacturing processes that result from the habits and customs of different countries and regions (Lebert et al. 2007). In rural households and on family farms in Croatia, Turkey, Argentina, Italia, according to old recipes, regional traditional meat products are produced, without the addition of starter cultures so that fermentation of these products depends on autochthonous microbiota originating from raw materials or from the environment (Bonomo et al. 2011, Babić et al. 2011, Kargozari et al. 2014, Palavecino Prpich et al. 2015). The composition of microbiota of each traditional product is specific, which means that the microbiota composition has an important role in the development of typical sensory characteristics. However, since the composition of the autochthonous microbiota is variable and its growth is under the influence of environmental conditions, the production of fermented products without the use of starter cultures can result in unequal product quality. Application of the commercial starter cultures is contributing to the achievement of standard sensory properties of products (flavor, color, texture), but specific or typical characteristics of traditional products, which are the result of autochthonous microbiota activity, are lost (Babić et al. 2011, Markov et al. 2013, Frece et al. 2014a, c , Kargozari et al. 2014, Palavecino Prpich et al. 2015). Besides, commercial starters cannot always compete with autochthonous microbial population, nor are adapted to growth conditions in a particular product, so sometimes a satisfactory level of growth is not achieved, resulting in loss of the desirable and specific sensory characteristics. Therefore, in order to preserve the typical characteristics of traditional products, a growing trend of finding and selecting autochthonous starter cultures from traditional products is present lately.

The presence of microorganisms in raw materials as an inevitable factor in the production of fermented foods has encouraged many researchers to study activity and composition of native microbiota on the ripening process. Raw materials in the production of autochthonous fermented products can be a source of different microorganisms that can contaminate the final product. Understanding the importance of microbiota in the production of fermented products has lead to the development of starter cultures as well as their application in the production of fermented food. Starter cultures are preparations containing live microorganisms and are applicable for the fermentation of different foods with the aim of enriching these foods with products of its metabolism (Šušković and Kos 2001, Frece et al. 2010c). Use of starter cultures has the influence on hygienic safety of production, it standardizes the process of ripening, balances and improves quality and provides better shelf life.

Depending on the type of fermented products, starter cultures can be LAB, Gram-positive catalase-positive cocci (GCC+) (*Staphylococcus, Kocuria*), yeasts and molds. In the selection of starter cultures it is necessary to choose the strains that do not have undesirable characteristics such as production of H_2O_2, CO_2, biogenic amines, mycotoxins and other compounds (Frece et al. 2010a,b, Markov et al. 2013). General criteria for the selection of starter cultures are shown in Table 1. One of the basic characteristics of starter cultures is their ability to compete with autochthonous microbiota, colonize the environment, dominate the microbial population and have such metabolic activities that enable them to grow and survive in the conditions of fermented product. Therefore, the knowledge about security and technological characteristics of autochthonous microbiota is of practical importance for the improvement of starter cultures technology and production of traditional fermented sausages (Comi et al. 2005, Drosinos et al. 2005, Fontana et al. 2005, Rantsiou et al. 2005, Lebert et al. 2007, Bonomo et al. 2011, Frece et al. 2014a, c, Palavecino Prpich et al. 2015). One of the main roles of starter cultures is to increase food safety by inactivation of pathogens and spoilage microorganisms through the production of acids and bacteriocins. Antagonistic cultures added due to inhibition of pathogens and/or extensions of food shelf life with minimum changes in sensory properties of products are called "protective cultures". One can make a difference between starter cultures and protective cultures

Autochthonous Starter Cultures

where metabolic activity (acid production, hydrolysis of proteins) and antimicrobial activity are the main difference (Castellano et al. 2008).

Table 1 General criteria for the selection of starter cultures

General criteria for the selection of starter cultures
1. Safety
– Starter microorganisms must not have pathogenic or toxic effects
– Preparation of starter cultures must be carried out under strictly controlled aseptic conditions
2. Technological characteristics
– Starter microorganisms must dominate over the autochthonous microbiota
– Starter microorganisms carry out specific metabolic activity
– During the technological process of starter cultures production must not be pollution
3. Economic aspects
– Cultivation of starter cultures must be easily feasible from an economic point of view
– Storage of the starter cultures can be carried out by methods of freezing or lyophilization with minimum loss of culture metabolic activity
– Important properties of starter cultures must be stable under defined storage conditions for several months
– Handling starter cultures must be facilitated as much as possible

Source: Šušković et al. 2001, Frece et al. 2010a, 2011

Although the technology of industrial production of meat products is standardized using commercial starter cultures, many manufacturers still apply traditional production technology without the addition of selected starter cultures so fermentation depends on autochthonous microbiota from raw materials and the environment. Due to specific composition and metabolic activities of autochthonous microbiota, together with activity of endogenous enzymes of raw materials, traditional fermented products in comparison with products manufactured with the addition of commercial starter cultures have specific organoleptic characteristics (Lebert et al. 2007, Babić et al. 2011, Frece et al. 2014a, c). The greatest problem of using commercial starter cultures in the production of traditional dry sausages are differences in technological microbiota of different types of sausages which are determined by the raw material and the specific microbiota of production environment. The consequence is that commercial starters, in which bacterial microbiota typical for other types of smoked sausages is dominant, speed manufacturing processes (economy), increase microbiological stability (safety), but on the other hand contribute to the formation of a typical sensory properties, particularly atypical flavor and taste (Aymerich et al. 2003, Babić et al.

274   *Fermented Meat Products: Health Aspects*

2011, Markov et al. 2011, Frece et al. 2014a, c). Therefore, the greatest potential for use as starter cultures have those microorganisms isolated and selected from autochthonous microbiota (autochthonous starter cultures), since the same naturally dominate in traditional fermentation processes, are adapted to micro conditions in the product, technological processes and technological parameters which are unique for each type of dry sausage, have greater metabolic capacities which improve quality and safety of the product and participate in the creation of typical sensory properties (color, texture, smell and taste). Besides, the use of different autochthonous types and strains could increase the diversity of markets with different types of starter cultures with whom it will be able to produce a variety of typical regional fermented products with specific properties and characteristics of the area they are produced in. The advantages of using autochthonous starter cultures are shown in Table 2.

Table 2 Advantages of using autochthonous starter cultures

Advantages of using autochthonous starter cultures
✓ Extending the shelf life of the product (production of antimicrobial compounds: lactic acid, acetic acid, diacetyl, hydrogen peroxide, bacteriocins) → bioconservation → reducing the use of additives
✓ Formation of desired aroma, color and consistency of the product
✓ Reduced ripening time
✓ With the use of functional starter cultures → improved product quality → functional foods
✓ Domination and adaptability of growth in a given raw material, since they are isolated within the same
✓ Getting the traditional product with characteristic aroma and improved properties

Source: Frece et al. 2011, 2014a

Talon et al. (2008) have developed autochthonous starter culture consisting of strains which have been isolated from traditional French fermented sausages. The aforementioned autochthonous starter culture consisted of three dominant strains: *Lactobacillus sakei*, *Staphylococcus equorum* and *St. succinus*, and was tested in French traditional facility from which these dominant strains were previously isolated. This is the first time that selection is conducted, development and validation of *in situ* autochthonous starter cultures, as well as the first time that the species *St. equorum* and *St. succinus* are used together as starter cultures for meat fermentation. Compared with the traditional fermentation of sausages, the addition of this autochthonous starter culture resulted in increase of product safety by inhibition of pathogenic bacteria

Autochthonous Starter Cultures

Listeria monocytogenes, by reducing the levels of biogenic amines and by preventing the oxidation of fatty acids and cholesterol. Besides, aforementioned autochthonous starter culture did not effect on the typical organoleptic characteristics of traditional sausages. Implementation of of *Lb. sakei* in Croatian traditional fermented sausages resulted in reduction of undesirable microbiota (*L. monocytogenes*, enterococci) and preserved sensorial features of final products (Zdolec et al. 2007a, Zdolec et al. 2008).

Several studies have shown that the microbiota of traditional sausages mainly consist of LAB and CoNS (Rantsiou and Cocolin 2006). The most commonly identified LAB species in traditional fermented sausages are *Lb. sakei*, *Lb. curvatus* and *Lb. plantarum* (Aymerich et al. 2006, Urso et al. 2006a). Among GCC+ (CoNS) isolates, *St. xylosus*, *St. equorum*, *St. succinus* and *St. saprophyticus* are identified, often with *St. xylosus* being predominant (Mauriello et al. 2004, Corbière Morot-Bizot et al. 2007, Drosinos et al. 2007).

Results of previous studies of autochthonous microbial population of traditional Croatian fermented sausages performed in the Laboratory for General Microbiology and Food Microbiology of Faculty of Food Technology and Biotechnology, University of Zagreb, showed that Croatian traditional meat products (Slavonian Kulen, horsemeat sausages, Varazdin and domestic Slavonian sausages, game sausages) have a different composition of autochthonous microbial population from other European traditional fermented sausages which can be explained by different climate, region, technology of production, different spices. Different composition of microbial populations than in traditional meat products we have identified in the game products, which can be explained by different type of nutrition (Table 3).

Table 3 shows the composition of autochthonous microbiota at the end of ripening period of some traditional fermented sausages, produced without the addition of starter cultures, i.e. diversity of LAB and CoNS identified from various European countries (see also Chapter 8 in this book).

Autochthonous bacterial isolates showed significant antimicrobial activity, adhesion properties, and inhibition of infection with bacteria from the genus *Salmonella*, and have satisfied some of the criteria for the selection of functional starter cultures (Babić et al. 2011, Frece et al. 2014a, b, c, Markov et al. 2013) (Table 4). By classical microbiological, phenotypic and genotypic methods (Slavonian sausage, horsemeat sausage, Varazdin and Slavonian sausage, game sausage) autochthonous starter cultures were isolated and identified: *Lb. plantarum*, *Lb. acidophilus*, *Lb. delbrueckii*, *Lactococcus lactis*, *St. xylosus*, *St. warneri*, *St. lentus* and *St. auricularis*.

Table 3 Autochthonous microbial population of traditional fermented sausages

Product	Microorganisms	Reference
Italy		
Traditional fermented sausages from Basilicata region	LACTIC ACID BACTERIA *Lb. sakei*[*] *Pd. pentosaceus, Ln. carnosum, Lb. plantarum, Lb. brevis, Ln. pseudomesenteroides*	Bonomo et al. 2008
	COAGULASE-NEGATIVE STAPHYLOCOCCI *St. xylosus*[*] *St. equorum, St. pulvereri/vitulus, St. succinus St. pasteuri, St. saprophyticus, St. warneri, St. caseolyticus*	Bonomo et al. 2009
Traditional fermented sausages from Friuli Venezia Giulia region, northeast Italy	LACTIC ACID BACTERIA *Lb. sakei*[*] *Lb.curvatus*[*] *Lb. plantarum, Lb. paraplantarum/pentosus, E. pseudoavium, Lc. lactis Lc. garvieae, Lb. brevis, Ln. mesenteroides, Ln. citreum, W. paramesenteroides/hellenica*	Comi et al. 2005, Urso et al. 2006b
	COAGULASE-NEGATIVE STAPHYLOCOCCI *St. xylosus*[*] *St. warneri, St. equorum, St. pasteuri, M. caseolyticus, St. epidermidis, St. carnosus, St. saprophyticus, St. pasteuri, St. cohnii*	Iacumin et al. 2006
"Ciauscolo"	LACTIC ACID BACTERIA *Lb. curvatus*[*] *Lb. plantarum*[*] COAGULASE-NEGATIVE STAPHYLOCOCCI *St. xylosus*[*]	Aquilanti et al. 2007
Spain		
"Androlla"	LACTIC ACID BACTERIA *Lb. sakei*[*] *Lb.curvatus*[*] *Lb. alimentarius, Lb. plantarum*	García Fontán et al. 2007a, b

"Botillo"	COAGULASE-NEGATIVE STAPHYLOCOCCI *St. xylosus** *St. epidermidis, St. equorum, St. capitis, St. saprophyticus* LACTIC ACID BACTERIA *Lb. sakei** *Lb. alimentarius, Lb. curvatus, Lb. plantarum, Lb. farcimini*	
Chorizo and salchichón	COAGULASE-NEGATIVE STAPHYLOCOCCI *St. saprophyticus** *St. xylosus, St. lentus, St. cohnii, St. epidermidis, St. sciuri, St. capitis* COAGULASE-NEGATIVE STAPHYLOCOCCI *St. xylosus** *St. warneri, St. epidermidis, St. carnosus, St. lugdunensis, St. chromogens, St. sciuri,* *St. saprophyticus, St. capitis*	Martín et al. 2006, 2007
Fuet and chorizo	LACTIC ACID BACTERIA *Lb. sakei** *Lb. curvatus, Lb. plantarum, E. faecium* COAGULASE-NEGATIVE STAPHYLOCOCCI *St. xylosus** *St. carnosus, St. epidermidis*	Aymerich et al. 2003
France Traditional fermented sausages	LACTIC ACID BACTERIA *Lb. sakei** *E. faecium, E. seriolicida, E. faecalis, Ln. mesenteroides subsp. mesenteroides/* *dextranicum, Lc. garvieae* COAGULASE-NEGATIVE STAPHYLOCOCCI *St. equorum** *St. succinus** *St. saprophyticus, St. xylosus, St. simulans, St. carnosus, St. warneri*	Ammor et al. 2005 Corbière Morot-Bizot et al. 2007, Leroy et al. 2010

Table 3 (*Contd.*)

Table 3 Autochthonous microbial population of traditional fermented sausages (*Contd.*)

Greece Traditional fermented sausages from the south of Greece	LACTIC ACID BACTERIA *Lb. plantarum*[*] *Lb. sakei*[*] *Lb. curvatus, Lb. pentosus, Lb. rhamnosus, Lb. brevis* COAGULASE-NEGATIVE STAPHYLOCOCCI *St. saprophyticus*[*] *St. xylosus*[*] *St. simulans, St. capitis, St. cohnii cohnii, St. equorum, St. gallinarum, St. hominis, St. lentus, St. sciuri , St. vitulinus, St.warneri, St. epidermitis, St. carnosus*	Drosinos et al. 2005, 2007
Croatia Homemade sausages (from Varaždin) Homemade sausages (Slavonian) Slavonian kulen Horsemeat sausages Venison Sausages	LACTIC ACID BACTERIA *Lb. plantarum* 1A*COAGULASE-NEGATIVE STAPHYLOCOCCI *St. carnosus**, St.sciuri* LACTIC ACID BACTERIA *Ln. mesenteroides* SL1* YEASTS *Candida famata* SLK LACTIC ACID BACTERIA *Lb. plantarum* 1K*, *Lb. delbrueckii* 2K, *Ln. mesenteroides* 6K1, *Lb. acidophilus* 7K2 COAGULASE-NEGATIVE STAPHYLOCOCCI *St. carnosus* 4K1*, *St. warneri* 3K1, *St. lentus* 6K2, *St. auricularis* 7K1 LACTIC ACID BACTERIA *Lc. lactis* ssp. *lactis* 5K1*, *Lb. plantarum* HS LACTIC ACID BACTERIA *Lc. lactis* ssp. *lactis* 5K1*, *Lb. plantarum* HS *Ln. mesenteroides* subsp. *mesenteroides* 1V, *Lb. plantarum* 1DV, *Lb. curvatus* 11SR, *E. faecalis* 2DK, *Pd. entosaceus* 6V	Babić et al. 2011, Frece et al. 2010a, b, c, d, 2011, 2014a, c, Markov et al. 2010a, 2013.

*Dominant species in fermented sausages; *Lb.* = *Lactobacillus*, *Ln.* = *Leuconostoc*, *Lc.* = *Lactococcus*, *E.* = *Enterococcus*, *St.* = *Staphylococcus*, *Pd.* = *Pediococcus*, *Source*: Babić et al. 2011, Frece et al. 2014a, c

Autochthonous Starter Cultures

Table 4 Selection criteria for LAB and CoNS used as starter cultures for fermented meat products

LAB	CoNS
✓ quick and adequate production of lactic acid	✓ rapid growth at different temperatures, salt concentrations and pH
✓ production of L (+) - lactic acid	✓ catalase activity and hydrolysis of hydrogen peroxide
✓ rapid growth at different temperatures, salt concentrations and pH	✓ reduction of nitrates and nitrites
✓ homofermentative species	✓ proteolytic and lipolytic enzyme activity
✓ catalase activity and hydrolysis of hydrogen peroxide	✓ conversion activity of branched amino acids
✓ reduction of nitrates and nitrites	✓ inhibition activity of free fatty acids oxidation
✓ lactose negative	✓ tolerance or synergism towards other microbial components of starter culture
✓ proteolytic and lipolytic enzyme activity	✓ production of antimicrobial compounds
✓ tolerance or synergism towards other microbial components of starter culture	✓ antagonism towards pathogenic microorganisms
✓ production of antimicrobial compounds	✓ antagonism towards technologically undesirable microorganisms
✓ antagonism towards pathogenic microorganisms	✓ no antimicrobial resistance
✓ antagonism towards technologically undesirable microorganisms	✓ no formation of biogenic amines
✓ no antimicrobial resistance	✓ economical factors
✓ no formation of biogenic amines	
✓ no mucus production	
✓ probiotic properties (tolerance to low pH, tolerance to bile salts, adhesion to human intestinal cells)	
✓ economical factors	

Source: Frece et al. 2011

3 AUTOCHTHONOUS FUNCTIONAL STARTER CULTURES

Functional starter culture can be defined as microbial cultures that possess at least one functional characteristic with the ultimate goal of improving the quality of the final product that will have a positive impact on the health and physiology of consumers (Leroy et al. 2006, Frece et al. 2010b, c, 2011). Autochthonous functional starter cultures provide additional functionality compared to commercial starter cultures and represent a way of improving and optimizing the fermentation process of foods and getting tastier, safer and healthier products. Functional starter cultures include microorganisms with properties: color development of the product, antimicrobial activity through the reduction of pH, production of antimicrobial compounds such as lactic acid and bacteriocins, catalase activity, reduction of nitrate, formation of flavor, bactericidal activity, reduction of ripening time, probiotic activity (with proper modification of probiotic bacteria may have properties of nutraceuticals – a positive effect on the health of consumers) etc. (Leroy et al. 2006, Ammor and Mayo 2007). Metabolism of most of LAB can lead to the H_2O_2 formation, which impairs sensory properties of meat products by oxidizing nitrosomyoglobin (NOMb) and lipids. The strains of bacteria used in the starter cultures (*Lactobacillus*, *Pediococcus* and *Leuconostoc*) can produce the enzyme catalase, or antioxidant enzyme that catalyzes degradation of H_2O_2 to O_2 and H_2. This enzyme is produced by aerobic forms of bacteria, such as CoNS (*St. carnosus* and *St. xylosus*). Besides its antimicrobial activity, useful probiotic bacteria exhibit and antioxidant properties, due to the activity of the enzyme catalase and/or superoxide dismutase (SOD), which may help in the inhibition of lipid oxidation, prevent the loss of desirable color and texture of food products and the occurrence of toxic compounds.

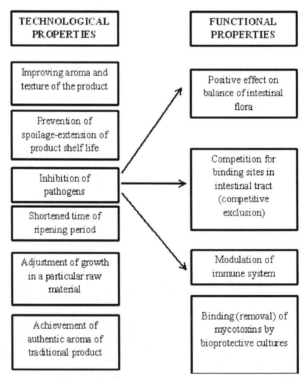

Figure 1 The role of autochthonous starter cultures in improving technological and functional properties in fermented foods production (Markov et al. 2010b, Frece et al. 2011).

Autochthonous Starter Cultures 281

Nowadays, there is an increasing trend of food preservation using autochthonous functional starter cultures, and traditional fermented foods of different origins represent a treasury of biodiversity, which can be used for the development of such starter cultures. For this purpose, there is more research focused on the isolation and identification of autochthonous functional starter cultures, in order to obtain new functional fermented products. The advantages of using autochthonous functional starter cultures in the production of fermented foods and traditional fermented products are shown in Figure 1 (Šušković et al. 2001, Markov et al. 2010b, Frece et al. 2010a, b, c, 2011, 2014a, c, Čvek et al. 2012).

In production technology of cured meats and smoked sausages, fermentation plays an important role. During production, in durable sausages occur numerous chemical, biochemical and microbial processes which contribute to the development of their characteristic smell and taste, and all thanks to the use of starter cultures consisted of certain types of bacteria, yeasts and molds.

3.1 Steps for Selection of Autochthonous Starter Cultures

The selection of a starter is a complex process, involving different steps, like identification of indigenous microbiota by culture independent methods, then phenotypic (Analytical Profile Index – API, Matrix-assisted laser desorption/ionization/Time of flight analizator – MALDI-TOF) and genotypic identification (Polimerase Chain Reaction – PCR, Random amplification of poliymorphic DNA – RAPD, Restriction Fragment Length Polymorphism – RFLP), then preliminary characterization under laboratory conditions (by selection criteria for starter cultures – Table 4), the selection of the most promising strains (selection by technological characteristics), a laboratory validation and final validation in a large-scale fermentation. The selection of a functional starter includes a further step (functional and probiotic characteristics), dealing with the assessment of the functional traits (Babić et al. 2011, Bevilacqua et al. 2012, Frece et al. 2014c).

Technological characterization of LAB and staphylococci isolates encompassed determination of influence of temperature of incubation and concentration of NaCl on their growth. The most important selection criteria for starter cultures are: viability of the cells in the presence of oxgall, proteolytic, lipolytic and antimicrobial activity of viable cells, antibiotic resistance and ability to produce lactic acid in relatively high concentrations. The growth rate at different temperatures of cultivation (from 2–4 to 24°C), the tolerance of salt concentrations of 2–15%, and of pH in the range from 4.2 to 6.0 are limiting factors affecting the viability of the starter cultures as well as competitiveness among it over the entire fermentation and ripening process (Ammor and Mayo 2007). Therefore, the ability of the isolates to grow at different temperatures of 12, 18 and 22°C, and in the presence of 5% of NaCl, that are usually applied during technological processes are one of the selection technological criteria for starter cultures. Technologically relevant properties such as proteolytic and lipolytic activity of LAB and staphylococci isolates are also most important. The LAB usually do not possess pronounced proteolytic or lipolytic activity, although a certain degree of peptidase and lipase activity has been defined for number of LAB strains isolated from meat products (Leroy et al. 2006). For *Staphylococcus* isolates one of the most important technological characteristic is that they must not produce enterotoxins (Dobranić et al. 2013). Catalase activity is one of the desired properties for starter cultures used in food fermentation as it can prevent rancidity and color defects in the products due to the effect of hydrogen peroxide. The production of lactic acid, and eventually decrease of the pH value of the medium, contributes to the development of texture, color and characteristic taste, prevents the growth of pathogenic microorganisms and thus improving food safety and stability of final meat products. Lactic acid and bacteriocin production are the main components responsible for antimicrobial activity of microorganisms (Zdolec et al. 2007b, 2009). One of the most important characteristics of functional starter cultures is inhibition of growth of pathogenic microorganisms which can be obtained by disk-diffusion method and turbidimetric method. In order to act as a probiotic in the gastrointestinal tract the bacteria must be able to survive the acidic conditions of the stomach and resist the bile acids at the beginning of the small intestine (Erkkilä and Petäjä 2000, Frece et al. 2005a, b, 2009). Therefore, investigation of probiotic activity of autochthonous starter cultures include: investigation resistance of lactic acid bacteria and staphylococci isolates to 1% oxgall bile at pH 2.5. Several species of the genus *Lactobacillus* possess surface S-layer protein (SlpA). Due to their structural regularity and the unique self-assembling properties S-layers have potential for many biotechnological applications (Åvall-Jääskeläinen and Palva 2005, Åvall-Jääskeläinen et al. 2008). *Lactobacillus helveticus* M92 possesses a SlpA. A gene coding for the SlpA protein from *Lb. helveticus* M92 was detected by Southern blot hybridization (Frece et al. 2005a). Various data suggested that some of the *Lb. helveticus* M92 probiotic strain could be mediated by its SlpA, notably data concerning strain adhesion to the host cells. Kos et al. (2003) and Frece et al. (2005a) have shown a protective role of S-layers during transit through the GIT and during freezedrying of cultures for probiotic applications. The role of the S-layer in the adherence of *Lb. helveticus* M92 to mouse and pig intestinal epithelial cells was demonstrated (Kos et al. 2003, Frece et al.

2005a). Adhesion is believed to be a requirement for the realisation of probiotic effects, such as pathogen exclusion and immunomodulation (Buck et al. 2005, Lebeer et al. 2008).

3.2 The Role of Autochthonous Starter Cultures in the Production of Fermented Sausages

Although fermented meat products can be produced without the addition of starter cultures, in industrial production starter cultures are used in order to ensure the safety and standardization of the product quality (flavor, color, texture) and shorter ripening period. LAB and GCC+ are commonly used as starters for the production of fermented meat products including fermented sausages. Molds and yeasts are also used as starter cultures. All these microorganisms affect separately or united with tissue enzymes, on the manner and level of changing sensory properties of the product (Drosinos et al. 2007, Ammor and Mayo 2007).

The LAB are the most commonly used starter cultures in industrial production of fermented dairy products, meat and vegetables. Due to the long tradition of using lactic acid bacteria without harmful effect on human health, LAB have GRAS (Generally Recognized As Safe) status under US FDA or QPS (Qualified Presumption of Safety) status under European Union legislation (Šušković et al. 2001, Frece et al. 2014a). The LAB have significant role in the production of fermented meat products, since they affect on technological properties and microbiological stability of the final product. They produce lactic acid, organic acids and bacteriocins, and thereby inhibit the growth of spoilage microorganisms and pathogens, and improve safety, stability and durability of the meat product. Furthermore, they have an important role in formation of aroma, texture and color of meat products, through production of small amounts of acetic acid, ethanol, diacetyl and CO_2 (Lücke 2000, Talon et al. 2002, Leroy et al. 2006, Zdolec 2012). LAB play a major role in sausages production because they produce lactic and acetic acid which lower the pH value. At reduced pH value (4.6 to 5.9) muscle proteins of meat coagulate and lose water, which leads to higher strength, cohesion and easier meat cutting. Such acidic conditions are also important for the ripening process and color formation. In addition, accumulation of lactic and acetic acid inhibits the growth of pathogens. GCC+ (CoNS) also play an important role in production of fermented meat products. They enhance the color stability, prevent rancidity and spoilage, reduce processing time and contribute to the development of flavor (Corbiere Morot-Bizot et al. 2007, Babić et al. 2011). Molds and yeasts are mainly used as starter cultures for the development of aroma, first through the fermentation of carbohydrates, and later through the oxidation of lactate, proteolysis, amino acid degradation and lipolysis (Mauriello et al. 2004). However, as in the case of bacterial starter cultures, selection of molds and yeasts starter cultures depends on the production technology, and must be done carefully, since the proteolytic and lipolytic activity, and therefore the impact on the final product, may vary significantly between different types of molds and yeasts.

While traditional production of fermented sausages relies on ripening of filling with naturally present microbiota or by inoculation with the products of good quality from the previous generation, in the sixties there has been commercialization of starter cultures and their intensive use in the meat industry (Leroy et al. 2006, Babić et al. 2011, Frece et al. 2014a). The use of starter cultures influences hygienic safety of production, ripening process, balances and improves quality and provides better product shelf life. The process of fermentation can take place at higher temperatures that leads to a shorter and safer fermentation and added desirable microorganisms quickly overgrow naturally present microorganisms which can cause undesirable changes in the filling or can produce toxins. Microorganisms commonly used as starter cultures for the production of meat products and their functional properties are summarized in Table 5.

Table 5 Microorganisms commonly used as starter cultures for the production of meat products and their functional properties

Species	Functional property for meat fermentation
Lactic acid bacteria (LAB)	
Lb. sakei	Acidification
	Catalase activity
	Aroma development
	Metabolism of amino acids
	Antioxidant properties: catalase and superoxide dismutase (SOD)
	Production of bacteriocins
Lb. curvatus	Acidification
	Proteolytic activity
	Antioxidant properties: catalase
	Production of bacteriocins
Lb. plantarum	Acidification
	Antioxidant properties: catalase
	Production of bacteriocins
Lb. rhamnosus	Probiotic properties
Pd. acidilactici	Acidification
	Production of bacteriocins
Pd. pentosaceus	Acidification
	Production of bacteriocins

Coagulase-negative staphylococci (CoNS)	
St. xylosus	Aroma development
	Antioxidant properties: catalase and superoxide dismutase (SOD)
	Catabolism of amino acids
	Metabolism of fatty acid
	Reduction of nitrates
St. carnosus	Aroma development
	Antioxidant properties: catalase and superoxide dismutase (SOD)
	Catabolism of amino acids
	Metabolism of fatty acid
	Reduction of nitrates
St. equorum	Aroma development
	Reduction of nitrates
Kocuria	
K. varians	Reduction of nitrates
Molds	
Penicillium nalgiovense	Does not produce toxins
	Good appearance - white color
P. chrysogenum	Non-toxogenic strains have proteases that contribute to the formation of characteristic aroma components
P. gladioli	Does not produce toxins
	Desirable color and appearance
Yeasts	
D. hansenii	Development of characteristic aroma
	Lipolytic activity
	Proteolytic activity
C. famata	Development of characteristic aroma
	Lipolytic activity
	Proteolytic activity

Lb. = *Lactobacillus*, *Pd.* = *Pediococcus*, *St.* = *Staphylococcus*, *K.* = *Kocuria*, *P.* = *Penicillium*, *D.* = *Debaryomyces*, *C.* = *Candida*
Source: Cocconcelli 2007, Babić et al. 2011

4 TRADITIONAL FERMENTED SAUSAGES – SOURCE OF AUTOCHTHONOUS STARTERS

The type or composition of microbiota which develops in traditional sausages is related to diversity of recipe and technology of fermentation and ripening. However, regardless of the different production technologies of traditional fermented sausages among countries or regions, autochthonous microbial population shows similar development. At the end of ripening phase of traditional sausages, LAB make the main population, followed by population of CoNS. It is usual that the number of LAB increases at the begining of fermentation and remains constant during the ripening stage and is 7–9 log cfu/g. The growth of LAB number often matches with a reduction of pH. The LAB are dominant microorganisms of traditional fermented sausages because of their good adaptation to the meat environment and rapid growth during fermentation and maturation. Domination of LAB is supported by anaerobic conditions, addition of curing salt and sugar, and by low initial pH of the mixture (<5.8). The GCC+ are second present microorganisms in traditional fermented sausages at the end of ripening phase, which populations in most sausage is 6-8 log cfu/g.

The LAB are usually present in raw meat in small amounts, but very quickly become dominant in fermentation due to the anaerobic environment and conditions that are favorable for their growth. Isolated from fermented meat products, LAB are well adapted to ecology of meat fermentation, salt concentration, production temperature, acidity of the product, and have the ability of growth and competitiveness in terms of sausages production. They play an important role in processes of conservation and fermentation of meat, are considered technologically fundamental, and can produce different bacteriocins.

The greatest contribution of LAB in flavor development is through their catabolism of carbohydrates, mostly through the production of organic acids, while CoNS are more suitable for formation of typical aroma compounds. LAB also increase spontaneous reduction of nitrite to nitric oxides, which react with myoglobin and form nitrosomyoglobin, which is responsible for the typical pink color of the sausages. For some LAB was discovered that they possess a nitrate reductase, haem-dependent and heme-independent nitrite reductase. These types of LAB are directly involved in nitrosomyoglobin formation mechanisms. LAB usually do not have strong proteolytic and lipolytic properties although for some species, a certain degree of peptidase and lipase activity is noticed.

The use of functional starter cultures may be helpful for reducing the levels of nitrate and nitrite, discussed because of their contribution to nitrosoamine formation. Nitrates and nitrites are necessary in fermentation technology of sausages as curing agents for microbial stability and color formation. Therefore, use of species that have the ability of generating nitrosylated derivatives of myoglobin i.e. the ability to convert brown metmyoglobin to red myoglobin derivatives can be useful in order to partially take over the function of stabilization and color enhancement from curing agents. Coagulase-negative staphylococci contribute to the development and stabilization of color through nitrate reductase activity i.e. they reduce nitrate to nitrite, which leads to the formation of nitrosomyoglobin which gives the product its characteristic red color.

Microorganisms often isolated from traditional fermented sausages are *Lb. sakei*, *Lb. curvatus*, *Lb. plantarum*, *St. xylosus* and *St. saprophyticus*.

These bacteria are autochthonous in the ecosystem of traditional fermented sausages and have the ability of survival and growth during fermentation. In traditional fermented sausages, facultative homofermentative lactobacilli are dominant microbiota during the entire fermentation process (Coppola et al. 2000, Cocconcelli 2007, Talon et al. 2008, Babić et al. 2011, Frece et al. 2014a, c, Danilović and Savić, Chapter 8 in this book).

Naturally fermented sausages are traditional Mediterranean products with a large diversity within different regions (Aymerich et al. 2003). A wide variety of traditional sausages is produced in Croatia of which the most famous one is "Slavonski kulen". Therefore, the aim of our recent research was to identify and characterize naturally present microbial population from the "Slavonski kulen", especially LAB and staphylococci on the basis of their important technological properties in order to select potential autochthonous functional starter cultures. By identification of isolated LAB, domination of *Leuconostoc mesenteroides* and *Lb. acidophilus* was determined, as well as *Lb. plantarum* whereas among the staphylococcal microbiota *St. xylosus* and *St. warneri* were found as the most numerous species (Babić et al. 2011). The high numbers of *Ln. mesenteroides* in Croatian dry sausage "Slavonski kulen", implicates even more the need for introducing competitive technologically suitable starter cultures, because that species is heterofermentative and can cause sensorial deviations of sausages (Zdolec et al. 2013). *Lactobacillus* and *Staphylococcus* species isolated from "Slavonski kulen" were considered as autochthonous functional starter cultures, which can be used in the industrial production, since they are natural microbiota of the "Slavonski kulen" and possess desirable technological and functional characteristics. Most of the isolated LAB and Staphylococcus species displayed good growth capability at 12, 18 and 22°C in the presence of 5% of NaCl. All LAB and most of the staphylococci possess proteolytic activity and only *St. xylosus* had lipolytic activity. All lactobacilli and staphylococci isolates produced significant concentration of lactic acid (as determined by HPLC) and showed antimicrobial activity against pathogenic test microorganisms. Dominant LAB and *Staphylococcus* species displayed growth in the presence of 1% of bile. Most of the staphylococci and all of lactobacilli showed sensitivity to all antibiotics tested (Babić et al. 2011). This work was preliminary study conducted in order to develop an autochthonous functional starter culture. Isolation and characterization of essential members of microbial mixed cultures from region-specific brand products is of great importance. Implementation of this culture in industrial production should improve quality and uniformity of final product as well as food safety by preserving typical sensory quality of the traditional fermented product and inhibiting the growth of undesirable microorganisms. Detailed knowledge about microbiological population that is responsible for ripening process, technological parameters and resulting sensory properties is needed. Variations in overall quality, especially organoleptic characteristics of the product, represents huge problem that remains to be solved before "Slavonski kulen" will meet new market opportunities (Babić et al. 2011). Therefore, besides isolation and characterization of dominant microorganisms, technological and functional characteristics of specific microbial population had to be defined in order to be implemented in production by small-to-medium enterprises. In this way the product will meet all regulations, as needed on a broad market, it will remain autochthonous, safe, and recognizable.

Furthermore, these defined autochthonous functional starter cultures (combination of *Lactobacillus* and *Staphylococcus* strains) were used to produce five different industrial sausages which were compared by a panel. The viability of introduced autochthonous *Lactobacillus* and *Staphylococcus* strains and their effect on the final product characteristics, namely microbiological, physicochemical and sensory properties were monitored. The obtained results indicate that autochthonous starter cultures survived industrial production of sausages and can be used for production of sausages under controlled conditions. Autochthonous starter cultures obtained better results in the organoleptic evaluation, microbial safety and prolonged shelf-life in comparison with commercial starter cultures (Frece et al. 2014a).

Further studies had been carried out to detail phenotypic, genotypic and physiological characterization of isolated strains of staphylococci and LAB, with additional purpose of creating a "Croatian bank" of autochthonous functional starter cultures for traditional fermented meat products.

5 CONCLUSION

Nowadays, in the line with the expanding trend of green living, safe food containing less synthetic additives and items produced by using technologies with less impact on the environment (so-called natural food) have to be offered to consumers. All this reasons opens up a huge space for new methods implemented in safe food production. Implementation of autochthonous starter cultures in industrial production should improve quality and uniformity of final product as well as food safety by preserving typical sensory quality of the traditional fermented product and inhibiting the growth of undesirable microorganisms as standardisation of traditional fermented meat products.

Key words: *autochthonous starter cultures, fermented meat, general and selection criteria*

REFERENCES

Ammor, M.S. and B. Mayo. 2007. Selection criteria for lactic acid bacteria to be used as functional starter cultures in dry sausage production: an update. Meat Sci. 76: 138–146.

Ammor, S., E. Dufour, M. Zagorec, S. Chaillou and I. Chevallier. 2005. Characterization and selection of *Lactobacillus sakei* strains isolated from traditional dry sausage for their potential use as starter cultures. Food Microbiol. 22: 529–538.

Aquilanti, L., S. Santarelli, G. Silvestri, A.Osimani, A. Petruzzelli and F. Clementi. 2007. The microbial ecology of a typical Italian salami during its natural fermentation. Int. J. Food Microbiol. 120: 136–145.

Åvall-Jääskeläinen, S. and A. Palva. 2005. Lactobacillus surface layers and their applications. FEMS Microbiol. Rev. 29: 511–529.

Åvall-Jääskeläinen, S., U. Hynönen, N. Ilk, D. Pum, U.B. Sleytr and A. Palva. 2008. Identification and characterization of domains responsible for self-assembly and cell wall binding of the surface layer protein of *Lactobacillus brevis* ATCC 8287. BMC Microbiol. 8: 1–15.

Aymerich, T., B. Martin, M. Garriga and M. Hugas. 2003. Microbial quality and direct PCR identification of lactic acid bacteria and nonpathogenic Staphylococci from artisanal low-acid sausages. Appl. Environ. Microbiol. 69: 4583–4594.

Aymerich, T., B. Martín, M. Garriga, M.C. Vidal-Carou, S. Bover-Cid and M. Hugas. 2006. Safety properties and molecular strain typing of lactic acid bacteria from slightly fermented sausages. J. Appl. Microbiol. 100: 40–49.

Babić, I., K. Markov, D. Kovačević, A. Trontel, A. Slavica, J. Đugum, D. Čvek, I.K. Svetec, S. Posavec and J. Frece. 2011. Identification and characterization of potential autochthonous starter cultures from a Croatian "brand" product "Slavonski kulen". Meat Sci. 88: 517–524.

Bevilacqua, A., M.R. Corbo and M. Sinigaglia. 2012. Selection of yeasts as starter cultures for table olives: a step-by-step procedure. Front. Microbiol. 31: 194.

Bonomo, M.G., A. Ricciardi, T. Zotta, E. Parente and G. Salzano. 2008. Molecular and technological characterization of lactic acid bacteria from traditional fermented sausages of Basilicata region (Southern Italy). Meat Sci. 80: 1238–1248.

Bonomo, M.G., A. Ricciardi, T. Zotta, M.A. Sico and G. Salzano. 2009. Technological and safety characterization of coagulase-negative staphylococci from traditionally fermented sausages of Basilicata region (Southern Italy). Meat Sci. 83: 15–23.

Bonomo, M.G., A. Ricciardi and G. Salzano. 2011. Influence of autochthonous starter cultures on microbial dynamics and chemical-physical features of traditional fermented sausages of Basilicata region. World J. Microbiol. Biotechnol. 27: 137–146.

Buck, B.L., E. Altermann, T. Svingerud and T.R. Klaenhammer. 2005. Functional analysis of putative adhesion factors in *Lactobacillus acidophilus* NCFM. Appl. Environ. Microbiol. 71: 8344–8351.

Castellano, P., C. Belfiore, S. Fadda and G.Vignolo. 2008. A review of bacteriocinogenic lactic acid bacteria used as bioprotective cultures in fresh meat produced in Argentina. Meat Sci. 79: 483–499.

Cocconcelli, P.S. 2007. Starter Cultures: bacteria. pp. 137–146. In: F. Toldrá, Y.H. Hui, I. Astiasaran, W.K. Nip, J.G. Sebranek, E.T.F. Silveira, L.H. Stahnke and R. Talon (eds.). Handbook of fermented meat and poultry. Blackwell Publishing, Oxford.

Comi, G., R. Urso, L. Iacumin, K. Rantsiou, P. Cattaneo and C. Cantoni. 2005. Characterisation of naturally fermented sausages produced in the North East of Italy. Meat Sci. 69: 381–392.

Coppola, S., G. Mauriello, M. Aponte, G. Moschetti and F. Villani. 2000. Microbial succession during ripening of Naples-type salami, a southern Italian fermented sausage. Meat Sci. 56: 321–329.

Corbière Morot-Bizot, S., S. Leroy and R. Talon. 2007. Monitoring of staphylococcal starters in two French processing plants manufacturing dry fermented sausages. J. Appl. Microbiol. 102: 238–244.

Čvek, D., K. Markov, J. Frece, M. Friganović, L. Duraković and F. Delaš. 2012. Adhesion of zearalenone to the surface of lactic acid bacteria cells. Croat. J. Food Technol. Biotechnol. Nutr. 7: 49–52.

Dobranić, V., N. Zdolec, I. Račić, A. Vujnović, M. Zdelar-Tuk, I. Filipović, N. Grgurević and S. Špičić. 2013. Determination of enterotoxin genes in coagulase-negative staphylococci from autochthonous Croatian fermented sausages. Vet. arhiv 83: 145–152.

Drosinos, E.H., M. Mataragas, N. Xiraphi, G. Moschonas, F. Gaitis and J. Metaxopoulos. 2005. Characterization of the microbial flora from a traditional Greek fermented sausage. Meat Sci. 69: 307–317.

Drosinos, E.H., S. Paramithiotis, G. Kolovos, I. Tsikouras and I. Metaxopoulos. 2007. Phenotypic and technological diversity of lactic acid bacteria and staphylococci isolated from traditionally fermented sausages in Southern Greece. Food Microbiol. 24: 260–270.

Erkkilä, S. and E. Petäjä. 2000. Screening of commercial meat starter cultures at low pH and in the presence of bile salts for potential probiotic use. Meat Sci. 55: 297–300.

Fontana, C., P.S. Cocconcelli and G. Vignolo. 2005. Monitoring the bacterial population dynamics during fermentation of artisanal Argentinean sausages. Int. J. Food Microbiol. 103: 131–142.

Frece, J., B. Kos, I.K. Svetec, Z. Zgaga, V. Mrša and J. Šušković. 2005a. Importance of S-layer proteins in probiotic activity of Lactobacillus acidophilus M92. J. Appl. Microbiol. 98: 285–292.

Frece, J., B. Kos, J. Beganović, S. Vuković and J. Šušković. 2005b. In vivo testing of functional properties of three selected probiotic strains. World J. Microbiol. Biotechnol. 21: 1401–1408.

Frece, J., B. Kos, I.K. Svetec, Z. Zgaga, J. Beganović, A. Leboš and J. Šušković. 2009. Synbiotic effect of Lactobacillus helveticus M92 and prebiotics on the intestinal microflora and immune system of mice. J. Dairy Res. 76: 98–104.

Frece, J., K. Markov and D. Kovačević. 2010a. Determination of indigenous microbial populations, mycotoxins and characterization of potential starter cultures in Slavonian kulen. Meso 12: 92–98.

Frece, J., D. Čvek, D. Kovačević, I. Gobin, T. Krcivoj and K. Markov. 2010b. Characterization of bacterial strain Lactobacillus plantarum 1K isolated from Slavonian "kulen", as probiotic functional starter culture. Meso 12: 208–214.

Autochthonous Starter Cultures

Frece, J., K. Markov, D. Čvek and D. Kovačević. 2010c. Staphylococci as potential indigenous starter cultures from „Slavonian kulen". Meso 12: 156–160.

Frece, J., J. Pleadin, K. Markov, N. Perši, V. Dukić, D. Čvek and F. Delaš. 2010d. Microbe population, chemical composition and mycotoxins in sausages from Varaždin County. Vet. Stn. 41: 189–198.

Frece, J., K. Markov and F. Delaš. 2011. Traditional fermented meat products - the source of autochthonous functional starter cultures. Proceedings of fourth symposium "Functional Foods in Croatia". Croatia. 6–11.

Frece, J., D. Kovačević, S. Kazazić, J. Mrvčić, N. Vahčić, F. Delaš, D. Ježek, M. Hruškar, I. Babić and K. Markov. 2014a. Comparison of sensory properties, shelf live and microbiological safety of industrial sausages produced with autochthonous and commercial starter cultures (Starter cultures for sausages production). Food Technol. Biotechnol. 52: 307–316.

Frece, J., J. Mrvčić, I. Uglešić, Z. Lončar, D. Kovačević, F. Delaš and K. Markov. 2014b. Industrial sausage produced with autochthonous starter culture. Proceedings of 8th International congress of food technologist, biotechnologists and nutritionists. Croatia. 6–11.

Frece, J., J. Cvrtila, I. Topić, F. Delaš and K. Markov. 2014c. *Lactococcus lactis* ssp. *lactis* as potential functional starter culture. Food Technol. Biotechnol. 52: 489–494.

García Fontán, M.C., J.M. Lorenzo, A. Parada, I. Franco and J. Carballo. 2007a. Microbiological characteristics of "androlla", a Spanish traditional pork sausage. Food Microbiol. 24: 52–58.

García Fontán M.C., J.M. Lorenzo, S. Martínez, I. Franco and J. Carballo. 2007b. Microbiological characteristics of Botillo, a Spanish traditional pork sausage. LWT 40: 1610–1622.

Iacumin, L., G. Comi, C. Cantoni and L. Cocolin. 2006. Ecology and dynamics of coagulase-negative cocci isolated from naturally fermented Italian sausages. Syst. Appl. Microbiol. 29: 480–486.

Kargozari, M., S. Moini, A.A. Basti, Z. Emam-Djomeh, H. Gandomi, I. Revilla Martin, M. Ghasemlou and A.A. Carbonell-Barrachina. 2014. Effect of autochthonous starter cultures isolated from Siahmazgi cheese on physicochemical, microbiological and volatile compound profiles and sensorial attributes of sucuk, a Turkish dry-fermented sausage. Meat Sci. 97: 104–114.

Kos, B., J. Šušković, S. Vuković, M. Šimpraga, J. Frece and S. Matošić. 2003. Adhesion and aggregation ability of probiotic strain *Lactobacillus acidophilus* M92. J. Appl. Microbiol. 94: 981–987.

Lebeer, S., J. Vanderleyden and S.C. De Keersmaecker. 2008. Genes and molecules of lactobacilli supporting probiotic action. Microbiol. Mol. Biol. Rev. 72: 728–764.

Lebert, I., S. Leroy, P. Giammarinaro, A. Lebert, J.P. Chacornac, S. Bover-Cid, M.C. Vidal-Carou and R. Talon. 2007. Diversity of microorganisms in the environment and dry fermented sausages of small traditional French processing units. Meat Sci. 76: 112–122.

Leroy, F., J. Verluyten and L. De Vuyst. 2006. Functional meat starter cultures for improved sausage fermentation. Int. J. Food Microbiol. 106: 270–285.

Leroy, S., P. Giammarinaro, J.-P. Chacornac, I. Lebert and R. Talon. 2010. Biodiversity of indigenous staphylococci of naturally fermented dry sausages and manufacturing environments of small-scale processing units. Food Microbiol. 27: 294–301.

Lücke, F.K. 2000. Utilization of microbes to process and preserve meat. Meat Sci. 56: 105–115.

Markov, K., J. Frece, D. Čvek, A. Trontel, A. Slavica and D. Kovačević. 2010a. Dominant microflora of fermented horse meat sausages. Meso 12: 217–221.

Markov, K., J. Frece, D. Čvek, N. Lovrić and F. Delaš. 2010b. Aflatoxin M1 in raw milk and binding of aflatoxin by lactic acid bacteria. Mljekarstvo 60: 244–251.

Markov, K., N. Perši, J. Pleadin, D. Čvek, V. Radošević, F. Delaš, L. Duraković and J. Frece. 2011. Characterization of natural microflora and chemical parameters in fresh domestic cheese. Vet. Stn. 42: 211–218.

Markov, K., J. Pleadin, M. Horvat, M. Bevardi, D. Sokolić Mihalak, F. Delaš and J. Frece. 2013. Microbiological and mycotoxicological safety risks and characterization of homemade sausages of game meat. Vet. Stn. 44: 177–186.

Martín, B., M. Garriga, M. Hugas, S. Bover-Cid, M.T. Veciana-Nogues and T. Aymerich. 2006. Molecular, technological and safety characterization of Gram-positive catalase-positive cocci from slightly fermented sausages. Int. J. Food Microbiol. 107: 148–158.

Martín, A., B. Colín, E. Aranda, M.J. Benito and M.G. Córdoba. 2007. Characterization of Micrococcaceae isolated from Iberian dry-cured sausages. Meat Sci. 75: 696–708.

Mauriello, G., A. Casaburi, G. Blaiotta and F. Villani. 2004. Isolation and technological properties of coagulase negative staphylococci from fermented sausages of southern Italy. Meat Sci. 67: 149–158.

Palavecino Prpich, N.Z., M.P. Castro, M.E. Cayre, O.A. Garro and G.M. Vignolo. 2015. Indigenous starter cultures to improve quality of artisanal dry fermented sausages from Chaco (Argentina). Int. J. Food Sci. article ID 931970.

Rantsiou, K., R. Urso, L. Iacumin, C. Cantoni, P. Cattaneo, G. Comi and L. Cocolin. 2005. Culture-dependent and -independent methods to investigate the microbial ecology of Italian fermented sausages. Appl. Environ. Microbiol. 71: 1977–1986.

Rantsiou, K. and L. Cocolin. 2006. New developments in the study of the microbiota of naturally fermented sausages as determined by molecular methods: a review. Int. J. Food Microbiol. 108: 255–267.

Šušković, J., B. Kos, J. Goreta and S. Matošić. 2001. Role of lactic acid bacteria and bifidobacteria in synbiotic effect. Food Technol. Biotechnol. 39: 227–235.

Talon, R., S. Leroy-Sétrin and S. Fadda. 2002. Bacterial starters involved in the quality of fermented meat products. pp. 175–191. In: F. Toldrá (ed.). Research advances in the quality of meat and meat products. Research Signpost, India.

Talon, R., S. Leroy, I. Lebert, P. Giammarinaro, J.P. Chacornac, M. Latorre-Moratalla, C. Vidal-Carou, E. Zanardi, M. Conter and A. Lebecque. 2008. Safety improvement and preservation of typical sensory qualities of traditional dry fermented sausages using autochthonous starter cultures. Int. J. Food Microbiol. 126: 227–234.

Urso, R., K. Rantsiou, C. Cantoni, G. Comi and L. Cocolin. 2006a. Technological characterization of a bacteriocin-producing *Lactobacillus sakei* and its use in fermented sausages production. Int. J. Food Microbiol. 110: 232–239.

Autochthonous Starter Cultures 293

Urso, R., K. Rantsiou, C. Cantoni, G. Comi and L. Cocolin. 2006b. Sequencing and expression analysis of the sakacin P bacteriocin produced by a *Lactobacillus sakei* strain isolated from naturally fermented sausages. Appl. Microbiol. Biotechnol. 71: 480–485.

Zdolec, N., M. Hadžiosmanović, L. Kozačinski, Ž. Cvrtila, I. Filipović, S. Marcinčak, Ž. Kuzmanović and K. Hussein. 2007a. Protective effect of *Lactobacillus sakei* in fermented sausages. Arch. Lebensmittelhyg. 58: 152–155.

Zdolec, N., L. Kozačinski, M. Hadžiosmaović, Ž. Cvrtila and I. Filipović. 2007b. Inhibition of *Listeria monocytogenes* growth in fermented sausages. Vet. arhiv 77: 507–514.

Zdolec, N., M. Hadžiosmanović, L. Kozačinski, Ž. Cvrtila, I. Filipović, M. Škrivanko and K. Leskovar. 2008. Microbial and physicochemical succession in fermented sausages produced with bacteriocinogenic culture of *Lactobacillus sakei* and semi-purified bacteriocin mesenterocin. Meat Sci. 80: 480–487.

Zdolec, N., L. Kozačinski, B. Njari, I. Filipović, M. Hadžiosmanović, B. Mioković, Ž. Kuzmanović, M. Mitak and D. Samac. 2009. The antimicrobial effect of lactobacilli on some foodborne bacteria. Arch. Lebensmittelhyg. 60: 115–119.

Zdolec, N. 2012. Lactobacilli – functional starter cultures for meat applications. pp. 273–289. *In*: A.I. Peres Campos and A.L. Mena (eds.). *Lactobacillus*: classification, uses and health implications. Nova Publishers, USA.

Zdolec, N., V. Dobranić, A. Horvatić and S. Vučinić. 2013. Selection and application of autochtonous functional starter cultures in traditional Croatian fermented sausages. Int. Food Res. J. 20: 1–6.

Chapter 13

Probiotics in Fermented Meat Products

Rodrigo J. Nova*, George Botsaris and Fabiola Cerda-Leal

1 INTRODUCTION

Several scientific reports suggest a positive correlation between the administration of certain microorganisms generally regarded as safe (GRAS) with health benefits. Microorganisms displaying these health properties are widely known as probiotics, which have been defined by FAO and WHO (2002) as "live microorganisms that, when administered in an adequate amount, confer a health benefit on the host". The administration of probiotics could be done through either, the direct ingestion of the probiotic cultures or through food products which could act as a carrier for these microorganisms.

Although the main reason for using microorganisms in food industry is related to the enhancement of technological characteristics of the food product (e.g.; organoleptic changes and extend shelf-life), the potential health benefits associated to the consumption of food products containing probiotic has been widely studied, albeit not fully understood. In general terms, the health benefits associated to probiotics are not only related to a direct health benefit in the host but also by preventing the harbouring and development of pathogenic microorganisms via competitive

---

*For Correspondence: School of Veterinary Medicine and Science, Sutton Bonington Campus, University of Nottingham, LE12 5RD, United Kingdom, Email: rodrigo.nova@nottingham.ac.uk

exclusion or through the production of exo-molecules (Salminen et al. 1996, Gänzle et al. 1999).

Health benefits associated to consumption of probiotics have been reported for a wide array of conditions, ranging from gastrointestinal (GI) infections, improvement in lactose metabolism, antimicrobial activity, immune system stimulation, reduction in serum cholesterol, anti-carcinogenic properties, prevention of several forms of diarrhoea, management of chronic inflammatory bowel disease, modulation of intestinal microflora prevention and treatment of food allergies and lowering plasma cholesterol level amongst others (Martini et al. 1991, Pool-Zobel et al. 1996, Hlivak et al. 2005, O'Mahony et al. 2005, Elli et al. 2006, Hickson et al. 2007, Nguyen et al. 2007, Francavilla et al. 2008, Huurre et al. 2008, Grandy et al. 2010, Rodríguez et al. 2010).

Most of the microorganisms characterised as probiotics are lactic acid bacteria (LAB), with several of them belonging to the genera *Bifidobacterium* and *Lactobacillus*. However, other genera of bacteria such as *Lactococcus*, *Enterococcus* and *Propionibacterium* as well as yeasts (*Saccharomyces*) have also been classified as probiotic cultures or suggested to display probiotic properties (Sanders and Huis in't Veld 1999, Blandino et al. 2003, Vinderola and Reinheimer 2003). Nevertheless, probiotic activity seems to be linked to specific strains rather than to a bacterial genus or species. Thus, further studies are needed in order to identify probiotic cultures suitable to be used in food production (Majamaa et al. 1995, Cotter et al. 2005, O'Mahony et al. 2005, Huurre et al. 2008).

Nowadays, in developed countries, there is an increasing consumer demand for food products, including meat and meat products that can offer health benefits (Kołożn-Krajewska and Dolatowski 2012). As a result there is an increasing number of food products available on the market claiming to grant specific health benefits (Arvanitoyannis and Van Houwelingen-Koukaliaroglou 2005). Prior to authorizing the commercialization of food products as beneficial to health, regulatory bodies require evidence of the claimed properties. The process starts with the selection of microbial strains that could be used as probiotics, and a number of criteria must be fulfilled so safety and production/manufacturing are not compromised when adding these microorganisms to a food product. The survival and colonization in the host must be also assessed to verify that the microorganisms can effectively colonize the host and are not only transient microflora. *In vitro* trials should be available to investigate whether the microbial strains fulfil all the previously described criteria. These trials should aid in the process of screening microorganisms to select probiotics. However, further validation tests with animal and human trials should follow (FAO and WHO 2002).

A large number of microorganisms that have been categorised as probiotics are already used as starter cultures in the dairy and meat industry for the production of fermented food products (Hill et al. 2014). However probiotics cultures and the health benefits associated to the food products containing them have been mostly studied, promoted and commercialized in dairy food products rather than in the meat sector (Heller 2001). It is important to highlight that this category of "healthy" is not related only to a functional value due to the presence of probiotic cultures, but also with a reduced level of fat, cholesterol, improved composition of fatty acid profile and lowering the levels of sodium chloride and nitrites in the food product (Casaburi et al. 2016).

Meat products have a major role in human nutrition because they represent an important source for protein, fat, essential amino acids, minerals and vitamins, amongst other nutrients (Biesalski 2005). Technologically, the target food products of meat origin which could be further developed to offer an added probiotic functional value are fermented meat dry products. In order to evaluate health benefits associated to the consumption of meat products containing probiotics, it is essential to assess and select the most suitable probiotic strains for the manufacturing of these fermented meat products, as probiotic strains should be able not only to withstand the environmental conditions found in a fermented meat product, but also must endure the processing and be able to compete and dominate over other micro microorganisms that could be present in the food product or in the starter mix (Kołożn-Krajewska and Dolatowski 2012).

Meat could be a suitable carrier to support and deliver probiotic microorganisms to the host (Klingberg et al. 2005, Muthukumarasamy and Holley 2007). The capacity of meat to carry and sustain probiotic cultures may be reduced due to the pH of the microenvironment and the activity water and carbohydrates available in the product. However, the meat matrix has been found to protect LAB strains against the bactericidal effect of the presence of bile salts in the intestinal track (Gänzle et al. 1999). Additionally, methods such as microencapsulation may allow probiotic cultures to reach the GI in a desirable number (Leroy et al. 2006, Muthukumarasamy et al. 2006, Amine et al. 2014).

As meat has been shown to be good vehicle for probiotics, it is of great interest to carry out further studies on the likely health benefits of probiotics cultures if they are used in the manufacturing of fermented meat products. This chapter will discuss the health and food technology benefit of probiotics as well as the potential of these microorganisms to be included in the manufacturing of meat fermented food products.

2 HEALTH BENEFITS ASSOCIATED TO THE CONSUMPTION OF PROBIOTICS

How changes in the human intestinal microbiota are associated to the human health status is not yet fully understood. However, it is widely accepted that bacterial species in the normal gut microflora of a healthy individual, amongst which there are several microorganisms that are considered as probiotic, do influence human health status. It is accepted that the supplementation of human diet with selected probiotic microorganisms could be beneficial for human health (Lee 2014). In fact, more than a hundred years ago Elie Metchnikoff, Nobel Laureate in Medicine and one of the pioneers of the study of immunology, linked human health and longevity to the consumption of bacteria present in fermented dairy products (Metchnikoff 1908, Shah 2007). Since then, several studies have demonstrated a strong association between various aspects of human health and the direct consumption of bacterial or yeast strains or food products that contain them or their exopolysaccharides (Rodríguez et al. 2010).

It has been proposed that for the host to obtain health benefits, probiotics must harbour and colonize the gastrointestinal mucosa. Although some reports assess the viability of a given microorganism as a probiotic by evaluating the number of potential probiotics shed and recovered in feces, probiotics must not only be able to stand the environmental stress to which they will be exposed during the gastrointestinal transit, but also to colonize the small and large intestines while maintaining the metabolic properties that make them probiotic (Tuomola et al. 2001, Shimakawa et al. 2003, Elli et al. 2006).

In vitro studies showed that the binding capacity of probiotic microorganisms to the intestinal epithelial cells and mucins is related to the expression of surface proteins (Granato et al. 2004). The capacity to adhere to mucosa can be measured indirectly by autoaggregation where a high autoaggregation capacity has a positive correlation to bacterial binding to the mucosa (Del Re et al. 1998). However, the relevance of these surface proteins is not only associated to gastrointestinal system (GI) binding, but also to immunomodulation as it has been demonstrated that surface proteins of probiotic cultures stimulate a proinflammatory response in the host (Granato et al. 2004).

Human health benefits associated to the consumption of probiotics could be achieved through modulation of the host immune response and competition for nutrients and niche spaces in the GI, thus limiting the harboring and colonization of pathogenic microorganisms (Marco et al. 2006, Cammarota et al. 2009). At the same time, the host could benefit from the presence of probiotic microorganism in many other different ways, such as a more effective digestion process, better absorption of nutrients and supporting and adequate epithelial cell proliferation and differentiation (Steenbergen et al. 2015). Most of the commonly associated health benefits due to the administration of probiotics are related to the GI system. In fact, it has been shown that prevention and treatment of acute infectious diarrhea and infantile colic; prevention of antibiotic-associated diarrhea, nosocomial diarrhea and traveler's diarrhea; prevention of necrotizing enterocolitis, could all be aided by the administration of probiotic microorganisms (Cruchet et al. 2015, Savino et al. 2015). However, beneficial effects to human health due to the consumption of probiotics are not limited to the GI.

Numerous species from Genus *Bifidobacterium* (e.g. *adolescentis, animalis, bifidum, breve* and *longum*) and *Lactobacillus* (e.g. *acidophilus, casei, fermentum, gasseri, johnsonii, paracasei, plantarum, rhamnosus and salivarius*) are well known probiotic microorganisms (Table 1). Hill et al. (2014) suggested categorizing the distribution of the probiotics' mechanisms associated to health benefits in three categories: widespread among commonly studied probiotic Genera, frequently observed among most strains of a probiotic species and rare and present in only a few strains of a given species. As seen in Table 1, a given probiotic strain may offer more than one health benefit however no strain has been described as being able to provide all the known benefits.

2.1 Gastrointestinal System

The mechanisms by which probiotics positively impact the GI system vary according the species and strain of the probiotics administrated. However some mechanisms, such as the process to reduce the effects of lactose intolerance are not only well understood, but also widely spread amongst probiotic LAB (Hill et al. 2014). LAB are able to produce β–galactosidase, allowing microorganisms to survive the GI transit by granting them resistance to bile acids. By modifying the colonic microbiota, where an increased number of bacteria are able to produce β–galactosidase will be present post-administration of probiotics. Thus increasing the amount of the enzyme present in the GI, therefore the host benefits with the alleviation of lactose intolerance symptoms (Martini et al. 1991, Elli et al. 2006, Montalto et al. 2006, He et al. 2008).

Irritable bowel syndrome can be caused or could be the result of an unbalanced microbiota (Fanigliulo et al. 2006). It has been shown that administration of probiotics could restore balance of the microbial population, resulting in the reduction of the severity of abdominal distension and abdominal pain, as well as decreasing bowel motility (Fanigliulo et al. 2006, Kajander et al. 2008, Rodríguez et al. 2010). It has been suggested that probiotics could provide these health benefits by stimulating an increase of the thickness of the gastric mucus layer in the intestine and modulating the host immune response, resulting in the recovery of the normal ratio between anti-inflammatory and pro-inflammatory cytokines (interleukins (IL)-10 and IL-12) (O'Mahony et al. 2005, Hickson et al. 2007, Rodríguez et al. 2010).

Colonisation of the GI by pathogens can also be prevented by probiotic strains, a beneficial effect that has also been described for other body systems such as the urogenital system (Shalev et al. 1996, Reid et al. 2004, Coudeyras et al. 2008). The effectiveness on preventing the harbouring of pathogens in the GI seems to be time dependant, where earlier administration of probiotics in infants results in a lower fecal anaerobic gram-negative viable count and higher total anaerobic gram-positive viable count; with the former including pathogens from the Enterobacteriaceae family, while the latter includes the ingested probiotic strains (Savino et al. 2015).

Prevention of colonisation of the GI by bacterial pathogens induced by probiotics can be achieved through the reinforcement of the intestinal barrier, where probiotics may reduce the permeability o the epitelium. Permeability of epithelial cells can be assessed *in vitro* by measuring the trans-epithelial electrical resistance (TER), with higher resistance resulting in lower permeability. Essays carried out cell culture demonstrated a higher trans-epithelial electrical resistance (TER) and a higher expression of heat shock protein (Hsp) after exposure to *Lactococcus* and *Bifidobacterium* probiotic cultures. It has been suggested that the over expression of Hsp may favour repair of the mucosal integrity after infection with *Salmonella enteritidis* (Koninkx et al. 2010). GI infection with pathogenic microorganisms can also be prevented by competitive exclusion through adherence of probiotics to the mucosa (Cruchet et al. 2015).

The severity and duration of symptoms associated to acute viral gastroenteritis (watery diarrhea), such as the one produced by rotavirus, can significantly decrease after administration of probiotics. However, these effects seem to be largely dose dependant and species/strain dependant (Majamaa et al. 1995, Shornikova et al. 1997, Grandy et al. 2010). Human health benefits related to the administration of probiotics to patients suffering of acute viral gastroenteritis include providing lasting immunity post probiotic administration which could prevent rotavirus reinfections, as probiotic microorganisms can stimulate the generation of IgA antibody-secreting cells resulting in high serum IgA antibody level during the recovery stage in the host (Majamaa et al. 1995).

Table 1 Probiotic microorganisms and the suggested mechanism associated to health benefit. Modified from Hill et al. 2014

Mechanism	Frequency	Microorganism	Reference
Colonization resistance	Widespread	Lb. acidophilus	Shalev et al. 1996,
		Lb. rhamnosus Lcr35	Coudeyras et al. 2008
Acid and short chain fatty acids production	Widespread	Lb. pentosus S3T60C	Pessione et al. 2015
		Lb. plantarum S11T3E	
Regulation of intestinal transit	Widespread	B. animalis subsp. lactis Bb12	Martini et al. 1991,
		B. infantis 35624	O'Mahony et al. 2005,
		B. longum W11	Elli et al. 2006,
		Lb. delbrueckii subsp. bulgaricus 1190	Fanigliulo et al. 2006,
		Lb. reuteri ATCC 55730	Francavilla et al. 2008,
		Lb. rhamnosus GG	Kajander et al. 2008,
		Lb. rhamnosus Lc705	Rodríguez et al. 2010
		P. freudenreichii subsp. shermanii JS	
		Str. salivarius subsp. thermophilus	
		Str. thermophilus	
		Str. thermophilus CRL	
Normalization of perturbed microbiota	Widespread	B. bifidum (W23)	Persborn et al. 2013
		B. lactis (W51)	
		B. lactis (W52)	
		Lb. acidophilus (W22)	
		Lb. casei (W56)	
		Lb. paracasei (W20)	
		Lb. plantarum (W62)	
		Lb. salivarius (W24)	
		Lb. lactis (W19)	
Increased turnover of enterocytes	Widespread	Lb. reuteri	Preidis et al. 2012
Competitive exclusion of pathogens	Widespread	Lb. bulgaricus	Hickson et al. 2007
		Lb. casei	
		Str. thermophilus	

Vitamin synthesis	Frequent	*B. adolescentis* *B. pseudocatenulatum* *Lb. plantarum*	Rossi et al. 2011
Gut barrier reinforcement	Frequent	*B. bifidum* W23 *B. infantis* W52 *Lb. acidophilus* W70 *Lb. casei* subsp. *casei* GG *Lb. casei* subsp. *rhamnosus* *Lb. casei* W56 *Lb. lactis* W58 *Lb. salivarius* W24	Majamaa et al. 1995, Shornikova et al. 1997, Grandy et al. 2010, Koninkx et al. 2010
Neutralization of carcinogens	Frequent	*B. breve* *B. longum* *Lb. acidophilus* CL1285 *Lb. casei* LBC80R *Lb. gasseri* P79 *Lb. confusus* DSM20196 *Lb. rhamnosus* CLR2 *Str. thermophilus*	Pool-Zobel et al. 1996, Desrouillères et al. 2015
Bile salt metabolism Enzymatic activity Direct antagonism Neurological effects	Frequent Frequent Frequent Rare	*Lb. reuteri* *Bacillus coagulans* NJ0516 *Lb. johnsonii* NCC533 *Lb. reuteri* ATCC 55730 *B. animalis* subsp. *lactis* CNCM I-2494 *B. longum* R0175 *Lb. bulgaricus* CNCM I-1632 *Lb. helveticus* R0052 *Lc. lactis* subsp. *lactis* CNCM I-1631 *Str. thermophilus* CNCM I-1630	Nollet et al. 1999, Wang and Gu 2010, Granato et al. 2004, Francavilla et al. 2008 Tillisch et al. 2013, Ait-Belgnaoui et al. 2014, Smith et al. 2014

Table 1 (*Contd.*)

Table 1 Probiotic microorganisms and the suggested mechanism associated to health benefit. Modified from Hill et al. 2014 (*Contd.*)

Production of specific bioactives	Rare	*Lb. acidophilus* ATCC 4356 *Lb. casei* ATCC 393 *Lb. paracasei subsp. paracasei* ATCC BAA52	Sah et al. 2014
Immunological effects	Rare	*B. animalis* subsp. *lactis* Bb12 *B. longum* *Lb. casei* *Lb. casei* Shirota *Lb. delbrueckii* subsp. *bulgaricus* *Lb. rhamnosus* GG *Str. thermophilus*	Kato et al. 1999, Kalliomäki et al. 2003, Meyer et al. 2006, Kukkonen et al. 2007, Medina et al. 2007, Huurre et al. 2008, Kekkonen et al. 2008
Endocrinological effects	Rare	*E. faecium* M-74 *Lb. curvatus* HY7601 *Lb. gasseri* SBT0270 *Lb. helveticus* *Lb. paracasei* *Lb. plantarum* KY1032 *Lb. plantarum* PH04 *Saccharomyces cerevisiae*	Hata et al. 1996, Hosono 2000, Jahreis et al. 2002, Nguyen et al. 2007, Hlivak et al. 2005, Ahn et al. 2015

Lb. = *Lactobacillus*, *B.* = *Bifidobacterium*, *Str.* = *Streptococcus*, *Lc.* = *Lactococcus*, *P.* = *Propionibacterium*, *E.* = *Enterococcus*

2.2 Other Suggested Health Benefits

2.2.1 Cancer Prevention

Reports suggest that the administration of probiotics could be associated with the potential prevention of cancer through the inhibition of genotoxic metabolites and by reducing the production of bacterial enzymes in the gastrointestinal track (Pool-Zobel et al. 1996, Desrouillères et al. 2015). However, the mechanisms responsible for the inhibition or enhancing of carcinogenic enzymatic activity have not been elucidated (Desrouillères et al. 2015). Current data on the potential role of probiotics in the prevention of cancer focus on *in vitro* experiments or animal trials; hence caution is needed when extrapolating these outcomes to humans.

2.2.2 Neurological Effects

It is recognised that the microbiota present in the GI system can modulate behaviour, neurophysiology and neurochemistry of an individual. This gut-brain axis, which involves the intestinal microbiota, is a bidirectional communication channel which, when out of balance, can result in the development of a number of pathologies (Cryan and Dinan 2012, Smith et al. 2014). For example, irritable bowel syndrome can lead to depressive disorders, however it has been demonstrated that administration of probiotics in experimental animals can restore normal behaviour. That effect could be associated to an increased production of serotonin (Desbonnet et al. 2008, Smith et al. 2014). Similarly, chronic stress can result in an unbalanced gut microbiota as well as chronic reduction of the abnormal brain plasticity and a reduction in neurogenesis. However, it has been suggested that the administration of probiotics could modulate neuroregulatory factors as well as signalling pathways at the central nervous system level reducing the negative impact of chronic stress (Ait-Belgnaoui et al. 2014).

2.2.3 Immunological Effects

Results of group intervention studies assessing probiotic microorganisms show that administration of probiotics can enhance the immune response in several different pathways. For example, the number of cytotoxic T lymphocytes and the cytotoxic activity significantly increase after ingesting dairy products containing probiotics and this effect can persist for months after ceasing probiotic administration (Meyer et al. 2006). Ingestion of probiotics could also result in a decrease of serum C-reactive protein levels in healthy adults (Kekkonen et al. 2008). Enhancement of immune response associated to administration of probiotics has also been demonstrated in *in vitro* trials, where it has been demonstrated that probiotics could induce the production of the cytokines interleukin (IL)-12 and interferon-gamma (IFN-gamma) (Kato et al. 1999, Medina et al. 2007). Another suggested effect on the modulation of immune response by probiotics are associated to the reduction of the risk of sensitization and eczema in high risk children (Kalliomäki et al. 2003, Kukkonen et al. 2007, Huurre et al. 2008).

2.2.4 Endocrinological Effects

The presence of LAB in the normal gut microbiota is reduced in individuals consuming diets with a high fat content (Sun et al. 2016). Reports suggest that administration of probiotics could result in a lower risk of obesity by inducing a reduction of fat mass and weight loss (Kang et al. 2013, Jung et al. 2015). Some of the described mechanisms involved in the prevention of obesity are a higher level of expression of fatty-acid oxidation genes and a reduction of serum levels of leptin and insulin (Kang et al. 2013). Clinical trials in humans have also shown that administration of probiotics can induce hypocholesterolemia (Hosono 2000, Hlivak et al. 2005, Ahn et al. 2015). Reduction of total serum cholesterol concentration has been associated to probiotic cultures enhancing the reabsorption of bile acids into the enterohepatic circulation as well as to increased excretion of acidic steroids in feces (Hosono 2000).

2.2.5 Production of Specific Bioactives

Several LAB, including probiotic cultures, are able to synthesize antimicrobial peptides known as bacteriocins. In fermented food products the natural presence of LAB able to produce bacteriocins is beneficial in regards of food preservation, additionally there is a potential use of bacteriocins as an alternative to antimicrobial treatment (Cotter et al. 2005). LAB belonging to the genera *Lactobacillus*, also normally present in fermented food products, can produce bioactive peptides which have shown a potential antioxidant and antimutagenic activity (Sah et al. 2014). However, the associated beneficial effects to bioactive substances is not necessarily linked to administration of the live microorganism, but also to subcellular components such as cell wall proteins or extra cellular proteins (Ramesh et al. 2015). Further research is needed in order to identify these bio-compounds and assess how their activity could have a beneficial effect in human health.

3 DEVELOPMENT PROCESS OF NOVEL FUNCTIONAL MEAT PRODUCTS

The aim of food processing is to transform raw food materials into edible, safe, wholesome, digestible, nutritious food products, with desirable physical and chemical properties, extended shelf life and optimal organoleptic characteristics. Further developing food products with novel functional features or optimized functional components can be difficult. Often the production of functional features increases the level of complexity and monitoring needed in the food product development process from the generation of the idea to the launch of the new product.Overall, the market share of functional food products is currently dominated by dairy products. To obtain a health benefit the probiotic strains added to the food product should be present in numbers higher than 10^6 cfu per gram of product (Kołożn-Krajewska and Dolatowski, 2012). Heat processed meat products would not fit in the category of a functional food. The heat treatment applied in this manufacturing process would decrease the activity or viability of any probiotic microorganism that could have been added to the food productresulting in a dramatic reduction in the expected health benefits. On the other hand, fermented meat products could potentially provide a good matrix to deliver probiotics to consumers demanding a higher availability of functional food products in the markets' shelves, (Saxelin et al. 2000, Heller 2001). Additionally, consumers are familiar with the fact that fermented food products contain living microorganisms.

Any attempt to expand to the production of functional food to the meat sector must be done in a systematic and reliable fashion; ensuring that the "health" status of the new food product is validated. It has been suggested that the validation process of the health benefits associated to a microorganism should be carried out in the following three steps (Reid et al. 2003):

- *In vitro* tests demonstrating the safety and the probiotic qualities of the proposed strain.
- *In vivo* tests in animal models to explain the mode of action of the probiotic.
- Clinical tests with volunteers.

At the same time, FAO and WHO (2002) proposed international guidelines on the selection of probiotic strains to be used in food products. These guidelines follow a series of assessments, where a set of stringent requirements must be fulfilled in order to validate health claims of a probiotic/functional food product (Figure 1). The process to develop a functional product starts with the conception of a general idea.

For example the concept could be a fermented sausage with a probiotic strain added in the starter culture mix. Later, the concept is developed further, firstly by evaluating the technical feasibility of producing such as a product then the question of when and at what numbers to add the probiotic strain. All of which are assessed through existing information and through practical challenge tests.

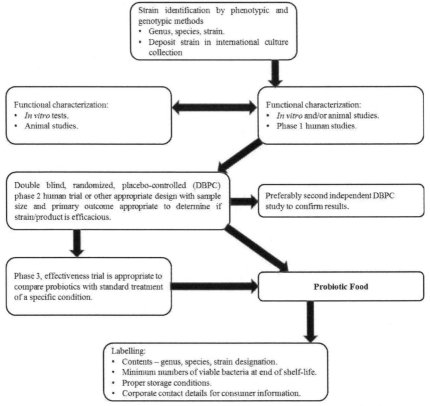

Figure 1 FAO/WHO guidelines for the evaluation of probiotics for food use (adapted from FAO and WHO, 2002).

Finally, before introducing the product into the market, trials must be performed in order to insure that the product is accepted by the consumers and that any legal aspect (health claims) have been verified properly (Hill et al. 2014). Achieving this requires reliable evidence and outcomes that could be equally applicable to all the major groups in the population, including those to whom their diet behaviour may compromise or reduce the expected health benefits from consuming the functional food. Additionally, the process might involve post-marketing monitoring in order to assess any potential effects on the dietary habits of the population. The production costs must not be forgotten when planning the development of a probiotic food, as the cost should be evaluated and optimized, if possible, in an attempt to produce an affordable product to the consumer.

In order to protect consumers, the authorisation of commercialisation of food products including health claims is strictly regulated in some countries. In the European Union (EU) for example, in addition to the required scientific evidence of the associated health benefits, the label or the advertisement of the product must include (European Commission 2006):

- A statement indicating the importance of a varied and balanced diet and a healthy lifestyle;
- The quantity of the food and pattern of consumption required to obtain the claimed beneficial effect;
- Where appropriate, a statement addressed to persons who should avoid using the food; and
- An appropriate warning for products that are likely to represent a health risk if consumed to excess.

3.1 Probiotics and Meat Products

Hammes and Hertel (1998) defined meat starter cultures as "preparations which contain living or resting microorganisms that develop the desired metabolic activity in the meat". Functional starter cultures offer additional beneficial properties compared to classical starter cultures. They represent a method to improve and optimize the meat fermentation process, thus resulting in the development of more complex flavors and aromas, as well as potentially achieving the production of safer and healthier food products (Leroy et al. 2006). There is an increasing interest in microorganisms that demonstrate probiotic properties however, data related to adding probiotic microorganisms to fermented meat products are scarce when compared with the existing reports on the use of probiotics strains in other food sectors such as dairy industry.

The development of fermented meat products with probiotic characteristics present some challenges. For example, in a fermented meat product several microorganisms will be present in the normal starter culture, where the resulting interaction between the starter culture and the probiotic strain may affect the viability of the latter. Thus the final food product may not be categorized as functional (Hill et al. 2014, Muthukumarasamy and Holley 2006). Additionally, the use of nitrites, high concentration of sodium chloride, low activity water and low concentration of carbohydrates in meat, may all have a detrimental impact on the proposed probiotic strain (Kołożn-Krajewska and

Dolatowski 2012). Not only the survival of the probiotic microorganism is of relevance, but also their metabolic activity, as it has been shown that probiotic cultures may not retain their probiotic properties until the end of the shelf-life of the food product (Saxelin et al. 1999, Antunes et al. 2005, Tamime et al. 2005, Vuyst et al. 2008, de Macedo et al. 2012).

The genetic and metabolic profile of several bacterial strains normally characterised as probiotics have been assessed. Neither technological nor sensory properties of the final product seem to be affected in fermented meat sausages when compared to similar meat product produced with a traditional starter culture (Erkkilä et al. 2001a, Klingberg et al. 2005). Adding probiotic cultures in meat products could result in a desirable inhibition of zoonotic food-borne pathogens such as *Escherichia. coli* O157, *E. coli* O111 or *Listeria monocytogenes*, which could potentially be present in the food product (Pidcock et al. 2002, Muthukumarasamy and Holley 2007). Additionally, it has been shown that some probiotic strains could survive both, the fermented meat product manufacturing process and the gastrointestinal transit. However, the harbouring of probiotics in the intestines of the host has been proved to be variable, depending not only of the probiotic strain used, but also of host related factors (Klingberg et al. 2005).

Methods such as bacterial microencapsulation have been suggested to be an effective intervention method in order to protect the probiotic cultures added to fermented meat products, thus the aim of delivering an adequate amount of microorganisms to the consumer could be achieved (Muthukumarasamy et al. 2006, Muthukumarasamy and Holley 2006, Muthukumarasamy and Holley 2007, Corona-Hernandez et al. 2013). Coating materials such as ionic polysaccharides, microbial exopolysaccharides, and milk proteins enhance microorganism survival rate and additionally improve microbial adhesion to the intestinal mucosa; however, the extent of the benefits of microencapsulation vary according the microencapsulation coating material used and the probiotic strain tested (Corona-Hernandez et al. 2013). Tests carried out in both, simulated gastric juice and dry fermented sausages, show a higher survival rate for microencapsulated probiotic microorganisms than for planktonic cells (Muthukumarasamy et al. 2006, Muthukumarasamy and Holley 2007). This higher viable count of microencapsulated probiotic microorganisms in a fermented meat product seems to lack of a detrimental impact in the organoleptic characteristics of the food product (Erkkilä et al. 2001b, Muthukumarasamy and Holley 2006). However, it has been shown that the coating of *Lb. reuteri and B. longum* could negatively affect desirable characteristics of the probiotics, such as impairing their capacity to inactivate pathogenic bacteria in the food product (Muthukumarasamy and Holley 2007).

4 CONCLUSION

Most of the research in probiotics and functional foods has been carried out in developed, countries where there is an increasing number of consumers expressing an avid interest in functional foods. However, in several developing countries fermented food products have played an important role in the daily diet of the most vulnerable groups of the population for centuries (Obodai and Dodd 2006, Franz et al. 2014). Hence the identification of probiotic strains and the transfer of this knowledge is of great relevance to these communities. The development and/or standardization of the production of fermented food products containing probiotics should be taken in consideration when implementing strategies for food security and could contribute to the efforts to achieve the eradication of extreme poverty and hunger (Franz et al. 2014).

The most common causes of morbidity and mortality in newborn children in developing countries are gastrointestinal illnesses. Adding probiotics cultures to products, such as fermented dairy and meat could not only improve their nourishment but also improve the health status of the whole population by potentially controlling and preventing human diseases (Drisko et al. 2003, Rodriguez et al. 2010). Animal trials show that feed enriched with some probiotic cultures can act as a non-medicated grow promoter, which may be related to increased protease and amylase activities (Wang and Gu 2010). Thus, food aimed to vulnerable and underweight children potentially could also be supplemented with probiotics in order to achieve a faster recovery of body condition (Majamaa et al. 1995, Hickson 2007, Rodriguez et al. 2010).

The suggestion of using of probiotics as therapeutic treatment rather than as a mean of prevention of illness must be justified with solid scientific evidence in order to avoid compromising the health of the patients or creating unrealistic expectations (Majamaa et al. 1995, Kalliomäki et al. 2003, O'Mahoney et al. 2005, Kukkonen et al. 2007, Huurre et al. 2008, Rodriguez et al. 2010). Although there are already promising outcomes in the use of probiotics, caution is needed when providing health recommendations, as there are several factors associated with the degree of effectiveness of the administration of probiotics against a condition or disease. Published research on probiotics tests carrying out human trials do not always report a clear health benefit or are lacking in descriptions of the biological basis of the suggested health benefit (Shornikova et al. 1997, Hickson et al. 2007, Kajander et al. 2008, Francavilla et al. 2008, Salazar-Lindo et al. 2014).

Intrinsic factors of the probiotic such as: expression of surface proteins that favour adhesion to the mucosa; factors of the host such as diet, physiological stage, pH of the GI; or severity of the condition amongst others, could result in a wide variation on the outcome on the health status of a patient after the administration. In order to decrease the variation of the observed positive health effects associated to probiotics, administration of mixed preparations seem to be more effective than the administration of a single strain (Majamaa et al. 1995, Kajander et al. 2008, Grandy et al. 2010).

Currently there is a limited amount of information available on the development of meat fermented products containing probiotics in the starter culture mix. Most of the existing research in probiotics and their use in meat products has been carried out in order to assess the viability of probiotic microorganisms during the food manufacturing process and during the gastrointestinal transit rather than focusing in the development of novel probiotic meat products (Papamanoli et al. 2003, Ruiz-Moyano et al. 2011).

A better understanding and characterisation of the natural microflora of fermented sausages may help to identify and select idoneous probiotic cultures, resulting in a more effective and standardised fermentation process, as well as a safer product containing beneficial live bacteria. As probiotics strains may not be able to survive in fermented food products in a numbers that could result in health benefits (Lücke 2000, Hill et al. 2014), alternative methods are available to make these microorganisms available to the consumer. Microencapsulation has been shown to reduce the negative effect of the environment of the food matrix on the viability of the probiotic cultures in the final product and at the same time may reduce required maturation time (Leroy et al. 2006, Muthukumarasamy et al. 2006, Amine et al. 2014).

Current issues in labelling of food products as "probiotics" with little or no evidence of the associated health benefit or with an unknown viable count number at the end of the shelf life of the product have resulted in a number of food products available in the market where expectation of the consumers may not be reached; such products should be best described as 'containing live and active cultures' instead of being commercialised as probiotic food products (Hill et al. 2014).

Nitrate reduction is an interesting and promising area in the technological development of meat products. Recent studies have shown that inoculation with different probiotic strains as starter cultures in the manufacturing of fermented meat products could contribute with nitrite depletion (Wang et al. 2013, Casaburi et al. 2016). These findings are of great relevance in view of the latest recommendations of the World Health Organization on the consumption of processed meat products (WHO 2015).

It is also necessary to establish a dietary recommendation for food products containing probiotic microorganisms (Hill et al. 2014).

Probiotics in Fermented Meat Products 311

However, to propose meat as a vehicle for delivering probiotics to humans will most likely be consider a conflictive proposal, as in order to achieve the maximum health benefits it would be necessary to consume high quantities of a given meat-based probiotic food, possibly resulting in adverse health effects associated to a higher than recommended dietary meat intake (e.g. obesity, hypertension and diabetes) (Katan 2012, Kołożn-Krajewska and Dolatowski 2012).

The use of probiotic bacteria in food manufacturing has resulted in the formulation of several doubts: how organoleptic characteristics are affected, how can their safety and effectiveness be assessed and reassured, which factors affect the viability of probiotic microorganisms and how could they be improved, The answers to most of these questions remain ambiguous and new products entering the functional food market should clearly address these concerns. Thus, it is of great interest to continue carrying out research on the use of probiotics in the manufacturing process of fermented meat products and more evidence is needed in order to elucidate the answers to these concerns.

Key words: *probiotics, meat, fermentation, starter culture, health benefits, functional foods, in vivo test, in vitro test.*

REFERENCES

Ahn, H.Y., M. Kim, J.S. Chae, Y.-T. Ahn, J-H. Sim, I-D. Choi, S.-H. Lee and J.H. Lee. 2015. Supplementation with two probiotic strains, *Lactobacillus curvatus* HY7601 and *Lactobacillus plantarum* KY1032, reduces fasting triglycerides and enhances apolipoprotein A-V levels in non-diabetic subjects with hypertriglyceridemia. Atherosclerosis 241: 649–656.

Ait-Belgnaoui, A., A. Colom, V. Braniste, L. Ramalho, A. Marrot, C. Cartier, E. Houdeau, V. Theodorou and T. Tompkins. 2014. Probiotic gut effect prevents the chronic psychological stress-induced brain activity abnormality in mice. Neurogastroent. Motil. 26: 510–520.

Amine, K.M., C.P. Champagne, Y. Raymond, D. St-Gelais, M. Britten, P. Fustier, S. Salmieria and M. Lacroix. 2014. Survival of microencapsulated *Bifidobacterium longum* in Cheddar cheese during production and storage. Food Control 37: 193–199.

Antunes, A.E.C., T.F. Cazetto and H.M. Abolini. 2005. Viability of probiotic micro-organisms during storage, postacidification and sensory analysis of fat-free yogurts with added whey protein concentrate. Int. J. Dairy Technol. 58: 169–173.

Arvanitoyannis, I.S. and M. Van Houwelingen-Koukaliaroglou. 2005. Functional foods: a survey of health claims, pros and cons, and current legislation. Crit. Rev. Food Sci. 45: 385–404.

Biesalski, H.K. 2005. Meat as a component of a healthy diet – are there any risks or benefits if meat is avoided in the diet? Meat Sci. 70: 509–524.

Blandino, A., M.E. Al-Aseeri, S.S. Pandiella, D. Cantero and C. Webb. 2003. Cereal-based fermented foods and beverages. Food Res. Int. 36: 527–543.

Cammarota, M., M. De Rosa, A. Stellavato, M. Lamberti, I. Marzaioli and M. Giuliano. 2009. In vitro evaluation of *Lactobacillus plantarum* DSMZ 12028 as a probiotic: emphasis on innate immunity. Int. J. Food Microbiol. 135: 90–98.

Casaburi, A., V. Di Martino, P. Ferranti, L. Picariello and F. Villani. 2016. Technological properties and bacteriocins production by *Lactobacillus curvatus* 54M16 and its use as starter culture for fermented sausage manufacture. Food Control 59: 31–45.

Corona-Hernandez, R.I., E. Alvarez-Parrilla, J. Lizardi-Mendoza, A.R. Islas-Rubio, L.A. de la Rosa and A. Wall-Medrano. 2013. Structural stability and viability of microencapsulated probiotic bacteria: a review. Compr. Rev. Food Sci. F. 12: 614–628.

Cotter, P.D., C. Hill and R.P. Ross. 2005. Bacteriocins: developing innate immunity for food. Nat. Rev. Microbiol. 3: 777-788.

Coudeyras, S., G. Jugie., M. Vernerie and C. Forestier. 2008. Adhesion of human probiotic *Lactobacillus rhamnosus* to cervical and vaginal cells and interaction with vaginosis-associated pathogens. Infect. Dis. Obstet. Gynecol. ID 549640.

Cruchet, S., R. Furnes, A. Maruy, E. Hebel, J. Palacios, F. Medina, N. Ramirez, M. Orsi, L. Rondon, V. Sdepanian, L. Xóchihua, M. Ybarra and R.A. Zablah. 2015. The use of probiotics in pediatric gastroenterology: a review of the literature and recommendations by Latin-American experts. Paediatr. Drugs 17: 199–216.

Cryan, J.F. and T.G. Dinan. 2012. Mind-altering microorganisms: the impact of the gut microbiota on brain and behaviour Nat. Rev. Neurosci. 13: 701-712.

de Macedo, R.E.F., S. Bertelli Pflanzer and C. Lugnani Gomes (2012). Probiotic meat products. Probiotic in animals. E. Rigobelo (ed.). InTech, DOI: 10.5772/50057. Available from: http://www.intechopen.com/books/probiotic-in-animals/probiotic-meat-products.

Del Re, B., A. Busetto, G. Vignola, B. Sgorbati and D. Palenzona. 1998. Autoaggregation and adhesion ability in a *Bifidobacterium suis* strain. Lett. Appl. Microbiol. 27: 307–310.

Desbonnet, L., L. Garrett, G. Clarke, J. Bienenstock and T.G. Dinan. 2008. The probiotic *Bifidobacteria infantis*: An assessment of potential antidepressant properties in the rat. J. Psychiatr. Res. 43: 164-174.

Desrouillères, K. M., Millette, K. Dang Vu, R. Touja and M. Lacroix. 2015. Cancer preventive effects of a specific probiotic fermented milk containing *Lactobacillus acidophilus* CL1285, *L. casei* LBC80R and *L. rhamnosus* CLR2 on male F344 rats treated with 1,2- dimethylhydrazine. J. Funct. Foods 17: 816–827.

Drisko, J. A., C.K. Giles and B.K. Bischoff. 2003. Probiotics in health maintenance and disease prevention. Altern. Med. Rev. 8: 143–155.

Elli, M., M.L. Callegari, S. Ferrari, E. Bessi, D. Cattivelli, S. Soldi, L. Morelli, N.G. Feuillerat and J.M. Antoine. 2006. Survival of yogurt bacteria in the human gut. Appl. Environ. Microbiol. 72: 5113–5117.

Probiotics in Fermented Meat Products   313

Erkkilä, S., M.L. Suihko, S. Eerola, E. Petäjä and T. Mattila-Sandholm. 2001a. Dry sausage fermented by *Lactobacillus rhamnosus* strains. Int. J. Food Microbiol. 64: 205–210.

Erkkilä, S., E. Petäjä, S. Eerola, L. Lilleberg, T. Mattila-Sandholm and M.L. Suihko. 2001b. Flavour profiles of dry sausages fermented by selected novel meat starter cultures. Meat Sci. 58: 111–116.

European Commission (2006). Regulation (EC) No 1924/2006 of the European Parliament and of the Council of 20 December 2006 on nutrition and health claims made on foods. Official Journal of the European Communities. L 404/9.

Fanigliulo, L., G. Comparato G. Aragona, L. Cavallaro, V. Iori, M. Maino, G.M. Cavestro, P. Soliani, M. Sianesi, A. Franzè and F. Di Mario. 2006. Role of gut microflora and probiotic effects in the irritable bowel syndrome. Acta Biomed. 77: 85–89.

FAO and WHO. 2002. Joint FAO/WHO Working group report on drafting guidelines for the evaluation of probiotics in food. Food and Agriculture Organization of the United Nations (FAO) and World Health Organization (WHO). London, Ontario, Canada, April 30 and May 1.

Francavilla, R., E. Lionetti, S.P. Castellaneta, A.M. Magista, G. Maurogiovanni, N. Bucci, A. De Canio, F. Indrio, L. Cavallo, E. Ierardi and V.L. Miniello. 2008. Inhibition of *Helicobacter pylori* infection in humans by *Lactobacillus reuteri* ATCC 55730 and effect on eradication therapy: a pilot study. Helicobacter 13: 127–134.

Franz, C.M.A.P., M. Hucha, J. Mathara, H. Abriouel, N. Benomar, G. Reid, A. Galvez and W.H. Holzapfel. 2014. African fermented foods and probiotics. Int. J. Food Microbiol. 190: 84–96.

Gänzle, M.G., S. Weber and W.P. Hammes. 1999. Effect of ecological factors on the inhibitory spectrum and activity of bacteriocins. Int. J. Food Microbiol. 46: 207–217.

Granato, D., G.E. Bergonzelli, R.D. Pridmore, L. Marvin, M. Rouvet and I.E. Corthésy-Theulaz. 2004. Cell surface-associated elongation factor Tu mediates the attachment of *Lactobacillus johnsonii* NCC533 (La1) to human intestinal cells and mucins. Infect. Imun. 72: 2160–2169.

Grandy, G., M. Medina, R. Soria, C.G. Terán and M. Araya. 2010. Probiotics in the treatment of acute rotavirus diarrhea. A randomized, double blind, controlled trial using two different probiotic preparations in Bolivian children. BMC Infect. Dis. 10: 253.

Hata, Y., M. Yamamoto, M. Ohni, K. Nakajima, Y. Nakamura and T. Takano. 1996. A placebo-controlled study of the effect of sour milk on blood pressure in hypertensive subjects. Am. J. Clin. Nutr. 64: 767–771.

Hammes, W.P. and C. Hertel. 1998. New developments in meat starter cultures. Meat Sci. 49: S125–S138.

He, T., M.G. Priebe, Y. Zhong, C. Huang, H.J.M. Harmsen, G.C. Raangs, J.M. Antoine, G.W. Welling and R.J. Vonk. 2008. Effects of yogurt and bifidobacteria supplementation on the colonic microbiota in lactose-intolerant subjects. J. Appl. Microbiol. 104: 595-604.

Heller, K.J. 2001. Probiotic bacteria in fermented foods: product characteristics and starter organisms. Am. J. Clin. Nutr. 73: 374s–379s.

314   Fermented Meat Products: Health Aspects

Hickson, M., A.L. D'Souza, N. Muthu, T.R. Rogers, S. Want, C. Rajkumar and C.J. Bulpitt. 2007. Use of probiotic *Lactobacillus* preparation to prevent diarrhea associated with antibiotics: randomized double blind placebo controlled trial. Brit. Med. J. 335: 80.

Hill, C., F. Guarner, G. Reid, G.R. Gibson, D.J. Merenstein, B. Pot, L. Morelli, R. Berni Canani, H.J. Flint, S. Salminen, P.C. Calder and M.E. Sanders. 2014. Expert consensus document: The International Scientific Association for Probiotics and Prebiotics consensus statement on the scope and appropriate use of the term probiotic. Nat. Rev. Gastroenterol. Hepatol. 11: 506–514.

Hlivak, P., J. Odraska, M. Ferencik, L. Ebringer, E. Jahnova and Z. Mikes. 2005. One-year application of probiotic strain *Enterococcus faecium* M-74 decreases serum cholesterol levels. Bratisl. Med. J. 106: 67–72.

Hosono, A.U. 2000. Effect of administration of *Lactobacillus gasseri* on serum lipids and fecal steroids in hypercholesterolemic rats. J. Dairy Sci. 83: 1705–1711.

Huurre, A., K. Laitinen, S. Rautava, M. Korkeamaki and E. Isolauri. 2008. Impact of maternal atopy and probiotic supplementation during pregnancy on infant sensitization: a double-blind placebo-controlled study. Clin. Exp. Allergy 38: 1342–1348.

Jahreis, G., H. Vogelsang, G. Kiessling, R. Schubert and W.P. Hammes. 2002. Influence of probiotic sausage (*Lactobacillus paracasei*) on blood lipids and immunological parameters of healthy volunteers. Food Res. Int. 35: 133–138.

Jung, S., J.L Young, K. Minkyung, K. Minjoo, H. Hyun, L. Ji-Won, A. Young-Tae, S. Jae-Hun and H.L. Jong. 2015. Supplementation with two probiotic strains, *Lactobacillus curvatus* HY7601 and *Lactobacillus plantarum* KY1032, reduced body adiposity and Lp-PLA2 activity in overweight subjects. J. Funct. Foods. 19: 744-752.

Kajander, K., E. Myllyluoma, M. Rajilic-Stojanovic, S. Kyronpalo, M. Rasmussen, S. Jarvenpaa, E.G. Zoetendal, W.M. de Vos, H. Vapaatalo and R. Korpela. 2008. Clinical trial: multispecies probiotic supplementation alleviates the symptoms of irritable bowel syndrome and stabilizes intestinal microbiota. Aliment. Pharmacol. Ther. 27: 48–57.

Kalliomäki, M., S. Salminen, T. Poussa, H. Arvilommi and E. Isolauri. 2003. Probiotics and prevention of atopic disease: 4-year follow-up of a randomised placebo-controlled trial. Lancet 361: 1869–1871.

Kang J.H, S.I. Yun, M.H. Park, J.H. Park, S.Y. Jeong and H.O. Park. 2013. Anti-Obesity Effect of *Lactobacillus gasseri* BNR17 in High-Sucrose Diet-Induced Obese Mice. PLoS ONE. 8(1): e54617. doi:10.1371/journal.pone.0054617.

Katan, M.B. 2012. Why the European Food Safety Authority was right to reject health claims for probiotics. Benef. Microbes. 3: 85–89.

Kato, I., K. Tanaka and T. Yokokura. 1999. Lactic acid bacterium potently induces the production of interleukin-12 and interferon-Ⅾ by mouse splenocytes. Int. J. Immunopharmacol. 21: 121–131.

Kekkonen, R.A., N. Lummela, H. Karjalainen, S. Latvala, S. Tynkkynen, S. Järvenpää, H. Kautiainen, I. Julkunen, H. Vapaatalo and R. Korpela. 2008. Probiotic intervention has strain-specific anti-inflammatory effects in healthy adults. World J. Gastroenterol. 7: 2029–2036.

Probiotics in Fermented Meat Products

Klingberg, T.D., L. Axelsson, K. Naterstad, D. Elsser and B.B. Budde. 2005. Identification of potential probiotic starter cultures for Scandinavian-type fermented sausages. Int. J. Food Microbiol. 105: 419–431.

Kołożn-Krajewska, D. and Z.J. Dolatowski. 2012. Probiotic meat products and human nutrition. Process Biochem. 47: 1761–1772.

Koninkx, J.F.J.G., P.C.J. Tooten and J.J. Malago. 2010. Probiotic bacteria induced improvement of the mucosal integrity of enterocyte-like Caco-2 cells after exposure to *Salmonella enteritidis* 857. J. Funct. Foods 2: 225–234.

Kukkonen, K., E. Savilahti, T. Haahtela, K. Juntunen-Backman, R. Korpel, T. Poussa, T. Tuure and M. Kuitunen. 2007. Probiotics and prebiotic galacto-oligosaccharides in the prevention of allergic diseases: a randomized, doubleblind, placebo-controlled trial. J. Allergy Clin. Immunol. 119: 192–198.

Lee, Y.K. 2014. What could probiotic do for us? Food Science and Human Wellness 3: 47–50.

Leroy F., J. Verluyten and L. Vuyst. 2006. Functional meat starter cultures for improved sausage fermentation. Int. J. Food Microbiol. 106: 270–285.

Lücke, F.K. 2000. Utilization of microbes to process and preserve meat. Meat Sci. 56: 105–115.

Majamaa, H., E. Isolauri, M. Saxelin and T. Vesikari. 1995. Lactic acid bacteria in the treatment of acute rotavirus gastroenteritis. J. Pediatr. Gastroenterol. Nutr. 20: 333–338.

Marco, M.L., S. Pavan and M. Kleerebezem. 2006. Towards understanding molecular modes of probiotic action. Curr. Opin. Biotechnol. 17: 204–210.

Martini, M.C., E.C. Lerebours, W.J. Lin, S.K. Harlander, N.M. Berrada, J.M. Antoine and D.A. Savaiano. 1991. Strains and species of lactic acid bacteria in fermented milks (yogurts): effect on in vivo lactose digestion. Am. J. Clin. Nutr. 54: 1041–1046.

Medina, M., E. Izquierdo, S. Ennaharand and Y. Sanz. 2007. Differential immunomodulatory properties of *Bifidobacterium logum* strains: relevance to probiotic selection and clinical applications. Clin. Exp. Immunol. 150: 531–538.

Metchnikoff, E.E. 1908. Prolongation of life: Optimistic studies. pp. 161-183 *In*: P. Chalmers Mitchell (ed.). Putnam's Sons. New York & London.

Meyer, A.L., M. Micksche, I. Herbacek and I. Elmadfa. 2006. Daily intake of probiotic as well as conventional yogurt has a stimulating effect on cellular immunity in young healthy women. Ann. Nutr. Metab. 50: 282–289.

Montalto M., V. Curigliano, L. Santoro, M. Vastola, G. Cammarota, R. Manna, A. Gasbarrini and G. Gasbarrini. 2006. Management and treatment of lactose malabsorption. World J. Gastroenterol. 12: 187–191.

Muthukumarasamy, P., P. Allan-Wojtas and R.A. Holley. 2006. Stability of *Lactobacillus reuteri* in different types of microcapsules. J. Food Sci. 71: M20-M24.

Muthukumarasamy, P. and R.A. Holley. 2006. Microbiological and sensory quality of dry fermented sausages containingalginate-microencapsulated *Lactobacillus reuteri*. Int. J. Food Microbiol. 111: 164–169.

Muthukumarasamy, P. and R.A. Holley. 2007. Survival of *Escherichia coli* O157:H7 in dry fermented sausages containing micro-encapsulated probiotic lactic acid bacteria. Food Microbiol. 24: 82–88.

Nguyen, T.D., J.H. Kang and M.S. Lee. 2007. Characterization of *Lactobacillus plantarum* PH04, a potential probiotic bacterium with cholesterol-lowering effects. Int. J. Food Microbiol. 113: 358–361.

Nollet, L.J.A., D.I. Pereira and W. Verstraete. 1999. Effect of a probiotic bile salt hydrolytic *Lactobacillus reuteri* on the human gastrointestinal microbiota as simulated in the SHIME reactor system. Microb. Ecol. Health Dis. 11: 13–21.

Obodai, M. and C.E.R. Dodd. 2006. Characterization of dominant microbiota of a Ghanaian fermented milk product, nyarmie, by culture- and non culture-based methods. J. Appl. Microbiol. 100: 1355–1363.

O'Mahony, L., J. McCarthy, P. Kelly, G. Hurley, F. Luo, K. Chen, G.C. O'Sullivan, B. Kiely, J.K. Collins, F. Shanahan and E.M. Quigley. 2005. *Lactobacillus* and *Bifidobacterium* in irritable bowel syndrome: symptom responses and relationship to cytokine profiles. Gastroenterology 128: 541–551.

Papamanoli, E., N. Tzanetakis, E. Litopoulou-Tzanetaki and P. Kotzekidou. 2003. Characterization of lactic acid bacteria isolated from a Greek dry-fermented sausage in respect of their technological and probiotic properties. Meat Sci. 65: 859–867.

Persborn, M., J. Gerritsen, C. Wallon, A. Carlsson, L.M.A. Akkermans and J.D. Söderholm. 2013. The effects of probiotics on barrier function and mucosal pouch microbiota during maintenance treatment for severe pouchitis in patients with ulcerative colitis. Aliment. Pharmacol. Ther. 38: 772–783.

Pessione, A., G. Lo Bianco, E. Mangiapane, S. Cirrincione and E. Pessione. 2015. Characterization of potentially probiotic lactic acid bacteria isolated from olives: Evaluation of short chain fatty acids production and analysis of the extracellular proteome. Food Res. Int. 67: 247–254.

Pidcock, K1, G.M. Heard and A. Henriksson. 2002. Application of nontraditional meat starter cultures in production of Hungarian salami. Int. J. Food Microbiol. 76: 75–81.

Pool-Zobel, B.L., C. Neudecker, I. Domizlaff, S. Ji, U. Schillinger, C. Rumney, M. Moretti, I. Vilarini, R. Scassellati-Sforzolini and I. Rowland. 1996. *Lactobacillus*-and *Bifidobacterium*-mediated antigenotoxicity in the colon of rats. Nutr. Cancer 263: 365–380.

Preidis, G.A., D.M. Saulnier, S.E. Blutt, T.A. Mistretta, K.P. Riehle, A.M. Major, S.F. Venable, J.P. Barrish, M.J. Finegold, J.F. Petrosino, R.L. Guerrant, M.E. Conner and J. Versalovic. 2012. Host response to probiotics determined by nutritional status of rotavirus-infected neonatal mice. J. Pediatr. Gastroenter. Nutr. 55: 299–307.

Ramesh, D., A. Vinothkanna, A. K. Rai and V.S. Vignesh. 2015. Isolation of potential probiotic *Bacillus* spp. and assessment of their subcellular components to induce immune responses in *Labeo rohita* against *Aeromonas hydrophila*. Fish Shellfish Immun. 45: 268-276.

Reid, G., J. Jass, M.T. Sebulsky and J.K. McCormick. 2003. Potential uses of probiotics in clinical practice. Clin. Microbiol. Rev. 16: 658-672.

Reid, G., J. Burton and E. Devil. 2004. The rationale for probiotics in female urogenital healthcare. Med. Gen. Med. 6: 49.

Rodríguez, C., M. Medici, F. Mozzi, and G.F. De Valdez. 2010. Therapeutic effect of *Streptococcus thermophilus* CRL 1190-fermented milk on chronic gastritis. World J. Gastroenterol. 16: 1622–1630.

Rossi, M., A. Amaretti and S. Raimondi. 2011. Folate production by probiotic bacteria. Nutrients 3: 118–134.

Ruiz-Moyano S., A. Martín, M.J. Benito, A. Hernández, R. Casquete and M.G. Córdoba. 2011. Application of *Lactobacillus fermentum* HL57 and *Pediococcus acidilactici* SP979 as potential probiotics in the manufacture of traditional Iberian dry-fermented sausages. Food Microbiol. 28: 839–847.

Sah, B.N.P., T. Vasiljevic, S. McKechnie and O.N. Donkor. 2014. Effect of probiotics on antioxidant and antimutagenic activities of crude peptide extract from yogurt. Food Chem. 156: 264–270.

Salazar-Lindo, E., P. Miranda-Langschwager, M. Campos Sanchez, E. Chea-Woo and R. Bradley Sack. 2004. *Lactobacillus casei* strain GG in the treatment of infants with acute watery diarrhea: A randomized, double-blind, placebo controlled clinical trial [ISRCTN67363048]. BMC Pediatrics. 4: 18.

Salminen, S., E. Isolauri and E. Salminen. 1996. Clinical uses of probiotics for stabilising the gut mucosal barrier: successful strains and future challenges. Anton. Leeuw. Int. J. G. 70: 251–262.

Sanders, M.E. and J. Huis in't Veld. 1999. Bringing a probiotic-containing functional food to the market: microbiological, product, regulatory and labelling issues. Anton. Leeuw. Int. J. G. 76: 293–315.

Savino, F., S. Fornasero, S. Ceratto, A. De Marco, N. Mandras, J. Roana, V. Tullio and G. Amisano. 2015. Probiotics and gut health in infants: A preliminary case–control observational study about early treatment with *Lactobacillus reuteri* DSM 17938. Clin. Chim. Acta 451: 82–87.

Saxelin, M., B. Grenov, U. Svensson, R. Fondén, R. Reniero and T. Mattila-Sandholm. 1999. The technology of probiotics. Trends Food Sci. Tech. 10: 387–392.

Shah, N.P. 2007. Functional cultures and health benefits. Int. Dairy J. 17: 1262–1277.

Shalev, E., S. Battino, E. Weiner, R. Colodner and Y. Keness. 1996. Ingestion of yogurt containing *Lactobacillus acidophilus* compared with pasteurized yogurt as prophylaxis for recurrent Candidal vaginitis and bacterial vaginosis. Arch. Fam. Med. 5: 593–596.

Shimakawa, Y., S. Matsubara, N. Yuki, M. Ikeda and F. Ishikawa. 2003. Evaluation of *Bifidobacterium breve* strain Yakult-fermented soymilk as a probiotic food. Int. J. Food Microbiol. 81: 131–136.

Shornikova, A.V., I.A. Casas, H. Mykkänen, E. Salo and T. Vesikari. 1997. Bacteriotherapy with *Lactobacillus reuteri* in rotavirus gastroenteritis. Pediatr. Infect. Dis. J. 16: 1103–1107.

Smith, C.J., J.R. Emge, K. Berzins, L. Lung, R. Khamishon, P. Shah, D.M. Rodrigues, A.J. Sousa, C. Reardon, P.M. Sherman, K.E. Barrett and M.G. Gareau. 2014. Probiotics normalize the gut-brain-microbiota axis in immunodeficient mice. Am. J. Physiol. Gastrointest. Liver Physiol. 307(8): 793-802.

Steenbergen, L., R. Sellaro, S. van Hemert, J.A. Bosch and L.S. Colzato. 2015. A randomized controlled trial to test the effect of multispecies probiotics on cognitive reactivity to sad mood. Brain Behav. Immun. 48: 258–264.

Sun, J., Y. Qiao, C. Qi, W. Jiang, H. Xiao, Y. Shi and G.W. Le. 2016. High fat diet induced obesity is associated with decreased anti-inflammatory *Lactobacillus reuteri* sensitive to oxidative stress in mice Peyer's patches. Nutrition 32: 262–272.

318   *Fermented Meat Products: Health Aspects*

Tamime, A. Y., M. Saarela, A.K. Sondergaard, V.V. Mistry and N.P. Shah. 2005. Production and maintenance of viability of probiotic micro-organisms in dairy products. pp. 39–72. *In*: A.Y. Tamime (ed.). Probiotic dairy products. Oxford, Blackwell Publishing.

Tuomola, E., R. Crittenden, M. Playne, E. Isolauri and S. Salminen. 2001. Quality assurance criteria for probiotic bacteria. Am. J. Clin. Nutr. 73: 393s–398s.

Vinderola, C.G. and J.A. Reinheimer. 2000. Enumeration of *Lactobacillus casei* in the presence of *L. acidophilus*, bifidobacteria and lactic acid starter bacteria in fermented dairy products. Int. Dairy J. 10: 271–275.

Vuyst L.D., G. Falony and F. Leroy. 2008. Probiotics in fermented sausages. Meat Sci. 80: 75–78.

Wang, Y. and Q. Gu. 2010. Effect of probiotic on growth performance and digestive enzyme activity of Arbor Acres broilers. Res. Vet. Sci. 89: 163–167.

Wang, X.H., H.Y. Ren, D.Y. Liu, W.Y. Zhu and W. Wang. 2013. Effects of inoculating *Lactobacillus sakei* starter cultures on the microbiological quality and nitrite depletion of Chinese fermented sausages. Food Control 32: 591–596.

WHO. 2015. IARC Monographs evaluate consumption of red meat and processed meat. International Agency for Research on Cancer, World Health Organization. Press Release N° 240.

Chapter 14

Antimicrobial Resistance of Lactic Acid Bacteria in Fermented Meat Products

Nevijo Zdolec*, Slavica Vesković-Moračanin, Ivana Filipović and Vesna Dobranić

1 INTRODUCTION

Antimicrobial resistance (AMR) is one of the leading public health issues which is closely related to farm animals, environment and food of animal origin. In this respect, production of ready-to-eat products, such as fermented meat products requires strict hygienic and technological procedures to prevent potential risk to consumers. The connection of primary production at the farm level and food processing is particularly evident in spreading of resistant bacteria or resistance determinants along the agri-food chain, including thermally untreated products such as fermented meat products (Zdolec 2016).

Safety aspects of microbiota related to fermented food products are evaluated during last decades (Resch et al. 2008, EFSA 2013). In general, low occurrence of hazards determinants was found in food lactic acid bacteria (LAB), however the transfer of antimicrobial resistance should be more studied *in vitro* and *in vivo*. In any case, the absence of mobile

---

*For Correspondence: University of Zagreb, Faculty of Veterinary Medicine, Department of Hygiene, Technology and Food Safety, Heinzelova 55, 10000 Zagreb, Croatia, Tel: ++38512390199, Email: nzdolec@vef.hr

AMR determinants in LAB must be the prerequisite for introducing LAB strains in sausage production as starter cultures (Talon and Leroy 2011). LAB have a long history of safe use in fermented food production and consumption that support their GRAS (generally recognized as safe) and QPS (qualified presumption of safety) status provided by FDA (US Food and Drug Administration) and EFSA, respectively. Predominant LAB species in dry-fermented sausages are *Lactobacillus sakei*, *Lb. curvatus*, *Lb. plantarum*, *Leuconostoc mesenteroideus*, *Pediococcus* spp. and *Enterococcus* spp. (Kozačinski et al. 2008, Vesković Moračanin et al. 2013, Federici et al. 2014, Wanangkarn et al. 2014, Vesković and Đukić 2015). Among LAB, 'Qualified Presumption of Safety' (QPS) status have 35 *Lactobacillus* species (e.g. *Lb. sakei*, *Lb. curvatus*, *Lb. plantarum*, *Lb. fermentum*, *Lb. brevis*, *Lb. rhamnosus* and *Lb. alimentarius* found in dry-fermented sausages), four *Leuconostocs* (*Ln. citreum*, *Ln. lactis*, *Ln. mesenteroides* and *Ln. pseudomesenteroides*), three *Pediococci* (*Pd. acidilactici*, *Pd. dextrinicus* and *Pd. pentosaceus*), *Lc. lactis* and *Streptococcus thermophilus* (EFSA 2013).

Testing of antibiotic resistance of LAB has not been extensively investigated until recently, in contrast to pathogenic species. However, interest in LAB and their antimicrobial resistances has increased in recent years, due to the observed horizontal transmission of antibiotic resistant determinants that can be transferred between bacterial species and also from beneficial bacteria to pathogens (Devirgiliis et al. 2013). The determination of AMR of LAB as the dominant microbial community in fermented meat products is highly important in terms of identifying and understanding mechanisms of persistence and spread of resistance genes in the microbial world (Levy and Miller 1989). Therefore, in addition to the need for their complete physiological and technological characterization, the demand to test LAB starter and probiotic strains for transmissible antimicrobial resistance is fully justified. It is vital to clearly define limit values used to classify strains as resistant or susceptible. Testing is particularly important in view of the results that contribute to distinguishing between natural (non-specific, non-transmissible) and acquired resistance using a procedure that involves comparison of antimicrobial resistance profiles for a large number of LAB originating from different sources (Teuber et al. 1999).

Therefore, both the qualitative and quantitative composition of epiphytic microorganisms in sausages and, hence, the presence of resistant LAB strains are directly correlated with the microorganisms naturally present in raw meat and other ingredients used in sausage production. Interestingly, bacteria resistant to antimicrobials occur under both home production (Toomey et al. 2010) and low temperature conditions. Environmental bacteria are inevitably found in unprocessed meat, even when slaughtering occurs under proper hygienic conditions (Devirgiliis et al. 2011). Antimicrobial drug resistance profiles in bacteria as commensal microorganisms in an ecosystem, such as lactic acid bacteria, serve as indicators of the selective pressure placed on these microorganisms under habitat contamination with antimicrobial agents. During the production process, raw milk and meat can frequently be contaminated with materials containing LAB resistant to antimicrobials. This is how resistance genes are transferred to end products, primarily raw-milk cheeses or fermented sausages.

2 ANTIMICROBIAL RESISTANCE: GENERAL ASPECTS

Antimicrobial resistance is the result of interactions between the microbial cell, its environment and the antimicrobial agent. In general, antimicrobial resistance is the capacity of a microorganism to resist the growth inhibitory or killing activity of an antimicrobial beyond the normal susceptibility of the specific bacterial species (Acar and Röstel 2001, Mathur and Singh 2005). Antimicrobial resistance in bacteria is the ability of bacteria to survive and grow in the presence of chemical molecules that would normally kill them or limit their growth. In other words, resistance in microorganisms is their acquired trait resulting from adaptation to changing environmental conditions (Rodríguez-Rojas et al. 2013).

In recent years, the use of antibiotics in human medicine has significantly increased (a 36% increase between 2000 and 2010), mostly in developing countries (Van Boeckel et al. 2014). However, apart from the justified use of antibiotics, there have been frequent records of antibiotic misuse which has reduced the numbers of susceptible bacterial strains, leading to the emergence of new resistant strains. Moreover, the use of antibiotics has been on the rise in animal husbandry as well, with overuse and improper use causing the emergence of resistant species/strains in the food chain (Teale 2002). The largest absolute increase in antibiotic use has been seen in cephalosporins, broad-spectrum penicillins and fluoroquinolones (Vesković et al. 2011). In fact, the last European Surveillance of Veterinary Antimicrobial Consumption (ESVAC 2013) report, regarding the overall sales in 2011 for 25 countries, states that the largest proportions, expressed as mg/PCU, were accounted for tetracyclines (37%), penicillins (23%), sulfonamides (11%) and polymyxins (7%). The World Health Organization report (WHO 2014) provided a comprehensive picture of increasing antibiotic resistance across an ever-increasing range of infectious agents and their growing threat to public health.

It should also be noted that the use of antimicrobials is not restricted to animal husbandry but also occurs in horticulture, for example the use of aminoglycosides in apple growing (Teale 2002). Therefore, antibiotic use for therapeutic, prophylactic and subtherapeutic purposes calls for an integrated approach in the fields of human medicine and animal health as well as in the environmental protection field (Janković et al. 2012, Berkner et al. 2014).

It is necessary to differentiate between natural or intrinsic and acquired (transmissible) resistance. Resistance to a given antimicrobial can be intrinsic to a bacterial species or genus (natural resistance), and it refers to the ability of a microorganism to survive in the presence of an antimicrobial agent, due to innate resistance characteristics i.e. due to the distinctive feature of a bacterial species found in all its strains. Since it is generally consistent and inheritable, intrinsic resistance is predictable.

Intrinsic resistance of a bacterial cell is presumed to present a minimal potential for horizontal spread of resistance (between different bacterial species), as was demonstrated for example with the chromosomal vancomycin resistance determinant of the *Lactobacillus rhamnosus* strain GG (Tynkkynen et al. 1998). In contrast, resistance is considered to be acquired when a strain of a normally susceptible species becomes resistant to an antimicrobial drug (European Commission 2008). In other words, acquired resistance is a characteristic of certain strains within the species. Acquired resistance to an antimicrobial results either from mutation in the bacterial genome or from the uptake of extra genes encoding resistance mechanism. It is not typical of most strains of certain species and is, thus, unpredictable (Gunell 2010). Acquired antimicrobial resistance is a specific attribute of microorganisms, chiefly those whose primary habitats include environments that are regularly challenged with antibiotics (human and animal intestines) (Teuber et al. 1999). Acquired resistance is considered as having a higher potential for horizontal spread of antibiotic resistance, since the resistance genes are present on mobile genetic elements (plasmids and transposons) (Devirgiliis et al. 2013, Đukić et al. 2015).

Antibiotic resistance genes can be transferred from one bacterial cell to another (horizontal transmission of genetic material) via a number of mechanisms. The transfers of DNA by transduction (via bacteriophages) or by transformation (when DNA is released from one bacterium and taken up by another) are not believed to be relevant mechanisms of antibiotic resistance transfer (Ammor and Mayo 2007, Đukić et al. 2015). By contrast, conjugation i.e. the direct cell-to-cell contact can potentially achieve horizontal gene transfer, as it has been shown to be a genetic information transfer mechanism with a broad host range (Courvalin 1994). Conjugation is thought to be the major mode of transfer of antibiotic resistance genes (Salyers 1995).

Since molecular analysis of resistance genes localized on plasmids and transposons shows identical genetic elements in humans and animals, it seems likely that food products of animal origin serve as transmission pathways for resistant bacteria i.e. antimicrobial resistance determinants. Over the last years, scientists have been seeking the answer to the question of whether commensal bacteria from food products can transfer resistance genes to intestinal bacteria in humans during their passage through the digestive tract. In certain pathogenic and potentially pathogenic bacteria, such as staphylococci and enterococci, the development of highly resistant bacterial clones has prompted the antimicrobial resistance crisis (Neu 1992). In the case of vancomycin resistant enterococci, there still remains no useful antimicrobial for a successful treatment (Jett et al. 1994).

2.1 Mechanisms of Antimicrobial Resistance

During more than 60 years of global antimicrobial use, several resistance mechanisms have been identified, viz. enzymatic degradation of antibiotics, antibiotic target modification, changing the bacterial cell wall permeability and alternative pathways to escape the activity (Levy 1997, Verraes et al. 2013).

In the presence of certain resistance genes, bacteria can avoid antimicrobial agents through any of the three mechanisms:

- direct inactivation of the active molecule;
- loss of bacterial susceptibility to the antimicrobial by modification of the target of action; and
- reduction of the drug concentration that reaches the target molecule without modification of the compound itself (efflux pump) (Fraqueza 2015).

The antibiotic defense mechanisms of intrinsic resistance are, in most of cases, related to the presence of low affinity targets, absence of targets, innate production of enzymes that inactivate the drug, inaccessibility of the drug into the bacterial cell by decreased drug uptake or extrusion by efflux of drug (Kumar and Schweizer 2005).

Bacteria that have resistance genes located on mobile genetic determinants pose a threat to public health (often referred to as "reservoirs") and enable the spread of these genes, especially if the environment contains numerous microbiota (Salyers et al. 2004, Haug et al. 2011). Today, many researchers have emphasized the hypothesis that commensal bacteria, primarily lactic acid bacteria, can serve as reservoirs of antibiotic resistance genes (Perreten et al. 1997, Levy and Salyers 2002). This is why the population of commensals is highly important in identifying mechanisms of persistence and spread of resistance genes in the microbial world. Accordingly, the food colonized by bacteria with transmissible antibiotic resistance genes has been specifically addressed. Antibioresistance of foodborne bacteria has aroused great interest because they may act as reservoirs for antibiotic resistance genes (Talon and Leroy 2011).

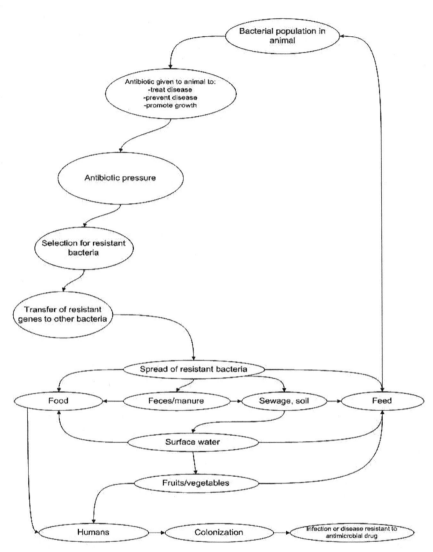

Figure 1 Possible transmission routes of antibiotic resistance bacteria from animals to humans (Modified from Khachatourians 1998, Claycamp and Hooberman 2004).

Generally, the food chain is considered a major transmission pathway for resistant bacteria between human and animal populations (Witte 1997, von Wright 2005). Antibiotic resistant bacteria in the food chain are commonly sustained through fecal contamination or recontamination due to incompetent or improper heat treatment during the food production process. After such a product is consumed, resistant bacteria colonize the digestive tract (Singer et al. 2003). In conclusion, the transmission of antibiotic resistance via the food chain is the same as for food borne pathogens (Figure 1).

Undisputedly, the presence of resistant strains of microorganisms isolated from food products, as either causative agents of alimentary diseases or commensal bacteria, and the potential to transfer resistant determinants to much more pathogenic species, pose a hazard. Therefore, the European Food Safety Authority (EFSA) has asked its Panel on Biological Hazards to identify, from a public health perspective, the extent to which food serves as a source of antimicrobial-resistant microorganisms or antimicrobial resistance genes, to rank the identified risk and to identify potential control options for reducing the risk (EFSA 2013). The WHO report on global surveillance of antimicrobial resistance (2014) claims that there is an association between antibiotic use in feed production and the emergence of resistance in mutual pathogenic microorganisms.

3 ANTIBIOTIC RESISTANCE OF NON-ENTEROCOCAL LACTIC ACID BACTERIA

Resistance in LAB is enchanced by the large numbers of LAB in fermented products and in the gastro-intestinal tract, but also from other bacteria in the environment. Once a LAB becomes resistant, amplified determinant can be transmitted to another host. Therefore, checking for signs of transferable antibiotic resistance in starter strains and bacteria used as feed and food additives is essential.

The determination of the antibiotic resistance profiles of LAB is mostly based on the use of numerous phenotypic methods. However, there is no consensus on breakpoints for most antimicrobials. The confusion in this area is primarily due to the fact that different methods are used to define resistance: Etest – based on antibiotic diffusion (Danielsen and Wind 2003), agar and broth dilution or agar dilution methods for the determination of minimum inhibitory concentration (MIC) (Herrero et al. 1996, Flórez et al. 2005), disk diffusion or Kirby-Bauer method (Charteris et al. 1998, Gevers et al. 2000), and microdilution (Kushiro et al. 2009, Klein et al. 2000), thus preventing direct comparison of results. The Etest (Epsilometer Testprinzip, Ellipse gradient test – AB Biodisk) is a popular quantitative technique for determining antimicrobial susceptibility. It is based on the combined concepts of *in vitro* dilution and diffusion tests. In the assay, there is an immediate and effective release of the antimicrobials in a continuous exponential gradient when they are applied to an agar surface (Ribeiro et al. 2005). The technique is accurate and reproducible because of the stability of the antibiotics (Sader et al. 1994).

Test results on antimicrobial resistance in bacteria can be affected by size of the inocula, temperature and incubation time, as well as by medium composition. Moreover, different growth media have been used for actual testing, including Iso-Sensitest, MRS, M17 and Müller-Hinton (Huys et al. 2002, Hummel et al. 2007). As found, most LAB species show relatively poor growth on common test media.

Table 1 Microbiological breakpoint values (mg/L) (EFSA 2012)

	Ampicillin	*Vancomycin*	*Gentamicin*	*Kanamycin*	*Streptomycin*	*Erythromycin*	*Clindamycin*	*Tetracycline*	*Chloramphenicol*
Lactobacillus obligate homofermentative[a]	1	2	16	16	16	1	1	4	4
Lactobacillus acidophilus group	1	2	16	64	16	1	1	4	4
Lactobacillus obligate heterofermentative[b]	2	n.r.	16	32	64	1	1	8	4
Lactobacillus reuteri	2	n.r.	8	64	64	1	1	16	4
Lactobacillus facultative heterofermentative[c]	4	n.r.	16	64	64	1	1	8	4
Lactobacillus plantarum/ pentosus	2	n.r.	16	64	n.r.	1	2	32	8
Lactobacillus rhamnosus	4	n.r.	16	64	32	1	1	8	4
Lactobacillus casei/paracasei	4	n.r.	32	64	64	1	1	4	4
Bifidobacterium	2	2	64	n.r.	128	1	1	8	4
Pediococcus	4	n.r.	16	64	64	1	1	8	4
Leuconostoc	2	n.r.	16	16	64	1	1	8	4
Lactococcus lactis	2	4	32	64	32	1	1	4	8
Streptococcus thermophilus	2	4	32	64	64	2	2	4	4
Propionibacterium	2	4	64	64	64	0.5	0.25	2	2

n.r.–not required.
[a]including *Lb. delbrueckii*, *Lb. helveticus*
[b]including *Lb. fermentum*
[c]including the homofermentative species *Lb. salivarius*

Breakpoint values (minimum inhibitory concentrations – MIC of antibiotic) that discriminate between resistant and susceptible strains should be clearly defined, as well as distinguishing between intrinsic (non-specific, non-transferable) and acquired, potentially transferable resistance (Ammor et al. 2007). In Table 1, breakpoint values are given for some LAB strains, and strains with MICs higher than the breakpoints

Antimicrobial Resistance of Lactic Acid Bacteria in Fermented Meat....   327

are considered as resistant. Breakpoints used for antibiotic susceptibility profiling are harmonized in Europe by the EUCAST but unfortunately no harmonized guidelines regarding the resistance–susceptibility breakpoints for non-enterococcus LAB are available, and therefore results are not well comparable (Flórez et al. 2008, Patel et al. 2012).

Phenotypic assays have now been complemented by molecular methods in which bacterial strains are directly screened for the presence of antibiotic resistance determinants. Generally, these methods include amplification by PCR with specific primers for single or multiplex antibiotic resistance genes (Strommenger et al. 2003), real time PCR (Volkmann et al. 2004) or the use of DNA microarrays containing large collections of antibiotic resistance genes (Perreten et al. 2005). The existing genetic studies used to confirm the transmission of known resistance determinants are hampered by many experimental factors and thus show variable results. Moreover, positive control strains for conjugation and/or transposition experiments are not available, and standard protocols for gene transfer demonstration are lacking. Two of the most commonly observed resistance genes in LAB found so far are *tet*(M) for tetracycline resistance and *erm*(B) for erythromycin, followed with cat genes coding for chloramphenicol resistance (Lin et al. 1996, Danielsen 2002, Gevers et al. 2003, Cataloluk and Gogebakan 2004).

3.1 Occurrence of Resistant Non-Enterococcal LAB in Fermented Meat Products

LAB possesses a broad spectrum of intrinsic and aquired antibiotic resistance (Table 2). An overview of antibiotic resistances reported together with potentially transferable resistance determinants in the food-associated LAB is given in Table 3. The antibiotic resistance profiles of *Lactobacillus, Lactococcus, Streptococcus, Pediococcus, Leuconostoc* and *Bifidobacterium* are quite different (although clear-cut species-specific patterns have not been observed) (Ammor et al. 2007). As stated before, lactobacilli, pediococci and leuconostocs are of particular importance in fermented meat production, hence their resistance will be further considered.

Lactobacilli originated from meat have been found to harbor the resistance to ampicillin, chloramphenicol, erythromycin, penicillin G, and tetracycline (Ahn et al. 1992, Tannock et al. 1994, Lin et al. 1996, Gevers et al. 2003, Essid et al. 2009, Toomey et al. 2010). Most common resistance genes detected in lactobacilli isolated from dry-fermented sausages have been *tet*(M), *tet*(W), *tet*(S) for tetracycline, and *erm*(B) and *erm*(C) for erythromycin resistance (Gevers et al. 2003, Zonenschain et al. 2009, Fraqueza 2015). The predominat *Lactobacillus* species isolated from dry-fermented sausages, such as *Lb. sakei* and *Lb. plantarum* had a resistance rate to tetracycline of 12–70% and 75–80%, respectively (Gevers et al. 2000, Aymerich et al. 2006, Zonenschain et al. 2009, Landeta et al. 2013, Federici et al. 2014). Furthermore, tetracycline resistance can be transferred from lactobacilli (originated from fermented sausages) to *Enterococcus faecalis* or *Lc. lactis* subsp. *lactis in vitro* (Gevers et al. 2003). A chloramphenicol-resistance *(cat)* gene has been found in many lactobacilli species of food origin, but not from lactobacilli isolated from dry-fermented sausages (Hummel et al. 2007).

Table 2 Intrinsic antibiotic resistance profile of LAB (modified from Teuber et al. 1999)

Type of Bacteria	Intrinsic Antibiotic Susceptibility	Intrinsic Antibiotic Resistance
Bifidobacterium	ampicillin, penicillin G, bacitracin, cephalosporin, chloramphenicol, erythromycin, clindamycin, nitrofurantoin, tetracycline.	vancomycin, gentamicin, fusidic acid, streptomycin, polymyxin B, trimethoprim, aminoglycosides, colistin, metronidazole
Lactococcus lactis	amikacin, ampicillin, 1st generation cephalosporine, chloramphenicol, erythromycin, gentamicin, penicillin, imipenem, oxacillin, sulfonamide, tetracycline, vancomycin	colistin, fosfomycin, pipemidic acid and rifamycin
Lactobacillus spp.	chloramphenicol, streptomycin, gentamycin, penicillin G, tetracycline and erythromycin	aminoglycosides, fluoro-quinolones, glycopeptides and vancomycin

Pediococci are usually susceptible to penicillin G, imipenem, gentamicin, netilmicin, erythromycin, clindamycin, rifampin, chloramphenicol, daptomycin and ramoplanin (Swenson et al. 1990, Tankovic et al. 1993, Zarazaga et al. 1999, Temmerman et al. 2003, Danielsen et al. 2007). On the contrary, *Pediococcus* species are intrinsically resistant to glycopeptides (vancomycin and teicoplanin) and to ciprofloxacin, sulphamethoxazole and trimethoprim-sulphamethoxazole (Swenson et al. 1990, Danielsen et al. 2007). Considering fermented sausages, Federici et al. (2014) tested the antimicrobial susceptibility of *Pd. pentosaceus* (n=9) which showed the resistance to streptomycin (100%), gentamicin (100%), tetracycline (100%), ampicillin (100%), chloramphenicol (98%), erythromycin (33%) and clindamycin (11%). In these isolates antibiotic-resistant genes for tetracycline $-tet$(M) and for erythromycin $-erm$(B) have been detected (Federici et al. 2014).

Leuconostoc species are mostly susceptible to rifampicin, chloramphenicol, erythromycin, clindamycin and tetracycline (Swenson

Antimicrobial Resistance of Lactic Acid Bacteria in Fermented Meat.... 329

et al. 1990, Zarazaga et al. 1999, Katla et al. 2001, Flórez et al. 2005). On contrary, *Leuconostoc* spp. are resistant to glycopeptides (e.g. vancomycin), cefoxitin and metronidazole, and usually (at least partially) to nalidixic acid, gentamicin, kanamycin, streptomycin, nitrofurantoin, sulphadiazine and trimethoprim (Swenson et al. 1990, Katla et al. 2001, Flórez et al. 2005, Toomey et al. 2010). Zdolec et al. (2011) isolated from traditionally fermented sausages *Ln. mesenteroides* strain resistant to streptomycin, enrofloxacin, trimethoprim, nalidixic acid, metronidazole, gentamicin and kanamycin. Antibiotic susceptibility of *Ln. mesenteroides* isolates (n=12) from dry-fermented sausages was tested using a modified disc diffusion technique (Aymerich et al. 2006). Isolates were resistant to vancomycin (100%), gentamicin (66.7%), penicillin G (50%), ampicillin (41.7%) and tetracycline (25%). All isolates were susceptible to chloramphenicol, erythromycin, linezolid and quinupristin/dalfopristin (Aymerich et al. 2006). Tetracycline resistance and the associated *tet*(S) determinant has been identified in one *Ln. mesenteroides* strain isolated from pork and beef abattoirs in Ireland (Toomey et al. 2010). The *tet*(S) gene was also found in a *Ln. citreum* strain isolated from raw pork meat in a sausage processing line (Gevers et al. 2003).

4 ANTIMICROBIAL RESISTANCE OF ENTEROCOCCI IN FERMENTED MEAT PRODUCTS

Enterococci are widely distributed in the environment, and primarily are hosted in the gastrointestinal tract of humans and animals. Most important and dominant species are *Enterococcus faecalis* and *Enterococcus faecium* (Franz et al. 2003). The presence of *E. faecalis* in foods is not always connected with fecal contamination meaning that enterococci are not considered as indicators of hygiene in food production and processing (Birollo et al. 2001). Regarding meat production, meat can be initially contaminated by enterococci during the slaughter processing. Due to their good adaptability to wide environmental conditions and their resistance to pasteurization, enterococci are commonly found in fermented sausages or thermally-treated meat products (Aymerich et al. 1996, Marchesino et al. 1992). Enterococci are capable to produce bacteriocins (enterocins), peptides with antimicrobial activity towards Gram positive bacteria (De Vuyst and Vandamme, 1994, Mojsova et al. 2015) and possess probiotic characteristics (Laukova, 2012). Enterococci, as constituents of LAB group, show a favorable features related to development of sensorial properties of fermented meat products (glycolytic, proteolytic, lipolytic activity, reduction of methmioglobin), as well as enhance the safety and shelf-life of products by inhibiting

Table 3 Overview of antibiotic resistances in the food-associated LAB (modified from Patel et al. 2012, Fraqueza 2015)

Food	Species	Resistance	Detection and location of gene	References
Ciauscolo salami	*Lb. sakei,* *Pd. pentosaceus,* *Lb. plantarum,* *Lb. paraplantarum, E. faecalis, Lactococcus*	ampicillin, chloramphenicol, clindamycin, erythromycin, gentamicin, streptomycin, tetracycline, vancomycin	*tet*(M), *tet*(W), *tet*(K), *tet*(L), *tet*(S), *erm*(A), *erm*(B), *van*(A), *van*(B)	Federici et al. 2014
Indian vegetables and fermented foods	*Lb. plantarum,* *Lb. fermentum, Weissella* spp., *Pd. parvulus*	gentamicin, vancomycin, norfloxacin, kanamycin	nd	Patel et al. 2013
Dry-fermented sausages	*Lb. sakei,* *Lb. plantarum,* *Lb. paracasei,* *Lb. coryniformis,* *E. faecium*	vancomycin, rifamycin, amikacin, tetracycline	*tet*(M)	Landeta et al. 2013
Chinese fermented foods-pickles, sausages	*Lb. plantarum,* *Lb. fermentum,* *Lb. helveticus,* *E. faecium*	tetracycline, erythromycin, chloramphenicol, kanamycin	*tet*(M) and *erm*(B), – plasmid and chromosome; gene *aph* A3, – plasmid, gene *mef* A, – chromosome	Pan et al. 2011
Chinese fermented foods	*Lb. fermentum* NWL24 and *Lb. salivarius* NWL33; *Lb. plantarum* NWL22 and *Lb. brevis* NWL59 *Lb. brevis* and *Lb. kefiri*	erythromycin tetracycline gentamicin ciprofloxacin	*erm*(B), *tet*(M), *tet*(S)	Nawaz et al. 2010
Italian fermented products	*Lb. paracasei* 197 strains	tetracycline erythromycin	*tet*(M), *erm*(B)	Comunian et al. 2010

Product	Species	Phenotypic resistance	Resistance genes	Reference
Italian dry-fermented sausages	Lb. sakei, Lb. curvatus, Lb. plantarum	tetracycline erythromycin	tet(M), tet(W), tet(L), tet(S), erm(B), erm(C)	Zonenschain et al. 2009
Spanish dry-fermented sausages	Lb. sakei, Lb. curvatus, Ln. mesenteroides	ampicillin chloramphenicol gentamicin penicillin G tetracycline vancomycin	nd	Aymerich et al. 2006
Fermented dry sausages	Lactobacillus spp.	tetracycline gentamicin penicillin G kanamycin	tet(M)	Gevers et al. 2003
European probiotic products	Lb. acidophilus, Lb. rhamnosus, Lb. casei, Lb. reuteri, Lb. johnsonnii, Lb. plantarum	tetracycline penicillin G erythromycin chloramphenicol	nd	Temmerman et al. 2003
Nigerian fermented foods and beverages	Lb. pentosus, Lb. acidophilus, Lb. casei, Lb. brevis, Lb. plantarum, Lb. jensenii	tetracycline erythromycin ampicillin cloxacillin penicillin	nd	Olukoya et al. 1993

nd-not determined

332 *Fermented Meat Products: Health Aspects*

foodborne pathogens and/or spoilage microorganisms. However, some enterococcal species are at the same time food spoilers, biogenic amines producers or carriers of mobile determinants of antimicrobial resistance (Čanžek Majhenić 2006, Zdolec 2016). According to Giraffa (2002) and Hugas et al. (2003) enterococci are capable to survive and grow during the fermentation and drying of fermented meat products, especially when competitive starter cultures are absent. It is known that their numbers in fermented sausages are very variable, even between the batches of the same product (Hugas et al. 2003, Alagić et al. 2010).

In recent years the antimicrobial resistance among food-related LAB is intensively evaluated, especially in *E. faecalis* and *E. faecalis*. The highest risk for consumers could be revealed by the presence of multi-resistant enterococci in ready-to-eat food products (Mathur and Singh 2005). The assessment of antimicrobial resistance risk potential in enterococci and other food LAB is highly dependent on methodology used. Therefore many studies showed a comparably low correlation of phenotypic and genotypic detection of antibiotic resistance (Temmerman et al. 2003).

4.1 Occurrence of Resistant Enterococci in Fermented Meat Products

Ready-to-eat meat products, including fermented sausages and fermented cured meats, are potential carriers of resistant enterococci (Chajęcka-Wierzchowska et al. 2014). Enterococci of food origin are not direct cause of resistant enterococci in humans, but they can transfer resistance determinants to human-adapted bacteria (Economou and Gousia 2015). The presence of enterococci harboring transmissible resistance determinants in fermented meat products is regular finding; however the resistance toward clinically relevant agents such as ampicillin or vancomycin is quite rare (Jahan et al. 2013). Similarly, Fontana et al. (2009) presented a low occurrence of resistant enterococci in Argentinian fermented sausages, lacking vancomycin resistance genes. Vancomycin resistant enterococci (VRE) are of the highest importance in food production and transfer to humans by agri-food chain. Vancomycin is highly important antibiotic used in treatment of human bacterial infections, including enterococci or methicillin-resistant *St. aureus*. Linezolid is the second important agent in treatment of vancomycin-, oxacillin- or methicillin-resistant enterococci/staphylococci infections, and linezolid-resistant enterococci (LRE) in food should be considered (Deresinski 2009, Zdolec et al. 2016). As mentioned before, the prevalence of VRE in fermented meat products seems to be very rare (Barbosa et al. 2009, Fontana et al. 2009). However, Houben (2003) investigated the potential of VRE to persist in fermented meat products and showed

Antimicrobial Resistance of Lactic Acid Bacteria in Fermented Meat.... 333

that VRE tolerated a high salt and nitrite concentrations and reduced pH. Consequently, vancomycin resistant enterococci were successfully recovered from fermented sausages investigated. Previously, Cocconcelli et al. (2003) have warned that transfer of vancomycin resistance determinant *van*A to meat-related *E. faecalis* occurs during sausage fermentation. A recent study results provided by Gousia et al. (2015) showed that phenotypic resistance to vancomycin was present in 30% of *E. faecium* (n=88) and 1% of *E. faecalis* (n=53) isolated from raw meat,which was encoded by *van*A and *van*B genes. Therefore, findings of VRE in fresh meat, as raw material for fermented meat production, should be still considered as potential reservoir of spreading the resistance genes to other bacteria of food or human origin.

In general, enterococci possess intrinsic antibiotic resistance to cephalosporins, ß-lactams, sulphonamides, and to certain levels of clindamycin and aminoglycosides, while acquired resistance exists to chloramphenicol, erythromycin, clindamycin, aminoglycosides, tetracycline, ß-lactams, fluoroquinolones and glycopeptides (Giménez Pereira 2005). Intrinsic resistance to ß-lactams as a general rule is not always demonstrated in food-related enterococci, as already mentioned for ampicillin (Jahan et al. 2013). Similarly, Barbosa et al. (2009) reported the absence of penicillin or ampicillin resistance in enterococci from Portuguese fermented meat products. In general, as presented in Table 4 the majority of enterococci from fermented meat products harbor transmissible resistance determinants, but the resistance toward clinically relevant agents is rare.

Results of many studies support the potential risk of acquired antimicrobial resistance in food-related enterococci and transfer of mobile genetic material to other bacteria, even in the condition of low antimicrobial pressure (Cocconcelli et al. 2003). Recently, Jahan et al. (2015) demonstrated *in vitro* the transfer of tetracycline resistant determinant, *tet*(M), by *E. faecium* strain isolated from fermented sausage to clinical isolates of both *E. faecium* and *E. faecalis.* In the same study, streptomycin resistance was also transferred from sausage-originated *E. faecium* to the clinical strain *E. faecalis via* an integron. Previously, the transfer of tetracycline and erythromycin resistance was reported by Gazzola et al. (2012) from human *E. faecium* to meat-borne enterococci, lactobacilli, pediococci and staphylococci during the sausage fermentation. Mathur and Singh (2005) emphasized that there is no barrier to prevent a development of acquired resistance between pathogens (e.g. streptococci), opportunistic pathogens (e.g. enterococci) and commensal LAB (e.g. intestinal lactobacilli, lactococci), which is visible by the presence of identical resistance determinants in all microbial groups.

Table 4 Dominant AMR patterns of enterococci from fermented meat products (modified from Zdolec 2016)

Enterococci species	Meat products	Country	Dominant resistance	Reference
E. faecalis, E. faecium, E. gallinarum	Dry-fermented sausages, dry-cured hams	Canada	clindamycin, tetracycline hydrochloride, tylosin, erythromycin	Jahan et al. 2013
E. faecalis, E. faecium	Sausages, ham	Germany	enrofloxacin, erythromycin, avilamycin, quinupristin/dalfopristin (E. faecium); tetracycline, erythromycin (E. faecalis)	Peters et al. 2003
E. faecalis, E. faecium, E. durans, E. gallinarum	Ham	Italy	tetracycline, erythromycin	Pesavento et al. 2014
E. faecalis, E. faecium, E. casseliflavus	Fermented sausages Alheira, Salpica˜ode Vinhais, Chouriça de Vinhais	Portugal	rifampicin, tetracycline, erythromycin, ciprofloxacin	Barbosa et al. 2009
E. faecium	Dry-cured sausages	Spain	tetracycline, rifampicin, ciprofloxacin	Landeta et al. 2013
E. faecalis, E. faecium	Sausages	Turkey	erythromycin, tetracycline, kanamycin	Toğay et al. 2010
E. faecium, E. faecalis, E. durans, E. hirae, E. casseliflavus	Slightly fermented sausages – chorizo, fuet	Spain	rifampicin, ciprofloxacin	Martin et al. 2005
E. faecalis	Fermented sausage Chouriço	Portugal	tetracycline, erythromycin, vancomycin	Ribeiro et al. 2011

5 CONCLUSION

The complex issue of antimicrobial resistance requires a wide multidisciplinary approach to predict and avoid the undesirable public-health consequences along the whole food-producing chain. Fermented meat products are traditionally well accepted by consumers that are aware on potential hazards (primarily related to nutritional aspects; e.g. salt, fat content, additives etc.). Classical microbiological hazards (foodborne pathogens) are less expected due to known hurdles in fermented meat technology. Lactic acid bacteria are even the part of hurdle concept, and controversialy they have been recognized as part of antimicrobial resistance problem. Nevertheless, their significance in human nutrition (probiotics) and food technology (starter cultures) should not be questioned. Strategies for reduction of AMR in fermented food microbiota should be based on prudent use of antimicrobials in food animals and application of competitive starter cultures in food fermentation.

Key words: antimicrobial resistance, lactic acid bacteria, resistance genes, phenotypic and genotypic methods, antimicrobial agents, veterinary public health, resistance transfer

REFERENCES

Acar, J. and B. Röstel. 2001. Antimicrobial resistance: An overview. Rev. Sci. Tech. OIE 20: 797–810.

Ahn, C., D. Collins-Thompson, C. Duncan and M.E. Stiles. 1992. Mobilization and location of the genetic determinant of chloram-phenicol resistance from Lactobacillus plantarum caTC2R. Plasmid 27: 169–176.

Alagić, D., L. Kozačinski, I. Filipović, N. Zdolec, M. Hadžiosmanović, B. Njari, Z. Kozačinski and S. Uhitil. 2010. Microbial changes during ripening of fermented horsemeat sausages. Meso 10: 200–203.

Ammor, M.S. and B. Mayo. 2007. Selection criteria for lactic acid bacteria to be used as functional starter cultures in dry sausage production: an update. Meat Sci. 76: 138–146.

Ammor, M.S., A.B. Flórez and B. Mayo. 2007. Antibiotic resistance in non-enterococcal lactic acid bacteria and bifidobacteria. Food Microbiol. 24: 559–570.

Aymerich, T., B. Martin, M. Garriga, M.C. Vidal-Carou, S. Bover-Cid and M. Hugas. 2006. Safety properties and molecular strain typing of lactic acid bacteria from slightly fermented sausages. J. Appl. Microbiol. 100: 40–49.

Aymerich, T., H. Holo, L.S. Havarstein, M. Hugas, M. Garriga and I.F. Nes. 1996. Biochemical and genetic characterization of enterocin A from Enterococcus faecium, a new antilisterial bacteriocin in the pediocin family of bacteriocins. Appl. Environ. Microbiol. 62: 1676–1682.

336   *Fermented Meat Products: Health Aspects*

Barbosa, J., V. Ferreira and P. Teixeira. 2009. Antibiotic susceptibility of enterococci isolated from traditional fermented meat products. Food Microbiol. 26: 527–532.

Berkner, S., S. Konradi and J. Schönfeld. 2014. Antibiotic resistance and the environment – there and back again. EMBO Rep. 15: 740–744.

Birollo, G.A., J.A. Reinheimer and C.G. Vinderola. 2001. Enterococci vs. nonlactic acid microflora as hygiene indicators for sweetened yoghurt. Food Microbiol. 18: 597–604.

Cataloluk, O. and B. Gogebakan. 2004. Presence of drug resistance in intestinal lactobacilli of dairy and human origin in Turkey. FEMS Microbiol. Lett. 236: 7–12.

Chajęcka-Wierzchowska, W., A. Zadernowska, B. Nalepa, M. Sierpińska and L. Laniewska-Trokenheim. 2014. Retail ready-to-eat food as a potential vehicle for *Staphylococcus* spp. harboring antibiotic resistance genes. J. Food Protect. 77: 993–998.

Charteris, W.P., P.M. Kelly, L. Morelli and J.K. Collins. 1998. Antibiotic susceptibility of potentially probiotic *Lactobacillus* species. J. Food Protect. 61: 1636–1643.

Claycamp, H.G. and B.H. Hooberman. 2004. Antimicrobial resistance risk assessment in food safety. J Food Protect. 67: 2063–2071.

Cocconcelli, P.S., D. Cattivelli and S. Gazzola. 2003. Gene transfer of vancomycin and tetracycline resistances among *Enterococcus faecalis* during cheese and sausage fermentations. Int. J. Food Microbiol. 88: 315–323.

Comunian, R., E. Daga, I. Dupré, A. Paba, C. Devirgiliis, V. Piccioni and G. Perozzi. 2010. Susceptibility to tetracycline and erythromycin of *Lactobacillus paracasei* strains isolated from traditional Italian fermented foods. Int. J. Food Microbiol. 138: 151–156.

Courvalin, P. 1994. Transfer of antibiotic resistance genes between gram-positive and gram-negative bacteria. Antimicrob. Agents Chemother. 38: 1447–1451.

Čanžek Majhenič, A. 2006. Enterococci: yin-yang microbes. Mljekarstvo 56: 5–20.

Danielsen, M. 2002. Characterization of the tetracycline resistance plasmid pMD5057 from *Lactobacillus plantarum* 5057 reveals a composite structure. Plasmid 48: 98–103.

Danielsen, M. and A.A. Wind. 2003. Susceptibility of *Lactobacillus* spp. to antimicrobial agents. Int. J. Food Microbiol. 82: 1–11.

Danielsen, M., P.J. Simpson, E.B. O'Connor, R.P. Ross and C. Stanton. 2007. Susceptibility of *Pediococcus* spp. to antimicrobial angents. J. Appl. Microbiol. 102: 384–389.

De Vuyst, L. and E. J Vandamme. 1994. Antimicrobial potential of lactic acid bacteria. pp. 91–142. *In:* L. De Vuyst and E.J. Vandamme (eds.). Bacteriocins of lactic acid bacteria: microbiology, genetics and applications. Blackie Academic & Professional, London.

Deresinski, S. 2009. Vancomycin in combination with other antibiotics for the treatment of serious methicillin-resistant *Staphylococcus aureus* infections. Clin. Infect. Dis. 49: 1072–1079.

Devirgiliis, C., P. Zinno and G. Perozzi. 2013. Update on antibiotic resistance in foodborne *Lactobacillus* and *Lactococcus* species. Front. Microbiol. 4: 301.

Antimicrobial Resistance of Lactic Acid Bacteria in Fermented Meat.... 337

Devirgiliis, C., S. Barile and G. Perozzi. 2011. Antibiotic resistance determinants in the interplay between food and gut microbiota. Genes Nutr. 6: 275–284.

Đukić, D., S. Vesković and L. Mandić. 2015. Genetics of microorganisms. pp. 81–93. *In*: Đukić D., S. Vesković and L. Mandić (eds). The general and industrial microbiology. Čačak, Serbia. (in Serbian).

Economou, V. and P. Gousia. 2015. Agriculture and food animals as a source of antimicrobial-resistant bacteria. Infect. Drug Resist. 8: 49–61.

EFSA. 2012. Guidance on the assessment of bacterial susceptibility to antimicrobials of human and veterinary importance. EFSA J. 10: 2740.

EFSA. 2013. Scientific Opinion on the maintenance of the list of QPS biological agents intentionally added to food and feed (2013 update). EFSA J. 11: 3449.

Essid, I., M. Medini and M. Hassouna. 2009. Technological and safety properties of Lactobacillus plantarum strains isolated from a Tunisian traditional salted meat. Meat Sci. 81: 203–208.

ESVAC. 2013. European Medicines Agency, European Surveillance of Veterinary Antimicrobial Consumption. Sales of veterinary antimicrobial agents in 25 EU/EEA countries in 2011 (EMA/236501/2013). Canary Wharf London, UK. pp. 1–97.

European Commission. 2008. Technical guidance prepared by the Panel on Additives and Products or Substances used in Animal Feed (FEEDAP) on the update of the criteria used in the assessment of bacterial resistance to antibiotics of human or veterinary importance. EFSA J. 732: 1–15.

Federici, S., F. Ciarrocchi, R. Campana, E. Ciandrini, G. Blasi and W. Baffone. 2014. Identification and functional traits of lactic acid bacteria isolated from Ciauscolo salami produced in Central Italy. Meat Sci. 98: 575–584.

Flórez, A.B., L. Tosi, M. Danielsen, A. von Wright, J. Bardowski, L. Morelli and B. Mayo. 2008. Resistance-susceptibility profiles of *Lactococcus lactis* and *Streptococcus thermophilus* strains to eight antibiotics and proposition of new cut-offs. Int. J. Probio. Prebio. 3: 249–256.

Florez, A.B., S. Delgado and B. Mayo. 2005. Antimicrobial susceptibility of lactic acid bacteria isolated from a cheese environment. Can. J. Microbiol. 51: 51–58.

Fontana, C., S. Gazzola, P.S. Cocconcelli and G. Vignolo. 2009. Population structure and safety aspects of *Enterococcus* strains isolated from artisanal dry fermented sausages produced in Argentina. Lett. Appl. Microbiol. 49: 411–414.

Franz, C.M.A.P., M.E. Stiles, K.H. Schleifer and W.H. Holzapfel. 2003. Enterococci in foods – a conundrum for food safety. Int. J. Food Microbiol. 88: 105–122.

Fraqueza, M. J. 2015. Antibiotic resistance of lactic acid bacteria isolated from dry-fermented sausages. Int. J. Food Microbiol. 212: 76–88.

Gazzola, S., C. Fontana, D. Bassi and P.S. Cocconcelli. 2012. Assessment of tetracycline and erythromycin resistance transfer during sausage fermentation by culture-dependent and -independent methods. Food Microbiol. 30: 348–354.

Gevers, D., G. Buys, F. Devlieghere, M. Uyttendaele, O. Debevere and E. Swings. 2000. Isolation and identification of tetracycline resistant lactic acid bacteria from pre-packed sliced meat products. Syst. Appl. Microbiol. 23: 279–284.

Gevers, D., G. Huys and J. Swings. 2003. In vitro conjugal transfer of tetracycline resistance from *Lactobacillus* isolates to other gram-positive bacteria. FEMS Microbiol. Lett. 225: 125–130.

Giménez Pereira, M.L. 2005. Enterococci in milk products. M.S. Thesis, Massey University Palmerston North, New Zealand.

Giraffa, G. 2002. Enterococci from foods. FEMS Microbiol. Rev. 744: 1–9.

Gousia, P., V. Economou, P. Bozidis and C. Papadopoulou. 2015. Vancomycin-resistance phenotypes, vancomycin-resistance genes, and resistance to antibiotics of enterococci isolated from food of animal origin. Foodborne Pathog. Dis. 12: 214–220.

Gunell, M. 2010. *Salmonella enterica*: Mechanisms of fluoroquinolone and macrolide resistance. Thesis PhD. University of Turku, Finland.

Haug, M.C., S.A. Tanner, C. Lacroix, M.J. Stevens and L. Meile. 2011. Monitoring horizontal antibiotic resistance gene transferin a colonic fermentation model. FEMS Microbiol. Ecol. 78: 210–219.

Herrero, M., B. Mayo, B. Gonzales and J.E. Suarez. 1996. Evaluation of technologically important traits in lactic acid bacteria isolated from spontaneous fermentations. J. Appl. Bacteriol. 81: 565–570.

Houben, J.H. 2003. The potential of vancomycin-resistant enterococci to persist in fermented and pasteurized meat products. Int. J. Food Microbiol. 88: 11–18.

Hugas, M., M. Garriga and M.T. Aymerich. 2003. Functionality of enterococci in meat products. Int. J. Food Microbiol. 88: 223–233.

Hummel, A.S., C. Hertel, W.H. Holzapfel and C.M.A.P Franz. 2007. Antibiotic resistances of starter and probiotic strains of lactic acid bacteria. Appl. Environ. Microbiol. 73: 730–739.

Huys, G., K. D'Haene and J. Swings. 2002. Influence of the culture medium on antibiotic susceptibility testing of food-associated lactic acid bacteria with the agar overlay disc diffusion method. Lett. Appl. Microbiol. 34: 402–406.

Jahan, M., D.O. Krause and R.A. Holley. 2013. Antimicrobial resistance of Enterococcus species from meat and fermented meat products isolated by a PCR-based rapid screening method. Int. J. Food Microbiol. 163: 89–95.

Jahan, M., G.G. Shanel, R, Sparling and R.A. Holley. 2015. Horizontal transfer of resistance from *Enterococcus faecium* of fermented mear origin to clinical isolates of *E. faecium* and *Enterococcus faecalis*. Int. J. Food Microbiol. 199: 78–85.

Janković, V., Lj. Petrović, S. Vesković, D. Karan, T. Radičević, S. Janković and S. Stefanović. 2012. Investigations of residue of veterinary medicines and environmental contaminants during production cycle of Petrovska klobasa as part of compulsory parameters for food safety. Vet. glasnik 66: 243–257 (in serbian).

Jett, B.D., M.M. Huycke and M.S. Gilmore. 1994. Virulence of enterococci. Clin. Microbiol. Rev. 7: 462–478.

Katla, A.K., H. Kruse, G. Johnsen and H. Herikstad. 2001. Antimicrobial susceptibility of starter culture bacteria used in Norwegian dairy products. Int. J. Food Microbiol. 67: 147–152.

Khachatourians, G.G. 1998. Agricultural use of antibiotics and the evolution and transfer of antibiotic-resistant bacteria. Can. Medic. Assoc. J. 159: 1129–1136.

Klein, G., C. Hallmann, I.A. Casas, J. Abad, J. Louwers and G. Reuter. 2000. Exclusion of *van*A, *van*B and *van*C type glycopeptide resistance in strains of *Lactobacillus reuteri* and *Lactobacillus rhamnosus* used as probiotics by polymerase chain reaction and hybridization methods. J. Appl. Microbiol. 89: 815–824.

Kozačinski, L., E. Drosinos, F. Čaklovica, L. Cocolin, J. Gasparik-Reichardt and S. Vesković. 2008. Investigation of microbial association of traditionally fermented sausages – microflora in traditionally fermented sausages. Food Technol. Biotechnol. 46: 93–106.

Kumar, A. and H.P. Schweizer. 2005. Bacterial resistance to antibiotics: active efflux and reduced uptake. Adv. Drug Deliv. Rev. 57: 1486–1513.

Kushiro, A., C. Chervaux, S. Cools-Portier, A. Perony, S. Legrain-Raspaud, D. Obis, M. Onoue and A. van de Moer. 2009. Antimicrobial susceptibility testing of lactic acid bacteria and bifidobacteria by broth microdilution method and Etest. Int. J. Food Microbiol. 132: 54–58.

Landeta, G., J.A. Curiel, A.V. Carrascosa, R. Muñoz and B. de las Rivas. 2013. Technological and safety properties of lactic acid bacteria isolated from Spanish dry-cured sausages. Meat Sci. 95: 272–280.

Laukova, A. 2012. Potential applications of probiotic, bacteriocin-producing enterococci and their bacteriocins. pp. 39–61. *In:* S. Lahtinen, A.C. Ouwehand, S. Salminen, A. von Wright (eds.). Lactic Acid Bacteria: microbiological and functional aspects. 4th ed. CRC Press, Taylor & Francis Group, Boca Raton.

Levy, S.B. 1997. Antibiotic resistance an ecological imbalance. pp. 1–14. *In:* D.J. Chadwick and J. Good (eds.). Antibiotic resistance. Origins, evolution, selection and spread. John Wiley & Sons, Chichester.

Levy, S.B. and A.A. Salyers. 2002. Reservoirs of antibiotic resistance (ROAR) Network. http://www.healthsci.tufts.edu/apua/Roar/roarhome.

Levy, S.B. and R.V. Miller. 1989. Horizontal gene transfer in relation to environmental release of genetically engineered microorganisms. pp. 405–420. Gene transfer in the environment McGraw-Hill Publishing Company, New York.

Lin, C.F., Z.F. Fung, C.L. Wu and T.C. Chung. 1996. Molecular characterization of a plasmid-borne (pTC82) chloramphenicol resistance determinant (cat-TC) from *Lactobacillus reuteri* G4. Plasmid 36: 116–124.

Marchesino, B., A. Bruttin, N. Romailler and R.S. Moreton. 1992. Microbiological events during commercial meat fermentations. J. Appl. Bacteriol. 73: 203–209.

Martin, B., M. Garriga, M. Hugas and T. Aymerich. 2005. Genetic diversity and safety aspects of enterococci from slightly fermented sausages. J. Appl. Microbiol. 98: 1177–1190.

Mathur, S. and R. Singh. 2005. Antibiotic resistance in food lactic acid bacteria–a review. Int. J. Food Microbiol. 105: 281–305.

Mojsova, S., K. Krstevski, I. Dzadzovski, Z. Popova and P. Sekulovski. 2015. Phenotypic and genotypic characteristics of enterocin producing enterococci against pathogenic bacteria. Mac. Vet. Rev. 38: 209–216.

Nawaz, M., J. Wang, A. Zhou, C. Ma, X. Wu, J.E. Moore, B. Cherie Millar and J. Xu. 2010. Characterization and transfer of antibiotic resistance in lactic acid bacteria from fermented food products. Curr. Microbiol. 62: 1081–1089.

Neu, H.C. 1992. The crisis in antibiotic resistance. Science 257: 1064–1073.

Olukoya, D.K., S.I. Ebigwei, O.O. Adebawo and F.O. Osiyemi. 1993. Plasmid profiles and antibiotic susceptibility patterns of *Lactobacillus* isolated from fermented foods in Nigeria. Food Microbiol. 10: 279–285.

Pan, L., X. Hu and X. Wang. 2011. Assessment of antibiotic resistance of lactic acid bacteria in Chinese fermented foods. Food Control 22: 1316–1321.

Patel, A.R., N.P. Shah and J.B. Prajapati. 2012. Antibiotic resistance profile of lactic acid bacteria and their implications in food chain. World J. Dairy & Food Sci. 7: 202–211.

Perreten, V., F. Schwarz, L. Cresta, M. Boeglin, G. Dasen and M. Teuber. 1997. Antibiotic resistance spread in food. Nature 389: 801–802.

Perreten, V., L. Vorlet-Fawer, P. Slickers, R. Ehricht, P. Kuhnert and J. Frey. 2005. Microarray-based detection of 90 antibiotic resistance genes of Gram-positive bacteria. J. Clin. Microbiol. 43: 2291-2302.

Pesavento, G., C. Calonico, B. Ducci, A. Magnanini and A. Lo Nostro. 2014. Prevalence and antibiotic resistance of Enterococcus spp. isolated from retail cheese, ready-to-eat salads, ham, and raw meat. Food Microbiol. 41: 1–7.

Peters, J., K. Mac, H. Wichmann-Schauer, G. Klein and L. Ellerbroek. 2003. Species distribution and antibiotic resistance patterns of enterococci isolated from food of animal origin in Germany. Int. J. Food Microbiol. 88: 311–314.

Resch, M., V. Nagel and C. Hertel. 2008. Antibiotic resistance of coagulase-negative staphylococci associated with food and used in starter cultures. Int. J. Food Microbiol. 127: 99–104.

Ribeiro, M.D.P.M.A., M.D.T.F. Dellias, S.M. Tsai, A. Bolmströn, L.W. Meinhardt and C.D.M. Bellato. 2005. Utilization of the Etest assay for comparative antibiotic susceptibility profiles of citrus variegated chlorosis and Pierce's disease strains of Xylella fastidiosa. Curr. Microbiol. 51: 262–266.

Ribeiro, T., M. Oliveira, M.J. Fraqueza, A. Lauková, M. Elias, R. Tenreiro, A.S. Baretto and T. Semedo-Lemsaddek. 2011. Antibiotic resistance and virulence factors among enterococci isolated from Chouriço, a traditional Portuguese dry fermented sausage. J. Food Protect. 74: 465–469.

Rodríguez-Rojas, A., J. Rodríguez-Beltrán, A. Couce and J. Blázquez. 2013. Antibiotics and antibiotic resistance: a bitter fight against evolution. Int. J. Med. Microbiol. 303: 293–297.

Sader, H.S. and A.C. Pignatari. 1994. E test: a novel technique for antimicrobial susceptibility testing. Revista Paulista de Medicina 112: 635–638.

Salyers, A.A. 1995. Antibiotic resistance transfer in the mammalian intestinal tract: implications for human health, food safety and biotechnology. Springer-Verlag.

Salyers, A.A., A. Gupta and Y. Wang. 2004. Human intestinal bacteria as reservoirs for antibiotic resistance genes. Trends Microbiol. 12: 412–416.

Singer, R.S., R. Finch, H.C. Wegener, R. Bywater, J. Walters and M. Lipsitch. 2003. Antibiotic resistance – the interplay between antibiotic use in animals and human beings. Lancet Infect. Dis. 3: 47–51.

Strommenger, B., C. Kettlitz, G. Werner and W. Witte. 2003. Multiplex PCR assay for simultaneous detection of nine clinically relevant antibiotic resistance genes in Staphylococcus aureus. J. Clin. Microbiol. 41: 4089–4094.

Swenson, J.M., R.R. Facklam and C. Thornsberry. 1990. Antimicrobial susceptibility of vancomycin-resistant Leuconostoc, Pediococcus, and Lactobacillus species. Antimicrob. Agents Chemother. 34: 543–549.

Talon, R. and S. Leroy. 2011. Diversity and safety hazards of bacteria involved in meat fermentations. Meat Sci. 89: 303–309.

Tankovic, J., R. Leclercq and J. Duval. 1993. Antimicrobial susceptibility of Pediococcus spp. and genetic basis of macrolide resistance in Pediococcus acidilactici HM3020. Antimicrob. Agents Chemother. 37: 789–792.

Antimicrobial Resistance of Lactic Acid Bacteria in Fermented Meat.... 341

Tannock, G.W., J.B. Luchansky, L. Miller, H. Connell, S. Thode-Andersen, A. A. Mercer and T. R. Klaenhammer. 1994. Molecular characterization of a Plasmid-Borne (pGT633) erythromycin resistance determinant (ermGT) from *Lactobacillus reuteri* 100-63. Plasmid 31: 60–71.

Teale, C.J. 2002. Antimicrobial resistance and the food chain. J. Appl. Microbiol. (Suppl.) 92: 85S–89S.

Temmerman, R., B. Pot, G. Huys and J. Swings. 2003. Identification and antibiotic susceptibility of bacterial isolates from probiotic products. Int. J. Food Microbiol. 81: 1–10.

Teuber, M., L. Meile and F. Schwarz. 1999. Acquired antibiotic resistance in lactic acid bacteria from food. Anton. Leeuw. Int. J. G. 76: 115–137.

Toğay, S.O., A.C. Keskin, L. Açik and A. Temiz. 2010. Virulence genes, antibiotic resistance and plasmid profiles of *Enterococcus faecalis* and *Enterococcus faecium* from naturally fermented Turkish foods. J. Appl. Microbiol. 109: 1084–1092.

Toomey, N., D. Bolton and S. Fanning. 2010. Characterisation and transferability of antibiotic resistance genes from lactic acid bacteria isolated from Irish pork and beef abattoirs. Res. Microbiol. 161: 127–135.

Tynkkynen, S., Singh, K.V. and P. Varmanen. 1998. Vancomycin resistance factor of *Lactobacillus rhamnosus* GG in relation to enterococcal vancomycin resistance (van) genes. Int. J. Food Microbiol. 41: 195–204.

Van Boeckel, T.P., S. Gandra, A. Ashok, Q. Caudron, B.T. Grenfell, S.A. Levin and R. Laxminarayan. 2014. Global antibiotic consumption 2000 to 2010: an analysis of national pharmaceutical sales data. Lancet Infect. Dis. 14: 742–750.

Verraes, C., S. Van Boxstael, E. Van Meervenne, E. Van Coillie, P. Butaye, B. Catry, M-E. de Schaetzen, X. Van Huffe, H. Imberechts, K. Dierick, G. Daube, C. Saegerman, K. De Block, J. Dewulf and L. Herman. 2013. Antimicrobial resistance in the food chain: a review. Int. J. Environ. Res. Public Health 10: 2643–2669.

Vesković, S. and D. Đukić. 2015. Lactic acid bacteria in meat industry. pp. 172–178. *In:* Vesković S. and D. Đukić (eds). Bioprotectors in food production. Čačak, Serbia. (in serbian).

Vesković Moračanin, S., L. Turubatović, M. Škrinjar and D. Obradović. 2013. Antilisterial activity of bacteriocin isolated from Leuconostoc mesenteroides subspecies mesenteroides IMAU:10231 in production of Sremska sausages (traditional Serbian sausage): lactic acid bacteria isolation, bacteriocin identification and meat application experiments". Food Technol. Biotechnol. 51: 247–256.

Vesković, S., S. Stefanović and S. Janković. 2011. Chapter VIII: Veterinary Drugs Residues. pp. 203–222; 369–398. *In:* Švarc-Gajić, J. (ed). Nutritional insights and food safety. New York: Nova Science Publishers.

Volkmann, H., T. Schwartz, P. Bischoff, S. Kirchen and U. Obst. 2004. Detection of clinically relevant antibiotic-resistance genes in municipal wastewater using real-time PCR (TaqMan). J. Microbiol. Meth. 56: 277–286.

von Wright, A. 2005. Regulating the safety of probiotics – The European approach. Curr. Pharm. Design 11: 17–23.

Wanangkarn, A., D.-C. Liu, A. Swetwiwathana, A. Jindaprasert, C. Phraephaisarn, W. Chumnqoen and F.-J. Tan. 2014. Lactic acid bacterial population dynamics during fermentation and storage of Thai fermented sausage according to restriction fragment length polymorphism analysis. Int. J. Food Microbiol. 186: 61–67.

WHO. 2014. Antimicrobial resistance: global report on surveillance. pp. 1-232. World Health Organization, Geneva, Switzerland.

Witte, W. 1997. Impact of antibiotic use in animal feeding on resistance of bacterial pathogens in humans. pp. 61-71. *In*: Chadwick, D.J., Goode, J. (eds.), Antibiotic resistance: origins, evolution, selection and spread, Ciba Foundation Symposium 207. Wiley, Chichester.

Zarazaga, M., Y. Sáenz, A. Portillo, C. Tenorio, F. Ruiz-Larrea, R. Del Campo, F. Baquero and C. Torres. 1999. In vitro activities of ketolide HMR3647, macrolides, and other antibiotics against *Lactobacillus*, *Leuconostoc*, and *Pediococcus* isolates. Antimicrob. Agents Chemother. 43: 3039–3041.

Zdolec, N. 2016. Antimicrobial resistance of fermented food bacteria. pp. 263–281. *In*: D. Montet and R.C. Ray (eds.). Fermented Foods– Part I: Biochemistry and Biotechnology. CRC Press, Taylor & Francis Group, Boca Raton.

Zdolec, N., I. Filipović, Ž. Cvrtila Fleck, A. Marić, D. Jankuloski, L. Kozačinski and B. Njari. 2011. Antimicrobial susceptibility of lactic acid bacteria isolated from fermented sausages and raw cheese. Vet. arhiv 81: 133–141.

Zdolec, N., V. Dobranić, I. Butković, A. Koturić, I. Filipović and V. Medvid. 2016. Antimicrobial susceptibility of milk bacteria from healthy and drug-treated cow udder. Vet. arhiv 86, 163-172.

Zonenschain, D., A. Rebecchi and L. Morelli. 2009. Erythromycin- and tetracycline-resistant lactobacilli in Italian fermented dry sausages. J. Appl. Microbiol. 107: 1559–1568.

Chapter 15

Microbial Spoilage of Fermented Meat Products

Spiros Paramithiotis and Eleftherios H. Drosinos*

1 INTRODUCTION

The microbiological quality of meat depends upon several factors including the physiological status of the animal at slaughter, the spread of contamination during slaughter and processing, as well as the conditions of storage and distribution (Doulgeraki et al. 2011). Gram-negative rods (mainly pseudomonads) and micrococci usually dominate the initial microbiota of fresh meat; the secondary microbiota is usually formed by *Enterobacteriaceae* and Gram-positive bacteria including spore formers, lactic acid bacteria (LAB), *Brochothrix thermosphacta* as well as yeasts and molds (von Holy et al. 1992, Gill 2005, Nychas et al. 2008). These microorganisms may concomitantly contaminate the processing environment and if insufficient cleaning and disinfection procedures are followed may re-contaminate the products at various steps of the production procedure (Chevallier et al. 2006, Lebert et al. 2007). Additionally, casings and additives are important sources of contamination. Regarding the former, 53 positive samples to sulphite-reducing clostridia out of a total of 138 samples were reported by Houben (2005). Moreover, the average population in the positive samples ranged

---

*For Correspondence: Laboratory of Food Quality Control and Hygiene, Department of Food Science and Human Nutrition, Agricultural University of Athens, Iera Odos 75, GR-11855 Athens, Greece, Tel: +302105294713 Fax: +302105294683, Email: ehd@aua.gr

from 5.6 ± 3.75 to 187.1 ± 558.19 cfu/g. Regarding the additives, they may significantly increase the microbiological load of the batter. Paramithiotis and Drosinos (2010) reported on the microbiological quality of 13 spices, 38 spice mixtures and 15 additives. Although absence of foodborne pathogens was verified from all samples, total aerobic mesophilic counts of more than 7 log cfu/g as well as yeast-molds counts of more than 6 log cfu/g were reported. Additionally, presence of *Salmonella* spp., *Bacillus cereus*, *Clostridium perfringens*, *Staphylococcus aureus* and *Escherichia coli* has been reported in a wide range of herbs and spices (Schwab et al. 1982, De Boer et al. 1985, McKee 1995, Garcia et al. 2001, Banerjee and Sarkar 2003, Sagoo et al. 2009).

Fermented foods may be defined as food substrates overgrown by edible microorganisms whose metabolic activity results directly or indirectly in the production of non-toxic products with flavors, aromas and textures pleasant to human. Therefore, the borderline between fermentation and spoilage is very fine and depends upon the characteristics of the end product and the properties of the causative microorganism. In the case of fermented meat products, microorganisms with spoilage potential are yeasts-molds, LAB, staphylococci and several Gram-negative bacteria. With the exception of the latter, the rest belong to the technological biota of this type of products. Therefore, not properly timed or uncontrolled growth, even of the technological biota, may be perceived as spoilage. In the following paragraphs a short description of the hurdles (see also Chapter 7 in this book) employed during manufacture of fermented meat products as well as a discussion on the microorganisms with spoilage potential and preventive measures is presented.

2 THE HURDLE CONCEPT

The hurdle concept is as old as the fermentation itself. It was re-invented in the late '70s when the application of even more preservative factors as well as their accurate control was possible. The essence of the hurdle concept lies in the simultaneous application of several mild antimicrobial factors rather than less but more intense. The latter may have a negative impact on the physicochemical and organoleptic properties of the final products whereas the former offers better control over the properties of the product. In terms of antimicrobial action, the simultaneous disruption of e.g. pH, Eh and/or a_w microbial homeostasis systems that is obtained by the hurdle concept offers optimal preservation. In the case of fermented meat products a very specific combination and sequence of hurdles is applied. In the beginning of fermentation, NaCl, nitrite and temperature inhibit a great proportion of the initial batter microbiota. However, another part of it is unaffected and consumes

Microbial Spoilage of Fermented Meat Products

any oxygen available resulting in a drop of Eh inhibiting thus aerobic microorganisms. Under these conditions LAB dominate, resulting in the establishment of another hurdle, namely acidification and reduction of the pH value. In the case of long-ripened products, some of these hurdles may gradually diminish, i.e. nitrite is depleted, population of LAB may decrease, pH and Eh values may increase. In that case, the main hurdle is the low a_w, provided that adequately dehydration has occurred. The hurdles encountered during sausage fermentation are:

2.1 Acidity

Acidity is a very important hurdle, both in terms of pH value and the type of organic acid applied. In general, at pH values below the pKa value of an organic acid, most of it remains in the undissociated form that can penetrate the microbial cell membranes and enter the cytoplasm. Once inside, the acid dissociates due to the higher intracellular pH value releasing a proton. The cell consumes energy in order to remove this proton reducing the amount of energy available for other cellular processes including multiplication and concomitantly population growth. At the same time the remaining anion accumulates conferring toxicity. In fermented meat products, this role is typically reserved for the lactic acid produced by LAB during fermentation.

2.2 Water Activity

Reduction of the available water for metabolic reactions is a very important hurdle that is developed during manufacturing of fermented meat products. Dehydration that takes place during maturation is responsible for the loss of water accounting for approximately 25–50% of the total weight of dry sausages and 10–15% for semi-dry sausages. Additionally, binding to NaCl, carbohydrates as well as other ingredients also contribute to the final a_w value of less than 0.9 at the end of ripening.

2.3 Sodium Chloride

Sodium chloride has been used for centuries for food preservation as well as taste enhancement. The antimicrobial action is based on cellular dehydration obtained due to its water binding ability. An amount of 2.4–3.0% is used in the batter formulation that is capable to affect negatively growth of most pathogenic and spoilage microorganisms associated with fermented meat products without affecting the lactic acid microbiota (Sofos 1984). The latter may be inhibited at concentrations above 5% resulting in products with higher pH values (Chikthimmah et al. 2001, Heir et al. 2010). The antimicrobial activity may be enhanced

346 *Fermented Meat Products: Health Aspects*

when it is combined with refrigeration, low pH value, nitrite, spices, smoke and several antioxidants and is greatly affected by previous heat treatments of the raw materials as well as storage under refrigeration (Sofos 1984). However, sodium intake has been related to high blood pressure, therefore partial or complete substitution has been the epicenter of intensive research. Several compounds, such as potassium chloride, potassium lactate, glycine, calcium ascorbate and calcium chloride have been used as sodium chloride substitutes. Generally, the use of potassium salts is limited to less that 30% of the sodium chloride amount due to the resulting bitterness. Furthermore, Ibanez et al. (1995) reported that potassium chloride led to higher heterofermentative activity and nitrosation intensity while Gou et al. (1996) observed a delay in the reduction of the pH value due to the use of potassium lactate that affected negatively colour stability, product consistency and overall quality and safety. The same authors also reported that an unacceptable sweet taste limits the use of glycine to less than 40 percent. The use of calcium ascorbate resulted in enhanced LAB development and acidification that affected significantly the colour, the hardness and the gumminess of the product (Gimeno et al. 2001). Similarly, sensorially less acceptable products with significant colour differences were the result of the use of potassium and calcium chlorides mixture as well as potassium, magnesium and calcium chlorides mixture (Gimeno et al. 1999).

2.4 Nitrite

The role of nitrite in meat products includes a contribution in color and flavor development as well as antioxidant and antimicrobial activity. The latter is achieved mainly through inhibition of certain metabolic enzymes and proton-dependent active transport (Yarbrough et al. 1980). The amount of nitrite added should not exceed 150 mg/kg. The antimicrobial activity is strongly pH dependent and affected, as in the case of sodium chloride, by the processing parameters including formulation. A great health concern quells from the potential of nitrite to form nitrosamines, compounds that have been characterized as carcinogenic, either in the food matrix or inside the human body. In the case of fermented meat products, both precursors, i.e. secondary amines and nitrite are present but it appears that no formation of nitrosamines takes place. Nevertheless, fermented meats should not be fried, as severe heating at low a_w favours nitrosamine formation and minimization of residual nitrite is an important aspect, since nitrosamines can also be formed in the stomach, due to the acidic environment (Drosinos and Paramithiotis 2012). The importance of nitrite during manufacturing of fermented meat products was recently highlighted by Pichner et al. (2006), which reported that the recall of STEC contaminated salametti-products in

Microbial Spoilage of Fermented Meat Products 347

September 2004 in Germany was due to manufacturing of ecological (without nitrite) long fermented dry sausages without strict compliance to hygiene regulations. It has been stated that when using standard manufacturing technologies, long fermented dry sausages do not pose a health hazard to consumer.

2.5 Temperature

Accurate control of the temperature throughout production procedure is essential for ensuring dominance of the lactic acid microbiota and at the same time suppress growth of pathogenic and spoilage biota. Generally, the higher the fermentation temperature the shorter the fermentation time but the higher the risk of growth of undesired microorganisms. However, depending on the type of the product specific range of temperatures is required, e.g semi-dry sausages are usually fermented at 20–25°C while dry sausages at 18–22°C. Ripening usually takes place at 12–15°C.

2.6 Atmosphere

This hurdle refers mostly to the inhibition of aerobic microorganisms due to the consumption of oxygen and concomitant Eh drop by the microbial consortium; in that way development of lactic acid microbiota is favored. However, during ripening the Eh value increase and thus the relative value as a hurdle decreases. However, this could not pose any safety concerns since at that time the low a_w is the main hurdle.

2.7 Smoke

Pyrolysis of cellulose, hemicellulose and lignin, i.e. the major wood components, results in the formation of acids and aldehydes from the former two and phenols and tars from the latter (Rozum 2009). The main antimicrobial compounds are the phenols followed by formaldehyde. Smoking of fermented meat products may take place from hours to days depending on several manufacturing parameters. However, penetration of smoke compounds into the sausage depends upon several factors such as the type and quality of casings; therefore surface microbiota is mostly affected and to a lesser extend the microbiota located towards the center of the product.

2.8 Herbs and Spices

Herbs and spices may harbor significant microbial populations that may increase the microbial load of the batter (Paramithiotis and

Drosinos 2010). Moreover, several *Salmonella* – related outbreaks associated with herbs and spices, such as paprika, cilandro, aniseed, basil and pepper have been reported (Lehmacher et al. 1995, Campbell et al. 2001, Koch et al. 2005, Pezzoli et al. 2008, Pakalniskiene et al. 2009). Regarding fermented meat products, a multistate outbreak caused by contamination of fermented meat products by *S.* Montevideo that occurred through red and black pepper has been reported (Gieraltowski et al. 2013). At the same time, several of their constituents have been reported to exert antimicrobial action. The latter has been the subject of extensive study. The majority of the studies have reached the conclusion that the amount used for the production of fermented meat products is not capable to interfere with spoilage or pathogenic biota and therefore they should be regarded as an additional hurdle rather than a stand-alone preservative (Zaika 1988, Gonzalez-Fandos et al. 1996).

2.9 Other Hurdles

Another hurdle that has been heavily studied is the inclusion of bacteriocinogenic strains in the starter culture or the addition of the bacteriocins in the formulation. A vast number of bacteriocinogenic LAB strains have been isolated from spontaneously fermented meat products. Moreover, the effect of the ingredients used in the batter formulation as well as the manufacturing conditions in the production and activity of bacteriocins has also been extensively studied (Drosinos et al. 2008). In most of the cases, only a 1–3 log cfu/g reduction of the population of the pathogen under study is achieved. This, along with the narrow range of their action, due to their strain-specific dependence, limits their use as a preservative.

3 MICROORGANISMS WITH SPOILAGE POTENTIAL

Uncontrolled growth, even of the technological biota may result in spoilage of the final product. As already mentioned, microorganisms with spoilage potential are Gram-negative bacteria, mainly pseudomonads and *Enterobacteriaceae*, yeasts-molds, staphylococci and LAB. In the next paragraphs their spoilage potential is discussed.

3.1 Yeasts-Molds

Yeasts and molds are very often present in the raw materials; in fresh meat their population varies from 2 to 4 log cfu/g and may even reach 6 log cfu/g depending on the storage conditions. Regarding casings, they are usually preserved by salting, curing and/or drying (Fischer and

Microbial Spoilage of Fermented Meat Products 349

Schweflinghaus 1988) and are generally considered as microbiologically acceptable. However, quality deterioration due to microbial growth is likely to occur (Houben 2005, Wijnker et al. 2006, 2011). As far as spices and additives are concerned, yeast-mold population is usually below detection limit but in some cases may form a dominant microbiota reaching populations such as 6.72 and 5.8 log cfu/g that were reported in the case of whole pepper and paprika, respectively (Paramithiotis and Drosinos 2010).

Thus, the initial yeast-mold population in the batter is usually less than 5 log cfu/g (Samelis et al. 1993, 1998, Osei Abunyewa et al. 2000, Metaxopoulos et al. 2001, Drosinos et al. 2005, Aquilanti et al. 2007). During fermentation their population may remain stable, increase or diminish depending on product formulation and technology applied (Paramithiotis et al. 2010); in any case the population may hardly exceed 6 log cfu/g. The species most frequently isolated belong to the genera *Debaryomyces*, *Rhodotorula*, *Candida*, *Pichia* and *Yarrowia* (Encinas et al. 2000, Coppola et al. 2000).

The effect that yeast and mold growth has during meat fermentation on the quality of the final product is manifold. Yeasts and molds require oxygen for their growth. More accurately, molds may only grow on the surface of the product due to their strict aerobic metabolism. The mycelium formed prevents excessive dehydration of the surface allowing thus a more homogeneous dehydration of the product and is required in many products. Yeasts, on the other hand, may even use organic molecules as final electron acceptors, decreasing the amount of oxygen required for their metabolic functions and concomitantly increasing the depth in which they may proliferate. Oxygen consumption results in the enhancement of the curing reactions and the development of the desired color and at the same time delays lipid oxidation. Yeast-mold growth affects sensorial properties through catabolism of lactic acid as well as lipolytic and proteolytic activities. The latter have been extensively studied and a secondary role in terms of flavor development has been suggested (Rodriguez et al. 1998, Martin et al. 2002).

Yeasts and molds may be the cause of spoilage, as this is defined in the introduction, when the above mentioned actions are not adequately controlled or occur when they are not desired. Growth on the surface, if occur very early, i.e. during the initial fermentation stage, may result in the formation of a biofilm that limits dehydration and cause discoloration. Surface growth may also occur at a later stage that if not controlled may as well lead to undesirable sensorial characteristics and may even pose health risks upon growth of mycotoxinogenic molds. Osmophilic yeasts such as *Candida famata* and *Zygosaccharomyces rouxii* as well as xerophilic molds such as *Penicillium expasum*, *Aspergillus flavus* and *A. ochraceus* may develop during storage and result in slime formation, discoloration,

350   *Fermented Meat Products: Health Aspects*

off-flavours and may even lead to mycotoxin production (Iacumin et al. 2009, 2012). Indeed, occurrence of ochratoxin A (OTA) and aflatoxin B1 in fermented meat products has been reported by Markov et al. (2013) and Pleadin et al. (2015) and at maximum concentrations of 7.83 and 3.0 µg/kg, respectively. This may be attributed either to direct contamination through growth of mycotoxinogenic molds or indirectly through contaminated feed. In a study by Dall'Asta et al. (2010), the muscle carry-over through oral administration of OTA contaminated feed was generally estimated to be negligible.

Surface growth of molds may be prevented by controlling the RH during all stages of fermentation and applying smoke. Moreover, treatment with natamycin, potassium sorbate or other anti-mold ingredient may be applied.

3.2 Gram-negative Bacteria

The spoilage potential of pseudomonads and *Enterobacteriaceae* of meat and meat products has been the epicenter of intensive study. The former dominate the spoilage microbiota upon storage in high O_2 conditions while the latter under increased CO_2 (Casaburi et al. 2015) taking for granted that storage takes place at refrigerating temperatures.

Pseudomonads may rapidly convert meat glucose and glucose-6-P to gluconate and gluconate-6-P that may as well use as carbon and energy sources; in addition, upon depletion of the former they may also utilize lactate, pyruvate and amino acids for the same purpose (Nychas and Arkoudelos 1990, Paramithiotis et al. 2009). The end products of their catabolism extend to various organic acids, alcohols, aldehydes, ketones, esters, amines and sulfides (Edwards et al. 1987, Tsigarida et al. 2003, Ercolini et al. 2010, 2011, La Storia et al. 2012). The most frequently isolated species is *Pseudomonas fragi* followed by *Ps. lundensis* and *Ps. fluorescens* (Samelis 2006). The properties that distinguish the former and give an ecological advantage over the latter species are the ability to utilize creatine and creatinine (Drosinos and Board 1994) as well as the ability to utilize a variety of iron sources that concomitantly suppresses the need for siderophore synthesis, which in turn saves energy for the cell (Labadie 1999).

The pseudomonads population in the batter may range depending on the quality of the raw materials; they have been reported from below detection limit to 5.2 log cfu/g (Samelis et al. 1998, Aymerich et al. 2003, Drosinos et al. 2005, Comi et al. 2005, Lebert et al. 2007). However, at the end of fermentation they are mostly reported below or near detection limit (Talon et al. 2007) and in only some cases their population may remain at their initial level (Lebert et al. 2007). Since pseudomonads are oxidative microorganisms, presence of oxygen is prerequisite for their growth. Growth may thus occur in the surface or in crevices and hollows that may result from a number of reasons including low quality raw materials. The former occurs mostly at the beginning of fermentation and may result in the formation of biofilm that prevent adequate dehydration of the product. In both cases, off-odors are produced, mostly due to the secretion of proteolytic and lipolytic enzymes. Increase of the ripening temperature to above 12°C as well as of the air velocity within the ripening chamber, in order to avoid water condensation may be applied as preventive measures against biofilm formation.

Members of the *Enterobacteriaceae* family may preferentially utilize meat glucose and glucose-6-P and upon their depletion degradation of amino acids may occur (Gill 1986). The end products of their metabolism include ammonia, volatile sulphides, including H_2S, as well as malodorous amines. Although research has been rightfully focused on specific genera with pathogenic potential (e.g. *Salmonella*), significant amount of studies currently exists regarding the occurrence, monitoring and ecology of *Hafnia alvei*, *Serratia liquefaciens*, *Enterobacter agglomerans* and *Rahnella* spp. in meat and meat products (Stanbridge and Davies 1998, Lindberg et al. 1998, Nychas et al. 1998, Ercolini et al. 2006, Doulgeraki et al. 2011). *Enterobacteriaceae* counts in the batter have been reported to vary considerably from below detection limit to 4.4 log cfu/g. During fermentation, population may diminish, remain stable or even increase, depending on the properties of the developing microecosystem (Samelis et al. 1998, Aymerich et al. 2003, Drosinos et al. 2005, Fontana et al. 2005, Comi et al. 2005, Rantsiou et al. 2005, Lebert et al. 2007). *Enterobacteriaceae* are generally sensitive to the acidity and therefore are outcompeted by LAB during the course of lactic acid fermentation. However, when acidity fails to develop quickly, which is often in spontaneous fermentations, or the quality of the raw materials is significantly compromised, i.e. harbors more than 5 log cfu/g, then it is very likely that their metabolic activity becomes organoleptically noticeable and perceived as spoilage. The most common defects associated with *Enterobacteriaceae* growth are off-odors and discoloration. Through their metabolic activities a range of metabolites are produced; diamines and hydrogen sulfide contribute to the putrid and sulphury tones, the latter on the other hand converts the muscle pigment to the green sulphyoglobin. Ammonia is another compound with a distinctive odour that is perceived and cheesy odours that are associated with acetoin and diacetyl formation (Borch et al. 1996). Use of high quality raw materials as well as ensuring adequate acidification may prevent *Enterobacteriaceae* growth and associated defects.

3.3 Gram-positive Bacteria

3.3.1 Staphylococci

Staphylococci are not considered among the bacteria with high spoilage potential; especially in the case of fermented meat products, in which they constitute a part of the technological microbiota. However, in nitrate-cured spontaneously fermented meat products an inability for the development of the desired bright red color of nitrosylmyoglobin is observed and the brown color of metmyoglobin prevails. This may be due to the lack of effective nitrate reductase activities by the staphylococci. High nitrate reductase activity is a crucial property; the desired color should be developed during the initial fermentation stages before the accumulated acidity restricts growth of staphylococci. Indeed, in a study by Gotterup et al. (2008) staphylococcal strains with high nitrate-reductase activity exhibited significantly faster rate of pigment formation; however, the nitrate-reductase activity of the *Staphylococcus* strains did not allow a direct prediction of the nitrosylmyoglobin formation rate and other factors such as growth characteristics and acid tolerance should also be taken into consideration. Since high nitrate reductase activity is such an important property, selection of the appropriate strains and utilization in the form of starter cultures may ensure color development.

3.3.2 Lactic Acid Bacteria

Lactic acid bacteria is probably the most studied spoilage agent of fresh meat. Their population dynamics as well as their spoilage potential have been extensively studied in a variety of meat products under diverse storage conditions (Remenant et al. 2015, Casaburi et al. 2015). In the case of fermented meat products, spoilage may be considered when heterofermentative species prevail over the desired homofermentative ones. Such a situation may occur during spontaneous fermentation, triggering undesired physicochemical and organoleptic deviations. The former occur due to the production of CO_2 and the latter due to the production of a mixture of organic acids. Furthermore, production of peroxides may result in greening especially when insufficient production activity of catalases or peroxydases by staphylococci (Anifantaki et al. 2002). The use of strictly defined starter cultures and the respective manufacturing conditions will definitely prevent such situations.

4 CONCLUSION

Manufacturing of fermented meat products is a well-studied procedure; when what is referred to as 'common practice' is applied, then quality

Microbial Spoilage of Fermented Meat Products

and safety of the products is ensured. Regarding the former, which is the subject of the current chapter, any deviation from the quality standards may be specifically assigned to a proper cause and accurate preventing measures can be taken. Therefore spoilage of fermented meat products is not often.

Recent advances in the field of molecular biology may trigger studies on the molecular basis of the development of this microbial consortium. Although such studies may seem to possess restricted practical significance, they may improve our understanding of the microbial physiology and response strategies to environmental stimuli and therefore may lead to new developments regarding both quality and safety of these products.

Key words: *fermented meat products, spoilage, hurdle concept*

REFERENCES

Anifantaki, K., J. Metaxopoulos, M. Kammenou, E.H. Drosinos and M. Vlasi. 2002. The effect of smoking, packaging and storage temperature on the bacterial greening of frankfurters caused by *Leuconostoc mesenteroides* subsp. *mesenteroides*. Ital. J. Food Sci. 14: 135–144.

Aquilanti, L., S. Santarelli, G. Silvestri, A. Osimani, A. Petruzzelli and F. Clementi. 2007. The microbial ecology of a typical Italian salami during its natural fermentation. Int. J. Food Microbiol. 120: 136–145.

Aymerich, T., B. Martin, M. Garriga and M. Hugas. 2003. Microbial quality and direct PCR identification of lactic acid bacteria and nonpathogenic Staphylococci from artisanal low-acid sausages. Appl. Environ. Microbiol. 69: 4583–4594.

Banerjee, M. and P.K. Sarkar. 2003. Microbiological quality of some retail spices in India. Food Res. Int. 36: 469–474.

Borch E., M.L. Kant-Muemans and Y. Blixt. 1996. Bacterial spoilage of meat and cured meat products. Int. J. Food Microbiol. 33: 103–120.

Campbell, J.V., J. Mohle-Boetani, R. Reporter, S. Abbott, J. Farrar, M. Brandl, R. Mandrell and S.B. Werner. 2001. An outbreak of *Salmonella* serotype Thompson associated with fresh cilantro. J. Infect. Dis. 183: 984–987.

Casaburi, A., P. Piombino, G.J.E. Nychas, F. Villani and D. Ercolini. 2015. Bacterial populations and the volatilome associated to meat spoilage. Food Microbiol. 45: 83–102.

Chevallier, I., S. Ammor, A. Laguet, S. Labayle, V. Castanet, E. Dufour and R. Talon. 2006. Microbial ecology of a small-scale facility producing traditional dry sausage. Food Control 17: 446–453.

Chikthimmah, N., R.C. Anantheswaran, R.F. Roberts, E.W. Mills and S.J. Knabel. 2001. Influence of sodium chloride on growth of lactic acid bacteria and subsequent destruction of *Escherichia coli* O157:H7 during processing of Lebanon bologna. J. Food Protect. 64: 1145–1150.

Comi, G., R. Urso, L. Iacumin, K. Rantsiou, P. Cattaneo, C. Cantoni and L. Cocolin. 2005. Characterisation of naturally fermented sausages produced in the North East of Italy. Meat Sci. 69: 381–392.

Coppola, S., G. Mauriello, M. Aponte, G. Moschetti and F. Villani. 2000. Microbial succession during ripening of Naples-type salami, a southern Italian fermented sausage. Meat Sci. 56: 321–329.

Dall'Asta, C., G. Galaverna, T. Bertuzzi, A. Moseriti, A. Pietri, A. Dossena and R. Marchelli. 2010. Occurrence of ochratoxin A in raw ham muscle, salami and dry-cured ham from pigs fed with contaminated diet. Food Chem. 120: 978–983.

De Boer, E., W.M. Spiegelenberg and F.W. Janssen. 1985. Microbiology of spices and herbs. Anton. Leeuw. Int. J. G. 51: 435–438.

Doulgeraki, A.I., S. Paramithiotis and G.J.E. Nychas. 2011. Characterization of the *Enterobacteriaceae* community that developed during storage of minced beef under aerobic or modified atmosphere packaging conditions. Int. J. Food Microbiol. 145: 77–83.

Drosinos, E.H. and R.G. Board. 1994. Metabolic activities of pseudomonads in batch cultures in extract of minced lamb. J. Appl. Bacteriol. 77: 613–620.

Drosinos, E.H. and S. Paramithiotis. 2012. Effective strategies towards healthier fermented meat products. European Journal of Nutraceutical and Functional Foods 23: 42–44.

Drosinos, E.H., M. Mataragas and S. Paramithiotis. 2008. Antimicrobial activity of bacteriocins and applications. pp. 375–397. *In*: F. Toldrá (ed.). Meat biotechnology. Springer, New York.

Drosinos, E.H., M. Mataragas, N. Xiraphi, G. Moschonas, F. Gaitis and J. Metaxopoulos. 2005. Characterization of the microbial flora from a traditional Greek fermented sausage. Meat Sci. 69: 307–317.

Edwards, R.A., R.H. Dainty and C.M. Hibbard. 1987. Volatile compounds produced by meat pseudomonads and related reference strains during growth on beef stored in air at chill temperatures. J. Appl. Bacteriol. 62: 403–412.

Encinas, J.P., T.M. Lopez-Diaz, M.L. Garcia-Lopez, A. Otero and B. Moreno. 2000. Yeast populations on Spanish fermented sausages. Meat Sci. 54: 203–208.

Ercolini, D., A. Casaburi, A. Nasi, I. Ferrocino, R. Di Monaco, P. Ferranti, G. Mauriello and F. Villani. 2010. Different molecular types of *Pseudomonas fragi* have the overall behaviour as meat spoilers. Int. J. Food Microbiol. 142: 120–131.

Ercolini, D., F. Russo, E. Torrieri, P. Masi and F. Villani. 2006. Changes in the spoilage related microbiota of beef during refrigerated storage under different packaging conditions. Appl. Environ. Microbiol. 72: 4663–4671.

Ercolini, D., I. Ferrocino, A. Nasi, M. Ndagijimana, P. Vernocchi, A. La Storia, L. Laghi, G. Mauriello, M.E. Guerzoni and F. Villani. 2011. Monitoring of microbial metabolites and bacterial diversity in beef stored in different packaging conditions. Appl. Environ. Microbiol. 77: 7372–7381.

Fischer, A. and M. Schweflinghaus. 1988. Naturdaerme 1: Anatomie und Gewinnung. Fleischerei 39: 10–14.

Fontana C., P.S. Cocconcelli and G. Vignolo. 2005. Monitoring the bacterial population dynamics during fermentation of artisanal Argentinean sausages. Int. J. Food Microbiol. 103: 131–142.

Microbial Spoilage of Fermented Meat Products 355

Garcia, S., F. Iracheta, F. Galvan and N. Heredia. 2001. Microbiological survey of retail herbs and spices from Mexican markets. J. Food Protect. 64: 99–103.

Gieraltowski, L., E. Julian, J. Pringle, K. Macdonald, D. Quilliam, N. Marsden-Haug, L. Saathoff-Huber, D. Von Stein, B. Kissler, M. Parish, D. Elder, V. Howard-King, J. Besser, S. Sodha, A. Loharikar, S. Dalton, I. Williams and C.B. Behravesh. 2013. Nationwide outbreak of *Salmonella* Montevideo infections associated with contaminated imported black and red pepper: warehouse membership cards provide critical clues to identify the source. Epidemiol. Infect. 141: 1244–1252.

Gill, C.O. 1986. The control of microbial spoilage in fresh meats. pp. 49–88. *In*: A.M. Pearson and T.R. Dutson (eds.). Advances in meat research: meat and poultry microbiology. Macmillan. New York.

Gill, C.O. 2005. Sources of bacterial contamination at slaughtering plants. pp. 231–243. *In*: J.N. Sofos (ed.). Improving the safety of fresh meat. CRC/Woodhead Publishing Limited, Cambridge.

Gimeno, O., I. Astiasaran and J. Bello. 1999. Influence of partial replacement of NaCl with KCl and $CaCl_2$ on texture and color of dry fermented sausages. J. Agr. Food Chem. 47: 873–877.

Gimeno, O., I. Astiasaran and J. Bello. 2001. Calcium ascorbate as a potential partial substitute for NaCl in dry fermented sausages: effect on colour, texture and hygienic quality at different concentrations. Meat Sci. 57: 23–29.

Gonzalez-Fandos, M.E., M.L. Sierra, M.L. Garcia-Lopez, A. Otero and J. Sanz. 1996. Effect of the major herbs and spices in Spanish fermented sausages on *Staphylococcus aureus* and lactic acid bacteria. Arch. Lebensmittelhyg. 47: 43–47.

Gotterup, J., K. Olsen, S. Knochel, K. Tjener, L.H. Stahnke and J.K.S. Moller. 2008. Colour formation in fermented sausages by meat-associated staphylococci with different nitrite- and nitrate-reductase activities Meat Sci. 78: 492–501.

Gou, P., L. Guerrero, J. Gelabert and J. Arnau. 1996. Potassium chloride, potassium lactate and glycine as sodium chloride substitutes in fermented sausages and in dry-cured pork loin. Meat Sci. 42: 37–48.

Heir, E., A.L. Holck, M.K. Omer, O. Alvseike, M. Hoy, I. Mage and L. Axelsson. 2010. Reduction of verotoxigenic *Escherichia coli* by process and recipe optimisation in dry fermented sausages. Int. J. Food Microbiol. 141: 195–202.

Houben, J.H. 2005. A survey of dry-salted natural casings for the presence of *Salmonella* spp., *Listeria monocytogenes* and sulphite-reducing *Clostridium* spores. Food Microbiol. 22: 221–225.

Iacumin, L., L. Chiesa, D. Boscolo, M. Manzano, C. Cantoni, S. Orlic and G. Comi. 2009. Moulds and ochratoxin A on surfaces of artisanal and industrial dry sausages. Food Microbiol. 26: 65–70.

Iacumin, L., M. Manzano and G. Comi. 2012. Prevention of *Aspergillus ochraceus* growth on and Ochratoxin A contamination of sausages using ozonated air. Food Microbiol. 29: 229–232.

Ibanez, C., L. Quintanilla, A. Irigoyen, I. Garcia-Jalon, C. Cid, I. Astiasaran and J. Bello. 1995. Partial replacement of sodium chloride with potassium chloride in dry fermented sausages: influence on carbohydrate fermentation and the nitrosation process. Meat Sci. 40: 45–53.

Koch, J., A. Schrauder, K. Alpers, D. Werber, C. Frank, R. Prager, W. Rabsch S. Broll, F. Feil, P. Roggentin, J. Bockemuehl, H. Tschaepe, A. Ammon and K. Stark. 2005. *Salmonella* Agona outbreak from contaminated aniseed, Germany. Emerg. Infect. Dis. 11: 1124–1127.

La Storia, A., I. Ferrocino, E. Torrieri, R. Di Monaco, G. Mauriello, F. Villani and D. Ercolini. 2012. A combination of modified atmosphere and antimicrobial packaging to extend the shelf-life of beefsteaks stored at chill temperature. Int. J. Food Microbiol. 158: 186–194.

Labadie, J. 1999. Consequences of packaging on bacterial growth. Meat is an ecological niche. Meat Sci. 52: 299–305.

Lebert, I., S. Leroy, P. Giammarinaro, A. Lebert, J.P. Chacornac, S. Bover-Cid, M.C. Vidal-Carou and R. Talon. 2007. Diversity of microorganisms in environments and dry fermented sausages of French traditional small units. Meat Sci. 76: 112–122.

Lehmacher, A., J. Bockemuehl and S. Aleksic. 1995. Nationwide outbreak of human salmonellosis in Germany due to contaminated paprika and paprika-powdered potato chips. Epidemiol. Infect. 115: 501–511.

Lindberg, A.M., A. Ljunghb, S. Ahrne, S. Lofdahlc and G. Molin. 1998. *Enterobacteriaceae* found in high numbers in fish, minced meat and pasteurised milk or cream and the presence of toxin encoding genes. Int. J. Food Microbiol. 39: 11–17.

Markov K., J. Pleadin, M. Bevardi, N. Vahcic, D. Sokolic-Mihalak and J. Frece 2013. Natural occurrence of aflatoxin B1, ochratoxin A and citrinin in Croatian fermented meat products. Food Control 34: 312–317.

Martin, A., M.A. Asensio, M.E. Bermudez, M.G. Cordoba, E. Aranda and J.J. Cordoba. 2002. Proteolytic activity of *Penicillium chrysogenum* and *Debaryomyces hansenii* during controlled ripening of pork loins. Meat Sci. 62: 129–137.

McKee, L.H. 1995. Microbial contamination of spices and herbs: a review. Lebensm. Wiss. Technol. 28: 1–11.

Metaxopoulos, J., J. Samelis and M. Papadelli. 2001. Technological and microbiological evaluation of traditional processes as modified for the industrial manufacturing of dry fermented sausage in Greece. Ital. J. Food Sci. 13: 3–18.

Nychas, G.J.E. and J.S. Arkoudelos. 1990. Microbiological and physico-chemical changes in minced meat under carbon dioxide, nitrogen or air at 3°C. Int. J. Food Sci. Tech. 25: 389–398.

Nychas, G.J.E., E.H. Drosinos and R.G. Board. 1998. Chemical changes in stored meat. pp. 288–326. *In*: A.R. Davies and R.G. Board (eds.). The microbiology of meat and poultry. Blackie Academic and Professional, London.

Nychas, G.J.E., P.N. Skandamis, C.C. Tassou and K.P. Koutsoumanis. 2008. Meat spoilage during distribution. Meat Sci. 78: 77–89.

Osei Abunyewa, A.A., E. Laing, A. Hugo and B.C. Viljoen. 2000. The population change of yeasts in commercial salami. Food Microbiol. 17: 429–438.

Pakalniskiene, J., G. Falkenhorst, M. Lisby, S.B. Madsen, K.E. Olsen, E.M. Nielsen, A. Mygh, J. Boel and K. Molbak. 2009. A foodborne outbreak of enterotoxigenic *E. coli* and *Salmonella* Anatum infection after a high-school dinner in Denmark, November 2006. Epidemiol. Infect. 137: 396–401.

Microbial Spoilage of Fermented Meat Products 357

Paramithiotis, S. and E.H. Drosinos. 2010. Microbiological quality and aflatoxin B1 content of some spices and additives used in meat. Qual. Assur. Saf. Crop. 2: 41–45.

Paramithiotis, S., E.H. Drosinos, J. Sofos and G.J.E. Nychas. 2010. Fermentation: microbiology and biochemistry. pp. 185–198. In: F. Toldra (ed.). Handbook of meat processing. Springer, New York.

Paramithiotis, S., P.N. Skandamis and G.J.E. Nychas. 2009. Insights into fresh meat spoilage. pp. 55–82. In: F. Toldra (ed.). Safety of meat and processed meat. Springer, New York, USA.

Pezzoli, L., R. Elson, C.L. Little, H. Yip, I. Fisher, R. Yishai, E. Anis, L. Valinsky, M. Biggerstaff, N. Patel, H. Mather, D.J. Brown, J.E. Coia, W. van Pelt, E.M. Nielsen, S. Ethelberg, E. de Pinna, M.D. Hampton, T. Peters and J. Threlfall. 2008. Packed with Salmonella—investigation of an international outbreak of Salmonella Senftenberg infection linked to contamination of prepacked basil in 2007. Foodborne Pathog. Dis. 5: 661–668.

Pichner, R., H. Hechelmann, H. Steinrueck and M. Gareis. 2006. Shigatoxin producing Escherichia coli (STEC) in conventially and organically produced salami products. Fleischwirtschaft 86: 112–114.

Pleadin, J., M.M. Staver, N. Vahcic, D. Kovacevic, S. Milone, L. Saftic and G. Scortichini. 2015. Survey of aflatoxin B1 and ochratoxin A occurrence in traditional meat products coming from Croatian households and markets. Food Control 52: 71–77.

Rantsiou, K., R. Urso, L. Iacumin, C. Cantoni, P. Cattaneo, G. Comi and L. Cocolin. 2005. Culture-dependent and -independent methods to investigate the microbial ecology of Italian fermented sausages. Appl. Environ. Microbiol. 71: 1977–1986.

Remenant, B., E. Jaffres, X. Dousset, M.F. Pilet and M. Zagorec. 2015. Bacterial spoilers of food: Behavior, fitness and functional properties. Food Microbiol. 45: 45–53.

Rodriguez, M., F. Nunez, J.J. Cordoba, M.E. Bermudez and M.A. Asensio. 1998. Evaluation of proteolytic activity of microorganisms isolated from dry cured hams. J. Appl. Microbiol. 85: 905–912.

Rozum, J.J. 2009. Smoke flavor. pp. 211–226. In: Tarte R. (ed.) Ingredients in meat products. Springer, New York.

Sagoo, S.K., C.L. Little, M. Greenwood, V. Mithani, K.A. Grant, J. McLauchlin, E. de Pinna and E.J. Threlfall. 2009. Assessment of the microbiological safety of dried spices and herbs from production and retail premises in the United Kingdom. Food Microbiol. 26: 39–43.

Samelis, J. 2006. Managing microbial spoilage in the meat industry. pp. 213–286. In: C.D.W. Blackburn (ed.). Food spoilage microorganisms. Woodhead Publishing Limited, Cambridge.

Samelis, J., G. Aggelis and J. Metaxopoulos. 1993. Lipolytic and microbial changes during the natural fermentation and ripening of Greek dry sausages. Meat Sci. 35: 371–385.

Samelis, J., J. Metaxopoulos, M. Vlassi and A. Pappa. 1998. Stability and safety of traditional Greek salami –A microbiological ecology study. Int. J. Food Microbiol. 44: 69–82.

Schwab, A.H., A.D. Harpestad, A. Swartzentruber, J.M. Lanier, B.A. Wentz, A.P. Duran, R.J. Barnard and R.B. Jr. Read. 1982. Microbiological quality of some spices and herbs in retail markets. Appl. Environ. Microbiol. 44: 627–630.

Sofos, J.N. 1984. Antimicrobials effects of sodium and other ions in foods: A review. J. Food Safety 6: 45–78.

Stanbridge, L.H. and A.R. Davies. 1998. The microbiology of chill stored meat. pp. 174–219. In: A.R. Davies and R.G. Board (eds.). The microbiology of meat and poultry. Blackie Academic and Professional, London.

Talon, R., S. Leroy and I. Lebert. 2007. Microbial ecosystems of traditional fermented meat products: the importance of indigenous starters. Meat Sci. 77: 55–62.

Tsigarida, E., I.S. Boziaris and G.J.E. Nychas. 2003. Bacterial synergism or antagonism in a gel cassette system. Appl. Environ. Microbiol. 69: 7204–7209.

von Holy, A., W.H. Holzapfel and G.A. Dykes. 1992. Bacterial populations associated with Vienna sausage packaging. Food Microbiol. 9: 45–53.

Wijnker, J.J., E.A.W.S. Weerts, E.J. Breukink, J.H. Houben and L.J.A. Lipman 2011. Reduction of Clostridium sporogenes spore outgrowth in natural sausage casings using nisin. Food Microbiol. 28: 974–979.

Wijnker, J.J., G. Koop and L.J.A. Lipman. 2006. Antimicrobial properties of salt (NaCl) used for the preservation of natural casings. Food Microbiol. 23: 657–662.

Yarbrough, J.M., J.B. Rake and R.G. Eagon. 1980. Bacterial inhibitory effects of nitrite. Inhibition of active-transport, but not of group translocation, and of intracellular enzymes. Appl. Environ. Microbiol. 39: 831–834.

Zaika, L.L. 1988. Spices and herbs: their antimicrobial activity and its determination. J. Food Safety 9: 97–118.

Chapter 16

Chemical and Sensorial Properties of Fermented Meat Products

Tanja Bogdanović, Jelka Pleadin*, Nada Vahčić
and Sandra Petričević

1 INTRODUCTION

All over the world there exists a long tradition of fermented meats' consumption and production, both by industries and rural households. Fermented meat products are usually divided into two groups: a) whole meat products, such as hams, and b) chopped meat products, such as various sausages. Among complex factors that affect the choice of a fermented meat product, physicochemical and sensory properties are the major ones. These properties widely vary due to the number of types of fermented products and processing procedures applied. Differences in chemical and sensorial characteristics depend on the product's composition and complex protein, lipid and carbohydrate metabolism, which occurs both due to the meat- and bacteria-induced enzymatic activity and oxidation processes that influence the lipid and the protein fraction (Kozačinski et al. 2006, Milićević et al. 2014, Marušić et al. 2014). In dry-cured hams, the distribution of ingredients is limited to the product surface, while the technology of fermented sausages' production enables the distribution of ingredients both inside

---

*For Correspondence: Croatian Veterinary Institute, Laboratory for Analytical Chemistry, Savska Cesta 143, 10000 Zagreb, Tel: +385 1 6123 626, Email: pleadin@veinst.hr

the product and on its surface. Significant differences exist also among the namesake products produced in different countries or even in the same country, but at different locations (Dellaglio et al. 1996, Kovačević et al. 2009, Kos et al. 2009, Zanardi et al. 2010).

A wide variety of fermented meat products is characterized by the specific flavour and a lot of research has been focused on understanding the control mechanisms affecting the flavour development, as well as on the elucidation of pathways leading to the attainment of either desirable or undesirable compounds of fermented meat products (Toldrá 1998, Demeyer et al. 2000, Virgili and Schivazappa 2002, García-González et al. 2013). The greatest impact on sensory properties, aroma and taste of these products is that of smoking and ripening operations (Kovačević 2001, Kovačević et al. 2010, Jerković et al. 2010, Kovačević 2014). Of particular significance are the processes of fat lipolysis, free fatty acids' formation, and degradation and oxidation of short-chain fatty acids, since these key reactions taking place during the ripening process may affect the formation of specific odour and taste of the final product.

Besides meat products' quality, food safety and healthy nutrition aspects have also been emphasized (Vignolo et al. 2010). Recent studies have shown that, due to healthy trends in meat products' consumption in terms of low-fat and low-salt products preference, meat producers are facing a new challenge in terms of providing for fat and salt reduction without any loss of sensory properties (Jiménez-Colmenero et al. 2009, Flores et al. 2013). During the recent decades, studies devoted to fermented meat products have mainly focused on the evaluation of physicochemical, microbiological and sensory properties (Virgili and Schivazappa 2002, Comi et al. 2005, Rantsiou et al. 2005, Di Cagno et al. 2008) in order to contribute to better characterization of final products, the definition of unique quality markers and the improvement of product specification protocols essential when dealing with the products of the protected designation of origin (PDO) or geographical indication (PGI), as laid down under the EU Council Regulation (EC) No 510/2006.

Suitable chemical and sensory markers enable better linkage between raw matter and processing parameters, and thus result in higher uniformity and consistency of the products (Virgili and Schivazappa 2002, Zanardi et al. 2010). When it comes to fermented meat products, the main obstacle to that goal is the lack of product uniformity; for instance, cured hams and fermented sausages are heterogeneous in their nature depending on the sampling position (inner or outer, thinner or thicker part of the product) (García-González et al. 2013).

The aim of this Chapter is to present the typical chemical and sensorial properties of fermented meat products most frequently produced by meat industries and rural households and therefore often offered on the market and consumed by consumers.

2 CHEMICAL PROPERTIES OF FERMENTED MEAT PRODUCTS

2.1 Variability of Chemical Properties

Meat and meat products are important sources of proteins, vitamins, and minerals, which can therefore be considered as their functional compounds; however, they also contain fat, saturated fatty acids, cholesterol and salt that are often labelled as the ingredients having a negative impact on human health. Sugar content (natural carbohydrates and added glucose with derivates) and other nutrients like amino acids, fatty acids, minerals, vitamins, etc., are considered to be the cause of the fermentation process. The quality of fermented meats is influenced by various factors starting from the selection of raw materials and stuffing recipes to the production technology, that is to say, technological processes (conditioning, fermentation, drying, smoking and ripening) and technological parameters (temperature, relative humidity and air/smoke velocity) applied (Kovačević 2014).

Due to the application of various technological processes, the activity of technological microflora (especially during the fermentation process) and long-term maturation of sausage stuffing, during the sausage production a complex microbiological, physicochemical and biochemical changes take place in fundamental building materials (fats, proteins and carbohydrates), wherein the process is followed by water loss and increase in dry matter weight. Complex physicochemical, biochemical and microbiological changes witnessed throughout the processes of acidification, dehydration, and protein and fat degradation, shape the final characteristics of the fermented meat product.

The properties of fermented meats are also influenced by various factors that affect the quality of fresh meat, such as farm animals' genotype, methods of their keeping and feeding, procedures applied before slaughtering, and post-slaughtering conditions. A number of other factors, such as the selection of fresh meat and fat, the addition of salt and spices, hygiene and environmental conditions (e.g. temperature, humidity, air flow) witnessed during fermentation, smoking, curing and ripening, can further contribute to the diversity of characteristics of the finished product.

Regardless of the recipe, whenever producing a meat product it is necessary to abide to the basic principles of production, since the degree of conformance or non-conformance with the latter principles affects the final quality and safety of the product and gives importance to the quality of raw materials and additives. Unlike industrial production, the household technology of autochthonous meat products' production is not regulated, so that the variety of production conditions, such as uneven weight and quality of raw materials and differences in production technology resulting in the diversity of finished products' compositions can be encountered. Therefore, chemical properties of fermented meats show huge variability across individual producers and production periods. Industrial production of traditional dry sausages and dry-cured products is conducted in line with the traditional recipes also used in rural households, but, as opposed to the seasonal household production, is carried out under controlled processing conditions and is not seasonal in nature, thus allowing for the continuous market supply throughout a year.

2.2 Methodology of Chemical Composition Assessment

Basic chemical composition of a fermented meat product is essential for the verification of its final quality, its labelling and official national regulations, as well as for the quality of food composition databases. Variations observed in numerous meat products' composition tables can arise on the grounds of several factors including the production technology and sample preparation, but also on the grounds of analytical methodology applied in the determination of the product's composition (Jiménez-Colmenero et al. 2010a).

Standard methods of evaluation of chemical meat products' composition have been established and widely adopted. The preference should be given to methods recommended by international organizations, e.g. the International Organization for Standardization (ISO) and the Association of Official Analytical Chemists (AOAC), due to the precisely defined level of performance. It is essential for control laboratories to verify whether their methodology has been properly applied in everyday practice; to that goal, the parameters of certified and reference materials should be well determined and their mean values compared to the values certified by the manufacturer.

Methods most commonly applied for the determination of basic physicochemical parameters descriptive of fermented meat products are summarized in Table 1.

The content of macronutrients can be determined using biophysical methods, including optical, dielectrical and nuclear magnetic resonance measurements (Damez 2008). Near-infrared radiation (NIR) is applied as a fast method for at-line analysis/control of fat, water, protein and collagen content, but it must be extensively calibrated (Steinhart 2009). Reference methods must always be used if deviations from the quality standards occur.

Chemical and Sensorial Properties of Fermented Meat Products   363

Table 1 Methods of analysis of basic physicochemical properties of fermented meats

Chemical parameter	Method	Principle
Moisture	ISO 1442:1997 standard. AOAC 2000. Official Method 950.46.	Oven drying at 100°C
Protein	ISO 937:1978 standard. AOAC 2000. Official Method 928.08.	Acid digestion of organic matter with catalytic agent, automatic distillation and titration of total nitrogen
Fat	ISO 1443:1973 standard. AOAC 2000. Official Method 991.36.	Acid Hydrolysis/solvent Extraction
Ash	ISO 936:1998 standard. AOAC 2000. Official Method 920.153.	Incineration at 500°C
pH	Voltarimetric method (pH-metry)	Directly or indirectly in sample homogenized in distilled water (10/1 water/sample, w/w)
Sodium chloride	AOAC 2002. Official Method 935.47.	Volumetric/titrimetric method
Collagen	ISO 3496:1994 standard.	Acid hydrolysis, oxidation, colorimetry
Carbohydrates	Calculation method (Review of Methods of Analysis 2003)	The calculation of carbohydrates by difference using the Weende proximate system of analysis

2.3 Physicochemical Characterization of Fermented Meat Products

Previous studies have pointed towards a high variability of the properties of fermented meat products (Ferreira et al. 2007). The researchers in the field have made a clear distinction between the fermented sausages produced in shorter and those produced in longer ripening periods (Demeyer and Stahnke 2002) and have indicated the existence of apparent regional differences. Northern type-European products are characterized by ripening periods of up to three weeks (German Salami, Hungarian Salami, Nordic Salami), whereas Mediterranean type-products are ripened for several months (Italian Salami, Spanish Salchichon, Chorizo, Slavonian Kulen, French Saucisson).

Throughout the drying/ripening stage, chemical composition of a fermented meat product undergoes a significant change: the water content decreases, while the content of protein, fat and ash increases. Based on the ratio of the water content over the protein content, semi-dry and dry meat products can be mutually distinguished. Differences also exist between home-made and industrial fermented sausages due to the markedly different recipes applied in home-based production as compared to standardized commercial production. Changes in chemical properties of commercial meat products also take place due to the variations in meat and non-meat ingredients and the manufacturing conditions (Jiménez-Colmenero et al. 2010a). Variations in the proximate composition in terms of fatty acid profile, cholesterol, energy and mineral content representation, have also been observed over the year. Physicochemical properties determined in certain dry-fermented sausages and dry-cured meat products are shown in Tables 2 and 3, respectively.

Literature data suggest that, due to the longer drying and ripening of dry-fermented sausages and a high share of lean meat used in the preparation of their stuffing, water and protein content in finished ripened sausages are on an equal level (30–40% w/w), suggesting a high nutritional value of the finished product (Pleadin et al. 2014). In dry-fermented sausages, the water content mostly raises up to 40% w/w (Vignolo et al. 2010). The ratio of water over proteins of 1.2 to 1.3 is typical of dry sausages (Incze 2007). When analysing dry sausages produced in individual households, Kozačinski et al. (2008) obtained the water/protein ratio of 1.00/2.12, also indicating that these products are of a high nutritional value. The amount of total fat present in dry-fermented sausages generally varies very widely (14 to 47% w/w, Table 2) depending on the recipe and the producing household, but also on the origin of raw materials. Such variations can be attributed to the differences in the amount of added fatback and the choice of more or less fatty meat, made by individual manufacturers (Pleadin et al. 2014). The share of collagen, i.e. the major connective tissue protein, is associated to the textural properties of meat products, smaller amounts of collagen thereby being typical of higher-quality products. In short-ripen sausages, the amount of collagen ranges from 8–20% w/w, while in long-ripen sausages it spans from 1–6% w/w (Dellaglio et al. 1996, Kovačević et al. 2009). The ash content ranges from 3% to 6% w/w, and is in direct correlation with the mineral content (Ockerman and Basu 2007, Jiménez-Colmenero et al. 2010a, Karolyi and Čurić 2012). Sodium chloride (salt) is an essential ingredient of fermented meats, so that meat products are one of the richest sodium chloride food sources contributing to the increased water and fat binding capacity, the formation of colour, taste and texture of the product and its microbiological safety. Typical

Chemical and Sensorial Properties of Fermented Meat Products

Table 2 Physicochemical properties determined in certain dry-fermented sausages

Dry-fermented sausage	Water (%)	Ash (%)	Total fat (%)	Total proteins (%)	Collagen (%)	Sodium chloride (%)	pH	Reference
Chorizo de Pamplona	24.3	–	46.6	22.2	5.3	4.4	4.7	Gimeno et al. 2000
Chorizo	41.7–43.8	3.4–3.6	35.3–37.3	14.5–17.6	–	–	–	Jiménez-Colmenero et al. 2010a
Italian salami	39.8	–	23.1	31.3	–	4.3	6.1	Zanardi et al. 2010
Felino salami	38.0	–	28.1	29.7	1.7	4.3	5.9	Dellaglio et al. 1996
Northern type sausage	37.0 (N) 43.0 (B)	–	35.9(N) 34.8 (B)	17.6 (N) 17.7(B)	2.3–2.9	5.3 (N) 3.0 (B)	4.9 (N) 4.8 (B)	Demeyer et al. 2000
Mediterranean type sausage	33.0 (B) 39.0 (I)	–	40.9 (B) 28.1 (I)	18.8 (B) 21.3 (I)	2.4–3.1	4.1 (B, I)	5.5 (B) 5.7 (I)	Demeyer et al. 2000
Sicilian salami	29.8	–	31.0	31.7	–	6.0	6.3	Moretti et al. 2012
Slavonian Kulen	30.8–39.2	5.7	14.1–28.8	26.2–43.2	0.9–2.7	3.8–6.4	5.1–5.6	Kovačević et al. 2009, Pleadin et al. 2014

Types of sausages: (B) Belgium; (I) Italy; (N) Norway
Values are given as % w/w (w – weight portion)

levels of salt found in finished dry-fermented sausages are in the range of 3.0–6.4% w/w (Table 2). The salinity of a product depends on the amount of added salt and the duration of drying and ripening of the product (Wirth 1986), and has a significant impact on hardness and elasticity of meat products and their resistance to chewing (Kovačević et al. 2010). In fermented sausages' stuffing, an average share of salt ranges from 2.0 to 2.6%, whereas during the drying process the value keeps growing up to its final level found in the finished product (Ockerman and Basu 2007, Stahnke and Tjener 2007). Jiménez-Colmenero et al. (2001) demonstrated that larger amounts of salt exceeding the value of 4–6% have been established in numerous fermented meats studies. The pH value is an indicator of fermentation and ripening of a meat product (Salgado et al. 2005), and is commonly used in the assessment of their

366 *Fermented Meat Products: Health Aspects*

sustainability. In dry-fermented sausages, the pH value spans from 4.7 to 6.3 (Table 2).

As for the raw materials and processing conditions, final components of a dry-cured ham are strongly affected by the pig breed and rearing practices, where inter-animal variability has a greater impact on whole-cuts meat. Based on the origin of raw material and the ripening process, dry-cured hams are grouped into those made from rustic and free range-reared genotypes (Iberian, Corsica, Cinta Senese) and those made from intensely-reared white pigs (Serrano, Parma, Bayonne) (Jiménez-Colmenero et al. 2010b). The water content found in dry-cured hams

Table 3 Physicochemical properties determined in particular dry-cured meat products

Dry-cured meat product	Water (%)	Ash (%)	Total fat (%)	Total proteins (%)	Sodium chloride (%)	pH	Reference
Italian typical dry-cured ham	49.9	–	19.4	24.9	4.8	–	Benedini et al. 2012
Dry-cured ham Kraški pršut	57.2–59.3	–	3.3–4.5	28.7–29.4	7.0–8.6	5.7	Andronikov et al. 2013
Iberian dry-cured ham	49.0	–	5.5–19.2	17.9–30.6	4.0–5.9	–	García-Rey et al. 2004, Jiménez-Colmenero et al. 2010b
Serrano dry-cured ham	48.5	–	3.5–18.8	27.9–30.6	5.0–6.0	–	García-Rey et al. 2004, Jiménez-Colmenero et al. 2010b
Parma dry-cured ham	48.7–61.2	–	3.6–20.0	25.4–27.0	4.7	–	Pastorelli et al. 2003, Lo Fiego et al. 2005, D'Evoli et al. 2009, Jiménez-Colmenero et al. 2010b, Benedini et al. 2012
San Daniele dry-cured ham	54.7–60.4	–	23.0	27.3–30.8	4.5–6.9	–	D'Evoli et al. 2009
Corsican dry-cured ham	45.2	–	5.3–12.3	32.5	–	–	Virgili et al. 1999, Gandemer 2009
Bayonne dry-cured ham	60.8	–	–	29.3	–	–	Virgili et al. 1999
Istrian dry-cured ham	37.9–41.0	6.7–7.2	13.5–17.0	32.4–43.1	6.3–7.4	–	Marušić et al. 2014

Values are given as % w/w (w – weight portion)

ranges from approximately 38 to 62% w/w; the fat content spans from 3 to 23% w/w, while the protein content ranges from 18 to 43% w/w (Table 3). The salt content is mostly in the range of 4.5–7.4% w/w. Hams having a skin and subcutaneous fat contain higher water amounts (Italian Parma and San Daniele dry-cured hams, Spanish Iberian and Serrano dry-cured hams) due to the slower dehydration. The protein content of dry-cured hams varies depending on the extent of drying and the fat content (Toldrá 2002). As for the salt level, higher levels of salt are found in dry-cured hams having a larger muscle tissue surface uncovered by fat (Istrian ham), as well as in those lower-weighing and rapidly dried. The variations in salt content seen across various ham types arise as the consequence of differences in curing treatments. The *semimebranosus* (SM) and the *biceps femoris* (BF) muscles are characterized by the significant differences in salt content. The more salt in the interior BF ham muscle (5.5–8.5%) as compared to the exterior SM muscle (4.9–6.6%), the greater the internal diffusion of salt into the muscle (Andronikov et al. 2013, Pugliese et al. 2015). The pH values recommended for fresh (SM) hams intended for dry-curing are above 5.5 (normal pH hams), so as to be able to obtain the final product of the required quality (Arnau 2004, García-Rey et al. 2004). During the salting and the resting period, the pH values of ham muscles decrease, as opposed to the ageing period after which they increase due to the certain level of proteolytic activity, ultimately reaching the initial values seen in normal pH hams (Andronikov et al. 2013, García-Rey et al. 2004). The final pH of dry-cured hams having the initial pH value below 5.5 is greater than in normal pH hams due to the more extensive protein hydrolysis identified as one of the fundamental reasons behind the defective product texture, explained later under the section discussing the sensory properties of fermented meat products.

2.4 Fatty Acid Composition and the Cholesterol Content

One of the characteristics of the modern diet is an excessive intake of fat, particularly saturated fatty acids and, at the same time, a disturbed balance of polyunsaturated fatty acids' intake in terms of an increased intake of n-6 as compared to n-3 fatty acids. Research results have shown the nutritional composition of meat products in terms of fat mass fraction and fatty acid composition to be affected by many factors, starting from the animal breed selection and feeding and farming practices to technological processes and parameters used during production. During the ripening period, lipids undergo a series of transformations, which include hydrolytic processes, the release of fatty acids (those short-chained thereby directly intervening into the development of product aroma), and the oxidation of the latter acids together with the formation of peroxides and volatile compounds, which contribute to the aroma of the final product (Jiménez-Colmenero et al. 2001, Siciliano et al. 2013, Barbir et al. 2014).

Studies have shown the average fatty acid meat content to be the following: about 40% of SFA, 40% of MUFA and about 2–25% of PUFA fatty acids (Pleadin et al. 2014). Pork and pork products are generally characterised by a high proportion of saturated fatty acids (SFA) (Wood et al. 2004, Wood et al. 2008, Woods and Fearon 2009) and fatty acid proportions found in fermented sausages are usually consistent with the amounts of pork fat used to make the sausages.

Modified n-6/n-3 ratio is associated to the disorders in a number of physiological processes that increase the incidence of the so-called diet-related chronic diseases, primarily that of the heart disease and other cardiovascular diseases (Cordain et al. 2005). Since meat and meat products are generally rich in fat, especially in saturated fatty acids, consumers are nowadays advised to reduce the consumption of this type of food (Valsta et al. 2005, Fernandez et al. 2007), while the producers are trying to modify meat products in order to get them closer to the nutritionally acceptable values (Muguerza et al. 2004, Pelser et al. 2007, Valencia et al. 2006). Jiménez-Colmenero et al. (2001) have pointed out that the modification of carcass composition, the manipulation with raw materials and the reformulation of meat products are three fundamental strategies to obtain healthier meat and meat derivatives. Modification of the fatty acid profile of fermented sausages is achieved through genetic and feeding strategies (the selection of races and genetic lines, adapted dietary fatty acid composition) and partial substitution of pork fatback with polyunsaturated fatty acid oils of animal and vegetable origin (pre-emulsified soy, linseed oils and deodorised fish oil) (Ansorena and Astiasaran 2007). Strategies based on the production approaches (genetic selection in combination with feeding practices) have contributed to the production of dry-cured hams with PUFA/SFA and n-6/n-3 ratios characteristic for healthy fats.

The results of fatty acid analyses of dry-fermented sausages show the largest representation of oleic acid (C18:1n-9 C), followed by palmitic (C16:0), stearic (C18:0) and linoleic (C18:2n-6) acid (Pleadin et al. 2014). Oleic fatty acid was generally found to be the most dominant fatty acid in traditional dry-fermented pork sausages (Casaburi et al. 2007, Visessanguan et al. 2006). Karolyi and Čurić (2012) determined the ratios of fatty acid groups in the autochthonous dry-fermented sausage Slavonski Kulen and found no significant changes in their representation during the production process, with the exception of docosapentaenoic acid (C22:5n-3), whose proportion decreased during ripening. The research concerned with changes in fatty acids' representation during the ripening of indigenous fermented pork sausages from Italy (Moretti

Chemical and Sensorial Properties of Fermented Meat Products   369

et al. 2004) and Spain (Franco et al. 2002) did not show any statistically significant changes in fatty acid representation during ripening, while in some studies the reduction of the PUFA proportion was determined and explained by lipid oxidation (Gandemer 2002).

MUFAs most prevalent in dry-cured hams are oleic (C18:1n-9) and palmitoleic (C16:1n-7) fatty acid. The most represented SFAs are palmitic (C16:0), stearic C18:0 and myristic C14:0 acids (Fernández et al. 2007). An intense lipolysis takes place during the first five months of curing (salty environment and lower water activity thereby favouring the activity of acid muscle lipases), while the length of ripening bears no significant influence for the fatty acid profile of dry-cured ham muscles (Toldrá 1998, Pugliese et al. 2015). In the *semimebranosus* (SM) and the *biceps femoris* (BF) muscles, fatty acid levels may differ due to the different content of intramuscular fat, as well as due to the higher lipolytic activity of the SM muscle, resulting in higher fatty acid content (Pugliese et al. 2015).

In general, literature data have revealed that the main fatty acids present in fermented sausages and dry-cured hams are MUFA (41–59%), SFA (30–45%) and PUFA (9–18%), respectively (Table 4). The most prevalent MUFAs are oleic (C18:1n-9) (38–42%) and palmitoleic (C16:1n-7) (2–3%) fatty acid. The principal SFAs are palmitic (C16:0) (23–24%) and stearic (C18:0) (10–15%) acids. The main PUFA component is linoleic acid (C18:2n-6), with percentage-shares of up to 6–16%, lower in dry-cured hams (7–10%) as compared to dry-fermented sausages (10–16%) (Moretti et al. 2004, Karolyi 2006, Olivares et al. 2011, Jurado et al. 2008).

Literature data have revealed that the intake of fat should account for 15–30% of the total energy intake; the contribution of saturated fatty acids (SFA) should be up to 10%, that of polyunsaturated fatty acids (PUFA) 6–10% (n-6: 5–8%; n-3: 1–2%), that of monounsaturated fatty acids (MUFA) around 10–15%, and that of trans-fatty acids less than 1%. It has also been recommended to limit the cholesterol intake to 300 mg/day (WHO 2003). Nutritional properties of fermented meat products are assessed based on PUFA/SFA and n-6/n-3 PUFA ratios. PUFA/SFA values above 0.4–0.5 and n-6/n-3 below 4 are recommended (Pleadin et al. 2014). Blood LDL-cholesterol levels can be reduced by PUFA-rich diets, but SFAs have the opposite effect, resulting in the recommended PUFA/SFA ratio of above 0.4 (UK Department of Health 1994). According to these recommendations, dry-cured ham would not be within the desirable limits both when it comes to PUFA/SFA (0.19–0.3) and n-6/n-3 PUFA ratio (9–40). PUFA/SFA ratios found in dry-fermented sausages usually span from 0.3 to 0.4, while n-6/n-3 PUFA ratios range from 10 to 19.

370   *Fermented Meat Products: Health Aspects*

Table 4   Fatty acid composition in particular fermented meat products

Meat product	Fatty acids (% total fatty acids)					Reference
	SFA	MUFA	PUFA	PUFA/ SFA	n-6/n-3	
Chorizo (% total fatty acids)	35.9–37.3	47.6–48.6	13.9–16.2	0.4–0.5	11.5–15.1	Jiménez-Colmenero et al. 2010a
Sicilian salami (% total fatty acids)	36.3	55.1	8.5	0.2	7.8	Moretti et al. 2004
Northern type sausage (% total fatty acids)	37.4 (N) –41.1 (B)	41.9–42.0	10.3–14.4	–	–	Demeyer et al. 2000
Mediterranean type sausage (% total fatty acids)	36.9 (B) –37.6 (I)	41.3–43.5	12.5–15.1	–	–	Demeyer et al. 2000
Slavonski kulen (% total fatty acids)	44.6	46.7	8.9	0.2	7.61	Pleadin et al. 2014
Serrano ham (Jamon Serrano)	32.6–33.4	52.8–54.1	9.1–10.5	0.2	18.3–18.6	Santos et al. 2008, Jiménez-Colmenero et al. 2010b, Campo and Sierra 2011
Teruel ham (Jamón de Teruel)	35.7–37.4	54.6–54.7	7.4–8	0.1–0.2	17.4–17.6	Campo and Sierra 2011
Iberian ham	32.5–35.2	51.4–59.4	67.8–13.4	0.2–0.4	9.4–28.2	Fernández et al. 2007, Ventanas et al. 2007
Bayonne ham	36.5	52.9	10.7–15.3	0.3–0.4	14.1–29.6	Gandemer 2009
Corsican ham	34.9–35.0	53.8–55.4	9.7–11.2	0.3	8.7	Gandemer 2009
Parma ham	30.4–37.9	50.2–54.6	7.3–17.8	0.2–0.6	12.3–39.9	Pastorelli et al. 2003, Lo Fiego et al. 2005, D'Evoli et al. 2009
San Danielle	38.5	51.9	9.6	0.3	–	D'Evoli et al. 2009
Cinta Senese ham	33.3	51.4	15.4	0.5	14.2	Pugliese 2009
Jinhua ham	37.1	46.6	14.2	0.4	–	Du and Ahn 2001
Istrian dry-cured ham	38.9–40.5	44.7–53.5	7.5–12.5	0.2–0.3	12.9–16.6	Karolyi 2006, Marušić et al. 2013
Dalmatian dry-cured ham	41.4	50.7	7.9	0.2	14.7	Marušić et al. 2013

SFA – saturated fatty acids; PUFA – polyunsaturated fatty acids; MUFA – monounsaturated fatty acids

Chemical and Sensorial Properties of Fermented Meat Products 371

The average cholesterol levels determined in fermented meat products range from 62 to 76 mg/100 g in dry-cured hams and from 59 to 105 mg/100 g in dry-fermented sausages (Jiménez-Colmenero et al. 2010a, b, Pleadin et al. 2010). Higher fat contents indicate the presence of higher cholesterol contents, although not in a direct proportion. This is explained by the fact that the cholesterol content found in meat products depends on their fat content and the representation of different muscles (red, white) and ingredients other than muscle tissue (offal) (Chizzolini et al. 1999). In case of need for the cholesterol reduction, the appropriate approach would be to substitute the meaty raw materials with the ingredients of vegetable origin that do not contain cholesterol. When it comes to dry-fermented products, cholesterol may rather cause problems due to its oxidation than due to its excessive intake. The oxidation of cholesterol can be prevented by α-tocopherol administered during the feeding period (Jiménez-Colmenero et al. 2010a).

2.5 Amino Acids and Other Nutrients

Dry-cured and fermented meats represent a significant source of proteins of high biological value. Proteolysis that results in peptides broken into free amino acids by endogenous and microbial enzymes, contributes to the nascence of both volatile and non-volatile flavours. The main amino acids derived from the muscle tissue decomposition that takes place in fermented sausages are alanine, leucine, isoleucine and glutamic acid. Small amounts of branched-chain amino acids, particularly leucine, isoleucine and valine, are metabolized, yielding the major aroma compounds that include aldehydes, branched alcohols and methyl acids (Ockerman and Basu 2007).

All types of dry-cured hams are characterized with high concentrations of glutamic acid, alanine, leucine, lysine, valine, and aspartic acid. Depending of the level of proteolysis, free amino acids (FAA) can positively affect dry-cured hams' flavour. An intense proteolysis resulting in the higher release of amino acids, leads to the formation of branched aldehydes via oxidative deamination going through the Strecker degradation (García-González et al. 2013, Marušić et al. 2014).

The representation of trace elements (zinc, iron, and selenium) and vitamins (niacin, pyridoxin, thiamine-B1, and riboflavin-B2) exceeds 20%, indicating a high energetic density of these nutrients in meat and meat products (Leveille and Cloutier 1986). From the nutritional value standpoint, fermented meat products contain essential mineral nutrients, some of them even in substantial amounts, which are readily bio-available. Dry-cured hams represent a natural source of iron (1.8–3.3 mg/100 g), zinc (2.2–3.0 mg/100g), phosphorus (157–180 g/100 g), vitamin B1 (0.57–0.84 mg/100 g), vitamin B2 (0.20–0.25 mg/100 g), vitamin B6 (0.22–0.42 mg/100 g), vitamin B12 (15.68 µg/100 g) and niacin (4.5–11.8 equivalents) (Jiménez-Colmenero et al. 2010b). The results of several studies examining into the fermented sausages' content have indicated that the magnesium content ranges from 17.9 to 53 mg/100 g, the zinc content from 1.6 to 12.5 mg/100 g, the iron content from 0.5 to 3.1 mg/100 g, the phosphorus content from 143 to 468 mg/100 g, the potassium content from 256 to 468 mg/100 g, and the calcium from 8.1 to 155 mg/100 g (Zanardi et al. 2010, Jiménez-Colmenero et al. 2010a, b, González-Tenorio et al. 2012). Data showed the hem iron to be a much more available iron source than the vegetable iron. Markedly low manganese meat levels (0.4 mg/kg) hamper the fermentation, while low calcium values are reflected in the Ca/P ratio of about 0.1 and contribute to calcium intakes below the recommended levels (Lombardi-Boccia et al. 2003).

Excessive meat consumption–based intakes of some minerals like sodium and phosphorus, the latter being attributed to the consumption of phosphate-added meat products, limit the suitability of the consumption of fermented meat products by certain population groups due to the health risks in terms of hypertension and anomalous bone metabolism (González-Tenorio et al. 2012). Sodium chloride (NaCl) is the major constituent of fermented meat products providing 20–30% of NaCl dietary intake, i.e. between 9 and 12 g per day per person (3.5–5 g of sodium/day/person). For the sake of comparison, WHO (2004) recommended the maximum intake of 5 g per day per person (2 g of sodium per day per person). In dry-cured hams, sodium contents are in the range of 1.1–1.8 g/100 g, while in fermented sausages this range goes from 0.9 to 2.4 mg/100 g (Jiménez-Colmenero et al. 2010a, Zanardi et al. 2010, González-Tenorio et al. 2012). Many strategies to formulate healthier dry-fermented sausages not only in terms of NaCl reduction, but also in terms of simultaneous addition of minerals such as magnesium and/or calcium required for the maintenance of an adequate bone health, are currently in place (Flores et al. 2013).

It has been reported that, when it comes to meat and meat products, pork and poultry are the main sources of niacin and selenium, respectively (Lombardi-Boccia et al. 2004), the first one being very low in beef because of the specificity of the ruminant metabolism. Meat and meat products are the major providers of vitamin B12. Animal diets are often supplemented by α-tocopherol to control oxidation and colour stability of pork and beef, its amounts thereby clearly exceeding those in use as a food additive, but improving the vitamin E status of the meat (McCarthy et al. 2001).

3 SENSORY PROPERTIES OF FERMENTED MEAT PRODUCTS

3.1 Sensory Quality and Methods Applied

The sensory quality of fermented meat products depends on the adequacy of the selected sensory attributes that include flavour (taste, aroma or odour), appearance, texture and sensations. The principal sensory attributes are described by a wide range of specific descriptors. As for the flavours of dry-fermented meat products, experts in the field recommend their origin-based classification into meat-related (meaty, cured, aged, mature, rancid, fat, butter) and process-related (hot, spice, paprika, smoky) flavours as appropriate for dry-fermented sausages. The flavour is a complex term defined as the combination of taste, aroma and texture (ISO 5492:2008). The major descriptors of dry-fermented sausages' taste mainly include four basic tastes, i.e. acid/sour, bitter, salty and sweet. In case of dry-cured hams, the main taste descriptors are salty, bitter, acid, and, rarely, sweet. Odour descriptors used with dry-fermented sausage profiling include lactic acid, black pepper and other spices, mouldy, and green odour. Dry-cured ham odour descriptors include cured/matured ham odour, fresh meat odour, smoke odour, mould odour and rancid odour. The usual texture descriptors applied both with dry-fermented sausages and cured hams are hardness, juiciness, and fibrousness. Additional descriptors of dry-fermented sausages' texture are chewiness, cohesiveness, and elasticity, while with dry-cured hams tenderness, toughness, dryness and pastiness are used to the same end. The major descriptors of dry-cured ham appearance used within the sensory assessment address the appearance of lean meat (colour intensity-redness, colour homogeneity, and brightness), fat content (the presence of intramuscular fat, fat colour) and tyrosine crystals. As for the appearance descriptors used within the frame of dry-fermented sausages' sensory assessments, besides the colour (expressed as the colour homogeneity and redness intensity), fat content (fat/lean ratio, fat/lean cohesiveness) (Lorenzo et al. 2012) and cross-section appearance (particle size, encrustation, slice cohesion) are also used as means to an end. The overall acceptance of fermented meat products depends on their main sensory attributes. Although some of the attributes seem to be more important than the other, they do not replace them, but rather complement to each other and to the overall description of the product.

Sensory methods applied in sensory analyses of food, fermented meat products included, are divided into two groups, that is to say, objective (analytical) methods and subjective (affective) methods. Depending on the selected task, objective methods include the difference testing (searching for differences between the two given samples using

374   *Fermented Meat Products: Health Aspects*

the triangular test, the paired-comparison test, or the duo-trio test) (ISO 4120:2004, ISO 5495:2005, ISO 10399:2004), the ranking tests (aiming to establish whether the products can be ranked based on intensity) (ISO 8587:2006), and the descriptive techniques (profiling, such as the Flavour Profile, the Texture Profile, the Quantitative Descriptive Analysis (QDA), or the Spectrum) (ISO 11035:1994). Affective methods include the measurement of the degree of liking of the product, the hedonic scale (nine- and seven-point scale) and the paired preference test (forced-choice method not allowing for the 'no preference' response) thereby being most commonly applied in sensory assessments (ASTM 2005).

A sensory assessment needs to adhere to the recommended practices that call for a panel performance and the observation of general guidance for the design of the test room and the testing conditions described in the ISO 8589:2007. The selection and training of panel members that are going to apply objective (analytical) sensory methods is carried out in several sessions according to the ISO 5492:2008 and the ISO 8586:2012. Prior to testing, it is important to organize a preparatory session in order to thoroughly discuss and clarify each attribute to be evaluated. An equal preparation time should be ensured for all fermented meat samples before serving them to the panel members in order to avoid the loss of complex flavour perception.

Contrary to some other, sensory assessment of fermented meat products (dry-cured hams and fermented sausages) is not standardized, since no consensus on sensory attributes to be evaluated has been reached yet. In several scientific papers, the appropriate attributes have been selected and applied through complex procedures involving the generation of the sensory vocabulary, the selection of the lexicon to be used to describe a fermented meat product, and the application of quantitative-descriptive analysis (Ruiz Pérez-Cacho et al. 2005, García-González et al. 2006, 2008, Benedini et al. 2012). Sensory assessment of solid food is complex and includes the aroma perception affected by the release of volatile compounds during mastication of fermented meat products (García-González et al. 2008). Sensory evaluation of dry-cured hams is even more difficult to deal with due to the presence of different muscles in the ham and the heterogeneity of the tissue. Ruiz et al. (1998) have pointed out that the texture and the appearance of the product are highly influenced by the slice location, but that the flavour is more influenced by the length of the process.

3.2 Flavour Compounds

Flavour is a complex sensory reaction that involves the taste, the odour and the texture of fermented meat products, developed in the final ripening/drying phase. In both fermented sausages and dry-cured hams,

Chemical and Sensorial Properties of Fermented Meat Products 375

biochemical changes of carbohydrates, proteins and lipids run in parallel. Sensory characteristics of fermented meat products are closely related to the lipid breakdown and the transformations seen during the ripening (Olivares et al. 2010). While lipolytic reactions taking place in dry-cured hams are mainly attributed to the endogenous enzymatic systems due to the fact that the microbial counts are usually low and do not exceed 10^4 or 10^5 cfu/g, in dry-fermented sausages lipases are derived from microorganisms naturally present in the product and added as starter cultures (Toldrá 1998, Olivares et al. 2010).

Lipolysis and proteolysis represent the principal biochemical reactions responsible for the generation of flavour and flavour precursors. The flavour and the odour of fermented meat products can partially be attributed to the presence of both volatile and non-volatile compounds with sapid and textural properties. Volatile compounds derive from lipids, proteins and carbohydrates. Amino acids and free fatty acids are the source of the majority of volatile compounds that shape the flavour profile of fermented meat products. Non-volatile compounds mainly consist of inorganic salts, nucleotide metabolites, sugars, organic acids, amino acids, and peptides.

3.2.1 Volatile Compounds Responsible for the Flavour of Meat Products

The assessment of odour (aroma) represents a very important component of the sensory assessment due to the high sensitivity of nasal receptors to numerous volatile compounds released during chewing and ingestion (Demeyer et al. 2000). The study of the aroma and the volatile compounds present in dry-fermented products implies the examination of the factors that lead to characteristics sensory notes (García-González et al. 2009). The principal chemical compounds responsible for the odour of fermented sausages and dry-cured hams are straight- and branched-chain aldehydes, acids and their methyl- and ethyl-esters, aromatic, heterocyclic and sulphur compounds, methyl ketones, and terpenes.

Based on the flavour studies that used volatile compounds' extraction techniques, it has been determined that the major volatile compounds found in fermented meat products (aldehydes, ketones, carboxylic acids, alkanes, alkenes) come from the fatty acid oxidation. Due to the low odour thresholds and high concentrations, aldehydes have an important role in the generation of flavour of fermented meat products. Depending on its concentrations, the saturated aldehyde hexanal can be associated with a pleasant and grassy aroma, fatty aroma, or rancid aroma, should the hexanal concentrations be higher (Demeyer et al. 2000, García-González et al. 2013). The aromatic aldehyde benzaldehyde, produced in amino acid reactions or during lipid oxidation, contributes to the dry-cured ham aroma with a bitter almond sensory note (Huan et al. 2005). The origin of benzaldehyde present in the Istrian dry-cured ham comes from laurel used in the production process (Marušić et al. 2013). Methyl ketones (2-nonanone, 2-heptanone) contribute to the fruity, musty and cheesy aroma of both dry-cured hams and dry sausages. Branched-chain aldehydes (2- and 3-methylbutanal) deriving from several reactions (the main pathway thereby being oxidative deamination-decarboxylation of amino acids going through the Strecker degradation) are usually tagged as molecules having the central role in the ripened flavour of dry-cured hams and fermented sausages, and are the largest contributor to the flavour of Italian dry-cured hams (Careri et al. 1993).

Compounds generated from the metabolism of carbohydrates present in dry-fermented sausages include the acetic acid contributing to the vinegar aroma, diacetyl, and acetoin that contributes to the buttery aroma (Demeyer et al. 2000, Ockerman and Basu 2007). In general, alcohols (linear and branched), coming mainly as the products of lipid oxidation and the Strecker degradation of amino acids (methyl-branched alcohols), are associated with the pleasant green and woody notes (3-methylbutanol), fruity and green notes (hexanol) and mushroom-like, earthy and dust notes (1-octen-3-ol) of dry-cured hams (García-González et al. 2013). The alcohols most abundantly present in dry-fermented sausages are 1-octen-3-ol contributing to the mushroom aroma (Sunesen et al. 2001) and 2-methyl-1-butanol and 2-methyl-1-propanol contributing to the malty aroma (Demeyer et al. 2000, Ockerman and Basu 2007). Ethyl-esters of carboxylic acids have always been reported to be present in fermented meat products and are associated with the fruity note (mainly those formed from the short-chain acids) (Demeyer et al. 2000, Ockerman and Basu 2007). The recent research of ester compounds present in dry-cured hams showed these compounds to be present only in low amounts due to the antimicrobial activity of sodium chloride and a long curing period, as well as a negligible contribution of the determined esters e.g. butyl-acetate) to the odour of dry-cured hams due to the high odour threshold (Gasparado et al. 2008, Marušić et al. 2013, García-González et al. 2013).

Unlike dry-cured hams, dry-fermented sausages are characterised by the prevalence of sulphides, terpenes, and phenolic compounds due to the role of spices and smoking in the generation of their flavour and odour (Ordóñez et al. 1999, Demeyer et al. 2000, Marušić et al. 2013). Pepper, garlic, onion, paprika, nutmeg, and rosemary are only the most widespread spices out of the variety of aromatic plant substances used in the manufacturing of not only dry-fermented sausages, but also certain dry-cured hams. Monoterpene hydrocarbons derived from pepper have been determined as the compounds most abundantly present in salami (Berger et al. 1990). The hydrocarbon limonene has been determined

Chemical and Sensorial Properties of Fermented Meat Products

in dry-cured hams and fermented sausages (García-González et al. 2013). Marušić et al. (2013) attributed its presence to the added spices and classified it as a terpene. The presence of limonene is associated with the lemon odour of dry-cured hams maturing longer than twelve months, while when analysing dry-fermented sausages using gas chromatography with olfactometry (GCO), the GCO-descriptor is citrus, orange (Flores 2011, García-González et al. 2013). Some terpenes have been found in Bayonne, Corsican and Istrian hams due to the black pepper treatment during processing (Sabio et al. 1998, Marušić et al. 2013). Phenol compounds diallyl disulphide and eugenol, derived from garlic and nutmeg, have been determined in the study of Mediterranean and Northern types of fermented sausages (Demeyer et al. 2000).

3.2.2 Non-Volatile Compounds Responsible for the Flavour of Meat Products

Amino acids and peptides represent the non-volatile fraction responsible for the taste of fermented meat products. Besides their contribution to the taste of such a meat product, amino acids contribute to its texture, as well, since they represent the origin of volatile compounds like branched alcohols and aldehydes, sulphide compounds or thiols arising on the grounds of the Strecker degradation (Toldrá 1998). L-amino acids including glycine, alanine, serine, threonine, proline, and hydroxiproline, are associated to the sweet taste (Kirimura et al. 1969) of fermented meat products. Arginine, histidine, isoleucine, leucine, methionine, valine, phenylalanine, and tryptophan contribute to the bitter taste (Virgili and Schivazappa 2002). Dipeptides also held responsible for the bitter taste of fermented meat products, are anserine and carnosine (Ordóñez et al. 1999). A salty taste derives from the sodium salts of the glutamic and the aspartic acid (Careri et al. 1993, Toldrá 1998, Ordóñez et al. 1999). The aspartic acid, asparagines, glutamic acid and histidine contribute to the acidic taste (Ordóñez et al. 1999). Toldrá (1998) reported that phenylalanine and isoleucine favourably affect the acid taste of dry-cured hams, while tyrosine does just the opposite. Free amino acids (FAA) like tyrosine and lysine contribute to the aged/dry-cured taste (Careri et al. 1993) of dry-cured hams. The free amino acid content is closely correlated with the length of the ripening process and positively affects dry-cured hams' taste. However, an excess in the degree of proteolysis and the increase in FAA and peptides lead to the unpleasant taste of enhanced bitterness and metal aftertaste of dry-cured hams (Careri et al. 1993, Virgili et al. 1995). Non-protein nitrogen affects sensory properties of dry-cured hams in terms of bitterness arising on the grounds of an uncontrolled extent of proteolysis (Virgili and Schivazappa 2002).

378   *Fermented Meat Products: Health Aspects*

Organic acids determined in fermented sausages including lactic acid, acetic acid and gluconic acid, induce the nascence of the acid taste and D-lactates (Ockerman and Basu 2007). As compared to dry-cured hams, where their contribution to the flavour in the capacity of odour-active aroma compounds is negligible due to their low concentrations as compared to the odour threshold (García-González et al. 2013), in dry-fermented sausages acids like butanoic, pentanoic, and hexanoic acid enhance their cheesy, sweaty and putrid notes (Schmidt and Berger 1998). Sodium chloride is responsible for the salty taste. In view of the general tendency to reduce NaCl content of dry-cured hams and fermented sausages so as to avoid unwanted health effects (hypertension, cardiovascular diseases and cancer) (WHO 2003), a partial replacement of NaCl with alternative salts (KCl, CaCl$_2$ and MgCl$_2$) has been attempted. It has been claimed that 30 to 50% of the NaCl content can be replaced without any detrimental effects on sensory quality (Flores et al. 2013). Gelabert et al. (2003) have determined that a 40+%-substitution of NaCl by KCl provides a bitter and metallic taste of fermented meat products. Zanardi et al. (2010) found that a 50%-reduction of NaCl attained in the Italian salami using a mixture of monovalent KCl and divalent CaCl$_2$ and MgCl$_2$ salts, significantly lowers the sodium content with limited detrimental effects on sensory attributes.

Depending on the fermented meat product, the texture develops in several stages, with the significant effects of technological parameters and physicochemical characteristics of the product and the raw materials. Dry-fermented sausages are characterized by a three-step texture development that includes the extraction of proteins using salt, the formation of protein gel during acidification (a pH decline) and the strengthening of the protein gel. Endogenous proteases calpains and catephsins lead to the meat tenderness during dry-cured ham processing. The sodium chloride (salt) content, the moisture content and the meat pH are physicochemical parameters that affect the texture of fermented meat products. In fermented sausages, salt favours the formation of gel and leads to the desirable texture. The reduction/substitution of salt affects the cohesiveness and chewiness, but not the hardness (Corral et al. 2013). The fermented sausage texture is under the influence of salt types used in salt substitution. Guàrdia et al. (2008) determined that KCl used alone increases the sausage hardness. Lower ham salt contents contribute to proteolysis, resulting in a softer texture of the final product (Gou et al. 2008, Andronikov et al. 2013). Lower salt and higher water contents may increase enzymatic activity and proteolysis going on in hams, which would increase the pastiness and the adhesiveness of these hams (Parolari et al. 1994). The initial meat pH (established 24 hours *post mortem*) has an important influence on texture characteristics of

Chemical and Sensorial Properties of Fermented Meat Products 379

dry-cured hams reflecting in softer, pastier, more crumbly and more adhesive texture descriptors (Guerrero et al. 1999, Virgili and Schivazappa 2002). A higher level of moisture together with a higher pH result in the defective texture (pasty and adhesive mouth feel) of dry-cured hams. The rise in non-volatile low molecular weight-nitrogen molecules as a measure of proteolysis going on in meat has been reported to increase the pastiness (Careri et al. 1993, Guerrero et al. 1996).

3.3 Colour

The colour of fermented meat products represents one of the major appearance descriptors. It is strongly affected by the processing aspects involving chemical, enzymatic and ultra-structural reactions the product undergoes. The colour typical of dry-fermented sausages is produced by the interaction between the meat pigments and the products resulting from the reduction of nitrates and nitrites added (Olivares et al. 2010). The formation of nitric oxide by nitrate- and nitrite-reducing bacteria (of the *Staphylococcus* and the *Micrococcus* genera) (Fischer and Schleifer 1980) that subsequently reacts with the meat myoglobin (Mb) leads to the stable red colour of nitrosomyoglobin (NOMb).

In dry-cured hams, the formation of colour is strongly affected by the diffusion of salt through the meat and is related to changes in water-holding capacity if taking about the earlier production stages, while during maturation it is mainly affected by the modification of the Mb structure (Pérez-Alvarez and Fernández-López 2011). Dry-cured ham production involves the approved use of nitrates, nitrites and salt during the curing process (Spanish Serrano ham) resulting in the formation of nitrosylmyoglobin (MbFe^{II}NO), hence the slices have a characteristic pink to purple colour. On the other hand, into some hams (Parma ham, the majority of Iberian hams, Dalmatian ham, Krk ham, Istrian ham) only salt is added. In the most investigated among them, the Parma dry-cured ham, zinc-porphyrin has been found to determine the stable bright red colour (Moller et al. 2007, Estévez et al. 2007). A more intense red colour of the Iberian dry-cured hams as compared to other hams of its kind is explained by the high myoglobin concentration in the muscles the Iberian hams are made of, which apparently comes as a consequence of the free range-rearing of the Iberian pigs.

The CIELAB method, which is based on the measuring of light reflected from the surface of the food and the colour coordinates including L^* (lightness), a^* (redness), b^* (yellowness), and a*/b* ratio, has been proven very useful in monitoring of the colour evolution going on in fermented meat products. The study by Pérez-Alvarez (1996), investigating into the colour evolution going on in fermented meat products (fermented sausages with and without paprika, dry-cured hams) during the fermentation stage (using CIELAB), colour coordinates L^* (lightness), b^* (yellowness), hue, and chroma showed a decreasing trend, as oppose to a^* (redness) and a^*/b^* ratio, whose trend was an increasing one. The ham fat portion contributes to the colour in terms of marbling (inter-muscular fat) and fat colour, as well as in terms of the white spots of tyrosine crystals formed due to the precipitation of the amino-acid tyrosine during long ripening periods of certain hams.

4 CONCLUSION

Physicochemical and sensory properties characteristic for fermented meats widely vary in dependance on the product's composition and complex protein, lipid and carbohydrate metabolism, which occurs due to enzymatic activity and oxidation processes. Analysis of fermented meat products in terms of food quality and safety should include its technological, nutritional, safety and sensorial value. The modern analytical methods applied enables the detection of large number of substances in trace amounts ensuring dynamic fermented meat quality approach in terms of analysis of biochemical pathways of minor food ingredients as well as hazardous compounds. Further studies should be focused on the evaluation of physicochemical, microbiological and sensory properties of fermented meat products, using sensitive analytical and standardized sensorial methods, in order to contribute to their better characterization, improvement of product specification protocols and the definition of unique quality markers. Suitable chemical and sensory markers should be used to enable higher uniformity and consistency of these products. Due to healthy trends in consumption and protection of human health, future trends should be linked to fat and salt reduction in fermented meat products without any loss of sensory properties.

Key words: *fermented meat products, physico-chemical characterization, sensorial properties, basic chemical properties, fatty acids composition, cholesterol, vitamins, minerals, flavour compounds, volatile compounds, non-volatile compounds, colour*

REFERENCES

American Society for Testing and Material (ASTM) 2005. Affective testing. pp. 73–79. *In:* E. Chambers and M. Baker-Wolf (eds.). Sensory Testing Methods Second Edition. ASTM manual series; MNL 26, Lancaster.

Andronikov, D., L. Gašperlin, T. Polak and B. Žlender. 2013. Texture and quality parameters of Slovenian dry-cured ham Kraški pršut according to mass and salt levels. Food Technol. Biotechnol. 51: 112–122.

Ansorena, D. and I. Astiasarán. 2007. Functional meat products. pp. 257–266. *In:* F. Toldrá (ed.). Handbook of fermented meat and poultry. Blackwell Publishing, Iowa.

AOAC. 2000. Ash in meat (920.153) in meat. In Official Methods of Analysis, 17th ed. Maryland, USA. Association of Official Analytical Chemists.

AOAC. 2000. Fat (crude) in meat and meat products (991.36). In Official Methods of Analysis, 17th ed. Maryland, USA. Association of Official Analytical Chemists.

AOAC. 2000. Moisture content (950.46) in meat. In Official Methods of Analysis, 17th ed. Maryland, USA. Association of Official Analytical Chemists.

AOAC. 2000. Protein content (928.08) in meat. In Official Methods of Analysis, 17th ed. Maryland, USA. Association of Official Analytical Chemists.

AOAC. 2002. Salt (chlorine as sodium chloride) in meat. Official Method 935.47. Official Methods of Analysis (17th ed). Arlington Va: Association of Official Analytical Chemists.

Arnau, J. 2004. Ham production. pp. 557–567. *In:* W.K. Jensen, C. Devine and M. Dikeman (eds.). Encyclopedia of meat sciences. Elsevier Academic Press, Oxford.

Barbir, T., A. Vulić and J. Pleadin. 2014. Fats and fatty acids in foods of animal origin. Vet. Stn. 2: 97–110.

Benedini, R., G. Parolari, T. Toscani and R. Virgili. 2012. Sensory and texture properties of Italian typical dry-cured hams as related to maturation time and salt content. Meat Sci. 90: 431–437.

Berger, R.G., C. Macku, J.B. German and T. Shibamoto. 1990. Isolation and identification of dry salami volatiles. J. Food Sci. 55: 1239–1242.

Campo, M.M. and I. Sierra. 2011. Fatty acid composition of selected varieties of Spanish dry-cured ham. Surveys from 1995 and 2007. Span. J. Agric. Res. 9: 66–73.

Careri, M., A. Mangia, G. Barbieri, L. Bolzoni, R. Virgili and G. Parolari. 1993. Sensory property relationships to chemical data of Italian-type dry-cured ham. J. Food Sci. 58: 968–972.

Casaburi, A., M.C. Aristoy, S. Cavella, R. Di Monaco, D. Ercolini, F. Toldra and F. Villani. 2007. Biochemical and sensory characteristics of traditional fermented sausages of Vallodi Diano (Southern Italy) as affected by the use of starter culture. Meat Sci. 76: 295–307.

Chizzolini, R., E. Zanardi, V. Dorigoni and S. Ghidini. 1999. Calorific value andcholesterol content of normal and low-fat meat and meat products. Trends Food Sci. 10: 119–128.

Comi, G., R. Urso, L. Iacumin, K. Rantasiou, P. Cattaneo, C. Cantoni and L. Cocolin. 2005. Characterization of naturally fermented sausages produced in the North East of Italy. Meat Sci. 69: 381–392.

Cordain, L., B.S. Eaton, A. Sebastian, N. Mannine, S. Lindeberg, B.A. Watkins, J.H. O'Keefe and J. Brand-Miller. 2005. Origins and evolution of the Western diet: health implications for the 21st century. Am. J. Clin. Nutr. 81: 341–354.

Corral, S., A. Salvador and M. Flores. 2013. Salt reduction in dry cured sausages affects aroma generation. Meat Sci. 93: 776–785.

Council Regulation (EC) No 510/2006 of 20 March 2006 on the protection of geographical indications and designations of origin for agricultural products and foodstuffs. Off. J. Eur. Union. L 93/12.

D'Evoli, L., M. Lucarini, S. Nicoli, A. Aguzzi, P. Gabrielli, G. Lombardi-Boccia. 2009. Nutritional profile of traditional Italian hams. In: Proceeding of 5th world congress of dry-cured ham. Aracena, Spain.

Damez, J.L. 2008. Review: Meat quality assessment using biophysical methods related to meat structure. Meat Sci. 80: 132–149.

Dellaglio, S., E. Casiraghi and C. Pompei. 1996. Chemical, physical and sensory attributes for the characterization of an Italian dry-cured sausage. Meat Sci. 42: 25–35.

Demeyer, D. and L. Stahnke. 2002. Quality control of fermented meat products. pp. 359–393. In: J. Kerry and D. Ledward (eds.). Meat processing, improving quality. Woodhead Publishing Ltd., Cambridge.

Demeyer, D., M. Raemaekers, A. Rizzo, A. Holck, A. De Smedt, B. ten Brink, B. Hagen, C. Montel, E. Zanardi, E. Murbrekk, F. Leroy, F. Vandendriessche, K. Lorentsen, K. Venema, L. Sunesen, L. Stahnke, L. De Vuyst, R Talon, R. Chizzolini and S. Eerola. 2000. Control of bioflavour and safety in fermented sausages: First results of a European project. Food Res. Int. 33: 171–180.

Di Cagno, R., C. Cháves López, R. Tofalo, G. Gallo, M. De Angelis, A. Paparella, W. Hammes and M. Gobbetti. 2008. Comparison of the compositional, microbiological, biochemical and volatile profile characteristics of three Italian PDO fermented sausages. Meat Sci. 79: 224–235.

Du, M. and D.U. Ahn. 2001. Volatile substances of Chinese traditional Jinhua ham and Cantonese sausage. J. Food Sci. 66: 827–831.

Estévez, M., D. Morcuende, J. Ventanas and S. Ventanas. 2007. Mediterranean products. pp. 393–405. In: F. Toldrá (ed.). Handbook of fermented meat and poultry. Blackwell Publishing, Iowa.

Fernández, M., J.A. Ordóñez, I. Cambero, C. Santos, C. Pin and L. de la Hoz. 2007. Fatty acids compositions of selected varieties of Spanish dry ham related to their nutritional implications. Food Chem. 101: 107–112.

Ferreira, B., J. Barboza, J. Silva, S. Vendeiro, A. Mota, F. Silva, M. Monteiro, T. Hogg, P. Gibbs and P. Teixeira. 2007. Chemical and microbiological characterization of "Salpic ã o de Vinhais" and "Chouri ç a de Vinhais": Traditional dry sausages produced in the North of Portugal. Food Microbiol. 24: 618–623.

Fischer, U. and K.H. Schleifer. 1980. Presence of staphylococci and micrococci in dry sausage. Fleischwirtschaft 60: 1046–1048.

Flores M., A. Olivares and S. Corral. 2013. Healthy trends affect the quality of traditional meat products in Mediterranean area. Acta Argic. Slov. Suppl. 4: 183–188.

Flores, M. 2011. Sensory descriptors for dry-cured meat products. pp. 173–196. In: M.L. Leo Nollet and F. Toldrá (eds.). Sensory analysis of foods of animal origin. CRC Press, Boca Raton.

Franco I., A. Martínez, B. Prieto and J. Carballo. 2002. Total and free fatty acids content during the ripening of artisan and industrially manufactured "Chorizo de cebolla". Grasas Aceites 53: 403–413.

Gandemer, G. 2002. Lipids in muscles and adipose tissues, changes during processing and sensory properties of meat products. Meat Sci. 62: 309–321.

Gandemer, G. 2009. Dry cured ham quality as related to lipid quality of raw material and lipid changes during processing: A review. Grasas Aceites 60: 297–307.

Chemical and Sensorial Properties of Fermented Meat Products 383

García-González, D.L., N. Tena and R. Aparicio. 2009. Contributing to interpret sensory attributes qualifying Iberian hams from the volatile profile. Grasas Aceites. 60: 277–283.

García-González, D.L., N. Tena, R. Aparicio-Ruiz and M.T. Morales. 2008. Relationship between sensory attributes and volatile compounds qualifying dry-cured hams. Meat Sci. 80: 315–325.

García-González, D.L., P. Roncales, I. Cilla, S. Río, J.P. Poma and R. Aparicio. 2006. Interlaboratory evaluation of dry-cured hams (from France and Spain) by assessors from two different nationalities. Meat Sci. 73: 521–528.

García-González, D.L., R. Aparicio and R. Aparicio-Ruiz. 2013. Volatile and amino acid profiling of dry cured hams from different swine breeds and processing methods. Molecules. 18: 3927–3947.

García-Rey, R.M., J.A. Garcia-Garrido, R. Quiles-Zafra, J. Tapiador and M.D. Luque de Castro. 2004. Relationship between pH before salting and dry-cured ham quality. Meat Sci. 67: 625–632.

Gasparado, B., G. Procida, B. Toso and B. Stefanon. 2008. Determination of volatile compounds in San Daniele ham using headspace GC–MS. Meat Sci. 80: 204–209.

Gelabert, J., P. Gou, L. Guerrero and J. Arnau. 2003. Effect of sodium chloride replacement on some characteristics of fermented sausages. Meat Sci. 65: 833–839.

Gimeno, O., D. Ansorena, I. Astiasarán and J. Bello. 2000. Characterization of chorizo de Pamplona: instrumental measurements of colour and texture. Food Chem. 69: 195–200.

González-Tenorio, R., A. Fernández-Diez, I. Caro and J. Mateo. 2012. Comparative assessment of the mineral content of a Latin American raw sausage made by traditional or non-traditional processes, atomic absorption spectroscopy. M. Akhyar Farrukh (ed.), ISBN: 978-953-307-817-5, InTech, USA.

Gou, P., R. Morales, X. Serra, M.D. Guàrdia and J. Arnau. 2008. Effect of a 10-day ageing at 30°C on the texture of dry-cured hams processed at temperatures up to 18°C in relation to raw meat pH and salting time. Meat Sci. 80: 1333–1339.

Guàrdia, M.D., L. Guerrero, J. Gelabert, P. Gou and J. Arnau. 2008. Sensory characterisation and consumer acceptability of small calibre fermented sausages with 50% substitution of NaCl by mixtures of KCl and potassium lactate. Meat Sci. 80: 1225–1230.

Guerrero, L., P. Gou and J. Arnau. 1999. The influence of meat pH on mechanical and sensory textural properties of dry-cured ham. Meat Sci. 52: 267–273.

Guerrero, L., P. Gou, P. Alonso and J. Arnau. 1996. Study of the physicochemical and sensorial characteristics of dry-cured hams in three pig genetic types. J. Sci. Food Agri. 70: 526–530.

Huan, Y., G. Zhou, G. Zhao, X. Xu and Z. Peng. 2005. Changes in flavor compound of dry-cured Chinese Jinhua ham during processing. Meat Sci. 71: 291–299.

Incze, K. 2007. European products. pp. 307–318. In: F. Toldrá (ed.). Handbook of fermented meat and poultry. Blackwell Publishing, Iowa.

ISO 10399:2004 standard. Sensory analysis – Methodology – Duo-Trio test. International Organization for Standardization, Geneve.

ISO 11035:1994 standard. Sensory analysis – Identification and selection of descriptors for establishing sensory profile by a multidimensional approach. International Organization for Standardization, Geneve.

ISO 1442:1997 standard. Meat and meat products – Determination of moisture content. International Organization for Standardization, Geneve.

ISO 1443:1973 standard. Meat and meat products – Determination of total fat content. International Organization for Standardization, Geneve.

ISO 3496:1994 standard. Meat and meat products – Determination of hydroxyproline content. International Organization for Standardization, Geneve.

ISO 4120:2004 standard. Sensory analysis – Methodology – Triangle Test. International Organization for Standardization, Geneve.

ISO 5492:2008 standard. Sensory analysis – Vocabulary. International Organization for Standardization, Geneve.

ISO 5495:2005 standard. Sensory analysis – Methodology – Paired comparison test. International Organization for Standardization, Geneve.

ISO 8586:2012 standard. Sensory analysis – General guidance for the selection, training and monitoring of selected assessors and expert sensory assessors. International Organization for Standardization, Geneve.

ISO 8587:2006 standard. Sensory analysis – Methodology – Ranking. International Organization for Standardization, Geneve.

ISO 8589:2007 standard. Sensory analysis – General guidance for the design of test rooms. International Organization for Standardization, Geneve.

ISO 936: 1998 standard. Meat and meat products – Determination of total ash. International Organization for Standardization, Geneve.

ISO 937:1978 standard. Meat and meat products – Determination of crude protein content. International Organization for Standardization, Geneve.

Jerković, I., D. Kovačević, D. Šubarić, Z. Marijanović, K. Mastanjević and K. Suman. 2010. Authentication study of volatile flavour compounds composition in Slavonian traditional dry fermented salami "kulen". Food Chem. 119: 813–822.

Jiménez-Colmenero, F, J. Carballo and S. Cofrades. 2001. Healthier meat and meat products: their role as functional foods. Meat Sci. 59: 5–13.

Jiménez-Colmenero, F., J. Ventanas and F. Toldrá. 2009. El jamón curado en una nutrición saludable. In Proceeding of 5th world congress of dry-cured ham. May 6–8, Aracena, Spain.

Jiménez-Colmenero, F., J. Ventanas and F. Toldrá. 2010b. Nutritional composition of dry-cured ham and its role in a healthy diet. Meat Sci. 84: 585–593.

Jiménez-Colmenero, F., T. Pintado, S. Cofrades, C. Ruiz-Capillas and S. Bastida. 2010a. Production variations of nutritional composition of commercial meat products. Food Res. Int. 43: 2378–2384.

Jurado, Á., C. García, M.L. Timón and A.I. Carrapiso. 2008. Improvement of dry-cured Iberian ham sensory characteristics through the use of a concentrate high in oleic acid for pig feeding Irish J. Agric. Food Res. 47: 195–203.

Karolyi, D. and T. Čurić. 2012. Total fatty acids composition of raw and ripe Slavonian kulen in relation to raw material used. Acta Agric. Slov. Suppl. 3: 231–234.

Karolyi, D. 2006. Chemical properties and quality of Istrian dry-cured ham. Meso. 7: 224–228.

Kirimura, J., A. Shimizu, A. Kirizuka, T. Ninomiya and N. Katsuya. 1969. The concentration of peptides and amino acids to the taste of foodstuffs. J. Agr. Food Chem. 17: 689–695.

Kos, I., R. Božac, A. Kaić, N. Kelava, M. Konjačić and Z. Janječić. 2009. Sensory profiling of Dalmatian dry-cured ham under different temperature conditions. Ital. J. Anim. Sci. 8: 216–218.

Kovačević, D. 2001. Chemistry and technology of meat and fish. University of J.J. Strossmayer. Faculty of food technology, Osijek, Croatia.

Kovačević, D. 2014. Technology of Kulen and other fermented sausages. University of J.J. Strossmayer. Faculty of food technology, Osijek, Croatia.

Kovačević, D., K. Mastanjević, D. Šubarić and K. Suman. 2009. Physico-chemical and colour properties of homemade Slavonian Sausage. Meso. 11: 280–284.

Kovačević, D., K. Mastanjević, D. Šubarić, I. Jerković and Z. Marijanović. 2010. Physico-chemical, colour and textural properties of Croatian traditional dry sausage (Slavonian Kulen). Meso. 12: 270–275.

Kozačinski, L., M. Hadžiosmanović, Ž. Cvrtila Fleck, N. Zdolec, I. Filipović and Z. Kozačinski. 2008. Quality of dry and garlic sausages from individual households. Meso. 10: 74–80.

Kozačinski, L., N. Zdolec, M. Hadžiosmanović, Ž. Cvrtila, I. Filipović and T. Majić. 2006. Microbial flora of the Croatian traditionally fermented sausage. Arch. Lebensmittelhyg. 57: 141–147.

Leveille, G.A. and P.F. Cloutier. 1986. Role of the food industry in meeting the nutritional needs of the elderly. Food Technol. 40: 82–88.

Lo Fiego, D.P., P. Macchioni, P. Santoro, G. Pastorelli and C. Corino. 2005. Effect of dietary conjugated linoleic acid (CLA) supplementation on CLA isomers content and fatty acid composition of dry-cured Parma ham. Meat Sci. 70: 285–291.

Lombardi-Boccia, G., A. Aguzzi, M. Cappelloni, G. Di Lullo and M. Lucarino. 2003. Total-diet study: Dietary intakes of macro elements and trace elements in Italy. Br. J. Nutr. 90: 1117–1121.

Lombardi-Boccia, G., S. Lanzi and M. Lucarini. 2004. Meat and meat products consumption in Italy: Contribution to trace elements, heme iron and selected B vitamins supply. Int. J. Vitam. Nutr. Res. 74: 247–251.

Lorenzo, J.M., S. Temperán, R. Bermúdez, N. Cobas and L. Purriños. 2012. Changes in physico-chemical, microbiological, textural and sensory attributes during ripening of dry-cured foal salchichón. Meat Sci. 90: 194–198.

Marušić, N., M. Petrović, S. Vidaček, T. Janči, T. Petrak and H. Medić. 2013. Fat content and fatty acid composition in Istrian and Dalmatian dry-cured ham. Meso. 15: 279–284.

Marušić, N., S. Vidaček, T. Janči, T. Petrak and H. Medić. 2014. Determination of volatile compounds and quality parameters of traditional Istrian dry-cured ham. Meat Sci. 96: 1409–1416.

McCarthy, T.L., J.P. Kerry, J.F. Kerry, P.B. Lynch and D.J. Buckley. 2001. Evaluation of the antioxidant potential of natural food/plant extracts as compared with synthetic antioxidants and vitamin E in raw and cooked pork patties. Meat Sci. 58: 45–52.

386 Fermented Meat Products: Health Aspects

Milićević, B., B. Danilović, N. Zdolec, L. Kozačinski, V. Dobranić and D. Savić. 2014. Microbiota of the fermented sausages: Influence to product quality and safety. Bulg. J. Agric. Sci. 20: 1061–1078.

Moller, J.K.S., C.E. Adamsen, R.R. Catharino, L.H. Skibsted, M.N. Eberlin. 2007. Mass spectrometric evidence for a zinc-porphyrin complex as the red pigment in dry-cured Iberian and Parma ham. Meat Sci. 75: 203–210.

Moretti, V.M., G. Madonia, C. Diaferia and T. Mentasti. 2004. Chemical and microbiological parameters and sensory attributes of a typical Sicilian salami ripened in different conditions. Meat Sci. 66: 845–854.

Moretti, V.M., G. Madonia, G. Diaferia, T. Mentasti, M.A. Paleari, S. Panseri, G. Pikrone and G. Gandini. 2012. Chemical and microbiological parameters and sensory attributes of typical Sicilian salami ripened in different conditions. Meat Sci. 66: 845–854.

Muguerza, E., D. Ansorena and I. Astiasaran. 2004. Functional dry fermented sasusages manifactured with high levels of n-3 fattyacids: nutritional benefits and evaluation of oxidation. J Sci. Food Agric. 84: 1061–1068.

Ockerman, H.W. and L. Basu. 2007. Production and consumption of fermented meat products. pp. 9–15. In: F. Toldrá (ed.). Handbook of fermented meat and poultry. Blackwell Publishing, Iowa.

Olivares, A., J.L. Navarro and M. Flores. 2011. Effect of fat content on aroma generation during processing of dry fermented sausages. Meat Sci. 87: 264–273.

Olivares, A., J.L. Navarro, A. Salvador and M. Flores. 2010. Sensory acceptability of slow fermented sausages based on fat content and ripening time. Meat Sci. 86: 251–257.

Ordóñez, J.A., E.M. Hierro, J.M. Bruna and L. Hoz. 1999. Changes in the components of dry-fermented sausages during ripening. Crit. Rev. Food Sci. 39: 329–367.

Parolari, G., R. Virgili and C. Schivazappa. 1994. Relationship between cathepsin B activity and compositional parameters in drycured hams of normal and defective texture. Meat Sci. 38: 117–122.

Pastorelli, G., S. Magni, R. Rossi, E. Pagliarini, P. Baldini, P. Dirinck, F. Van Opstaele and C. Corino. 2003. Influence of dietary fat, on fatty acid composition and sensory properties of dry cured Parma ham. Meat Sci. 65: 571–580.

Pelser, W.M., J.P.H. Linssen, A. Legger and J.H. Houben. 2007. Lipid oxidationin n-3 fatty acid enriched Dutch style fermented sausages. Meat Sci. 75: 1–11.

Pérez-Alvarez, J.A. 1996. Contribución al estudio objetivo del color en productos cárnicos crudo-curados. PhD Thesis, Universidad Politécnica de Valencia, Valencia, Spain.

Pérez-Alvarez, J.A. and J. Fernández-López. 2011. Color measurement on muscle-based foods. pp. 3-19. In: M.L. Nollet and F. Toldrá (eds.). Sensory analysis of foods of animal origin. CRC Press, Boca Raton.

Pleadin, J., G. Krešić, T. Barbir, M. Petrović, I. Milinović and D. Kovačević. 2014. Changes in basic nutrition and fatty acid composition during production of "Slavonski kulen". Meso. 16: 514–519.

Pleadin, J., N. Vahčić, N. Perši, A. Vulić, M. Volarić and I. Vraneš. 2010. Cholesterol levels in homemade and industrial sausages. Meso. 12: 156–161.

Chemical and Sensorial Properties of Fermented Meat Products 387

Pugliese, C. 2009. Effect of genetic type and rearing conditions on characteristicsof Italian quality hams: The instance of Tuscan ham. In: Proceeding of 5th world congress of dry-cured ham, Aracena, Spain.

Pugliese, C., F. Sirtori, M. Škrlep, E. Piasentier, L. Calamai, O. Franci and M. Čandek-Potokar. 2015. The effect of ripening time on the chemical, textural, volatile and sensorial traits of *Biceps femoris* and *Semimembranosus* muscles of the Slovenian dry-cured ham Kraški pršut. Meat Sci. 100: 58–68.

Rantsiou, K., E. Drosinos, M. Gialitaki, R. Urso, J. Krommer, J. Gasparik - Reichardt, S. Tóth, I. Metaxopoulos, G. Comi and L. Cocolin. 2005. Molecular characterization of *Lactobacillus* species isolated from naturally fermented sausages produced in Greece, Hungary and Italy. Food Microbiol. 22: 19–28.

Review of Methods of Analysis. 2003. Food compositional data. pp. 97–149. *In:* H. Greenfield and D.A.T. Southgate (ed.). Food and Agriculture Organizations of the United Nations (FAO), Rome.

Ruiz Pérez-Cacho, M.P., H. Galán-Soldevilla, F. León Crespo and G. Molina Recio. 2005. Determination of the sensory attributes of a Spanish dry-cured sausage. Meat Sci. 71: 620–633.

Ruiz, J., J. Ventanas, R. Cava, M.L. Timon and C. García. 1998. Sensory characteristics of Iberian ham: Influence of processing time and slice location. Food Res. Int. 31: 53–58.

Sabio, E., M.C. Vidal-Aragon, M.J. Bernalte and J.L. Gata. 1998. Volatile compounds present in six types of dry-cured ham from south European countries. Food Chem. 61: 493–503.

Salgado, A., M.C. García Fontán, I. Franco, M. López and J. Carballo. 2005. Biochemical changes during the ripening of Chorizo de cebolla, a Spanish traditional sausage. Effect of the system of manufacture (homemade or industrial). Food Chem. 92: 413–424.

Santos, C., L. Hoz, M.I. Cambero, M.C. Cabeza and J.A. Ordoñez. 2008. Enrichment of dry-cured ham with α-linolenic acid and α-tocopherol by use of linseed oil and α-tocopheryl acetate. Meat Sci. 80: 668–674.

Schmidt, S. and R.G.Berger. 1998. Aroma compounds in fermented sausages of different origins. Lebensm. Wiss. Technol. 31: 559–567.

Siciliano, C., E. Belsito, R. De Marco, M.L. DiGioia, A. Leggio and A. Liguori. 2013. Quantitative determination of fatty acid chain composition in pork meat products by high resolution 1H NMR spectroscopy. Food Chem. 136: 546–554.

Stahknke, L.H. and K. Tjener. 2007. Influence of processing parameters on cultures performance. pp. 187–194. *In:* F. Toldrá (ed.). Handbook of fermented meat and poultry. Blackwell Publishing, Iowa.

Steinhart, H. 2009. Introduction: Importance of analysis in meat products. pp. 3–7. *In:* L. M.L. Nollet and F. Toldrá (eds.). Handbook of processed meats and poultry analysis. CRC Press, Boca Raton.

Sunesen, L.O., V. Dorigoni, E. Zanardi and L. Stahnke. 2001. Volatile compounds released during ripening in Italian dried sausage Meat Sci. 58: 93–97.

Toldrá, F. 1998. Proteolysis and lipolysis in flavour development of dry cured meat products. Meat Sci. 49: 101–110.

Toldrá, F. 2002. Dry-cured meat products (3rd ed.), Food and Nutrition Press, Inc. Connecticut.

UK Department of Health. 1994. Nutritional aspects of cardiovascular disease. Report on Health and Social Subject No. 46. London: Her Majesty's Stationery Office.

Valencia, I., D. Ansorena and I. Astiasaran. 2006. Nutritional and sensory properties of dry fermented sasusages enriched with n-3 PUFAs. Meat Sci. 72: 727–733.

Valsta, L.M., H. Tapanainen and S. Mannisto. 2005. Meat fats in nutrition. Meat Sci. 70: 525–530.

Ventanas, S., J. Ventanas, J. Tovar, C. Garcia and M. Estevez. 2007. Extensive versus oleic acid and tocopherol enriched mixed diets for the production of Iberian dry-cured hams: Effect on chemical composition, oxidative status and sensory traits. Meat Sci. 77: 246–256.

Vignolo, G., C. Fontana and S. Fadda. 2010. Semidry and dry fermented sausages. pp. 379–398. In: F. Toldrá (ed.). Handbook of meat processing. Wiley-Blackwell, Oxford.

Virgili, R. and C. Schivazappa. 2002. Muscle traits for long matured dried meats. Meat Sci. 62: 331–343.

Virgili, R., G. Parolari, C. Schivazappa, C. Soresi Bordini and M. Borri. 1995. Sensory and texture quality of dry-cured ham as affected by endogenous cathepsin B activity and muscle composition. J. Food Sci. 60: 1183–1186.

Virgili, R., G. Parolari, C. Soresi Bordini, C. Schivazappa, M. Cornet and G. Monin. 1999. Free amino acids and dipeptides in drycured ham. J. Muscle Foods. 10: 119–130.

Visessanguan, W., S. Benjakul, S. Riebroy, M. Yarchai and W. Tapingkae. 2006. Changes in the lipid composition and fatty acid profile of Nham, a Thai fermented pork sausage, during fermentation. Food Chem. 94: 580–588.

WHO. 2003. Diet, nutrition and the prevention of chronic diseases. WHO Technical Report Series 916.

WHO. 2004. Food and health in Europe – a new basis for action. WHO Regional Publication European Series, No. 96 (pp. 1–385).

Wirth, F. 1986. Zur Technologie bei rohen Fleischerzengniseen. Fleischwirtschaft 66: 531–536.

Wood, J.D., M. Enser, A.V. Fisher, G.R. Nute, P.R. Sheard, R.I. Richardson, S.I. Hughes and F.M. Whittington. 2008. Fat deposition, fatty acids composition and meat quality: A review. Meat Sci. 78: 343–358.

Wood, J.D., R.I. Richardson, G.R. Nute, A.V. Fisher, M.M. Campo, E. Kasapidou, P.R. Sheard and M. Enser. 2004. Effects of fatty acids on meat quality: a review. Meat Sci. 66: 21–32.

Woods, V.B. and A.M. Fearon. 2009. Dietary sources of unsaturated fatty acids for animals and their transfer into meat, milk and eggs: a review. Livestock Sci. 126: 1–20.

Zanardi, E., S. Ghidini, M. Conter and A. Ianieri. 2010. Mineral composition of Italian salami and effect of NaCl partial replacement on compositional, physic-chemical and sensory parameters. Meat Sci. 86: 742–747.

Chapter 17

Fermented Meats Composition— Health and Nutrition Aspects

Peter Popelka

1 INTRODUCTION

The emphasis in any discussion of food composition in relation to nutrient content differs considerably with the overall level of food supply. This is especially true when meat and meat products are discussed in the environment of either developing or industrialized countries (WHO 2003).

The largest amounts of macronutrients supplied by meat and meat products are by far the protein and fat of animal muscle (19 and 2.5%, respectively) and adipose tissue (90% of fat). Actual protein and fat contents, as well as the corresponding amino and fatty acid composition, vary considerably depending on several endogenous (e.g., anatomic location, breeds, sex, species) and exogenous (mainly feeding regime) factors.

Fermented dry sausages are defined as a mixture of comminuted fat and lean meat, salt, nitrate and/or nitrite, sugar and spices (mostly oregano and black pepper), which is stuffed into casings, subjected to fermentation and then allowed to dry (Hugas and Monfort 1997). The quality of the final product is closely related to the ripening that takes place during drying. This process, which confers on the product its

---

For Correspondence: Department of Food Hygiene and Technology, University of Veterinary Medicine and Pharmacy in Košice, Komenského 73, 041 81 Košice, Email: peter.popelka@uvlf.sk

particular slice ability, firmness, color and flavor, is characterized by a complex interaction of chemical and physical reactions associated with the microbiological development of the batter flora (Ordóñez et al. 1999).

The many types of fermented sausages have been narrowed down by industrial production and a major distinction can be made between those produced in shorter and longer ripening periods (Demeyer and Stahnke 2002). In both, pork and/or beef are mixed and comminuted (cuttered) with pork backfat and salt, processes allowing for the incorporation of large amounts of pork fat and low-value meat cuts, including rind. For most fermented sausage types, such practice is reflected in a high calorific value, a high fat content, a high salt content, and a limited nutritional quality of the sausage crude protein. A major action, characteristic for the manufacture of meat products in general, is the addition of 2–3% of salt (NaCl) containing about 0.5% of $NaNO_2$. For the longer ripened products, nitrate rather than nitrite may be used. It should be noted that the amount of collagen in the sausage protein may reach 35%, and this affects price and protein quality but not digestibility as measured in humans (Reuterswärd et al. 1985).

The effect of the presence or absence of various factors: fat, fatty acid composition, cholesterol, calorific value, salt, nitrite or lipid oxidation products that can cause health problems must be analyzed. Bearing in mind these considerations, it then describes the strategies used in animal production, treatment of meat raw material and reformulation of meat products to obtain healthier meat and meat products. Meat and meat products are important sources of proteins, vitamins and minerals, but they also contain fat, saturated fatty acids, cholesterol, salt, etc. In order to produce "healthier" meat products we need to fully understand their positive and negative effects on health. Only then shall we be able to devise suitable strategies to effectively control and adjust their characteristics to suit our needs (Jiménez-Colmenero et al. 2001).

Food consumption is clearly related to the rapidly increasing incidence of chronic non communicable disease, in both industrialized and developing countries (WHO 2003). Over the past 50 years this association has been related to a steadily increasing consumption of animal fat and, more specifically, to the increased consumption of saturated fat present in dairy products and beef. This has resulted in nutritional guidelines advocating a decrease in fat consumption and in various proposals to decrease fat content and alter fat composition in animal production (Givens 2005). However, although available statistics are based on production and trade rather than on actual consumption of foods, they clearly indicate a shift from animal fat to sugar and vegetable oil intake in most industrialized countries.

Like any other food, meat and meat products contain elements which in certain circumstances have a negative effect on human health. Some of these are constituents (natural or otherwise) present in live animals, others are added to the product during processing for technological, microbiological or sensory reasons. Some of them are produced by technological treatment and finally, there are those that develop particularly in the storage/commercialization phase, notably the growth of some pathogenic bacteria, the formation of certain lipid oxidation products and the migration of compounds from the packing material to the product.

2 HEALTH ASPECTS OF FERMENTED MEAT PRODUCTS

The elements which in certain circumstances and in inappropriate proportions have a negative effect on human health presented in live animals are, for instance, fat, cholesterol, residues from environmental pollution or the use of pharmaceuticals, etc. Others are added to the product during processing for technological, microbiological or sensory reasons (salt, nitrite, phosphate, etc.). There is a third group that is produced by technological treatment (toxic compounds formed during smoking, polycyclic aromatic hydrocarbons, nitrosamines, etc.).

2.1 Salt and Reduction of Salt Content

Sodium intake exceeds the nutritional recommendations in many industrialized countries. Excessive intake of sodium has been linked to hypertension and consequently to increased risk of stroke and premature death from cardiovascular diseases (Law et al. 1991). The main source of sodium in the diet is sodium chloride. It has been established that the consumption of more than 6 g NaCl per day per person is associated with an age-increase in blood pressure. Therefore, it has been recommended that the total amount of dietary salt should be maintained at about 5–6 g per day (Nishida et al. 2004). Currently the daily sodium adult intake is approximately three times the recommended daily allowance (Ireland and UK) and therefore public health and regulatory authorities are recommending reducing dietary intake of sodium to 2.4 g (6 g salt) per day (Desmond 2006). Genetically salt susceptible individuals and hypertensive would particularly benefit from low sodium diets, the salt content of which should range between 1 and 3 g per day. In industrialized countries, meat products and meat meals at home and in catering comprise one of the major sources of sodium, in the form of sodium chloride. As a final note, the role of meat products in sodium intake should be considered. Meat as such is relatively poor in sodium; meat products, however, contain about 2% of added salt, values increasing up to 6% in dried products. It can be estimated that meat and meat products provide 20–30% of salt (NaCl) intake (Jiménez-Colmenero et al. 2001). The latter value ranges between 9–12 g per day (3.5 and 5 g per day Na) for industrialized countries, a value well above the dietary goal of 5 g per day (2 g per day Na). The recent finding of a close relationship between consumption of cured meat products and the incidence of hypertension (Paik et al. 2005) highlights the importance of meat processing in sodium overconsumption.

The salt intake derived from meat dishes can be lowered by, whenever possible, adding the salt, not during preparation, but at the table. In most cases, salt contents of over 2% can be markedly lowered without substantial sensory deterioration or technological problems causing economical losses. Salt contents down to 1.4% NaCl in cooked sausages and 1.75% in lean meat products should be enough to produce a heat stable gel with acceptable perceived saltiness as well as firmness, water-binding and fat retention. A particular problem with low-salt meat products is not only the perceived saltiness, but also the intensity of the characteristic flavor decreases. Sodium chloride affects also texture and shelf life of meat products.

The potential sodium chloride reduction depends on aspects connected with the type of the product, its composition, the type of processing required and the preparation conditions. These factors determine the type of product that can be modified and the technological limitations of salt reduction (Ruusunen and Puolanne 2005). Increased meat protein content (i.e. lean meat content) in meat products reduces perceived saltiness. When phosphates are added or the fat content is high, lower salt additions provide a more stable gel than in non-phosphate and in low-fat products.

One of the biggest barriers to salt replacement is cost as salt is one of the cheapest food ingredients available. Also, consumers have grown accustomed to salt through processed foods so in some cases it has being difficult to remove as previously discussed. Another issue is that although there are alternatives to salt in term of functionality some consumers and retailers may not be comfortable with these new ingredients on the label (Searby 2006). There are several approaches for reducing the sodium content in processed meats (Sofos 1989):

- lowering the level of sodium chloride (NaCl) added;
- replacing all or part of the NaCl with other chloride salts (KCl, $CaCl_2$, and $MgCl_2$);
- replacing part of the NaCl with non-chloride salts, such as phosphates, or with new processing techniques or process modifications; and
- combinations of any of the above approaches.

An excessive intake of meat products, particularly dry fermented sausages, is not recommended from a health point of view, at least for some population groups, due to their high level of sodium and animal fat. KCl, CaCl$_2$, and/or calcium ascorbate, among others, have been assessed as partial substitutes of NaCl, giving products with acceptable sensory quality, smaller amounts of sodium and being sometimes a significant source of potassium or calcium (Muguerza et al. 2004). Firstly, and probably the most widely used, is the use of salt substitutes, in particular, potassium chloride (KCl). Masking agents are commonly used in these products. Secondly, the use of flavor enhancers which do not have a salty taste, but enhance the saltiness of products when used in combination with salt. This allows less salt to be added to the products. Thirdly, optimizing the physical form of salt so that it becomes more taste bioavailable and therefore less salt is needed (Angus et al. 2005). Cimeno et al. (2001) replaced NaCl with potassium, magnesium and calcium ascorbates and were able to reduce the NaCl content by about 50 percent. The only relevant difference was the lower consistency which is to be expected when the chloride ions are replaced with ascorbate ions that do not effectively react with myofilaments. A wider application and development of such technologies involving other meat products is recommended (Fernández-López et al. 2004). Substitution of salt is the more delicate intervention because of its universal role in meat technology to improve texture and taste as well as providing a hurdle for bacterial development (Ruusunen and Puolanne 2005). Nevertheless, partial substitution using calcium ascorbate has lowered sausage salt level from 2.3 down to 1.4 percent, with limited damage to sensory quality while ensuring a simultaneous increase of the sausage calcium and ascorbate content, nutrients known to be very low and absent in meat, respectively (Gimeno et al. 2001).

Salt, nitrite, pH and temperature control the fermentation as well as the safety and the quality of dry fermented meat products (Leistner et al. 1971). They are all inter-related, and if one of these factors is reduced it should be compensated with an increase of one or more of the other factors to keep the same safety and technological quality. Salt is such an essential constituent of dry fermented products that there are not so many studies on its reduction. Petäjä et al. (1985) concluded that 2.5% NaCl is the lower limit for a good quality salami-type fermented sausage, but with 2.25% the sausages are less firm and the typical aroma is weaker and the yield lower than with higher salt contents. Also, Stahnke (1995) found that a low NaCl content favors pH decrease.

2.2 Nitrites and Reduction of Nitrites

Nitrate and nitrite are used for the purpose of curing meat products. In most countries the use of both substances, usually added as potassium or sodium salts, is limited. Either the ingoing or the residual amounts are regulated by laws. The effective substance is nitrite acting primarily as an inhibitor for some microorganisms.

Nitrite added to a batter of meat is partially oxidized to nitrate by sequestering oxygen – thus it acts as an antioxidant – a part of nitrite is bound to myoglobin, forming the heat stable NO-myoglobin, a part is bound to proteins or other substances in meat. Nitrate may be reduced to nitrite in raw meat products by microorganisms. As oxidation and reduction may occur the concentrations of nitrite plus nitrate in a product has to be controlled and measured especially if the residual amounts are regulated. This sum of both compounds is important for the human body. Intake of nitrate with food leads to its absorption over the digestive tract into the blood. In the oral cavity nitrate appears again where it is reduced to nitrite. With the saliva the nitrite is mixed with food, having the same effect as nitrite in a batter (inhibiting growth of some pathogenic microorganisms) and swallowed. In the stomach nitrite can eventually form carcinogenic nitrosamines in the acidic environment (Honikel 2008).

After it was discovered that nitrite was the genuine curing agent it took only a few years until nitrite was introduced into meat products manufacturing. But nitrite itself is rather toxic in comparison to nitrate. With Regulation 1333/2008/EC of 16 December 2008 the use of nitrates is limited to non-heated meat products with 150 mg per kg (ingoing amount must be calculated as sodium nitrite) but with quite a number of exemptions.

Issues that have been raised concerning the safety of using nitrate and nitrite for cured meat have included chemical toxicity, formation of carcinogens in food or after ingestion, and reproductive and developmental toxicity. None of these issues represent relevant concerns for nitrate or nitrite in light at the current regulated levels of use in processed meats. While nitrite is recognized as a potentially toxic compound, and there have been cases where nitrite was mistakenly substituted for other compounds in food or drink at concentrations great enough to induce toxicity symptoms, the normally controlled use of nitrite in processed meats represents no toxicity risk. However, the issue of carcinogenic nitrosamines formed from nitrite in meat products was a very serious concern in the 1970s. Fortunately, changes in manufacturing practices and reduced levels of nitrite used in curing solved the problem of nitrosamine formation in meat products. It was recognized that carcinogenic nitrosamines could be formed in the stomach following ingestion. Subsequent work has shown that less than 5 percent of the nitrite and nitrate typically ingested comes from cured meat, the rest coming from vegetables and saliva (Archer 2002). Nitrite

Fermented Meats Composition—Health and Nutrition Aspects 395

has been suspected to be a carcinogen for several decades, but numerous epidemiologic studies have failed to support consistently a link between nitrate or nitrite and cancer. Recent chronic feeding studies in two rodent species failed to link nitrite, even at extremely high oral dose levels, to cancer. Recent suspicions that nitrite might be a developmental toxicant were also found to lack foundation. Since 93% of ingested nitrite comes from normal metabolic sources, if nitrite caused cancers or was a reproductive toxicant, it would imply that humans have a major design flaw. Despite all evidence to the contrary, questions about the safety of nitrate and nitrite will likely persist until definitive prospective studies in humans are conducted.

There are two basic strategies for reducing the potential health risks of nitrites in meat products. One is to reduce or eliminate the addition of nitrite, and the other is to use N-nitrosamine inhibitors. N-nitrosamine production depends on the residual nitrite level. Reducing this level will lower the risk of these carcinogenic compounds forming. In fact, residual nitrite has been substantially reduced (as much as 80%) in the last few years. This change has come about thanks to the addition of less nitrite, the increased use of ascorbates, improvements in manufacturing processes and changes in composition (e.g. larger proportions of ingredients (Cassens 1997). Nevertheless, N-nitrosamine production cannot be totally eliminated while its precursors (nitrites, amines and amino acids) are still present. Alternatives must therefore be found, but this is not easy because of the numerous reactions of N-nitrosamine with the complex biological systems present in meat. Indeed, it is impossible to find any single compound capable of replacing the functions of nitrite. The solution must therefore be to combine several compounds, which together have a cumulative effect on color, flavor, and antioxidant and antimicrobial activity.

2.3 Polycyclic Aromatic Hydrocarbons

Polycyclic aromatic hydrocarbons (PAHs) comprise the largest class of chemical compounds known to be cancer causing agents. Some, while not carcinogenic, may act as synergists. PAHs may form directly in food as a result of some heat processes as charcoal grilling, roasting, smoke drying, smoking (Šimko 2002). Smoke is generated by thermal pyrolysis of a certain kind of wood when there is limited access of oxygen. Temperature of smoke generally plays a very important role, because the amount of PAHs in smoke formed during pyrolysis increases linearly with the smoking temperature within the interval 400–1000°C. By lowering the temperature of the smoldering pile of wood shavings or sawdust to 300–400°C and using filters, the content of PAH in the smoke can be decreased about 10-fold (Stołyhwo and Sikorski 2005).

The highest concentration of PAHs in smoked products is immediately after finishing the smoking then it decreases due to light decomposition and interaction with present compounds. However, PAHs also penetrate into smoked products, where they are protected from light and oxygen, and after some time, the concentration stabilizes at a certain constant level (Šimko 2002).

In recent years significant progress has been made in the understanding of the biological action of PAHs. These compounds enter the organism by inhalation, ingestion or penetration with a following distribution to the various organs. These compounds form covalent adducts with proteins and nucleic acids. The DNA adducts are thought to initiate cell mutation and eventual malignancy (Bartle 1991).

Fermented meat products belong to the group of special meat products that are not treated by a thermal procedure during their technological production. The PAH in fermented meat products did not express the real situation because of the water content, which is successively lowered during ripening. In principle, these losses in weight should result in the increase in PAH content. But, the content is decreased also due to photodegradation of PAH. The final result of these processes was a more or less constant PAH content in the salami. However, a far more objective view of changes in PAH contents in the salami can be obtained after recalculation of the PAH content on a dry weight basis. This recalculation eliminates the effect of variable water content on changes in PAH content during production. The changes took place in the period of ripening, while during the whole interval of storage the PAH content was practically constant (Šimko 2005). Smoking of fermented meat products under mild conditions, in modern smokehouses supplied with filtered smoke from external generators, does not lead to significant contamination of the products with carcinogenic PAHs (Stołyhwo and Sikorski 2005).

The alternative for reducing of PAH are smoke flavorings, which have been produced commercially since about the middle of the last century for use in the meat and fish industries contain only trace amounts of PAHs. These flavorings are generally smoke extracts, filtered and separated from the resinous material that contains most of the PAHs. Some flavorings are obtained by distillation of pyroligneous liquids. They are available as aqueous solutions or in dry form on different supports, e.g., salt, yeast, or other free-flowing material (Miler and Sikorski 1990).

2.4 Reduction of Other Undesirable Compounds

It is important during strain selection (starter cultures) that no undesirable compounds such as toxins and biogenic amines, that could adversely affect health, are formed. If surface mold growth is desirable, it must be

Fermented Meats Composition—Health and Nutrition Aspects 397

checked if the mold starter culture produces mycotoxins or antibiotics (Holzapfel 2002). The use of molds that are free of mycotoxin production as starter cultures could be useful in outcompeting mycotoxin-producing strains from the house flora. Molds may also produce green or dark spots that are not acceptable to most consumers or have a negative impact on flavor and taste (Sunesen and Stahnke 2003).

During the ripening of fermented sausages, biogenic amines such as tyramine, histamine, tryptamine, cadaverine, putrescine, and spermidine, may be formed by the action of microbial decarboxylases on amino acids that originate from meat proteolysis (Komprda et al. 2004, Pleadin and Bogdanović, Chapter 18 in this book). Microbial decarboxylation reactions may be ascribed to both the microorganisms that were introduced via the starter culture and the ones that constitute part of the natural population of the meat. In general, starter bacteria have limited tyrosine-decarboxylating activity, but contaminant non-starter LAB, in particular enterococci, are believed to be responsible for tyramine production (Ansorena et al. 2002). The use of decarboxylase-negative starter cultures that are highly competitive and fast acidifiers prevents the growth of biogenic amine producers and leads to end-products nearly free of biogenic amines (González-Fernández et al. 2003), as long as the raw material is of sufficient quality (Bover-Cid et al. 2001). Also, the introduction of starter strains that possess amine oxidase activity might be a way of further decreasing the amount of biogenic amines produced in situ (Gardini et al. 2002). Although it is known that the superficial inoculation with *Penicillium camemberti* in cheeses increases the concentration of certain amines, the production of amines by molds in fermented sausages does not appear significant but has not been fully studied yet (Bruna et al. 2003).

The nature of the lactic isomer produced by the LAB strains is of concern, since high levels of the d(–)-lactic acid isomer are not hydrolysed by lactate dehydrogenase in humans and are thus capable of causing acidosis. Therefore, strains producing l(+)-lactic acid should be preferably selected (Holzapfel 2002).

2.5 Strategies for Achieving Healthier Meat and Meat Products

A promising approach to improving health care would be to produce a healthier food supply as a preventive health care strategy. The food supply could be improved by producing functional foods that have nutritional profiles that are healthier than conventional products. However, production of functional foods is not always easily accomplished since they must also taste good, be convenient and reasonably priced so that consumers will regularly purchase and use the products. Meats have great potential for delivering important nutrients such as fatty acids, minerals, dietary fiber, antioxidants and bioactive peptides into the diet. However, to produce successful products with these ingredients, technologies must be developed to increase their stability and decrease their flavor impact on muscle foods (DHHS 2010).

Probiotic foods receive market interest as health-promoting, functional foods. They have been introduced in a wide range of food industries (Nova et al. Chapter 13 in this book). However, commercial application of probiotic microorganisms in fermented sausages is not common yet. There are both advantages and disadvantages connected to fermented meat matrices. They are adequate for the carriage of probiotic bacteria since they are usually not or only mildly heated and may promote the survival of probiotic bacteria in the gastrointestinal tract. In contrast, bacterial viability may be reduced due to the high content in curing salt and the low water activity and pH. Therefore, results are expected to be strain-dependent. Up till now, several approaches have been followed but most results are too preliminary to be able to evaluate the effect of probiotic fermented meats on human health (De Vuyst et al. 2008).

The evaluation of the end-products needs to deal with both health effects and technological characteristics, for instance through human intervention studies and taste panels, respectively. Although the concept is not new, only few manufacturers consider the use of fermented sausages as carriers for probiotic lactic acid bacteria (LAB) (Arihara 2006). Since meat products are seldom perceived as "healthy foods", due to the perceived image of meat and its controversial nutrient profiling clause with respect to the presence of nitrite, salt, and fat, their marketing potential may be compromised (Lücke 2000). In addition, the more artisan orientation of sausage manufacturers as compared to the dairy industry, the larger variety of products, and a number of uncertainties concerning technological, microbiological, and regulatory aspects seem to be problematic (Kröckel 2006). If any health claims are to be taken seriously, the application of probiotic LAB must in all cases be based on a careful selection procedure. This requires enhanced scientific research efforts dealing with the use of probiotics in fermented meats, taking into account that probiotics are and remain food constituents that are in no case to be considered as drugs or therapeutic agents. Much attention has been paid to the contribution of probiotic strains as meat starter cultures to improve food safety and little to the introduction of beneficial health effects (Leroy et al. 2006).

In relation to dry fermented sausages major difficulties are associated with developing low-salt and low-fat products than in the traditional products, because of the important functions that these two ingredients, salt and fat, have in the quality of those products. NaCl has an important influence on the final taste of dry fermented sausages and also plays an

Fermented Meats Composition—Health and Nutrition Aspects 399

important role in assuming microbiological stability. Fat is necessary for the satisfactory development of those properties particularly related to texture, juiciness and flavor.

There are diverse possible strategies for developing healthier meat and meat products, including functional foods. Items listed below are strategies suggested by Jiménez-Colmenero et al. (2001).

- Modification of carcass composition.
- Manipulation of meat raw materials.
- Reformulation of meat products.
  - Reduction of fat content.
  - Modification of the fatty acid profile.
  - Reduction of cholesterol.
  - Reduction of calories.
  - Reduction of sodium content.
  - Reduction of nitrites.
  - Incorporation of functional ingredients.
- Anticarcinogenic and antitumour activity.

Although all aspects from animal production to product processing should be considered for designing healthier products, the following paragraphs focuses on functional meat products mainly from the viewpoint of food processing.

2.5.1 Reformulation of Meat Products

Depending on the product type (composed of identifiable pieces of meat, coarsely or finely ground, emulsions, cooked, cured, etc.), one of the best moments at which to alter the composition of foods is perhaps during one of the preparation stages. At this stage reformulation is used as far as possible to develop a range of derivatives with custom-designed composition and properties. For this there are two possible types of complementary intervention. The first involves reducing some compounds normally present in these foods to appropriate amounts, for example, fat, SFAs, cholesterol and so on. The second is to incorporate ingredients that are potentially health-enhancing (functional), for example, fibre, certain types of vegetable protein, MUFAs and PUFAs, antioxidants, etc.

2.5.1.1 Reduction of fat content

Health organizations all over the world have promoted the choice of a diet low in saturated fat and cholesterol and moderated in total fat, as a means of preventing cardiovascular disease (AHA 2000), which constitutes one of the main causes of mortality in the world.

400   *Fermented Meat Products: Health Aspects*

Hypertension, which is one of the main risk factors for cardiovascular disease, is known to be correlated with an excessive sodium intake from the diet (Truswell 1994).

Generally speaking, fat reduction is achieved by reformulation. This consists of combining pre-selected meat raw materials (leaner and of known composition) with appropriate amounts of water, fat (animal or vegetable depending on the new composition), flavorings (with modified formulation) and other ingredients (fat replacements or substitutes), which coupled with technological processes give the product certain desirable characteristics, such as improved composition, sensory and technological properties, safety, nutritional value, convenience, etc.

Although there is a lot of research work for reduced and low-fat frankfurter type sausages, ground beef and hamburger, fresh pork and coarse ground sausages, there are only a few papers concerning fermented sausages (Keeton 1994). According to Wirth (1988), if the climatic conditions are properly controlled during fermentation and drying fat-reduced salami and saveloy products of acceptable standard can be made with fat contents in the raw material of about 15%. Papadima and Bloukas (1999) found that in traditional Greek sausages the fat can be reduced to 20%. Bloukas et al. (1997) have found that up to 20% of pork backfat can be replaced by olive oil in the form of pre-emulsified fat with soy protein isolate without negatively affecting the processing and quality characteristics of dry fermented sausages. Muguerza et al. (2001) have also found that up to 25% of pork backfat can be replaced with pre-emulsified olive oil in the production of Chorizo de Pamplona fermented sausages, resulting in nutritional advantages related to cholesterol reduction and an increase in MUFA and PUFA fractions.

Consequently, the development of low-fat products means that factors associated with meat raw materials, non-meat ingredients, and manufacture and preparation procedures (Jiménez-Colmenero 2000) together with other factors such as the characteristics of the new derivative must be taken into account. In relation to fat, recent research has focused on the use of different types of fibres and vegetable oils as partial substitutes of pork backfat. The use of fibres results in low-fat and low-energy products. The nutritional value of meat products is mainly due to the energy supplied by these products, and also to their high biological value proteins, vitamins and minerals.

2.5.1.2 *Modification of the fatty acid profile*

A higher level of polyunsaturated FA can be achieved in pork meat and adipose tissue by altering the fatty acid profile of the diet (Realini et al. 2010) and can also be achieved in meat products by the direct application of vegetable oils (Del Nobile et al. 2009). It is well known that in contrast

Fermented Meats Composition—Health and Nutrition Aspects

to beef adipose tissue, the PUFA content of pork backfat can easily be increased by dietary means (Warnants et al. 1998). As summarized by Raes et al. (2004) and De Smet et al. (2004), the intramuscular fat content may vary between about 1% (high value cuts of very lean pork and beef breeds) and 6% and such variation is also affecting fatty acid composition. The contents of saturated (SFA) and monounsaturated (MUFA) fatty acids increase faster with increasing fatness than does the content of polyunsaturated (PUFA) fatty acids, resulting in a decrease in the relative proportion of PUFA. In beef, fat level and fat composition are to a large extent determined by breed, whereas in pork, nutrition will have a larger impact. The fat level also influences the n-6/n-3 PUFA ratio, due to the difference of this ratio in polar and neutral lipids. However, these effects are much smaller than the effects that can be achieved by dietary means. Although linseed or linseed oil inclusion in animal diet supplies -linolenic acid (LNA), the conversion of LNA to its longer chain metabolite eicosapentanoic acid (EPA) seems to be limited, resulting in only a small increase in their intramuscular fat. For both ruminants and monogastrics, the only effective way to increase docosahexanoic acid (DHA) levels in intramuscular fat seems to be the feeding of fish oil or fish meal.

2.5.1.3 *Reduction in serum cholesterol*

In the last few decades there has been a clear drop in the fat content of carcasses, but there is no easy way of determining whether the same is also true of cholesterol content (Chizzolini et al. 1999). The amount of intra- and intermuscular fat is not always directly related to the cholesterol level. In dry matter, the amount of cholesterol in beef, pork, lamb and poultry lean tissue may be as much as twice that present in adipose tissue, but in wet matter, the cholesterol content of lean tissues is slightly lower than that of adipose tissue (Mandigo 1991).

The level of serum cholesterol is a major factor for coronary heart disease, and elevated levels of serum cholesterol, particularly LDL-cholesterol, have been linked to an increased risk (Liong and Shah 2006). There is a high correlation between dietary saturated fat or cholesterol intake and serum cholesterol level. Although dietary goals recommend cholesterol intakes less than 300 mg per day, and meat and meat product intake may provide up to half of that amount, cholesterol intake and serum cholesterol levels are recently considered of less importance for the incidence of cardiovascular disease (CVD). Other factors relating to genetics and dietary antioxidant intake appear to be very important. The latter could be involved in the protection against oxidation of both dietary and endogenous cholesterol. Cholesterol oxides or oxysterols group about 70–80 compounds, and a number of those have been

402   *Fermented Meat Products: Health Aspects*

identified in meats and in atherosclerotic lesions. Their toxicity probably involves interference with the gene expression related to cell viability (Chizzolini et al. 1999).

Probiotic bacteria used also for production of fermented meat products are reported to de-conjugate bile salts: deconjugated bile acid does not absorb lipid as readily as its conjugated counterpart, leading to a reduction in cholesterol level. *Lactobacillus acidophilus* is also reported to take up cholesterol during growth and this makes it unavailable for absorption into the blood stream (Klaver and Meer 1993). In all studies conducted thus far, the real factor responsible for a reduction in cholesterol levels remains unknown. Klaver and Van der Meer (1993) suggested that the reduction of cholesterol is not due to assimilation or to a direct interaction between the bacteria and cholesterol, but rather due to the co-precipitation of cholesterol with deconjugated bile salts at pH values below 6.0. This would not explain reduction of cholesterol in vivo as the pH of the lower gastrointestinal tract (GIT) is neutral to alkaline. Marshall and Taylor (1995) also observed co-precipitation of cholesterol with deconjugated bile salts, but also reported cholesterol removal in the absence of bile. However, cholesterol and bile salt metabolism is closely linked, where cholesterol is the precursor for synthesis and bile salts the water-soluble excretory end product. Bile salts are deconjugated during enterohepatic circulation by bile salt hydrolase (BSH). The free bile acids as well as glycine and taurine, are not so easily reabsorbed and are excreted in the faeces (De Smet et al. 1994). The loss in bile salts increases the catabolism of cholesterol to bile acids, resulting in lower cholesterol levels (De Rodas et al. 1996). BSH activity has been shown in *Lactobacillus, Enterococcus, Peptostreptococcus, Bifidobacterium, Clostridium* and *Bacteroides* spp. (Bateup et al. 1995). The BSH hypothesis has not definitely been proved. Recent observations indicate that free bile acids are less effectively absorbed by the active transport system in the ileum, but are more effectively reabsorbed in the intestine and colon by passive diffusion (Marteau et al. 1997).

De Smet et al. (1998) have shown that feeding pigs with cells of *Lactobacillus reuteri* containing active BSH resulted in significant lowering of serum total and LDL-cholesterol concentrations, accompanied by a gradual increase in *Lactobacillus* cell numbers. No change in HDL-cholesterol concentration was observed. The authors have also shown that during the final three weeks of changing from a high fat diet to a regular diet, the cholesterol levels significantly decreased and the differences in total and LDL-cholesterol concentrations between the treated and untreated animals largely disappeared. *Lb. reuteri* cells were gradually washed out and they did not succeed in permanently colonizing the intestinal tract.

Fermented Meats Composition—Health and Nutrition Aspects   403

The another approach could be products with less cholesterol obtained by replacing fat and lean meat raw materials (since dietary cholesterol is strictly linked to animal cells) with other vegetable materials containing no cholesterol, but not in fermented meat products when vegetable material should not be used.

2.5.2 Anticarcinogenic and Antitumour Activity

Another shift in emphasis concerns the substitution of food based for nutrient-based nutritional guidelines. Based on a "probable increasing risk" of colon cancer incidence, "preserved meat (e.g., sausages, salami, bacon, ham)" is one of the rare food groups for which a moderate consumption is recommended (WHO 2003). Since 1975, meat consumption has indeed been associated with colorectal cancer reflected in recommendations for healthy eating from the American Cancer Association and the U.S. National Cancer Institute. These involved less frequent eating of red meat to be substituted with chicken or fish and moderation in the eating of salt cured, smoked, and nitrite-cured foods. Recent work (Norat et al. 2005) has confirmed the association of red and processed meat consumption with colorectal cancer risk. The hazard ratio significantly increased from 0.96 to 1.35 when increasing average consumption from 20 g per day to >160 g per day. The association was stronger for processed than for unprocessed red meat, and risks were reduced by increased fiber intake. The mechanism underlying the association may relate to endogenous nitrosation of heme iron or colonic formation of toxic heme metabolites (Sesink et al. 1999) and/or to heterocyclic amines (HCAs) and/or polycyclic aromatic hydrocarbons (PAHs) produced during the preparation of meat or production of meat products. The first mechanism is considered to be more likely because chicken is a major contributor to HCA intake and no association was observed between poultry intake and colorectal cancer risk in the study (Norat and Riboli 2001). Evidence has been presented for the multifactorial causes for colon cancer (Emmons et al. 2005) as well as for genetic differences in its association with meat consumption (Brink et al. 2005). Recent reports also highlight the association of red meat and meat products both with the incidence of various diseases as well as with the dietary supply of functional food components. The former suggest association of red meat and/or meat products consumption with type 2 diabetes (Fung et al. 2004), ovarian cancer (Kiani et al. 2006), prostate cancer, and degenerative arthritis (Hailu et al. 2006).

Microbial enzymes such as azoreductase, β-glucuronidase and nitroreductase may convert procarcinogens into carcinogens and cause colon cancer (Fernandes and Shahani 1990). LAB may also retard or

404   *Fermented Meat Products: Health Aspects*

prevent the initiation and promotion of tumours. *Lactobacillus* acidophilus and *Lb. bulgaricus* and/or *Lb. casei* suppressed Ehrlich ascites tumour or Sarcoma 180 in mice (Goldin 1990). Tumour suppression is associated with intact viable cells, intact dead cells, and cell wall fragments of lactobacilli and bifidobacteria.

Nitrites used in food processing are converted to carcinogenic nitrosamines in the GIT. Cellular uptake of nitrites by lactobacilli and bifidobacteria has been shown in vitro (Grill et al. 1995).

Lb. acidophilus reduced the biotransformation of primary to secondary bile salts, thus reducing the possible initiation of cancer (Fernandes and Shahani 1990). Modler et al. (1990) suggested that the reduction of intestinal pH, through metabolic activities of LAB, could inhibit the growth of putrefactive bacteria and thus prevent large bowel cancer.

3   NUTRITION ASPECTS OF FERMENTED MEAT PRODUCTS

Regarding obesity, it is very important to understand how meat or meat products affects biological and physiological mechanisms of appetite, satiety, and long-term behavior. Meat and meat products show highly satiating characteristics and, in this respect, functional foods could be a food-related solution because these types of products could be designed to be less calorifically dense and while remaining more highly satiating and tasty. In this way, the food industry in general, the meat and related products industry in particular, could contribute to making lives easier and more active.

The nutritional value of fermented meat products is mainly due to the energy supplied by these products, and also to their high biological value proteins, vitamins and minerals. However, from a health point of view, an excessive intake of these products cannot be recommended especially for certain population groups, because of their significant sodium and fat content. Concerning fat supply, it is well known that the animal raw material used in the elaboration of these products contains cholesterol and a higher proportion of saturated fatty acids (SFA) than polyunsaturated fatty acids (PUFA).

Besides contributing to human health, probiotic fermented meats need to be of sufficient commercial value. Therefore, a primary requirement remains – as for any fermented meat – the sensory and technological aspects of the end-product. Negative side effects on the overall quality of the obtained fermented meats cannot be tolerated and evaluation of such effects is primordial (De Vuyst et al. 2008).

Meat and poultry products are a food category with both positive and negative nutritional attributes. Muscle foods are major sources for

Fermented Meats Composition—Health and Nutrition Aspects 405

many bioactive compounds including iron, zinc, conjugated linoleic acid (mainly ruminants) and B vitamins (Jiménez-Colmenero et al. 2001). However, meats and processed meats are also associated with nutrients and nutritional profiles that are often considered negative including high levels of saturated fatty acids, cholesterol, sodium and high fat and caloric contents (Whitney and Rolfes 2002). Some of these negative nutrients in meats can be minimized by selection of lean meat cuts, removal of adipose fat, dietary manipulation to alter fatty acid composition and proper portion control to decrease fat consumption and caloric intake. In addition, the nutritional profile of meat products could be further improved by addition of potentially health promoting nutrients. These products would be categorized as functional foods which are defined as foods with nutritional profiles that exceed conventional products. In deciding proper nutrients for functional foods, several factors should be considered including the bioactive compound's current intake level in the diet (e.g. would the consumer benefit from an increase in the bioactive compound in the diet), biological efficacy in humans, stability in the food product and impact on quality parameters such as color, flavor and texture. The major nutrients currently under consumed by adults in the U.S. include calcium, potassium, magnesium, fiber as well as vitamins A, C and E (Dietary Guidelines for Americans, 2005). In addition, omega-3 fatty acids are currently under consumed according to recommended intake levels set by associations such as the American Heart Association (Harris 2007). Finally, several newly recognized health promoting bioactive compounds have potential as functional food components such as conjugated linoleic acid and bioactive peptides.

3.1 Unsaturated Fatty Acids

In many countries, consumers are over consuming saturated fatty acids and under consuming polyunsaturated fatty acids especially the omega-3 fatty acids (Dietary Guidelines for Americans 2005). Fatty acid intake is a major problem because of the ability of fatty acids to impact low density lipoprotein (LDL) cholesterol levels which are associated with cardiovascular disease. In general, saturated fatty acids increased LDL-cholesterol levels in the plasma and thus increase cardiovascular disease risk while polyunsaturated fatty acids decrease LDL-cholesterol levels (Whitney and Rolfes 2002). There is also much interest in incorporating omega-3 fatty acids into functional foods. This is because many consumers are currently under consuming omega-3 fatty acids so increased consumption could be beneficial by decreasing blood triacylglycerols, sudden cardiac death, depression and arthritis (Harris 2007).

The fatty acid composition of meat from ruminants is generally more saturated due to the fact that unsaturated fatty acids are subjected to biohydrogenation in the rumen. The lack of a rumen means that the fatty acid composition of muscle foods from animals such as pigs, poultry and fish can be altered by diet as many papers have been published on increasing the unsaturated fatty acids composition of pigs and poultry (Bou et al. 2009). This practice already occurs in specialty products such as Iberian hams which are high in oleic acid due to consumption of a corns (Narvaez-Rivas et al. 2008) and aquaculture salmon which are fed fish oils high in omega-3 fatty acids (Blanchet et al. 2005). However, these practices are limited by the fact that increasing levels of unsaturated fatty acids decreases the oxidative stability of the meat product. Increasing unsaturated fatty acids is especially a problem in muscle foods since they are high in prooxidative metals, generally low in endogenous antioxidants and subjected to processing operations that greatly increase oxidative stress (e.g. cooking to produce warmed-over flavor). Thus, alteration of dietary fatty acids to change muscle composition is most easily accomplished with oleic acid since it is at least 10 times more oxidatively stable than polyunsaturated fatty acids such as linolenic (McClements and Decker 2008). However, if dietary manipulation is performed to increase polyunsaturated fatty acids, then antioxidant technologies must also be employed to minimize oxidative deterioration. Inhibition of oxidative deterioration could be accomplished by increasing muscle antioxidants by diet or food ingredients, decreasing storage temperature and/or oxygen exclusion by vacuum packaging.

As mentioned earlier, the fatty acid composition of ruminants is difficult to change because unsaturated fatty acids are biohydrogenated in the rumen. Attempts have been made to use protected dietary fats that are unavailable to ruminant bacterial enzymes but are then released in the small intestine where they are absorbed and incorporated into muscle or milk. Protected lipids have been largely unsuccessful due to their higher cost especially in relation to the low level of fatty acid change they produce in tissue. There are also many references of improved fatty acid compositions in grass fed beef. These references relate to the increased levels of omega-3 fatty acids (especially α-linolenic acid) in beef on pasture. However, the nutritional importance of increased α-linolenic acid concentration is not clear since α-linolenic acid is not as bioactive as longer chain omega-3 fatty acids such as EPA and DHA and since the total increase in omega-3 fatty acids in an 85 g portion of grass feed beef is generally less than 50 mg, which is less than 10% of current recommendation of 500 mg omega-3 fatty acids per day (Harris 2007).

Another approach to altering the fatty acid composition is direct addition of oils to processed meat products. This approach is most promising for omega-3 fatty acids which only need to be added at low levels (50–100 mg, approximately 150–300 mg fish oil) to be nutritionally significant. An advantage of this approach is that the omega-3 fatty acids

Fermented Meats Composition—Health and Nutrition Aspects

can be encapsulated in delivery systems that inhibit oxidation without decreasing bioavailability.

3.2 Dietary Fiber

Dietary fiber can be classified as soluble and insoluble fiber. Both types of fiber have numerous health benefits including maintaining bowel integrity and health, lowering blood cholesterol levels, controlling blood sugar levels and providing a non-caloric bulking agent that can aid in weight loss by replacing caloric food components such as fat. According to the Dietary Guidelines for Americans (2005), dietary fibers are under consumed by most adults indicating that fiber fortification in meat products could have health benefits. Numerous papers have shown that fiber fortification into sausages at nutritionally significant levels (2–3 g per serving) can be accomplished without adverse impact on sensory quality (Salazar et al. 2009, Yilmaz and Gecgel 2009). In addition to the benefit of increased fiber consumption, dietary fibers in meat products also have other advantages such as fat replacement, increased water-holding capacity and improved oxidative stability when the fiber source is associated with phenolic antioxidants (Sayago-Ayerdi et al. 2009).

3.3 Minerals

Dietary mineral are essential for bone health, hypertension, muscle and nerve function, regulation of blood sugar levels and thus are important in diseases such as hypertension, cardiovascular disease, osteoporosis and diabetes (Whitney and Rolfes 2002). Calcium, potassium and magnesium are the most commonly under consumed minerals in the diet (Dietary Guidelines for Americans, 2005). In some populations, iron is also under consumed but its addition to meat is problematic as it will rapidly promote lipid oxidation and discoloration (McClements and Decker 2008). Fortification of processed meats with calcium, potassium and magnesium can be accomplished without major changes in sensory profile (Moon et al. 2008, Selgas et al. 2009). An additional advantage of calcium, potassium and magnesium fortification is that these minerals can be used to reduce sodium levels in processed meats thus further improving the nutritional profile of the product (Moon et al. 2008). Salt such as potassium lactate can also be advantageous since it inhibits lipid oxidation and microbial growth (Moon et al. 2008). Calcium can activate the protease, calpain, which increases meat tenderness. Research has shown that calcium infusion into beef cattle can both increase muscle calcium concentrations (Dikeman et al. 2003) and improve beef tenderness after 14 days of aging (Diles et al. 1994). However, in some processed

408   *Fermented Meat Products: Health Aspects*

meats, calcium, potassium and magnesium inhibit proteolytic enzymes important in the flavor development of dry-cured meats although they do not negatively impact salty flavor (Armenteros et al. 2009).

There has also been considerable interest in the dietary supplementation of livestock with selenium. Muscle selenium concentrations can be increased by dietary selenium supplementation in beef (Juniper et al. 2008), pork (Morel et al. 2008) suggesting that supplemented livestock could be beneficial for human health. Increases in muscle selenium concentrations correlate with increases in the activity of glutathione peroxidiase, a selenium containing enzyme that decomposes lipid hydroperoxides. However, increases in the activity of glutathione peroxidase in beef did not increase the oxidative stability of the muscle (Juniper et al. 2008).

Increases in dietary selenium for humans has been of interest because of epidemiological evidence that selenium decreases prostate cancer risk and beneficial for thyroid function (Combs et al. 2009).

3.4   Antioxidant Vitamins

Dietary antioxidants have been suggested to be beneficial to immune function, heart disease and cancer. Vitamins A, C and E are consumed at levels below their recommended dietary intake levels by many consumers (Dietary Guidelines for Americans 2005). Addition of vitamin C to meats is difficult because it is not very stable at meat pH and it tends to promote lipid oxidation (Haak et al. 2009). In addition, vitamin C addition to meats is often prohibited since it can stabilize meat color and thus is considered adulteration. β-carotene is an important source of dietary Vitamin A. In general, meats are not a good source of β-carotene or other carotenoids with the exception of chicken. β-carotene concentrations can be increased by dietary supplementation (Bou et al. 2009). Addition of β-carotene to meats as a food ingredient is often difficult due to its chemical instability. Both exogenous addition and dietary fortification of β-carotene is also limited since it will alter the color of the muscle (Torrissen 2000).

Muscle foods could be even better sources of vitamin E through meat fortification or dietary supplementation with α-tocopherol (McClements and Decker 2008, Schaefer et al. 1995). Incorporation of α-tocopherol via the diet allows the biological system to efficiently incorporate α-tocopherol into the cell membranes (the other tocopherol homologs are not efficiently incorporated into biological membranes when incorporated into the diet). Cell membranes with high levels of tocopherol are significantly more resistant to lipid oxidation since membrane phospholipids are the primary site of oxidation. Thus, α-tocopherol has a major benefit to protecting meat flavor. In addition, lipid oxidation in beef muscle causes myoglobin discoloration so α-tocopherol supplementation also protects meat color. Overall, α-tocopheryl acetate supplementation of livestock could have major benefits to both meat quality and nutritional composition.

4 CONCLUSION

Fermented meats are often studied by food technologists and microbiologists with respect to their safety and quality properties. They have originated as the products of empirical methods for meat preservation in a distant past and have evolved over many centuries towards a large assortment of varieties with strong territorial and socio-cultural connotations. To be successful, innovative strategies should respect the "original" make-up of the products, a feature that is hard to pin down. As an example, changes at the recipe level may be more acceptable than changes at the level of the raw materials (Leroy et al. 2015).

Most innovative strategies are to be found on the level of technology, often driven by a need to meet prescribed standards of food safety and hygiene. Besides achieving higher food safety levels, novel technologies include the use of non-traditional ingredients to reduce costs and to mask suboptimal sensory quality, as exemplified by the use of chemical acidulants, liquid smoke, phosphates, glutamate, and colourants. Moreover, the use of novel but carefully selected industrial starter cultures with specific functionalities could contribute to the improvement of quality and safety of fermented meats, to the benefit of both producers and consumers.

Several contemporary expectations and notions have been added to the traditional setup during the last decades, leading to products with increased convenience, "naturalness" (e.g., organic meat), or "functionality" (e.g., "sport salami" with low-fat content). Again, innovation strategies that aim at the creation of additional levels of meaning should make sure not to be inconsistent with traditional conceptions (Cotillon et al. 2013). This approach is thus not always straight forward, despite the many efforts aiming at the development of new types of fermented meats, for instance when adding potential probiotic value (De Vuyst et al. 2008). Other attempts of creating "nutritional" meaning, mostly as a reaction to the negative communication around the health hazards coupled to cured meat, consist of adding or increasing levels of fibres, micronutrients, or antioxidants, or even using meat replacers (Muguerza et al. 2004). Generally, the use of meat substitutes in food markets is not straightforward and more insight is needed into the way consumers classify them within different food categories (Hoek et al. 2011), which

410   *Fermented Meat Products: Health Aspects*

could be particularly challenging for the category of "traditional fermented meats".

Key words: *salt, nitrite, nitrosamine, PAHs, fat, cholesterol, fatty acids, vitamins, minerals*

REFERENCES

[AHA] American Heart Association 2000. Dietary guidelines revision. A statement for Healthcare Professionals from the Nutrition Committee of the Americans Heart Association. Circulation 102: 2284–2299.

Angus, F., T. Phelps, S. Clegg, C. Narain, C. den Ridder and D. Kilcast. 2005. Salt in processed foods: collaborative research project. Leatherhead Food International.

Ansorena, D., M.C. Montel, M. Rokka, R. Talon, S. Eerola, A. Rizzo, M. Raemaekers and D. Demeyer. 2002. Analysis of biogenic amines in northern and southern European sausages and role of flora in amine production. Meat Sci. 61: 141–147.

Archer, D.L. 2002. Evidence that ingested nitrate and nitrite are beneficial to health. J. Food Protect. 65: 872–875.

Arihara, K. 2006. Strategies for designing novel functional meat products. Meat Sci. 74: 219–229.

Armenteros, M., M. Aristoy and F. Toldrà. 2009. Effect of sodium, potassium, calcium and magnesium chloride salts on porcine muscle proteases. Eur. Food Res. Technol. 229: 93–98.

Bartle, K.D. 1991. p. 41. *In:* C. Creaser and R. Purchase (eds.). Food contaminants, sources and surveillance. The Royal Society of Chemistry, Cambridge.

Bateup, J.M., M.A. McConnell, H.F. Jenkinson and G.W. Tannock. 1995. Comparison of *Lactobacillus* strains with respect to bile salt hydrolase activity, colonization of the gastrointestinal tract, and growth rate of the murine host. Appl. Environ. Microbiol. 61: 1147–1149.

Blanchet, C., M. Lucas, P. Julien, R. Morin, S. Gingras and E. Dewailly. 2005. Fatty acid composition of wild and farmed Atlantic salmon (*Salmo salar*) and rainbow trout (*Oncorhynchus mykiss*). Lipids 40: 529–531.

Bloukas, J.G., E.D. Paneras and G.C. Fournitzis. 1997. Effect of replacing pork backfat with olive oil on processing and quality characteristics of fermented sausages. Meat Sci. 45: 133–144.

Bou, R., R. Codony, A. Tres, E.A. Decker and F. Guardiola. 2009. Dietary strategies to improve nutritional value, oxidative stability, and sensory properties of poultry products. Crit. Rev. Food Sci. 49: 800–822.

Bover-Cid, S., M. Izquierdo-Pulido and M.C. Vidal-Carou. 2001. Effectiveness of a *Lactobacillus sakei* starter culture in the reduction of biogenic amine accumulation as a function of the raw material quality. J. Food Protect. 64: 367–373.

Brink, M., M.P. Weijenberg, A.F.P.M. de Goeij, G.M.J.M. Roemen, M.H.F.M. Lentjes, A.P. de Bruine, R.A. Goldbohm and P.A. van den Brandt. 2005. Meat consumption and K-ras mutations in sporadic colon and rectal cancer in The Netherlands. Cohort Study. Br. J. Cancer 92: 1310–1320.

Fermented Meats Composition—Health and Nutrition Aspects

Bruna, J.M., E.M. Hierro., L. de la Hoz, D.S. Mottram, M. Fernández and J.A. Ordóñez. 2003. Changes in selected biochemical and sensory parameters as affected by the superficial inoculation of *Penicillium camemberti* on dry fermented sausages. Meat Sci. 85: 111–125.

Cassens, R.G. 1997. Residual nitrite in cured meat. Food Technol. 51: 53–55.

Chizzolini, R., E. Zanardi, V. Dorigoni and S. Ghidini. 1999. Calorific value and cholesterol content of normal and low-fat meat and meat products. Trends in Food Sci. Tech. 10: 119–128.

Cimeno, O., I. Astiasarán and J. Bello. 2001. Calcium chloride as a potential partial substitute of NaCl in dry fermented sausages: effects on colour, texture and hygienic quality at different concentrations. Meat Sci. 57: 23–29.

Combs, G.F., Jr, D.N. Midthune, K.Y. Patterson, W.K. Canfield, A.D. Hill and O.A. Levander. 2009. Effects of selenomethionine supplementation on selenium status and thyroid hormone concentrations in healthy adults. Am. J. Clin. Nutr. 89: 1808–1814.

Cotillon, C., A.-C. Guyot, D. Rossi and M. Notarfonso. 2013. Traditional food: a better compatibility with industry requirements. J. Sci. Food Agric. 93: 3426–3432.

Demeyer, D. and L. Stahnke. 2002. Quality control of fermented meat products. pp. 359–393. *In:* J. Kerry, J. Kerry and D. Ledward (eds.). Meat processing, improving quality. Cambridge, Woodhead Publishing Ltd.

Del Nobile, M.A., A. Conte, A.L. Incoronato, O. Panza, A. Sevi and R. Marino. 2009: New strategies for reducing the pork back-fat content in typical Italian salami. Meat Sci. 81: 263–269.

De Rodas, B.Z., S.E. Gilliland and C.V. Maxwell. 1996. Hypocholesterolemic action of *Lactobacillus acidophilus* ATCC 43121 and calcium in swine with hypercholesterolemia induced by diet. J. Dairy Sci. 79: 2121–2128.

De Smet, S., K. Raes and D. Demeyer. 2004. Meat fatty acid composition as affected by fatness and genetic factors: A review. Anim. Res. 53: 81–98.

De Smet, I., P. De Boever and W. Verstraete, W. 1998. Cholesterol lowering in pigs through enhanced bacterial bile salt hydrolase activity. Brit. J. Nutr. 79: 185–194.

De Smet, I., L. Van Hoorde, N. De Saeyer, M. Van de Woestyme and W. Vestraete. 1994. In vitro study of bile salt hydrolase (BSH) activity of BSH isogenic *Lactobacillus plantarum* 80 strains and estimation of lowering through enhanced BSH activity. Microb. Ecol. Health D. 7: 315–329.

Desmond, E. 2006. Reducing salt: A challenge for the meat industry. Meat Sci. 74: 188–196.

De Vuyst, L., G. Falony and F. Leroy. 2008. Probiotics in fermented sausages. Meat Sci. 80: 75–78.

[DHHS] Department of Health and Human Services 2010. Chronic disease cost. http://health.nv.gov/CD_ChronicDisease_Costs.htm.

Dietary guidelines for Americans 2005. U.S. Department of Health and Human Services, and U.S. Department of Agriculture. http://www.health.gov/dietaryguidelines/dga2005/document/.

Dikeman, M.E., M.C. Hunt, P.B. Addis, H.J. Schoenbeck, M. Pullen and E. Katsanidis. 2003. Effects of postexsanguination vascular infusion of cattle with a solution of saccharides, sodium chloride, and phosphates or with

412 *Fermented Meat Products: Health Aspects*

calcium chloride on quality and sensory traits of steaks and ground beef. J. Anim. Sci. 81: 156–166.

Diles, J., M.F. Miller and B.L. Owen. 1994. Calcium-chloride concentration, injection time, and aging period effects on tenderness, sensory, and retail color attributes of loin steaks from mature cows. J. Anim. Sci. 72: 2017–2021.

Emmons, K.M., A.M. Stoddard, R. Fletcher, C. Gutheil, E.G. Suarez, R. Lobb, J. Weeks and J.A. Bigby. 2005. Cancer prevention among working class, multiethnic adults: Results of the healthy directions – Health centers study. Am. J. Publ. Health 95: 1200–1205.

Fernandes, C.F. and K.M. Shahani. 1990. Anticarcinogenic and immunological properties of dietary lactobacilli. J. Food Protect. 53: 704–710.

Fernández-López, J., J.M. Fernandez-Gines, L. Aleson-Carbonell, E. Sendra, E. Sayas-Barberá and J.A. Pérez-Alvarez. 2004. Application of functional citrus by-products to meat products. Trends Food Sci. Tech. 15: 176–185.

Fung, T.T., M. Schulze, J.E. Manson, W.C. Willett and F.B. Hu. 2004. Dietary patterns, meat intake, and the risk of type 2 diabetes in women. Arch. Intern. Med. 164: 2235–2240.

Gardini, F., M. Matruscelli, M.A. Crudele, A. Paparella and G. Suzzi. 2002. Use of Staphylococcus xylosus as a starter culture in dried sausages: effect on biogenic amine content. Meat Sci. 61: 275–283.

Giese, J. 1996. Fats, oils and fat replacers. Food Technol. 50: 78–83.

Gimeno, O., I. Astiasarán and J. Bello. 2001. Calcium ascorbate as a potential partial substitute for NaCl in dry fermented sausages: Effect on colour, texture and hygienic quality at different concentrations. Meat Sci. 57: 23–29.

Givens, D.I. 2005. The role of animal nutrition improving the nutritive value of animal-derived foods in relation to chronic disease. Proc. Nutr. Soc. 64: 395–402.

Goldin, B.R. 1990. Intestinal microflora: Metabolism of drugs and carcinogens. Ann. Med. 22: 43–48.

González-Fernández, C., E. Santos, I. Jaime and J. Rovira. 2003. Influence of starter cultures and sugar concentrations of biogenic amine contents in chorizo dry sausage. Food Microbiol. 20: 275–284.

Grill, J.-P., J. Crociani and J. Ballongue. 1995. Effect of bifidobacteria on nitrites and nitrosamines. Lett. Appl. Microbiol. 20: 328–330.

Haak, L., K. Raes and S. De Smet. 2009. Effect of plant phenolics, tocopherol and ascorbic acid on oxidative stability of pork patties. J. Sci. Food Agr. 89: 1360–1365.

Hailu, A., S.F. Knutsen and G.E. Fraser. 2006. Associations between meat consumption and the prevalence of degenerative arthritis and soft tissue disorders in the Adventist health study, California USA. J. Nutr. Health Aging 10: 7–14.

Harris, W.S. 2007. International recommendations for consumption of long-chain omega-3 fatty acids. J. Cardiovasc. Med. 8(Suppl 1): 50–52.

Hoek, A.C., M.A.J.S. van Boekel, J. Voordouw, and P.A. Luning. 2011. Identification of new food alternatives: how do consumers categorize meat and meat substitutes? Food Qual. Prefer. 22: 371–383.

Holzapfel, W.H. 2002. Appropriate starter culture technologies for small-scale fermentation in developing countries. Int. J. Food Microbiol. 75: 197–212.

Fermented Meats Composition—Health and Nutrition Aspects 413

Honikel, K.O. 2008. The use and control of nitrate and nitrite for the processing of meat products. Meat Sci. 78: 68–76.

Hugas, M. and J.M. Monfort. 1997. Bacterial starter cultures for meat fermentation. Food Chem. 59: 547–554.

Jiménez-Colmenero, F., J. Carballo and S. Cofrades. 2001. Healthier meat and meat products: their role as functional foods. Meat Sci. 59: 5–13.

Jiménez-Colmenero, F. 2000. Relevant factors in strategies for fat reduction in meat products. Trends Food Sci. Tech. 11: 56–66.

Juniper, D.T., R.H. Phipps, E. Ramos-Morales and G. Bertin. 2008. Effect of dietary supplementation with selenium-enriched yeast or sodium selenite on selenium tissue distribution and meat quality in beef cattle. J. Anim. Sci. 86: 3100–3109.

Keeton, J.T. 1994. Low-fat meat products-technological problems with processing. Meat Sci. 36: 261–276.

Kiani, F., S. Knutsen, P. Singh, G. Ursin and G. Fraser. 2006. Dietary risk factors for ovarian cancer: The Adventist Health Study. United States. Cancer Cause Contol. 17: 137–146.

Klaver, F.A.M. and R. Van der Meer. 1993. The assumed assimilation of cholesterol by lactobacilli and *Bifidobacterium bifidum* is due to their bile salt-deconjugating activity. Appl. Environ. Microbiol. 59: 1120–1124.

Komprda, T., D. Smělá, P. Pechová, L. Kalhotka, J. Štencl and B. Klejdus. 2004. Effect of starter culture, spice mix and storage time and temperature on biogenic amine content of dry fermented sausages. Meat Sci. 67: 607–616.

Kröckel, L. 2006. Use of probiotic bacteria in meat products. Fleischwirtschaft 86: 109–113.

Law, M.R., C.D. Frost and N.J. Wald. 1991. III – Analysis of data from trials of salt reduction. Brit. Med. J. 302: 819–824.

Leistner, L., F. Herzog and F. Wirth. 1971. Untersuchungenu¨ber die Wasseraktivität (aw-Wert) von Rohwurst. Fleischwirthschaft 51: 213–216.

Leroy, F., P. Scholliers and V. Amilien. 2015. Elements of innovation and tradition in meat fermentation: Conflicts and synergies. Int. J. Food Microbiol. 212: 2–8.

Leroy, F., J. Verluyten and L. De Vuyst. 2006. Functional meat starter cultures for improved sausage fermentation. Int. J. Food Microbiol. 106: 270–285.

Liong, M.T. and N.P. Shah. 2006. Effects of *Lactobacillus casei* symbiotic on serum lipoprotein, intestinal microflora, and organic acids in rats. J. Dairy Sci. 89: 1390–1399.

Lücke, F.-K. 2000. Utilization of microbes to process and preserve meat. Meat Sci. 56: 105–115.

Mandigo, R.W. 1991. Meat processing: Modification of processed meat. pp. 119–132. *In:* C. Huberstroh and C.E. Morris (eds.). Fat and cholesterol reduced foods. Technologies and strategies. PPC Portfolio Publishing Company, Houston.

Marshall, V.M. and E. Taylor. 1995. Ability of neonatal human *Lactobacillus* isolates to remove cholesterol from liquid media. Int. J. Food Sci. Technol. 30: 571–577.

Marteau, P., J.P. Vaerman, J.P. Bord, D. Brassart, P. Pochart and J.F. Desjeux. 1997. Effects of intrajejunal perfusion and chronic ingestion of *Lactobacillus johnsonii* strain LA1 on serum concentrations and jejunal secretions of

414   *Fermented Meat Products: Health Aspects*

immunoglobulins and serum proteins in healthy humans. Gastroen. Clin. Biol. 21: 293–298.

McClements, D.J. and E.A. Decker. 2008. Lipids. pp. 155–216. *In:* S. Damodaran, K.L. Parkin, and O.R. Fennema (eds.), Food chemistry. Boca Raton, FL: CRC Press.

Miler, K.M.B. and Z.E. Sikorski. 1990. Smoking. pp. 163–180. *In:* Z.E. Sikorski (ed.), Seafood: resources nutritional composition and preservation. Boca Raton: CRC Press.

Modler, H.W., R.C. McKellar and M. Yaguchi. 1990. Bifidobacteria and bifidogenic factors. Can. I. Food Sc. Tech. J. 23: 29–41.

Moon, S.S., Y.T. Kim, S. Jin and I. Kim. 2008. Effects of sodium chloride, potassium chloride, potassium lactate and calcium ascorbate on the physico-chemical properties and sensory characteristics of sodium-reduced pork patties. Korean J. Food Sci. An. 28: 567–573.

Morel, P.C., J.A. Janz, M. Zou, R.W. Purchas, W.H. Hendriks and B.H. Wilkinson. 2008. The influence of diets supplemented with conjugated linoleic acid, selenium, and vitamin E, with or without animal protein, on the composition of pork from female pigs. J. Anim. Sci. 86: 1145–1155.

Muguerza, E., O. Gimeno, D. Ansorena and I. Astiasarán. 2004. New formulations for healthier dry fermented sausages: a review. Trends Food Sci. Tech. 15: 452–457.

Muguerza, E., O. Gimeno, D. Ansorena, J. Bloukas and I. Astiasaran. 2001. Effect of replacing pork backfat with pre-emulsified olive oil on lipid fraction and sensory quality of Chorizo de Pamplona— a traditional Spanish fermented sausage. Meat Sci. 59: 251–258.

Narvaez-Rivas, M., I.M. Vicario, E.G. Constante and M. Leon-Camacho. 2008. Changes in the fatty acid and triacylglycerol profiles in the subcutaneous fat of Iberian ham during the dry-curing process. J. Agr. Food Chem. 56: 7131–7137.

Nishida, Ch., R. Uauy, S. Kumanyika and P. Shetty. 2004. The Joint WHO/FAO Expert Consultation on diet, nutrition and the prevention of chronic diseases: process, product and policy implications. Public Health Nutr. 7: 245–250.

Norat, T., S. Bingham, P. Ferrari and et al. 2005. Meat, fish, and colorectal cancer risk: the European prospective investigation into cancer and nutrition. J. Natl. Cancer Inst. 97: 906–916.

Norat, T. and E. Riboli. 2001. Meat consumption and colorectal cancer: a review of epidemiologic evidence. Nutr. Rev. 59: 37–47.

Ordóñez, J.A., E.M. Hierro, J.M. Bruna and L. Hoz. 1999. Changes in components of dry-fermented sausages during ripening. Crit. Rev. Food Sci. Nutr. 39: 329–367.

Paik, D.C., T.D. Wendel and H.P. Freeman. 2005. Cured meat consumption and hypertension: An analysis from NHANES III (1988–94). Nutr. Res. 25: 1049–1060.

Papadima, S.N. and J.G. Bloukas. 1999. Effects of fat level and storage conditions on quality characteristics of traditional Greek sausages. Meat Sci. 51: 103–113.

Petäjä, E., E. Kukkonen and E. Puolanne. 1985. Einfluss des Salzgehaltes auf die Reifung von Rohwurst. Fleischwirtschaft, 65: 189–193.

Raes, K., S. De Smet and D. Demeyer. 2004. Effect of dietary fatty acids on

Fermented Meats Composition—Health and Nutrition Aspects 415

incorporation of long chain polyunsaturated fatty acids and conjugated linoleic acid in lamb, beef and pork meat: A review. Anim. Feed Sci. Technol. 113: 199–221.

Realini C.E., P. Duran-Montgé, R. Lizardo, M. Gispert., M.A. Oliver and E. Esteve-Garcia. 2010. Effect of source of dietary fat on pig performance, carcass characteristics and carcass fat content, distribution and fatty acid composition. Meat Sci. 85: 606–612.

Regulation (EC) No 1333/2008 of the European Parliament and of the Council of 16 December 2008 on food additives, OJ L 354, p. 16.

Reuterswärd, A.L., H. Andersson and N.G. Asp. 1985. Digestibility of collagenous fermented sausage in man. Meat Sci. 14: 105–121.

Ruusunen, M. and E. Puolanne, E. 2005. Reducing sodium intake from meat products. Meat Sci. 70: 531–541.

Salazar, P., M.L. Garcia and M.D. Selgas. 2009. Short-chain fructo oligosaccharides as potential functional ingredient in dry fermented sausages with different fat levels. Int. J. Food Sci. Technol. 44: 1100–1107.

Sayago-Ayerdi, S.G., A. Brenes and I. Goni. 2009. Effect of grape antioxidant dietary fiber on the lipid oxidation of raw and cooked chicken hamburgers. LWT-Food Sci. Technol. 42: 971–976.

Schaefer, D.M., Q.P. Liu, C. Faustman and M.C. Yin. 1995. Supranutritional administration of vitamin-E and vitamin-C improves oxidative stability of beef. J. Nutr. 125: 1792–1798.

Searby, L. 2006. Pass the salt. Int. Food Ingredients (February/March): 6–8.

Sesink, A.L.A., D.S.M.L. Termont, J.H. Kleibeuker and R. Van der Meer. 1999. Red meat and colon cancer: The cytotoxic and hyperproliferative effects of dietary heme. Cancer Res. 59: 5704–5709.

Selgas, M.D., P. Salazar and M.L. Garcia. 2009. Usefulness of calcium lactate, citrate and gluconate for calcium enrichment of dry fermented sausages. Meat Sci. 82: 478–480.

Sofos, J.N. 1989. Phosphates in meat products. pp. 207–252. *In:* S. Thorne (ed.). Developments in food preservation – 5. New York: Elsevier Applied Science.

Stahnke, L.H. 1995. Dried sausages fermented with *Staphylococcus xylosus* at different temperatures and with different ingredient levels – Part I. Chemical and bacteriological data. Meat Sci. 41: 179–191.

Stołyhwo, A. and Z.E. Sikorski. 2005. Polycyclic aromatic hydrocarbons in smoked fish – a critical review. Food Chem. 91: 303–311.

Sunesen, L.O. and L.H. Stahnke. 2003. Mould starter cultures for dry sausages – selection, application and effects. Meat Sci. 65: 935–948.

Šimko, P. 2005. Factors affecting elimination of polycyclic aromatic hydrocarbons from smoked meat foods and liquid smoke flavorings. Mol. Nutr. Food Res. 49: 637–647.

Šimko, P. 2002. Determination of polycyclic aromatic hydrocarbons in smoked meat products and smoke flavouring food additives. J. Chromatogr. B 770: 3–18.

Torrissen, O.J. 2000. Chapter 11: Dietary delivery of carotenoids. pp. 289–313. *In:* E.A. Decker, C. Faustman and C.J. Lopez-Bote (eds.). Antioxidants in muscle foods. Wiley-Interscience.

Truswell, A.S. 1994. The evolution of diets for western diseases. *In:* B. Harris-

While and R. Hoffenderg (eds.). Multidisciplinary perspectives. Oxford: Basil Blackwell.

Valencia, I., D. Ansorena and I. Astiasarán. 2006. Nutritional and sensory properties of dry fermented sausages enriched with n-3 PUFAs. Meat Sci. 72: 727–733.

Warnants, N., M.J. Van Oeckel and C.V. Boucque. 1998. Effect of incorporation of dietary polyunsaturated fatty acids in pork backfat on the quality of salami. Meat Sci. 49: 435–445.

WHO. 2003. Diet, Nutrition and the Prevention of Chronic Diseases. Report of a joint WHO/FAO expertconsultation, WHO Technical Report Series 919; 148 pp. http://www.who.int/dietphysicalactivity/publications/trs916/summary/en/print.html.

Whitney, E.N. and S.R. Rolfes. 2002. Understanding nutrition, Ninth ed. Belmont, CA: Wadsworth.

Wirth, F. 1988. Technologies for making fat-reduce meat products. What possibilities are there? Fleischwirtschaft 68: 1153–1156.

Yilmaz, I. and U. Gecgel. 2009. Effect of inulin addition on physico-chemical and sensory characteristics of meatballs. J. Food Sci. Tech. Mys. 46: 473–476.

Chapter 18

Chemical Hazards in Fermented Meats

Jelka Pleadin* and Tanja Bogdanović

1 INTRODUCTION

During the production, storage and distribution, food of animal origin may be contaminated by a number of toxic compounds of chemical origin, thus representing a growing concern of public authorities, food chain stakeholders and consumers. Meat and other raw materials and spices that are used in the production of fermented meat products from "farm to table" may be contaminated by various chemical toxicants originating from the environment; the contamination in question can also arise due to the treatment of animals during meat production or final processing. Chemical toxicants entering into the food chain may exhibit biological activity in humans, causing adverse effects on their health.

Different studies have linked an increased risk of cancer attributable to meat and meat products' consumption to the presence of carcinogenic chemical substances that are tagged as a health concern and have been the subject of extensive studies and international regulations. Hazardous chemical compounds present in fermented meat products include those still used on farms during meat production in order to reduce production losses (antibacterial substances and other veterinary

---

*For Correspondence: Croatian Veterinary Institute, Laboratory for Analytical Chemistry, Savska Cesta 143, 10000 Zagreb, Tel: +385 1 6123 626, Email: pleadin@veinst.hr

drugs) and those used or misused so as to increase animal production yields (substances having an anabolic effect). Furthermore, there exist environmental contaminants present in all production stages arising as a consequence of inappropriate conditions under which fermented meats are processed either in industrial or household settings. It is well known that most of the fee.d components are digested and absorbed by farm animals during their passage through the intestinal tract. Cases of carry-over effects and the persistence of residues of toxic chemical substances in meat products have been evidenced. Therefore, research of their distribution, accumulation and persistence in farm animal tissues used as raw materials and their possible presence in a variety of foods of animal origin, including fermented meats, is very important.

Furthermore, in order to prevent the possible presence of chemical toxicants in meat products, systematic implementation of all legal provisions applicable to certain types or groups of substances through the appropriate enforcement of Hazard Analysis and Critical Control Points (HACCP) system at all critical points of the production and storage is very important. It is necessary to implement continuous sampling of various biological materials (fluids and tissues) of animals bred for meat production to be carried out while the animals are still fattened on farms, as well as in slaughterhouses; final products intended for marketing are to be controlled, too. Systematic monitoring of toxic chemical substances should be carried out in accordance with the adopted annual monitoring state plans, not only state-wide, but also by each and every individual involved in the handling of food; such a monitoring also implies the application of validated screening and confirmatory analytical methods exercised by the control laboratories. Generally speaking, foodstuffs should be analysed for the presence of growth promoters, antibiotics, environmental contaminants and toxic substances that may be found during the processing of fermented meats, either to ensure their absence or to determine whether their levels are in accordance with the maximum permitted ones stipulated by the law.

This chapter gives an overview of the main representatives of toxic compounds of chemical origin that can be present in fermented meats and enter these meats at any given point from animal farms to the final stages of the industrial or household production process, representing thereby chemical hazards and, consequently, a threat to the health of consumers. The substances discussed under this Chapter shall be the following: (i) substances having an anabolic effect; (ii) other veterinary drugs, such as antibiotics and carcass disinfectants; (iii) environmental contaminants, such as heavy metals, dioxins, organophosphorous and organochlorine compounds including polychlorinated biphenyls (PCBs), pesticides and mycotoxins; and (iv) substances that may be generated

Chemical Hazards in Fermented Meats

during the processing of fermented meats, such as nitrites and nitrates, N-nitrosamines and polycyclic aromatic hydrocarbons (PAHs).

2 VETERINARY DRUGS

Using different routes, veterinary drugs may enter the food chain and therefore be present in meat and meat products (including fermented ones), thus posing a risk for food safety and human health. Many compounds are used to treat animals against diseases; these fall into the category of veterinary drugs and are represented, for instance, by antibiotics and anthelmintics. Other substances are used to improve animal breeding; those are growth-promoting agents usually called the growth hormones. Veterinary pharmaceutical drugs have been used in animals as therapeutic agents so as to control infectious diseases, as well as prophylactic agents administered so as to prevent disease outbreaks and control parasitic infections. Growth-promoting agents, like steroid hormones, stilbenes and b-agonists, are intended to improve feed efficiency by increasing the lean/fat ratio. Classification of veterinary drugs and substances having an anabolic effect, laid down under the Council Directive 96/23/EC (Anonymous 1996), is shown in Table 1.

Table 1 List of veterinary drugs and substances having an anabolic effect, provided for and classified under the Council Directive 96/23/EC

Group of substances	*Main representatives*
Group A: *Substances having anabolic effect and unauthorized substances*	
A1 Stilbenes, stilbene derivatives, and their salts and esters	Diethylstilbestrol, dienestrol, hexestrol
A2 Antithyroid agents	Thiouracil, methylthiouracil, propylthiouracil, tapazole
A3 Steroids	17β-Estradiol, progesterone, testosterone, trenbolone, 19-nortestosterone, boldenone, methyltestosterone, stanozolol
A4 Resorcyclic acid lactones including zeranol	Zeranol, zearalanone
A5 Beta-agonists	Clenbuterol, brombuterol, mabuterol, cimaterol, isoxsuprine, ractopamine, salbutamol, zilpaterol
A6 Other substances	Nitroimidazoles, chloramphenicol, nitrofurans, dapsone, chlorpromazine
Group B: *Veterinary drugs and contaminants*	
B1 Antibacterial substances including sulphonomides, quinolones	Sulfonamides, tetracyclines, quinolones, β-lactam, macrolides (tylosin), aminoglycosides, carbadox, olaquindox

B2	Other veterinary drugs	
a)	Anthelmintics	Benzimidazoles, probenzimidazoles, piperazines, imidazothiazoles, avermectins, tetrahydropyrimidines, anilides
b)	Anticoccidials	Nitroimidazoles, carbanilides, 4-hydroxyquinolones, pyridinols, ionophores
c)	Carbamates and pyrethroids	Esters of carbamyc acid, type 1 and 2 pyrethroids
d)	Sedatives	Acepromazine, propiopromazine, haloperidol
e)	Nonsteroideal anti-inflammatory drugs	Phenylbutazone, oxyphenbutazone, ibuprofen, naproxen, mefenamic acid, diclofenac
f)	Other pharmacologically active substances	Dexamethasone
B3	Other substances and environmental contaminants	
a)	Organochloride compounds including PCBs	PCBs, compounds derived from aromatic ciclodiene, or terpenic hydrocarbons
b)	Organophosphorus compounds	Malathion, phorate
c)	Chemical elements	Heavy metals
d)	Mycotoxins	Aflatoxin B_1, ochratoxin A
e)	Dyes	
f)	Others	

2.1 Substances Having an Anabolic Effect

Substances having an anabolic effect represent organic chemical substances that stimulate tissue growth by virtue of their impact on metabolic processes involved in protein synthesis, the latter stimulation being particularly pronounced when it comes to skeleton muscle cells (Lone 1997). In animals, their action mechanisms can be either direct or indirect, ultimately resulting in an enhanced nitrogen retention and protein synthesis, that is to say, in an enhanced growth. The above is the exact reason why livestock production started to make use of anabolic substances as early as in the 1950-ties. The efficiency of animal growth improvement depends on animal species/breed, animal age, reproductive status and hormone administration route; the growth improvement attained by virtue of anabolic hormones' use is estimated at over 20% (Meyer 2001).

The application of these substances has been a point of international disputes about the safety of meat originating from animals treated with such anabolics. As opposed to the total ban of all hormone-based growth promoters ("hormones") in livestock production imposed by the EU, the USA allows for the use of five such hormones (17β-estradiol, testosterone, progesterone, trenbolone and zeranol) in form of small solid ear implants and for the use of two hormones (melengestrol acetate and ractopamine) as feed additives intended for feedlot heifers and swine, respectively, leading to sharp differences in regulatory mechanisms enforced by the two. In the EU, anabolic treatment of slaughter animals is considered to be an offence, therefore being the subject of inspection programs. In the USA, inspection programmes aim at testing for the compliance of the level of anabolic hormones found in edible animal tissues (muscles, fat, liver or kidney) with the maximum residue level stipulated by the law, while the EU inspection programs are more focused on sample materials such as urine, faeces or hair, which are more suitable for banned substances testing, especially if the animals are still on the farm. Should the biological materials be retrieved from slaughtered animals, the more preferred sample tissues are bile, blood, eyes, liver and, in rare occasions, muscles (Stephany 2010). The latter tissues are sampled only within the frame of import controls or monitoring programs dealing with meat sampled in butcher shops or supermarkets. As a result, data on hormone concentrations in muscle meat circulating on the EU market are very rare and mostly obtained within the frames of small-scale monitoring programs on an *ad hoc* basis. The EU data on natural hormones in meat are even rarer because of the absence of "legal natural levels" of these hormones to be compared against during the compliance testing. With the exception of samples retrieved from the application sites – in the EU the site of injection of liquid hormone preparations or the site of application of "pour on" preparations – hormone concentrations observed in meat samples of illegally treated animals are typically in the range of a few micrograms per kilogram down to a few tenths of a microgram per kilogram. In the EU, dozens of illegal hormones are used to the above purpose, with the number of such active compounds still rising. On top of estrogenic, androgenic and progestagenic compounds, thyreostatic, corticosteroid and beta-adrenergic compounds are also used, either alone or in "smart" combinations (Stephany 2010). However, the presence of these substances or their residua is also to be expected in meat products in case of farm animal's treatment with anabolic doses of these substances, since the latter cumulate and persist in edible animal tissue subsequently used in meat products' production (Pleadin et al. 2010). Of note, thermal meat processing is incapable of either inactivating or removing the substances in question (Rose et al. 1995). Should such products be consumed by humans, they can cause alimentary intoxications and bear serious health consequences.

2.1.1 Stilbenes

Stilbenes and their derivatives (salts and esters) are non-steroidal synthetic estrogenic compounds with anabolic properties. The most representative stilbene is diethylstilbestrol (DES), an endocrine disruptor with carcinogenic properties and one of the first growth promoters used in veal production, administered to livestock either orally or via injection, due to its effects in terms of body mass increase and fat tissue reduction. However, later studies have shown that DES exhibits a strong mutagenic, teratogenic and carcinogenic effect, so that the use of this anabolic substance may have a serious adverse impact on human health. On top of DES, stilbenes most commonly used worldwide were hexestrol and dienestrol (Payne et al. 1999). Nevertheless, due to their evidenced toxicity manifested in mutagenic, teratogenic and carcinogenic effects (Robboy et al. 1982), the administration of stilbenes has been banned since 1981 (Anonymous 1981), though literature data have indicated that DES has continued to be misused as an anabolic in meat production (Lone 1997).

2.1.2 Thyreostats

Thyreostats or anti-thyroid agents are a complex group of substances, which inhibit the thyroid function and, as a consequence, reduce the circulation of thyroid hormones. Weight gain obtained by thyreostatic treatment should mainly be attributed to the increased filling of the animal gastrointestinal tract and the increased water retention (Stolker and Brinkman 2005). Oral thyrostatics are active compounds, which may be used as growth promoters in end-stage cattle breeding, approximately 4 weeks prior to slaughtering. These compounds are potentially harmful to humans (carcinogenic and teratogenic) and have therefore, together with the stilbenes, been banned in the EU. Thiouracil (2-thiouracil), as a particularly strong drug of its kind, was one of the most frequently abused thyrostatic agents at that time.

2.1.3 Steroids

Anabolic steroids can be distinguished based on their chemical structure and origin (estrogens, gestagens, androgens and corticosteroids). These compounds stimulate growth, leading to improved feed metabolism and protein deposition increase. This group of substances includes both natural compounds and their synthetic derivatives. Among natural hormones, 17β-estradiol is known for its indirect and direct animal growth enhancement estimated at roughly 5–15% and has therefore been used in livestock industry (Meyer 2001). The impact of this hormone on animal growth depends on animal species, animal age, gender and the applied hormone dose, and exhibits in direct stimulation of muscles through oestrogen receptors. It used to be administered in combination with other androgenic and gestagen compounds. If given to animals in the approved therapeutic dose, its residua in meat are low-levelled and not dangerous for consumers. The most important natural androgen is testosterone, most commonly given to farm animals in combination with 17β-estradiol. In livestock production settings, it used to be applied in form of ear implants. Long-term progesterone use results in toxic effects in terms of ovarian, breast, vaginal and uterine tumours. Proper therapeutic use of natural hormones does not bear adverse human health consequences (FAO/WHO 2000). Of all synthetic steroids, trenbolone and 19-nortestosteron have deserved most research attention. Trenbolone is a steroid having androgenic and anabolic properties. It indirectly affects muscle growth by virtue of changing concentrations and metabolic paths of certain hormones, and is usually applied in form of ester derivatives, such as trenbolone acetate (Meyer 2001). 19-nortestosteron is a steroid having more potent anabolic and androgenic effects as compared to testosterone; in humans, it stimulates muscle growth and enhances appetite, increases the production of erythrocytes and bone density, and may exhibit hepatotoxic effects, as well as cause gynecomastia, decreased libido, cardiovascular symptoms and the reduction of luteinising hormone (Noé et al. 1999).

2.1.4 Resorcylic Acid Lactones

Resorcylic acid lactones are represented by taleranol, zearalenol and zeranol, the latter being the best known representative of the group. Zeranol is produced by molds of the *Fusarium* genus and represents a derivative of the mycotoxins termed zearalenone. The effect of zeranol on humans and animals hasn't been fully investigated yet, but the research conducted insofar has indicated that, as compared to other estrogens, oral zeranol is less active (Lamming et al. 1987). Literature data have suggested variations in zeranol's affinity to bind to estrogenic receptors (Meyer 2001). Due to its estrogenic effects, zeranol used to be applied as an anabolic drug (Kennedy et al. 1998). It affects animal growth both directly, by virtue of binding to estrogenic receptors, and indirectly, by virtue of increasing the concentrations of growth hormone and insulin-like growth factor IGF-I in animal blood (Thomas et al. 2000). Toxic effects of zeranol seen in humans include developmental impairments, immune toxicity, genotoxicity, possible carcinogenic effects, and changes in blood concentrations of various hormones (Wilson et al. 2002).

2.1.5 Beta-Adrenergic Agonists

Beta-adrenergic agonists (β-agonists) are chemical substances used for over 40 years in human and veterinary medicine for the treatment of chronic bronchitis, chronic obstructive pulmonary disease and asthma, as well as animal tocolytics (Barnes 1999, Anderson et al. 2005). They represent the derivatives of catecholamines, i.e. the hormones epinephrine and norepinephrine. Beta-agonists promote growth in many animal species used for meat production, attaining that effect through binding to specific β-adrenergic receptors located on cell membranes of the target tissue (Mersmann 1998). Application of β-agonists in anabolic doses result in a significant increase in lean body mass and a significant decrease in the amount of body fat, as well as in better utilization of feed and increased animal growth (Meyer 2001, Courtheyn et al. 2002, Anderson et al. 2005), improving thereby the sensory properties of the final product in terms of fat reduction and muscle tissue share increase (Armstrong et al. 2004). Former misuse of these substances in the livestock industry had significant negative implications on human health. However, beta-agonists still represent the most important group of anabolic drugs and the major grounds for misuse, since certain substances belonging to this group are allowed to be used in some countries (the European Union excluded) up to the stipulated Maximum Residue Limits (MRLs). Beta-adrenergic agonists are represented by dozens of compounds, among which the long-acting substance clenbuterol has been the one most studied in the recent period. Data also indicate the presence of many short-acting substances of this class on the market, such as salbutamol, ractopamine, cimaterol, zilpaterol, terbutaline, mabuterol and others, estimated to be able to achieve an increase in meat protein share and to reduce meat fat content by about 40% (Courtheyn et al. 2002). However, illegal use of clenbuterol has been evidenced to lead to numerous cases of poisoning of consumers who had consumed the meat coming from treated animals, the symptoms of poisoning thereby being rapid heartbeat, tremor, nervousness, general weakness, dizziness and headache (Martinez-Navarro 1990, Kuiper et al. 1998, Woodward 2005). When it comes to the justification of the application of certain β-agonists (clenbuterol, ractopamine, zilpaterol) in animals used for meat production in order to achieve better utilization of feed and improve meat sensory characteristics, as well as when it comes to possible adverse consequences that may result from the use of these substances, scientists and legislative frames adopted in different countries substantially vary in their standing. As a consequence of these different views, the use of these substances is regulated by a range of different legislations spanning from the ban enforced by the European Union to the approval of several hormone-containing preparations used for growth-promoting purposes or dairy cattle milk production increase.

2.2 Therapeutic and Prophylactic Agents

2.2.1 Antibiotics

Antibiotics include naturally-occurring, semi-synthetic and synthetic chemical compounds with antimicrobial activity. Antibiotics used to treat and prevent farm animal diseases may be administered for the purposes of therapy (to animals exhibiting clinical signs of a disease), control (of a disease spread) or prevention (of diseases); depending on their chemical nature, the administration route can be oral, parenteral or topical. Their use to the end of growth promotion involves the administration of an antibiotic agent usually as a feed additive over a period of time and results in an improved physiological performance of livestock, particularly in the increase in average daily weight gain and increase in feed efficiency (Gaskins et al. 2002, Lefebvre et al. 2006). An incorrect use of antibiotics in Veterinary Medicine practice may leave behind their residues in edible tissues that may have direct toxic effects on consumers, e.g. allergic reactions in hypersensitive individuals, or may cause problems indirectly through the nascence of resistant bacterial strains. Antibiotics comprise the following sub-groups: aminoglycosides, tetracyclines, β-lactams, macrolides, peptides, sulphonamides (and trimethoprim), quinolones, and miscellaneous (chloramphenicol and malachite green) (Stolker and Brinkman 2005). Some of the most commonly used antibiotics are benzathine penicillin, procaine penicillin, ampicillin, amoxicillin, cloxacillin, cefuroxime, cephalonium dihydrate, cefuroxime, ceftiofur, erythromycin, oxytetracycline, sulfadiazine, sulphadimidine, sulphadoxine, dihydrostreptomycin, novobiocin, trimethoprim, fluorphenicol, neomycin, tylosin and flunixin (Khan et al. 2008). Codex MRLs in muscle and fat tissue and the Acceptable Daily Intakes (ADIs) of some of the most important antimicrobial drugs are shown in Table 2. Polyether ionophores (lasalocid, monensi, narasin and salinomycin) are used to improve feed metabolism efficiency in farm animals (Page 2003) and are toxic to many bacteria, protozoa, fungi and higher organisms (Khan et al. 2008). The lipolytic nature of polyether ionophores enables easy penetration through cell membranes and an uncontrolled influx and/or efflux of the selected ions such as potassium and sodium from the cell, leading to its death. Tylosin is a macrolide produced by *Streptomyces* spp. bacteria that is able to penetrate host cells and be therapeutically active, therefore being particularly valuable in the treatment of cell-associated pathogens such as mycoplasmas. Sulphonamides represent broad-spectrum antibiotics active against gram-positive and gram-negative bacteria, acting on specific targets in bacterial DNA synthesis. Aminoglycosides are a large and diverse class of antibiotics that characteristically contain two or more aminosugars linked by glycosidic bonds to an aminocyclitol component. Well-known aminoglycosides are gentamicin, lincomycin, neomycin and streptomycin (Stolker and Brinkman 2005). Beta-lactams are probably the class of antibiotics most widely used in Veterinary Medicine for the treatment of bacterial infections of livestock bred on farms and used for bovine milk production. The presence of antibiotics or their residues in food of animal origin that goes above the maximum permissible level has been recognized worldwide. When it comes to the presence of antibiotic residues in meat and meat products, the fact that raises special concern is the increasing incidence of resistance to antibiotics among a wide variety of pathogenic bacteria and potential deleterious effects that these compounds may have on natural soil bacterial populations. The meat processing industry also faces some problems when using antibiotic-treated meats as raw materials in the fermented meats' production, including a lower quality of the resulting products, problems in fermentation arising on the grounds of antibiotic residues' presence and a lower meat fat content with a subsequent loss of juiciness and a poorer flavour development (Toldrá and Reig 2006). Therefore, careful management of antibiotic residues is necessary in order to minimise the potential public health risks coming from resistant bacterial strains (Khan et al. 2008).

Table 2 Codex Maximum Residue Limits (MRLs) in muscle and fat tissue and the Acceptable Daily Intakes (ADI) of some important antimicrobial drugs

Antimicrobial substance	Acceptable daily intake (µg/kg)	Muscle and fat (mg/kg)
Ampicillin	–	0.050
Benzyl/procaine, benzyl penicillin	0–30	0.050
Cloxacillin	–	0.300
Erythromycin	–	0.400
Gentamicin	0–20	0.100
Sulfadimidine	0–50	0.100
Sulfonamides (combinations)	0–50	0.100
Streptomycin	0–50	0.600
Oxytetracycline	0–30	0.100
Tetracyclines (OTC+CTC+TC)	0–30	0.100
Trimethoprim	–	0.050

2.2.2 Other Veterinary Drugs

Drugs most commonly used to treat a wide range of parasitic infestations known as the ectoparasiticides include benzimidazoles such as thiabendazole, albendazole, febendazole, mebendazole, and oxfendazole, mostly used in cattle and sheep, while levamisole and ivermectine are used against gastrointestinal nematodes in cattle, pigs and sheep. Organophosphates and synthetic pyrethroids are widely used

Chemical Hazards in Fermented Meats 427

against external parasites such as fleas, lice, and ticks (Deshpande 2002). Anthelmintics are used for the treatment of livestock against internal parasites. Many antifungal drugs are also available, mostly for topical treatment (e.g. nystatin for the treatment of gastrointestinal infections).

3 ENVIRONMENTAL CONTAMINANTS

Environmental contaminants such as heavy metals and organic chemicals may enter the meat production chain through stock feed and water or through direct soil consumption. Stock feed and water are the integral factors of meat production, the quality of which has an influence on the quality of meat produced. Contamination through stock feed and water may also come from the presence of endogenous plant toxicants or mycotoxins. Biehl and Buck (1987) concluded that, of all possible sources of contamination, at least 80% of all residues in food of animal origin are estimated to come from feed. Data have revealed that different environmental contaminants can be present in meat and meat products as the consequence of contamination of raw materials used in production or the contamination of final products. This group of meat-contaminating substances is represented by dioxins, organophosphorous and organochlorine compounds including polychlorinated biphenyls (PCBs), pesticides, mycotoxins, heavy metals and so forth. Bottom-line, different environmental contaminants can be present in fermented meats due to animal consumption of contaminated feed and water (a carry-over effect) or may enter these meats directly through contaminated ingredients (spices) or under inadequate processing conditions.

3.1 Toxic Heavy Metals

Significant amounts of metals in feed can be a result of agricultural or industrial production or a consequence of accidental or deliberate misuse. These pollutants often have direct toxic effects on mammals because they are stored or incorporated in tissues, sometimes even permanently (Mariam et al. 2004). Heavy metals are considered particularly dangerous for human health because they do not decompose during food preparation and therefore are prone to bio-accumulation. Their representatives of the highest importance for meat and meat products' contamination are cadmium (Cd), lead (Pb), arsenic (As) and mercury (Hg), out of which the presence of lead and cadmium in meat products has been confirmed by numerous studies (Ferreira et al. 2006, González-Weller 2006, Hoha et al. 2014).

Cadmium is a contaminant primarily toxic to the kidney, especially to the proximal tubular cells, where it accumulates over time and

428 Fermented Meat Products: Health Aspects

may therefore cause renal dysfunction. The International Agency for Research on Cancer has classified Cd as a Group 1 human carcinogen. The percentages of Cd exceeding the MRLs determined in various meat and meat product were 3.6% in bovine, sheep and goat meat, 1.6% in pork meat, 3.7% in liver and 1.0% in kidneys of different farm animals (Andrée et al. 2010). Lead is present in most foods in low concentrations, meat products included, whereas offal and molluscs have been shown to contain higher lead levels (Kan and Meijer 2007). It accumulates in plants and animals, so that its concentration gets to be magnified throughout the food chain (Halliwell et al. 2000). Some chronic Pb effects involve colics, constipation, anaemia, increased blood pressure, etc. If residual levels exceed the maximal permissible ones due to the accumulation in the human body consequential to repetitive ingestion, adverse health effects shall be induced; hence, the duty of all farmers and meat processors is to minimize the possibilities and probabilities of contamination (Hoha et al. 2014). Studies have shown higher levels of Cd and Pb in kidney and liver samples as compared to muscle samples that did not contain Cd or Pb in levels higher than MRLs established by many countries or international organizations (Andrée et al. 2010). The main sources of pork products' contamination are mineral components of commercial feeds used for animal feeding and spices added to processed pork. Lukáčová et al. (2014) determined higher average concentrations of Pb in homogenized samples of Malokarpatska and Lovecka Salami in which additives and spices are added (4.32–11.02 µg/kg). Nkansah and Amoako (2010) determined high Pb concentrations in black and white pepper (0.965 and 0.978 mg/kg, respectively), while Ozkutlu et al. (2006) found high Pb levels in garlic samples (0.99 mg/kg). The study conducted on processed meat including bacon, ham, sausage and salami demonstrated the highest Pb and Cd concentrations in salami (0.96 mg/kg and 0.21 mg/kg) and sausages (0.82 mg/kg and 0.16 mg/kg), as compared to bacon and ham, in which the obtained Pb and Cd values ranged from 0.58 to 0.65 mg/kg and from 0.11 to 0.13 mg/kg, respectively (Hoha et al. 2014). Cadmium levels significantly higher than the maximum permissible ones (0.1 mg/kg for pork products) were explained by the fact that the salami was abundantly spiced. Arsenic and mercury were found to be present in meat products only in negligible concentrations, often below the limit of detection of the applied analytical methods, their presence being attributed to the carry-over from feeds to edible tissues of mammalian species (Kan and Meijer 2007).

3.2 Dioxins and Polychlorinated Biphenyls (PCBs)

Dioxins are lipophilic compounds which accumulate in animal fat. Foodstuffs having the highest dioxin concentrations are dairy products,

Chemical Hazards in Fermented Meats

meat and poultry, eggs, fish, and animal fats. It is estimated that more than 90% of human exposure to dioxins occurs through the food supply chain, mainly from the consumption of meat and dairy products, fish and shellfish. Schecter et al. (1997) measured dioxins in pooled food samples that were collected in 1995 at supermarkets across the USA and determined the following toxic equivalency factors (TEQs): 0.38 in beef, 0.32 in chicken and 0.32 in pork. The authors concluded that the dietary intake of dioxins depends on the relative intake of foodstuffs with either high or low levels of dioxin contamination and on the quantity of food consumed. Patandin et al. (1999) investigated into the dietary dioxin intake of a group of Dutch preschool children and concluded that meat/ meat products and processed foods contribute to the dioxin intake by about 20% and 25%, respectively. In the late 2008, Ireland withdrew many tons of pork meat and pork products from the market when dioxin concentrations of up to 200 times the safe limit were detected in pork samples. This led to one of the largest food recalls related to chemical contamination in history. The contamination was also traced back to contaminated feed (WHO 2014).

Since the ingestion of dioxins present in contaminated vegetation and soil is considered to be the major pathway of exposure of farmed animals, different feeding practices, such as confinement feeding, grazing, and percentage grain feeding should have significant effects on the actual concentrations of dioxins in animal tissues. Study results have indicated that dioxin concentrations considerably vary across farmed animals, as well as that certain geographical areas may generate animals with higher dioxin concentrations, also pointing out that more research is needed to identify the potential risk factors associated with varying dioxin concentrations in animals intended for meat production (Feil and Ellis 1998).

Polychlorinated biphenyls (PCBs) are a group of toxic and highly persistent organic compounds that consist of 209 congeners differing in the number and position of chlorine atoms bound to the two coupled biphenyl rings (Costabeber et al. 2006). Various reports on toxicological effects of PCBs, which include skin rashes (chloracne), eye irritation, liver damage, neuropathy, endocrine disruption, immune suppression and an increased risk of cancer have been released (WHO 1993). Although most countries have prohibited PCBs' production, a considerable amount of these compounds is still in use. Environmental persistence of PCBs seems to increase with the degree of chlorination (WHO 1993). Raw materials and other ingredients used in meat processing may be a source of PCBs, together with meat and meat products that have been shown to contain residual PBC levels (Glynn et al. 2000). As compared to other foodstuffs, the estimated contribution of meat and meat products to the cumulative toxic effects of PCBs amounts to 14%-19% (Patandin et al. 1999).

430   *Fermented Meat Products: Health Aspects*

Other chlorinated congeners most frequently found in meat and meat products are hereby listed in the descending order: heptachlorobiphenyl PCB 180 > hexachlorobiphenyl PCB 153 > hexachlorobiphenyl PCB 138 > tetrachlorobiphenyl PCB 52 = trichlorobiphenyl PCB 28 = dichlorobiphenyl PCB 10). Costabeber et al. (2006) concluded that the frequency of PCB 180 and PCB 153 presence in meat and meat products (41% and 16%, respectively) was low as compared to most environmental contaminants. Literature data have revealed that, owing to their high liposolubility, PCBs bind to lipid components of animal tissues and accumulate in livestock throughout the food chain. Because of their high thermodynamic stability, degradation mechanisms are rather ineffective, so that environmental and metabolic degradation generally goes very slowly (Erickson 1997). In meat products, the highest PCB levels were found in hot dog and Bologna Sausage, which contain mechanically deboned meat; when compared to the salami that contains no mechanically deboned meat, these levels were much higher (Costabeber et al. 2006). Lower levels of PCBs determined in salami suggest a possibility of PCB degradation using a starter culture employed with fermentation. In general, PCB levels obtained in meat products were higher than those found in meat, suggesting the influence of meat processing on PCB levels (Schecter et al. 1997).

3.3 Pesticides

Pesticides are defined as substances or mixtures intended for preventing, destroying, repelling or mitigating pests and are globally applied to the goal of food protection (CFS 2014). It is estimated that, without pesticides, world production of food would be reduced by 30% (Biswas et al. 2010). Animal exposure to pesticides may originate from direct pesticide treatments, inhalation of contaminated air and the ingestion of contaminated feed, all of these contamination routes leading to intoxication of animals and bio-accumulation of pesticides in products of animal origin, meat and meat products included. Among pesticides most investigated for their presence in meat and meat products, five groups (persistent organic pollutants – POPs, including organochlorinated pesticides – OCPs, organophosphorus pesticides – OPPs, synthetic pyrethroid pesticides – PYRs, carbamates and triazines) have been of particular research interest (LeDoux 2011). OCPs were banned for agricultural and domestic use in most countries pursuant to the Stockholm Convention adopted in 1980s. They induce health problems and have been considered as endocrine-disrupting and carcinogenic substances (Amaral-Mendes 2002, Kavlock 1996). OCPs include pesticides such as DDT, lindane or hexachlorobenzene, heptachlor, heptachlor epoxide and aldrin/dieldrin and represent the most persistent pesticides to be found

Chemical Hazards in Fermented Meats 431

in the environment and therefore also a major health hazard for both humans and animals. PYRs stand for an effective and broad spectrum of insecticides, although of lower mammalian toxicity and shorter-term environmental persistence, as compared to the above-detailed OCPs, due to their rapid degradation. OPPs are mainly used as insecticides and are less persistent than OCPs, therefore used as efficacious, safe and cost- effective agents against a wide range of pests. They can be absorbed by inhalation, ingestion and dermal absorption and can also concentrate along the food chain. Triazines are among the most widely used herbicides in agriculture and are very stable and persistent compounds. There are also some other pesticides such as carbamates, benzoylureas, amines, quinoxalines and fluorides. The MRLs have been set by the European Commission to protect consumers from exposure to unacceptable levels of pesticide residues in food (29 applicable to food of animal origin) and feed, in most of the cases calculated as the sum of the parent compound and its relevant metabolites (EFSA 2011).

As for fermented meat products, pesticides occurring in higher-fatty products (> 20% of fat), i.e. in the product category into which most of the fermented meat products fall, are fat-soluble, whereas in low-fatty products (2–20% of fat) lipophilic and hydrophilic pesticides may occur. Due to their high thermodynamic stability and liposolubility, OCPs bind to lipid components of animal tissues, becoming a major route of human exposure when these tissues are used as food. LeDoux (2011) pointed out that, since diet represents the main source of chronic exposure to low doses of these substances, humans are mainly exposed to these chemicals through the ingestion of meat and meat products. According to the Report on Pesticide Residues in Food issued by the European Union, the majority of food of animal origin was free of detectable pesticide residues (99.7% of samples were reported to contain these residues in quantities below the limit of quantification (LOQ)). In most cases, such a contamination was predominantly attributed to environmental contamination with persistent pesticides, the most frequently found pesticides thereby being DDT and hexachlorobenzene, which were detected in 16.5% and 10.9% of the samples analysed for these pesticides, respectively. Studies devoted to the determination of pesticide contaminants in meat and meat products have been focused on OCPs, mainly in fatty tissue and muscles, pig and cattle liver tissue and meat products (meat pate, sausage, canned meat) (Covaci et al. 2004, Nougadère et al. 2012). Concentrations of OCPs found in pork fat samples were the following: p,p′–DDD (1.9 µg/kg in pork fat sample), p,p′–DDE (1.4 µg/kg in pork fat sample), alpha–HCH (7.65 µg/kg in pork fat sample) and tecnazene (8.25 µg/kg in pork fat sample), whereas in beef fat only p,p′–DDE (2.15 µg/kg) and hexachlorobenzene (1.41 µg/kg) were detected. OCP concentrations determined in pork,

432   *Fermented Meat Products: Health Aspects*

beef and meat products were usually below the LOQ (Garrido-Frenich et al. 2007). Data have shown that the differences in pesticide levels before and after processing are appreciable. After 72 hours, meat fermentation reduced DDT and lindane residues (in fermented sausages) by 10% and 18% as compared to the baseline level of 5 and 2 mg/kg, respectively, the reduction in question thereby being explained by the meat starter culture activity (Abou-Arab 2002).

3.4 Mycotoxins

Mycotoxins are secondary toxic metabolites produced by molds, which often contaminate different food. Literature data have shown that, due to the natural mycotoxins contamination of feed farm animals are fed on and consequent contamination of raw materials used in the production of meat products, fermented meat products may also be contaminated by mycotoxins in significant levels. The contamination may also arise on the grounds of contaminated spices used in production, as well as the consequence of the activity of molds overgrowing the surface of dry-fermented sausages and dry-cured meat products. Ripening that makes use of the surface growth of fungi is common for many European Mediterranean-style products. Several studies have shown that molds of *Penicillium* and *Aspergillus* genera isolated from the surface of fermented sausages and hams produce these mycotoxins under certain meat production and storage conditions, such as temperature, water activity, cover damage and the presence or absence of skin (ham) or cracks favouring mold growth (Asefa et al. 2011, Rodríguez et al. 2012, Pleadin et al. 2015b, Pleadin et al. 2015c). Table 3 shows the main parameters that influence muld growth and favour the production of mycotoxins in fermented meats. Mycotoxins most commonly contaminating fermented meat products are aflatoxin B_1 (AFB$_1$) and ochratoxin A (OTA) (Pleadin et al. 2013, Perši et al. 2014, Pleadin et al. 2015a).

Table 3   Parameters that influence mold growth and the production of mycotoxins in fermented meats (Asefa et al. 2011)

Parameter	Mold growth	Production of mycotoxins
Temperature	−12–55°C	+4 do +44°C, optimum < 20°C
pH	1.7–10	> 2.5; optimum 5–7
a_w	min. 0.62	min. 0.8–0.85
Redox potential	aerobic conditions	aerobic conditions
Mass fraction of salts	to 20%	to 14%
The impact of spices	inhibition	Inhibition

AFB$_1$ represents the most potent carcinogen affecting the mammalian liver and is classified by the International Agency for Research on Cancer

(IARC) as evidenced Group 1 human carcinogen (IARC 2002). The presence of AFB_1 in animal diet may cause reduced production of food of animal origin, causing a number of toxic effects in different animal species (Richard 2007). AFB_1 was determined in meat and meat products after farm animals' exposure to contaminated diet that contained significant AFB_1 levels (Richard 2007, Herzallah 2009). As for OTA, given that research has indicated a widespread distribution of this mycotoxin and with respect to its nephrotoxic properties and other toxic effects on humans and animals, it is included into the IARC 2B group of possible human carcinogens (IARC 1993). Elevated levels of OTA were determined in fermented and other meat products produced from contaminated raw materials (Pleadin et al. 2013, Perši et al. 2014), significant OTA levels thereby being found in dry-fermented sausages and dry-cured meat products obtained from the market (Dall'Asta et al. 2010, Pleadin et al. 2015a).

Markov et al. (2013) observed significant OTA contamination in fermented meat products; at the same time, the contamination of these products with AFB_1 was found to be very low. Pleadin et al. (2014a) evidenced an extremely high AFB_1-contamination of maize and feed produced from such maize and used in animal breeding, which certainly could significantly contribute to the contamination of meat and meat products. Many studies have pointed towards the possibility of high AFB_1 and OTA contamination of food and feed in many countries of the world, coming as a result of inadequate control of production and storage conditions, indicating the necessity of prevention of contamination with and the continuous control of these extremely toxic mycotoxins.

Flaws witnessed during the production process that favour mycotoxin contamination of the final products are the following: (i) Non-standardised production quality and technology, often also in form of unregistered production and illegal marketing that is not subjected to proper veterinary health surveillance and therefore jeopardises the safety of fermented meat products; (ii) Challenges faced during the ripening period, coming as a consequence of the non-use of automated ripening chambers protected from environmental mold (spore) contamination (by virtue of biological micro-filters, pressure barriers, etc.), which is especially the issue in facilities located in rural areas due to the vicinity of grain (maize, wheat, etc.) fields.; (iii) The use of raw meat contaminated through animal feedstuffs and contaminated spices, as well as the contamination arising as a consequence of the activity of toxin-producing molds overgrowing the product surface; and (iv) The inability to remove mycotoxins using standard production and preservation technologies. Study results have also shown that technological operations used with the production of these types of products, such as thermal processing, curing, drying and ripening, as

434   *Fermented Meat Products: Health Aspects*

well as storage conditions, do not have a significant impact on reducing the levels of these potent toxins in the final meat product (Pleadin et al. 2014b). Casing damaging occurring during long-term ripening may result in the entry of mycotoxins from the dry-fermented sausage's surface into the product's interior. Studies have therefore pointed out the need for casing preservation in each production stage, together with the need for continuous removal of molds from the product surfaces, all to the end of minimizing the risk of mycotoxin contamination of the final product (Pleadin et al. 2015a).

4 CONTAMINANTS OF MEAT PROCESSING

Studies have shown that food of animal origin is mainly contaminated during processing (Wenzl et al. 2006). Smoking of meat products has been exercised for centuries in many countries, not only so as to achieve particular sensorial meat products' profiles in terms of taste, colour and aroma, but also to improve their preservation based on antimicrobial, antioxidant and drying effects of this process (Djinovic et al. 2008a, b). During smoking of fermented meats, conditions favourable for the formation of compounds harmful to human health such as polycyclic aromatic hydrocarbons (PAHs) are witnessed. PAHs can also enter food products through contaminated packaging material (Wenzl et al. 2006). Formation of biogenic amines in all foods of animal origin having high protein contents has been reported. Such a formation can occur as a result of the activities of spoilage microbiota and/or intentionally added microorganisms. Microbial starter cultures should be selected and controlled so as to avoid the production of biogenic amines during the meat fermentation process. The production of fermented meat products also implies the addition of different additives, such as nitrites and nitrates widely used in the meat products' manufacturing so as to obtain specific sensory properties, stability and safety. However, traditional nitrate and nitrite use in the meat industry is being questioned due to the involvement of these substances into the formation of carcinogenic nitrosamines that may be formed during salting and the thermal treatment of fermented meat products (Walker 1990, Toldrá and Reig 2007).

4.1 Polycyclic Aromatic Hydrocarbons (PAH)

Smoke contains flavouring substances that lead to a typical smoke flavour of the product, but also contain some health-hazardous compounds like polycyclic aromatic hydrocarbons (PAHs), phenols and formaldehyde. Many traditional dry-fermented meat products are still produced by virtue of smoking processing. Yet, the smoke produced from wood

*Chemical Hazards in Fermented Meats*435

combustion in a low-oxygen environment contains considerable amounts of toxic PAHs (FAO/WHO 1998). PAH formation during smoking depends on the type (higher in soft wood) and composition (e.g. moisture content) of the wood in use, as well as on the oxygen concentration in the combustion chamber, wood temperature during combustion (higher if the combustion takes place in the temperature range of 500–700°C) and the duration of meat smoking (Šimko 2005). Wood pyrolysis generates smoke through different oxidation routes. Moist wood chips can also be used for the direct generation of smoke. Smoke is condensed and adsorbed on the surface of a meat product. To a certain depth, it can penetrate into the product depending on the processing conditions (Toldrá and Reig 2007).

The PAH group is represented by roughly 660 different compounds, among which the list of those possibly present in primary products used in the generation of smoke flavourings reads as follows: benzo[a] pyrene, anthanthrene, benz[a]anthracene, benzo[b]fluoranthene, benzo[j] fluoranthene, benzo[k]fluoranthene, benzo[ghi]perylene, chrysene, cyclopenta[cd]pyrene, dibenz[a,h]anthracene, dibenzo[a,e]pyrene, dibenzo[a,h]pyrene, dibenzo[a,i]pyrene, dibenzo[a,l]pyrene, indeno[cd] pyrene, 5-methylchrysene, perylene, phenanthrene and pyrene. PAHs containing up to four fused benzene rings are known as the light PAHs, while those containing more than four benzene rings are called the heavy PAHs. Heavy PAHs are more stable and more toxic than the light ones (Wenzl et al. 2006). The meat industry is interested in monitoring the presence of PAHs in smoked meat and smoked meat products in order to establish good smoking practices and improve food safety, in particular that of the products originating from protected geographic areas. Smoked meat products are important due to their high nutritional value, extensive production and wide variety of products and the fact that they represent a significant part of the human diet in certain EU countries (Lorenzo et al. 2011, Gomes et al. 2013, Škrbić et al. 2014).

The 15 + 1 EU-priority PAHs include benz[a]anthracene, benzo[c] fluroene, benzo[b]fluoranthene, benzo[j]fluoranthene, benzo[k] fluoranthene, benzo[ghi]perylene, chrysene, cyclopenta[cd]pyrene, dibenz[a,h]anthracene, dibenzo[a,e]pyrene, dibenzo[a,h]pyrene, dibenzo[a,i]pyrene, dibenzo[a,l]pyrene, indeno[1,2,3-cd]pyrene and 5-methylchrysene (SCF 2002). Of the 15 EU priority PAHs, 12 are identical to those that were reasonably anticipated by the International Agency for Research on Cancer (IARC) to be human carcinogens. Benzo[a]pyrene (BaP) is still used as a marker in wide-scale scientific investigations of its carcinogenic and mutagenic properties (Djinovic et al. 2009). Recently, dibenzo[a,l]pyrene has been in the spotlight of scientific interest, as toxicological investigations have indicated that its carcinogenic potential is probably much stronger than that of benzo[a] pyrene. In the position statement recently published by the European Food Safety Authority (EFSA 2008), 'PAH4' (BaP, CHR, BaA and BbF) and 'PAHs8' (BaP, BaA, BbF, BkF, BgP, CHR, DhA and IcP) are tagged as more suitable indicators of PAH presence in food than BaP alone, the latter being proven unsuitable for the investigation into the occurrence and toxicity of PAHs in food. This resulted in significant changes in the EU legislation and the setting of new maximum permissible levels of total 'PAHs4' and BaP alone.

Since PAHs are adsorbed on the surface of meat and do not significantly penetrate into the interior of smoked meat products (Jira et al. 2006), the surface/mass ratio substantially influences PAH contents in smoked meat products (Andrée et al. 2010). The authors stated that the analysis of representative samples of German smoked meat products clearly demonstrated that the production of smoked meat products at benzo[a]pyrene levels below 1 µg/kg is possible without any problems, their findings being in accordance with the outcomes of other studies performed in European countries in the last decades. Although the latest maximum permissible BAP levels (of 2 µg/kg as compared to the previous 5 µg/kg) adopted for smoked meat products and the total PAH4 levels (set at 12 µg/kg as compared to the previous 30 µg/kg) have been reduced pursuant to the EU Regulation 835/2011, the Commission Regulation (EU) 1327/2014 (Anonymous 2014) provides for a three-year derogation from the application of these recently-set lower maximum permissible PAH levels for PAHs in local production and consumption of traditionally smoked meat and meat products and/ or fish and fishery products in half of the Member States. Despite the application of good smoking practices to a possible extent, lower PAH levels in these foodstuffs commodities have not been achieved yet, since they require smoking practices that significantly change the organoleptic food characteristics. The Member States shall establish programmes to implement good smoking practices wherever possible, but within economically feasible limits and to the extent that shall not significantly affect typical organoleptic characteristics of these products. The Codex Alimentarius Commission (CAC 2009) has published the 'Code of Practice for the Reduction of Contamination of Food with PAH from Smoking and Direct Drying Processes' with the objective of lowering PAH in smoked meat products and other foods.

Due to the lipophilic nature of PAHs, some diffusion into the inner layers of meat products takes place, with the migration rate strongly affected by water activity and fat content. The presence of barriers like smoked sausage casings and skin in case of bacon can influence PAH diffusion into the product's internal layers (García-Falcón and Simal-Gándara 2005, Djinovic et al. 2008b). Available studies on PAH profiles

Chemical Hazards in Fermented Meats 437

in smoked meat products have shown the prevalence of light PAHs over the heavy ones. In two types of Spanish traditional sausages, light PAHs phenanthrene (49.78 and 53.54 µg/kg, respectively), naphthalene (41.61 and 25.56 µg/kg, respectively) and anthracene (9.96 and 15.39 µg/ kg, respectively) were determined to be most abundantly represented (Lorenzo et al. 2011). Light PAH content of Portuguese fermented sausages consists of acenaphthylene, acenaphthene, fluorene and naphthalene with their sum corresponding to at least 98% of the total PAHs highlighted by the US Environmental Protection Agency (EPA) (Gomes et al. 2013). The prevalence of phenanthrene, anthracene, fluoranthene, fluorene and pyrene has been found in *Petrovská Klobása*, with phenanthrene being represented in the highest average concentration of 31 µg/kg in traditionally smoked samples as compared to industrially smoked ones, in which its content amounted to 3 µg/kg (Škrbić et al. 2014). All of the abovementioned PAH studies have demonstrated the prevalence of light PAHs over the heavy ones. BaP concentrations were below or around 1 µg/kg, i.e. in line with the EU Commission recommendations. In the study by Lorenzo et al. (2011), BAP was determined to be a good marker of 15 SCF-PAH priority compounds. The analysis of cold-smoked meat products of Serbia (traditional and industrial) revealed the dependency between the PAH contents and the duration of smoking; industrial smoked meat products exhibited lower PAH contents in comparison to conventionally smoked ones (Djinovic et al. 2008a, b).

Some alternative processes have been implemented in order to reduce the contamination of smoked meat products with hazardous compounds. Some of these strategies, which can significantly reduce the PAH content in smoked meat products, are the filtration of particles, the use of cooling traps, smoking at lower temperatures and/or shortening of the process duration. The extended use strategy is characterised by the incorporation of smoke flavourings produced from primary products obtained from different woods under specific pyrolysis conditions and extraction protocols into the meat products in the range of 0.1–1.0% (Toldrá and Reig 2007).

4.2 Biogenic Amines

Meat and meat products have repetitively been reported to contain biogenic amines (Maijala et al. 1995, Bover-Cid et al. 2006). Biogenic amines are organic bases of low molecular weight containing aliphatic dyamines (cadaverine, putrescine) and polyamines (agmatine, spermidine and spermine), aromatic (tyramine, phenylethylamine) and heterocyclic (histamine, tryptamine) structures mainly produced by microbial decarboxylation of precursor amino acids (Suzzi and Gardini 2003). Some biogenic amines introduced via diet can cause distinctive pharmacological and toxic effects. High- dose human exposure to biogenic amines results in vomiting, respiratory problems, perspiration, palpitation, hypo- or hypertension and migraine (Hernández-Jover et al. 1997, Stadnik and Dolatowski 2010). A set of harmful reactions produced by biogenic amines is classified as intolerance reactions, i.e. a form of hypersensitivity (Jansen et al. 2003). Literature data have shown that in fermented meat products biogenic amines can be found due to the use of poor-quality raw materials or contamination occurring during processing and storage, in particular under inappropriate processing and storage conditions. Microorganisms responsible for the fermentation process may also contribute to their accumulation (Bover-Cid et al. 2006, Latorre-Moratalla et al. 2010). Biogenic amines most prevalent in meat and meat products are tyramine, cadaverine, putrescine and histamine (Ruiz-Capillas and Jiménez-Colmenero 2004). In South European sausages, the tyramine level ranged from 176 to 187 mg/kg, while the putrescine content amounted to 125 mg/kg, while in North European sausages 76 mg of tyramine/kg and 33 mg of putrescine/kg were determined. In South European sausages having high tyramine content, the highest ranges of phenylethylamine (27–39 mg/kg) were observed as well. Naturally occurring spermine was found in the range from 12 to 24 mg/kg, while the obtained spermidine ranges found both in the South and North European sausages spanned from 3 to 6 mg/kg (Ansorena et al. 2002).

Fermented products having comparable microbiological profiles may differ in their biogenic amine content, indicating that the production of such compounds depends on complex interaction between various factors, out of which raw material hygiene appears to be the most important one. Concentrations of some biogenic amines (tyramine, putrescine, cadaverine) increase during the processing and storage of meat and meat products as opposed to spermidine and spermine whose concentrations remain constant or even slightly decrease. Several authors have suggested the possibility of using biogenic amines as the hygienic index of raw materials and/or manufacturing practices (Ruiz-Capillas and Jiménez-Colmenero 2004, Min et al. 2007, Latorre-Moratalla et al. 2008). The amine index should include all biogenic amines responsible for meat spoilage, so that several authors have proposed the use of biogenic amine index (BAI) calculated as the sum of concentrations of histamine, cadaverine, tyramine and putrescine (Wortberg and Woller 1982, Hernández-Jover et al. 1996). BAI index below 5 mg/kg points to good quality fresh meat; values between 5 and 20 mg/kg indicate acceptable meat quality, while indices between 20 and 50 mg/kg indicate low meat quality and that above 50 mg/kg meat spoilage. The BAI index seems to be a more reliable indicator of the quality of fresh meat and heat-treated meat products than of fermented meat products;

Chemical Hazards in Fermented Meats 439

namely, low biogenic amine concentrations do not always vouch for good microbiological quality of the product (Wortberg and Woller 1982, Ruiz-Capillas and Jiménez-Colmenero 2004). Relatively less attention has been focused on the effect of chemical substances added during the manufacturing of fermented meat products (Stadnik and Dolatowski 2010).

Amine contents and profiles may also vary depending on various extrinsic and intrinsic factors of relevance for the manufacturing process, such as pH, redox potential, temperature, sodium chloride content, the size of sausage, hygienic conditions and manufacturing practices and the effect of starter cultures (Latorre-Moratalla et al. 2008). Maijala et al. (1995) concluded that pH reduction reduces the growth of amine-positive microorganisms. The presence of sodium chloride was found to activate tyrosine decarboxylase and to inhibit histidine decarboxylase (Karovičová and Kohajdová 2005). Suzzi and Gardini (2003) reported that the accumulation of biogenic amines markedly decreases with the increase in sodium chloride concentration, with higher proteolytic activity seen with medium-high salt concentrations, pointing out that the correlation between the two is not necessarily in place. Karovičová and Kohajdová (2005) determined that the reduction in redox potential stimulates histamine production. A relationship was also found between biogenic amine content and the size of dry-fermented sausages. The diameter of the sausage affects the environment in which microorganisms grow; for example, in larger-diameter sausages the concentration of salt is usually lower, while the activity of water is usually higher. A larger diameter may be one of the reasons behind the higher production of certain amines, such as tyramine and putrescine (Stadnik and Dolatowski 2010).

Generally speaking, biogenic amine levels in larger diameter sausages were higher than those found in thinner sausages, the central part of the sausages thereby being proven to harbour higher levels of biogenic amines than the edge portions (Suzzi and Gardini 2003, Ruiz-Capillas and Jiménez-Colmenero 2004). The selection of starter cultures used in the production of fermented meat products is fundamental for the assurance of quality of final fermented meat products and allow for the presence and relative activity of amine oxidases that are able to catalyse oxidative deamination of amines, thus resulting in their physiological inactivation (Suzzi and Gardini 2003). However, the implementation of measures focused on the hygienic quality of both raw material and processing units is critical for the avoidance of the development of contaminating aminogenic bacteria and the reduction of biogenic amine content, but is often not fully capable of preventing biogenic amines' formation should other technological measures fail to be applied (Latorre-Moratalla et al. 2010).

4.3 Nitrites, Nitrates and Nitrosamines

Nitrates and nitrites are additives widely used in the manufacturing of meat products to the end of obtaining specific sensory properties, stability and hygienic safety by virtue of inhibiting the growth of spores of *Staphylococcus aureus* and *Clostridium botulinum*. However, their traditional use in the meat industry is being questioned due to their involvement in the formation of nitrosamines that have been shown to have carcinogenic properties; at the same time, meat products have been recognized as important sources of microorganisms frequently causing human food-borne diseases.

Nitrates are sodium or potassium nitrate acid salts, while nitrites represent sodium or potassium salts of the nitrite acid (HNO_2). Nitrates and nitrites are added into meat products for two main reasons: to preserve them and to extend their shelf-life, as well as to create and stabilise their light red colour (Kovačević 2001). Owing to the activity of reducing (denitrifying) bacteria, following their addition into the stuffing and their full dissociation into metal cations (K^+ ili Na^+) and nitrate anions (NO_3^-), nitrates get to be reduced into nitrites, that is to say, NO_3^- transforms into NO_2^- (nitrite anion) active in meats. Therefore, when it comes to the production of slow-fermented sausages, the meat industry resorts to nitrates (that are first transformed into nitrites), while the production of quick-fermented sausages leans on nitrites (that have an immediate effect). Of note, nitrates do not act directly and do not destroy bacteria, but merely pose as a nitrite source. Antimicrobial activity comes as the result of the formation of non-dissociated nitrate acid (HNO_3) that passes through the ionic barrier of the bacterial cell wall and exhibits intracellular toxicity. Bacteriostatic activity is underpinned by the influence on ▯-amine groups and the sulphur metabolism (Kovačević 2014).

Literature data have shown that nitrites (or nitrates) should be added into the majority of dry-cured meat products so as to prevent microbial growth and the production of the *Cl. botulinum* toxin; namely, despite all research efforts, alternatives to the use of sodium nitrite to the effect of production of safe dry-cured meat products are yet to be found. The amount of nitrites spanning from 5 to 20 mg/kg has been proven sufficient for the maintenance of the red meat colour; the amount of 50 mg/kg enables the attainment of the characteristic product taste, while the amount of 100 mg/kg has a sufficient antimicrobial effect (EFSA 2003). The amount of nitrites necessary to inhibit *Cl. botulinum* varies across meat products. It has been demonstrated that some meat products can be produced even without the addition of nitrites (nitrates) provided that good hygienic practices are implemented, that HACCP principles are observed "to a tee" and that the products are stored only for a short period of time at well-controlled temperatures; however,

Chemical Hazards in Fermented Meats

the production of nitrate and nitrite-free meat products calls for ideal production technologies.

If present in higher amounts, nitrites pose as health hazards since they cause erythrocyte and vitamin A decomposition and having toxic reproductive effects; in addition, in combination with amides, some 17% of nitrites not converting into nitrosyl myoglobin (NOMb) form nitrosoamides, while their combination with ammonia derivatives amines (whose quantity rises along with the rise in bacterial and enzymatic activity seen during fermentation and ripening) yield carcinogenic compounds termed the nitrosamines. The amount of nitrites added into a meat product poses as the key factor in N-nitrosamine nascence, together with the processing conditions, the amount of lean meat used in production and the presence of catalysts and inhibitors. The nascence of nitrosamines is favoured by longer ripening and longer sausage storage, as well as by lower pH values and, in case of thermally processed meat products, higher processing temperatures (Hotchkiss and Vecchio 1985, Walker 1990). Some of the most important nitrosamines are N-nitrosodimethylamine, N-nitrosopirrolidine, N-nitrosopiperidine, N-nitrosodiethylamine, N-nitrosodi-n-propylamine, N-nitroso-morpholine and N-nitrosoethylmethylamine (Toldrá and Reig 2007). Nitrosodimethylamine and nitrosopiperidine were the main nitro-samines found at levels of above 1 µg/kg in European fermented sausages, in which the levels of nitrosamines were generally low or even negligible (Demeyer et al. 2000). The presence of some N-nitrosamines has also been reported in hams packaged in elastic rubber nettings, coming as the consequence of the interaction between nitrites and amine additives present in rubber nettings (Sen et al. 1987).

Commonly occurring N-nitrosamines found in different types of food are N-nitrosodimethylamine, N-nitrosopyrrolidine, N-nitrosothia-zolidine, N-nitrososarcosine, N-nitrosohydroxyproline, N-nitrosoproline and N-nitrosothiazolidine-4-carboxylic acid (Tricker 1997). The Committee in charge of food additives and nutrient sources added into food, established by the EFSA (2010), concluded that the exposure to transformed nitrosamines present in food should be reduced by virtue of implementation of appropriate technological practices capable of downsizing the level of nitrates and nitrites added into food to the maximal achievable point, allowing thereby for the necessary food preservation and microbiological safety effects. In line with the foregoing, there exists the need for continuous monitoring of the intake of these substances into the human body, especially through meat products, and for seeking alternative technological solutions allowing for the maximal achievable downsizing of nitrate and nitrite use without prejudice to safety and durability of the meat products concerned.

5 CONCLUSION

Chemical toxicants entering into the food chain may exhibit biological activity in humans, causing adverse effects on their health. In order to prevent their presence in meat and meat products systematic control of certain types or groups of substances through all critical points of the production and storage is absolutely necessary. The applied regulatory measures in monitoring plan of chemical contaminants over the last decades have demonstrated improved level of compliance with the regulations. Still, there is need to implement continuous control from animal farms to the final stages of the industrial or household production process. Systematic monitoring of toxic chemical substances should be carried out in accordance with the adopted annual monitoring state plans and include the application of sophisticated analytical methods, either to ensure their absence or to determine whether their levels are in accordance with the maximum permitted ones stipulated by the law. Future trends should include identification and determination of new generation substances that could be abused in meat industry or through different pathways contaminate fermented meat products, representing thereby chemical hazards and threat to the health of consumers.

Key words: *chemical hazards, veterinary drugs, substances having anabolic effect, therapeutic agents, prophylactic agents, environmental contaminants, contaminants of meat processing*

REFERENCES

Abou-Arab, A.A.K. 2002. Degradation of organochlorine pesticides by meat starter in liquid media and fermented sausage. Food Chem. Toxicol. 40: 33–41.

Amaral-Mendes, J.J. 2002. The endocrine disrupters: A major medical challenge. Food Chem. Toxicol. 40: 781–788.

Anderson, D.B., D.E. Moody, and D.L. Hancock. 2005. Beta adrenergic agonists. pp. 104–107. *In*: W. Pond and A. Bell (eds.) Encyclopedia of animal science. Marcel Dekker, USA.

Andrée, S., W. Jira, K.-H. Schwind, H. Wagner and F. Schwägele. 2010. Chemical safety of meat and meat products. Meat Sci. 86: 38–48.

Anonymous. 1981. Council Directive 81/602/EEC of 31 July 1981 concerning the prohibition of certain substances having a hormonal action and of any substances having a thyrostatic action. Off. J. Eur. Commun. L 222/32–33.

Anonymous. 2006. Council Directive 96/23/EC of 29 April 1996 on measures to monitor certain substances and residues thereof in live animals and animal products and repealing Directives 85/358/EEC and 86/469/EEC and Decisions 89/187/EEC and 91/664/EEC. Off J. Eur. Commun. L 125/10.

Chemical Hazards in Fermented Meats 443

Anonymous. 2014. Commission Regulation (EU) No 1327/2014 of 12 December 2014 amending Regulation (EC) No. 1881/2006 as regards maximum levels of polycyclic aromatic hydrocarbons (PAHs) in traditionally smoked meat and meat products and traditionally smoked fish and fishery products.

Ansorena, D., M.C. Montel, M. Rokka, R. Talon, S. Eerola, A. Rizzo, M. Raemaekers and D. Demeyer. 2002. Analysis of biogenic amines in northern and southern European sausages and role of flora in amine production. Meat Sci. 61: 141–147.

Armstrong, T.A., D.J. Ivers, J.R. Wagner, D.B. Anderson, W.C. Weldon and E.P. Berg. 2004. The effect of dietary ractopamine concentration and duration of feeding on growth performance, carcass characteristics, and meat quality of finishing pigs. J. Anim. Sci. 82: 3245–3253.

Asefa, D.T., C.F. Kure, R.O. Gjerde, S. Langsrud, M.K. Omer, T. Nesbakken, and I. Skaar. 2011. A HACCP plan for mycotoxigenic hazards associated with dry-cured meat production processes. Food Control. 22: 831–837.

Barnes, P.J. 1999. Effect of β-agonists on inflammatory cells. J. Aller. Clin. Immunol. 104: S10–S17.

Biehl, M.L. and W.J. Buck. 1987. Chemical contaminants: their metabolism and their residues. J. Food Protect. 50: 1058–1073.

Biswas, A.K., N. Kondaiah, A.S.R. Anjaneyulu and P.K. Mandal. 2010. Food safety concerns of pesticides, veterinary drug residues and mycotoxins in meat and meat products. Asian J. Anim. Sci. 4: 46–55.

Bover-Cid, S., M.J. Miguelez-Arrizado, L.L. Latorre Moratalla and M.C. Vidal Carou. 2006. Freezing of meat raw materials affects tyramine and diamine accumulation in spontaneously fermented sausages. Meat Sci. 72: 62–68.

Centre for Food Safety (CFS) of the Food and Environmental Hygiene Department (FEHD) of the Government of the Hong Kong Special Administrative Region 2014. Organochlorine pesticide residues, The first Hong Kong total diet study. Report No. 8.

Codex Alimentarius Commission (CAC) 2009. Code of practice for the reduction of contamination of food with PAH from smoking and direct drying processes. CAC/RCP 68-2009.

Costabeber, I., J. Sifuentes dos Santos, A.A.Odorissi Xavier, J. Weber, F. Leal Leães, S. Bogusz Junior and T. Emanuelli. 2006. Levels of polychlorinated biphenyls (PCBs) in meat and meat products from the state of Rio Grande do Sul, Brazil. Food Chem. Toxicol. 44: 1–7.

Courtheyn, D., B. Le Bizec, G. Brambilla, H.F. De Brabander, E. Cobbaert, M. Van De Wiele, J. Vercammen, and K. De Wasch. 2002. Recent developments in the use and abuse of growth promoters. Anal. Chim. Acta. 473: 71–82.

Covaci, A., A. Gheorghe and P. Schepens. 2004. Distribution of organochlorine pesticides, polychlorinated biphenyls and a-HCH enantiomers in pork tissues. Chemosphere 56: 757–766.

Dall'Asta, C., G. Galaverna, T. Bertuzzi, A. Moseriti, A. Pietri, A. Dossena and R. Marchelli. 2010. Occurence of ochratoxin A in raw ham muscle, salami and dry-cured ham from pigs fed with contaminated diet. Food Chem. 120: 978–983.

Demeyer, D., M. Raemaekers, A. Rizzo, A. Holck, A. De Smedt, B. ten Brink, B. Hagen, C. Montel, E. Zanardi, E. Murbrekk, F. Leroy, F. Vandendriessche,

K. Lorentsen, K. Venema, L. Sunesen, L. Stahnke, L. Vuyst, R. Talon, R. Chizzolini and S. Eerola. 2000. Control of bioflavour and safety in fermented sausages: First results of a European project. Food Res. Int. 33: 171–180.

Deshpande, S.S. 2002. Handbook of food toxicology, Marcel Dekker, Inc.

Djinovic, J., A. Popović and W. Jira. 2009. Polycyclic Aromatic Hydrocarbons (PAHs) in wood smoke used for production of traditional smoked meat products in Serbia. Mitteilungsblatt der Fleischforschung Kulmbach 48: 123–132.

Djinovic, J., A. Popovic and W. Jira. 2008a. Polycyclic aromatic hydrocarbons (PAHs) in traditional and industrial smoked beef and pork ham from Serbia. Eur. Food Res. Technol. 227: 1191–1198.

Djinovic, J., A. Popovic and W. Jira. 2008b. Polycyclic aromatic hydrocarbons (PAHs) in different types of smoked meat products from Serbia. Meat Sci. 80: 449–456.

European Food Safety Authority (EFSA). 2003. The effects of nitrites/nitrates on the microbiological safety of meat products. EFSA J. 14: 16.

European Food Safety Authority (EFSA). 2008. Polycyclic Aromatic Hydrocarbons in Food. Scientific Opinion of the Panel on Contaminants in the Food Chain (Question N° EFSA-Q-2007-136) Adopted on 9 June 2008, EFSA J. 724: 1–114.

European Food Safety Authority (EFSA). 2010. Scientific Opinion. Statement on nitrites in meat products, EFSA Panel on Food Additives and Nutrient Sources added to Food. EFSA J. 8: 1538.

European Food Safety Authority (EFSA). 2011. The 2009 European Union report on pesticide residues in food. EFSA J. 9: 2430. Parma, Italy.

Erickson, M.D. 1997. Analytical Chemistry of PCBs, second ed. CRC, Lewis publishers, Boca Raton.

Feil, V.J. and R.L. Ellis. 1998. The USDA perspective on dioxin concentrations in dairy and beef. J. Anim. Sci. 76: 152–159.

Ferreira, V., J. Barbosa, S. Vendeiro, A. Mota, F. Silva, M.J. Monteiro, T. Hogg, P. Gibbs and P. Teixeira. 2006. Chemical and microbiological characterization of *alheira*: A typical Portuguese fermented sausage with particular reference to factors relating to food safety. Meat Sci. 73: 570–575.

Food and Agriculture Organisation/World Health Organisation (FAO/WHO). 1998. Selected non-heterocyclic polycyclic aromatic hydrocarbons. Environmental Health Criteria 202. Geneva: International Programme on Chemical Safety, World Health Organization.

Food and Agriculture Organisation/World Health Organisation (FAO/WHO). 2000. Toxicological evaluation of certain veterinary drug residues in food. Estradiol-17β, progesterone and testosterone. The Fifty-second meeting of the Joint FAO/WHO Expert Committee in Food Additives (JECFA). WHO Food Aditives Series 43.

García-Falcón, M. S. and J. Simal-Gándara. 2005. Polycyclic aromatic hydrocarbons in smoke from different woods and their transfer during traditional smoking into chorizo sausages with collagen and tripe casing. Food Addit. Contam. 22: 1–8.

Garrido-Frenich, A., P.P. Bolaños, and J.L.M. Vidal. 2007. Multiresidue analysis of pesticides in animal liver by gas chromatography using triple quadrupole tandem mass spectrometry. J. Chromatogr. A 1153: 194–202.

Chemical Hazards in Fermented Meats 445

Gaskins, H.R., C.C. Collier, and D.B. Anderson. 2002. Antibiotics as growth promotants: Mode of action. Anim. Biotechnol. 13: 29–42.

Glynn, A.W., L. Wernroth, S. Atuma, C.-E. Linder, M. Aune, I. Nilsson and P.O. Darnerud. 2000. PCB and chlorinated pesticide concentrations in pork and bovine adipose tissue in Sweden 1991–1997: spatial and temporal trends. Sci. Total Environ. 246: 195–206.

Gomes, A., C. Santos, J. Almeida, M. Elias and L.C. Roseiro. 2013. Effect of fat content, casing type and smoking procedures on PAHs contents of Portuguese traditional dry fermented sausages. Food Chem. Toxicol. 58: 369–374.

González-Weller, D., L. Karlsson, A. Caballero, F. Hernández, A. Gutiérrez, T. González-Iglesias, M. Marino and A. Hardisson. 2006. Lead and cadmium in meat and meat products consumed by the population in Tenerife Island, Spain. Food Addit. Contam. 23: 757–763.

Halliwell, D., N. Turoczy and F. Stagnitti. 2000. Lead concentrations in *Eucalyptus* sp. in a small coastal town. Bull. Environ. Contam. Toxicol. 65: 583–590.

Hernández-Jover, T., M. Izquierdo-Pulido, M.T. Veciana-Nogués, A. Mariné-Font and M.C.Vidal-Carou. 1997. Biogenic amine and polyamine contents in meat and meat products. J. Agric. Food Chem. 45: 2098–2102.

Hernández-Jover, T., M. Izquierdo-Pulido, M.T. Vecina-Nogués and M.C. Vidal-Crou. 1996. Biogenic amine sources in cooked cured shoulder pork. J. Agric. Food Chem. 44: 3097–3101.

Herzallah, S.M. 2009. Determination of aflatoxins in eggs, milk, meat and meat products using HPLC fluorescent and UV detectors. Food Chem. 114: 1141–1146.

Hoha, G. V., E. Costăchescu1, A. Leahu and B. Păsărin. 2014. Heavy metals contamination level in processed meat marketed in Romania. Environ. Eng. Manag. J. 13: 2411–2415.

Hotchkiss, J.H. and A.L. Vecchio. 1985. Nitrosamines in firedout bacon fat and its use as a cooking oil. Food Technol. 39: 67–73.

International Agency for Research on Cancer (IARC). 2002. Aflatoxins. IARC Monograph on the evaluation of carcinogenic risks to humans. Some traditional herbal medicines, some mycotoxins, naphthalene and styrene. Vol. 82. IARC *Press*, Lyon, France.

International Agency for Research on Cancer (IARC). 1993. Some naturally occurring substances: food items and constituents, heterocyclic aromatic amines and mycotoxins. IARC monographs on the evaluation of carcinogenic risks to humans. Vol. 56. IARC, Lyon, France.

Jansen, S.C., M. van Dusseldorp, K.C. Bottema and A.E.J. Dubois. 2003. Intolerance to dietary biogenic amines: a review. Ann. Allerg. Asthma Im. 91: 233–241.

Jira, W., K. Ziegenhals and K. Speer. 2006. Values don't justify high maximum levels: PAH in smoked meat products according to the new EU standards. Fleischwirtschaft Int. 4: 11–17.

Lorenzo, J.M., L. Purriños, R. Bermudez, N. Cobas, M. Figueiredo and M.C. García Fontán. 2011. Polycyclic aromatic hydrocarbons (PAHs) in two Spanish traditional smoked sausage varieties: "Chorizo gallego" and "Chorizo de cebolla". Meat Sci. 89: 105–109.

446   *Fermented Meat Products: Health Aspects*

Lukáčová, A., J. Golian, P. Massányi and G. Formicki. 2014. Lead concentration in meat and meat products of different origin. Potravinarstvo 8: 43–47.

Kan, C.A. and G.A.L. Meijer. 2007. The risk of contamination of food with toxic substances present in animal feed. Anim. Feed Sci. Technol. 133: 84–108.

Karovičová, J. and Z. Kohajdová. 2005. Biogenic amines in food. Chem. Pap. 59: 70–79.

Kavlock, R.J. 1996. Research needs for the risk assessment of health and environmental effects of the US EPA-sponsored workshop. Environ. Health Perspect. 104: 714–740.

Kennedy, D.G., S.A. Hewitt, J.D. McEvoy, J.W. Currie, A. Cannavan, W.J. Blanchflower and C.T. Elliot. 1998. Zeranol is formed from *Fusarium* spp. toxins in cattle in vivo. Food Addit. Contam. 15: 393–400.

Khan, S.J., D.J. Roser, C.M. Davies, G.M. Peters, R.M. Stuetz, R. Tucker and N.J. Ashbolt. 2008. Chemical contaminants in feedlot wastes: Concentrations, effects and attenuation. Environ. Int. 34: 839–859.

Kovačević, D. 2001. Chemistry and technology of meat and fish. University of J.J. Strossmayer. Faculty od Food Technology, Osijek, Croatia.

Kovačević, D. 2014. Technology of Kulen and other fermented sausages. University of J.J. Strossmayer. Faculty od Food Technology, Osijek, Croatia.

Kuiper, H.A., M.Y. Noordam, M.M.H. Dooren-Flipsen, R. Schilt and A.H. Roos. 1998. Illegal use of β-adrenergic agonists: European Community. J. Anim. Sci. 76: 195–207.

Lamming, G.E., G. Ballarini, G. Baulieu, P. Brooks, P.S. Elias and R. Ferrando. 1987. Scientific report on anabolic agents in animal production. Vet. Rec. 121: 389–392.

Latorre-Moratalla, M.L., S. Bover-Cid, R. Talon, M. Garriga, E. Zanardi, A. Ianieri, M.J. Fraqueza, M. Elias, E.H. Drosinos and M.C. Vidal-Carou. 2010. Strategies to reduce biogenic amine accumulation in traditional sausage manufacturing. LWT-Food Sci. Technol. 43: 20–25.

Latorre-Moratalla, M.L., T. Veciana-Nogués, S. Bover-Cid, M. Garriga, T. Aymerich, E. Zanardi, A. Ianieri, M.J. Fraqueza, L. Patarata, E.H. Drosinos, A. Lauková, R. Talon and M.C. Vidal-Carou. 2008. Biogenic amines in traditional fermented sausages produced in selected European countries. Food Chem. 107: 912–921.

LeDoux, M. 2011. Analytical methods applied to the determination of pesticide residues in foods of animal origin. A review of the past two decades. J. Chromatogr. A 1218: 1021–1036.

Lefebvre, B., F. Malouin, G. Roy, K. Giguere and M.S. Diarra. 2006. Growth performance and shedding of some pathogenic bacteria in feedlot cattle treated with different growth-promoting agents. J. Food Protect. 69: 1256–1264.

Lone, K.P. 1997. Natural sex steroids and their xenobiotic analogs in animal production: growth, carcass quality, pharmacokinetics, metabolism mode of action, residues, methods and epidemiology. Crit. Rev. Food Sci. Nutr. 37: 193–209.

Maijala, R., E. Nurmi and A. Fischer. 1995. Influence of processing temperature on the formation of biogenic amines in dry sausages. Meat Sci. 39: 9–22.

Mariam, I., S. Iqbal and S.A. Nagra. 2004. Distribution of some trace and macro minerals in beef, mutton and poultry. Int. J. Agric. Biol. 6: 816–820.

Chemical Hazards in Fermented Meats 447

Markov, K., J. Pleadin, M. Bevardi, N. Vahčić, D. Sokolić-Mihalek and J. Frece. 2013. Natural occurrence of aflatoxin B1, ochratoxin A and citrinin in Croatian fermented meat products. Food Cont. 34: 312–317.

Martinez-Navarro, J.F. 1990. Food poisoning related to consumption of illicit beta-agonist in liver. Lancet 336: 1311.

Mersmann, H.J. 1998. Overview of the effects of β-adrenergic receptor agonists on animal growth including mechanisms of action. J. Anim. Sci. 76: 160–172.

Meyer, H.H.D. 2001. Biochemistry and physiology of anabolic hormones used for improvement of meat production. APMIS 109: 1–8.

Min, J.S., S.O. Lee, A. Jang, C. Jo, C.S. Park and M. Lee. 2007. Relationship between the concentration of biogenic amines and volatile basic nitrogen in fresh beef, pork, and chicken meat. Asian-Aust. J. Anim. Sci. 20: 1278–1284.

Nkansah, M.A. and C.O. Amoako. 2010. Heavy metal content of some common spices available in markets in the Kumasi metropolis of Ghana. Am. J. Sci. Ind. Res. 44: 215–222.

Noé, G., J. Suvisaari, C. Martin, A.J. Moo-Young, K. Sundaram, S.I. Saleh, E. Quintero, H.B. Croxatto and P. Lähteenmäki. 1999. Gonadotrophin and testosterone suppression by 7α-methyl-19-nortestosterone acetate administered by subdermal implant to healthy men. Hum. Reprod. 14: 2200–2206.

Nougadère, A., V. Sirot, A. Kadar, A. Fastier, E. Truchot, C. Vergnet, F. Hommet, J. Baylé, P. Gros and J.C. Leblanc. 2012. Total diet study on pesticide residues in France: Levels in food as consumed and chronic dietary risk to consumers. Environ. Int. 45: 135–150.

Ozkutlu, F., S.M. Kara and N. Sekeroglu. 2006. Heavy metals, micronutrients, mineral content, toxicity, trace elements. ISHS Acta Hortic. 22: 756–760.

Page, S.W. 2003. The role of enteric antibiotics in livestock production. Canberra, ACT, Avcare Limited.

Patandin, S., P.C. Dagnelie, P.G. Mulder, E. Op de Coul, J.E. van der Veen, N. Weisglas-Kuperus and P.J. Sauer. 1999. Dietary exposure to polychlorinated biphenyls and dioxins from infancy until adulthood: A comparison between breast-feeding, toddler, and long term exposure. Environ. Health Persp. 107: 45–51.

Payne, M.A., R.E. Baynes, S.F. Sundlof, A. Craigmill, A.I. Webb and J.E. Riviere. 1999. Drugs prohibited from extralabel use in food animals. JAVMA 215: 28–32.

Perši, N., J. Pleadin, D. Kovačević, G. Scortichini and S. Milone. 2014. Ochratoxin A in raw materials and cooked meat products made from OTA-treated pigs. Meat Sci. 96: 203–210.

Pleadin, J., M. Malenica Staver, N. Vahčić, D. Kovačević, S. Milone, L. Saftić and G. Scortichini. 2015a. Survey of aflatoxin B_1 and ochratoxin A occurrence in traditional meat products coming from Croatian households and markets. Food Control. 52: 71–77.

Pleadin, J., N. Perši, D. Kovačević, N. Vahčić, G. Scortichini and S. Milone. 2013. Ochratoxin A in traditional dry-cured meat products produced from sub-chronic-exposed pigs. Food Addit. Contam. 30: 1827–1836.

Pleadin, J., N. Perši, D. Kovačević, A. Vulić, J. Frece and K. Markov. 2014b. Ochratoxin A reduction in meat sausages using processing methods practiced in households. Food Addit. Contam. Part B. 7: 239–246.

Pleadin, J, D. Kovačević and I. Perković. 2015b. Impact of casing damaging on aflatoxin B_1 concentration during the ripening of dry-fermented meat sausages. J. Immunoassay Immunochem. 36: 655–666.

Pleadin, J., D. Kovačević and N. Perši. 2015c. Ochratoxin A contamination of the autochthonous dry-cured meat product "Slavonski Kulen" during a six-month production process. Food Control. 57: 377–384.

Pleadin, J., A. Vulić, N. Perši, M. Škrivanko, B. Capek and Ž. Cvetnić. 2014a. Aflatoxin B1 occurrence in maize sampled from Croatian farms and feed factories during 2013. Food Control. 40: 286–291.

Pleadin, J., A. Vulić, N. Perši and N. Vahčić. 2010. Clenbuterol residues in pig muscle after repeat administration in a growth-promoting dose. Meat Sci. 86: 733–737.

Richard, J.L. 2007. Some major mycotoxins and their mycotoxicoses – An overview. Int. J. Food Microbiol. 119: 3–10.

Robboy, S.J., O. Taguchi and G.R. Cunha. 1982. Normal development of the human female reproductive tract and alterations resulting from experimental exposure to diethylstilbestrol. Human. Pathol. 13: 190–198.

Rodríguez, A., M. Rodríguez, A. Martín, F. Nuñez and J.J. Córdoba. 2012. Evaluation of hazard of aflatoxin B1, ochratoxin A and patulin production in dry-cured ham and early detection of producing moulds by qPCR. Food Control. 27: 118–126.

Rose, M.D., G. Shearer and W.H.H. Farrington. 1995. The effect of cooking on veterinary drug residues in food: clenbuterol. Food Addit. Contam. 12: 67–76.

Ruiz-Capillas C. and F. Jiménez-Colmenero. 2004. Biogenic amines in meat and meat products. Crit. Rev. Food Sci. 44: 489–499.

Schecter, A., P. Cramer, K. Boggess, J. Stanley and J.R. Olson. 1997. Levels of dioxins, dibenzofurans, PCB and DDE congeners in pooled food samples collected in 1995 at supermarkets across the United States. Chemosphere 34: 1437–1447.

Scientific Committee on Food (SCF). 2002. Opinion of the Scientific Committee on Food on the risk to human of polycyclic aromatic hydrocarbons in food, SCF/CS/CNTM/ PAH/29 Final. Health and consumer protection directorategeneral, Brussels.

Sen, N.P., P.A. Baddoo and S.W. Seaman. 1987. Volatile nitrosamines in cured meats packaged in elastic rubber nettings. J. Agric. Food Chem. 35: 346–350.

Šimko, P. 2005. Factors affecting elimination of polycyclic aromatic hydrocarbons from smoked meat foods and liquid smoke flavourings. Mol. Nutr. Food Res. 49: 637–647.

Škrbić, B., N. Đurišić-Mladenović, N. Mačvanin, A. Tjapkin and S. Škaljac. 2014. Polycyclic aromatic hydrocarbons in smoked dry fermented sausages with protected designation of origin Petrovská klobása from Serbia. Maced. J. Chem. Chem. Eng. 33: 227–236.

Stadnik, J. and Z.J. Dolatowski. 2010. Biogenic amines in meat and fermented meat products. Acta Sci. Pol. Technol. Aliment. 9: 251–263.

Stephany, R.W. 2010. Hormonal growth promoting agents in food producing animals. Handb. Exp. Pharmacol. 195: 355–367.

Stolker, A.A.M. and U.A.Th. Brinkman. 2005. Analytical strategies for residue analysis of veterinary drugs and growth-promoting agents in food-producing animals—a review. J. Chromatogr. A, 1067: 15–53.

Suzzi, G. and F. Gardini. 2003. Biogenic amines in dry fermented sausages: a review. Int. J. Food Microbiol. 88: 41–54.

Thomas, M.G., J.A. Carroll, S.R. Raymond, R.L. Matteri and D.H. Keisler. 2000. Transcriptional regulation of pituitary synthesis and secretion of growth hormone in growing wethers and the influence of zeranol on these mechanisms. Domest. Anim. Endocrin. 18: 309–324.

Toldrá, F. and M. Reig. 2006. Methods for rapid detection of chemical and veterinary drug residues in animal foods. Trends Food Sci. Technol. 17: 482–489.

Toldrá, F. and M. Reig. 2007. Chemical origin toxic compounds. pp. 469–475. In: F. Toldrá (ed.). Handbook of fermented meat and poultry. Blackwell Publishing.

Tricker, A.R. 1997. N-nitroso compounds and man: Sources of exposure, endogenous formation and occurrence in body fluids. Eur. J. Cancer Prev. 6: 226–268.

Walker, R. 1990. Nitrates, nitrites and nitrosocompounds: A review of the occurrence in food and diet and the toxicological implications. Food Addit. Contam. 7: 717–768.

Wenzl, T., R. Simon, J. Kleiner and E. Anklam. 2006. Analytical methods for polycyclic aromatic hydrocarbons (PAHs) in food and the environment needed for new food legislation in the European Union. Trends Analyt. Chem. 25: 716–725.

World Health Organization (WHO). 1993. Polychlorinated Biphenyls and Terphenyls, second ed. WHO, Geneva.

World Health Organization (WHO). 2014. Dioxins and their effects on human health http://www.who.int/mediacentre/factsheets/fs225/en/

Wilson, T.W., D.A. Neuendorff, A.W. Lewis and R.D. Randel. 2002. Effect of zeranol or melengestrol acetat (MGA) on testicular and antler development and agression in farmed fallow bucks. J. Anim. Sci. 80: 1433–1441.

Woodward, K.N. 2005. Veterinary pharmacovigilance. Part 2. Veterinary pharmacovigilance in practice – the operation of a spontaneous reporting scheme in a European Union country – the UK, and schemes in other countries. J. Vet. Pharmacol. Therap. 28: 149–170.

Wortberg, B. and R. Woller. 1982. Quality and freshness of meat and meat products as related to their content of biogenic amines. Fleischwirtschaft 62: 1457–1463.

Chapter 19

Biogenic Amines in Fermented Meat Products

José M. Lorenzo, Daniel Franco and Javier Carballo*

1 INTRODUCTION

Biogenic amines are nitrogen compounds with physiological activity, generally formed in foods as a result of the protein metabolism by microorganisms. Due to their importance on the consumers' health, derived from their biological activities, and also on the food quality, as their presence indicates excessive growth of undesirable microorganisms, their presence, formation, quantification and control in foods have demanded considerable attention in the scientific literature of the last decades.

Fermentation, and subsequent ripening of fermented meat products offer very favorable conditions for the production of biogenic amines: (i) Intense proteolysis by action of autochthonous and microbial enzymes that supply abundant free amino acids, the main precursors of amines; (ii) Slightly acidification due to the carbohydrate metabolism by the lactic acid bacteria dominating in the first stages; (iii) Active growth of microbial populations coming from raw materials, processors, etc. As a consequence of this, and also due to the wide consumption of these meat

---

*For Correspondence: Área de Tecnología de los Alimentos, Facultad de Ciencias de Ourense, Universidad de Vigo, 32004 Ourense, Spain. Tel: +34-988-387052, Fax: +34-988-387001, Email: carbatec@uvigo.es

Biogenic Amines in Fermented Meat Products 451

products around the world, study of biogenic amines in these products achieves a special relevance.

The aim of this chapter in to expose the main concerns related to biogenic amines in fermented meat products. Fermented meat products are typically sausages, and to these products this chapter will be devoted. This chapter does not pretend to be a thorough scientific treatise, nor intended to be an exhaustive review of all the information present in the literature regarding this subject. As befits a chapter of a book, is intended to contain authoritative information, set out in an orderly and simple way so that it can be useful in teaching this concrete aspect of the food science.

2 BIOGENIC AMINES, CHEMICAL NATURE AND IMPORTANCE IN FOOD QUALITY AND HUMAN HEALTH

Biogenic amines are low molecular weight nitrogen compounds, possessing one or more amino groups, and having aliphatic, aromatic or heterocyclic structures. They are present in live animal and vegetable tissues and in microbial cells, where they have biological activities generally consisting in regulatory functions. Biogenic amines are usually formed by decarboxylation of the corresponding amino acids that act as precursors, although some aliphatic amines can be formed through amination or transamination reactions from aldehydes and amines. In foods, they can come from the raw materials, but they are above all formed by decarboxylation of amino acids, catalyzed by enzymes of microbial origin.

The main biogenic amines in foods, their chemical structures and their characteristics regarding origin and physiological activities are shown in Figure 1 and Table 1. The presence of biogenic amines in foods has a different significance depending on the nature of the products considered. Non-fermented foods such as meat and milk contain moderate levels of biogenic amines (above all spermine and spermidine, and putrescine, cadaverine, histamine and tyramine in lover quantities) as part of the non-protein nitrogen fraction (Silla Santos 1996). The presence of biogenic amines in non-fermented foods in amounts above certain levels is generally due to undesired microbial activity and is associated with poor quality of the raw materials and/or with deficiencies in the manufacture and storage conditions. In such cases, biogenic amine content can be used as indicator of microbial spoilage and undesirable hygienic practices. At this respect, not all the biogenic amines have the same significance; levels of histamine, putrescine and cadaverine usually increase along the spoilage of meat or fish (they are produced as products of the microbial metabolism), while spermidine and spermine decrease as a result of their degradation by the microorganisms. On the contrary, fermented foods are produced through microbial action, being the microorganisms the main agents responsible for the biochemical changes that give to these foods their unique and characteristic organoleptic and nutritional features. Fermented foods have particular environmental conditions that favour the formation of biogenic amines, e.g. presence of microorganisms (most of them having decarboxylase activities) as a part important in the manufacturing process, low pH values due to the formation of organic acids in the fermentation reactions, and presence of free amino acids at high levels as a consequence of proteolytic reactions that are concomitant with the fermentation processes. In such circumstances, the presence of biogenic amines at significant levels can be expected. Most of these products in which lactic acid bacteria have grown and developed their metabolic activities contain considerable amounts of putrescine, cadaverine, histamine and tyramine (Brink et al. 1990).

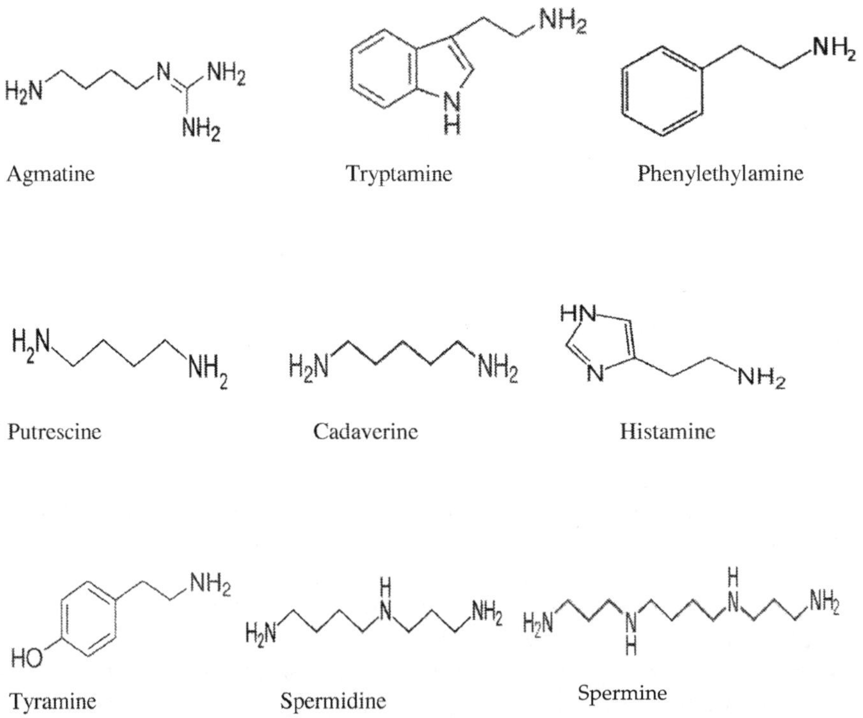

Agmatine   Tryptamine   Phenylethylamine

Putrescine   Cadaverine   Histamine

Tyramine   Spermidine   Spermine

Figure 1 Main biogenic amines found in fermented meat products.

Table 1 Main biogenic amines present in fermented meats: origins and physiological roles

Biogenic amine	Precursor	Reaction of formation	Enzyme involved	Physiological activities
Monoamines				
Phenylethylamine	Phenylalanine	Decarboxylation	Phenylalanine decarboxylase (EC 4.1.1.53)	Neurotrasmitter, neuromodulator, vasoactive
Tyramine	Tyrosine	Decarboxylation	Tyrosine decarboxylase (EC 4.1.1.25)	Vasoactive
Tryptamine	Tryptophan	Decarboxylation	Tryptophan decarboxylase (EC 4.1.1.28)	Neurotrasmitter, neuromodulator, vasoactive
Histamine	Histidine	Decarboxylation	Histidine decarboxylase (EC 4.1.1.22)	Modulator of immune responses, neuromodulator, vasoactive, modulator of gastric acid secretion, modulator of the smooth muscle tone
Diamines				
Putrescine	Ornithine	Decarboxylation	Ornithine decarboxylase (EC 4.1.1.17)	Growth factor, necessary for cellular division
Cadaverine	Lysine	Decarboxylation	Lysine decarboxylase (EC 4.1.1.18)	Growth factor, necessary for cellular division
Polyamines				
Agmatine	Arginine	Decarboxylation	Arginine decarboxylase (EC 4.1.1.19)	Modulatory action at several molecular targets (neurotransmitter systems, key ion channels, nitric oxide synthesis, polyamine metabolism)
Spermidine	Putrescine	Transfer of a propylamine group from S-adenosylmethioninamine	Spermidine synthase (EC 2.5.1.16)	Multifunctions in live cell: Maintaining membrane potential and controlling intracellular pH and volume, modulation of cell growth and proliferation, chromatin structure, gene transcription and translation, DNA stabilization, signal transduction
Spermine	Spermidine	Transfer of a propylamine group from S-adenosylmethioninamine	Spermine synthase (EC 2.5.1.22)	Multifunctions in live cell: Maintaining membrane potential and controlling intracellular pH and volume, modulation of cell growth and proliferation, chromatin structure, gene transcription and translation, DNA stabilization, signal transduction

454   *Fermented Meat Products: Health Aspects*

The presence of biogenic amines in excessive quantities in foods can compromise quality and safety of them. Biogenic amines commonly present in foods are necessary elements and develop very important functions in the human organism. However, consumption of foods containing high amounts of biogenic amines, particularly histamine, putrescine and tyramine, may have negative effects on the consumer, with episodes of intoxication that can become very serious in people debilitated, sick or being treated with some particular drugs. Common symptoms derived from consumption of high amounts of biogenic amines are vomiting, sweating, respiratory problems, palpitations, and hypo- or hypertension with their associated consequences (Kordiovská et al. 2006).

Histamine is, by far, the most toxic among biogenic amines. Histamine poisoning can occur after consumption of foods, generally fish or cheese, having very high levels of this amine. Histamine, due to its vasoactive condition, causes dilatation of the peripheral blood vessels, resulting in hypotension, flushing and headache (Stratton et al. 1991). By acting on the intestinal smooth muscle, it causes sudden and intense contractions, accounting for cramps, diarrhea and vomiting (Taylor 1986). In any case, the toxic effects depend on the histamine intake concentration, and also on the physiological state of the consumer (Bacellar Ribas Rodriguez et al. 2014); healthy individuals possess efficient detoxification systems (by amine oxidation or conjugation) in the intestinal tract that quickly detoxified the normal dietary intakes of biogenic amines. The detoxificant enzymes, called amine-oxidases, are induced in the presence of amines (Wendakoon and Sakaguchi 1993). Usually, two amine oxidases, mono amine oxidase (MAO) and diamine oxidase (DAO) are active and play a decisive role in the detoxification of the biogenic amines. However, some drugs inhibitors of the mono amine oxidase, widely used in treatments of depression, Parkinson's disease and some other related disorders, can act as inhibitors of these natural activities. In patients being treated with such drugs, and also in people hipersensitive to the amines, moderate concentration of these amines can cause a true intoxication with fatal outcomes in some cases. Also, some antihistamines, drugs used in the treatment of malaria, and other medications can inhibit the enzymes involved in the metabolism of histamine, thus potentiating its undesirable effects (Brink et al. 1990, Stratton et al. 1991).

Some amines (tyramine, tryptamine and 2-phenylethylamine) have vasoconstrictor effects. These amines, mainly tyramine, have been blamed onset hypertension crisis and of dietary-induced migraine (Stadnik and Dolatowski 2010). Negative effects of tyramine comprise peripheral vasoconstriction with its immediate consequences, increased heart rate and breathing, increased blood glucose, and release of norepinephrine

(McCabe-Sellers et al. 2006). In other cases, some amines present in foods as a result of the protein metabolism can act as potentiators of the toxic effects of other biogenic amines. Putrescine and cadaverine can act as inhibitors of the histamine-detoxifying enzymes; also, other amines can act as potentiators of the undesirable toxic actions of the histamine; these include tyramine (that can act as inhibitor of the MAO), and phenylalanine (which inhibit the enzyme histamine N-methyltransferase) (Stratton et al. 1991). Tryptamine and phenylethylamine also act as inhibitors of DAO (Stratton et al. 1991).

Adverse responses to biogenic amines in foods are, in their mildest forms, often considered as intolerance reactions, also named pseudoallergy. Intolerance is a hypersensitivity not mediated by the immune system, in contrast to allergic reactions that involve immune responses (Jansen et al. 2003). Another issue related to the negative effects of some biogenic amines deals with the formation, in concrete environmental conditions, of carcinogenic compounds. In cured meat products, nitrite in commonly added as anti-botulinic agent, and also to achieve a stable and desirable red color via the reaction of NO (coming from the nitrite reduction) with the muscle myoglobin, forming a typical pigment (nitrosomyoglobin) responsible for this desirable color. However, some amines such as putrescine and cadaverine can react with nitrite to give rise to the heterocyclic nitrosamines nitrosopyrrolidine and nitrosopiperidine (Huis in't Veld et al. 1990); the formation of nitrosamines through reaction with nitrite can occur also in the cases of agmatine, spermidine and spermine (Halász et al. 1994). These nitrosamines can be formed during storage and heat treatment of foods (Hotchkiss, 1987), but also endogenously in the stomach through nitrosation of nitrite with amines (Tricker and Preussmann 1991). N-nitrosodimethylamine (NDMA) is the most frequently occurring nitrosamine in foods (Chowdhury 2014). Consumption in diet of these nitrosamines as well as other N-nitroso compounds together with endogenous-formed nitrosamines, seems to increase the risk of oral cavity, oesophagus and stomach cancer (Crampton 1980, Bartsch and Montesano 1984, Larsson et al. 2006, Butler 2015).

Regarding the effect of amines on food quality, amines, together with peptides and other low molecular weight nitrogen compounds actively contribute to the genuine and desirable smell and taste of the fermented foods (Maga and Katz 1978). However, excessive amounts of amines, particularly putrescine and cadaverine, can confer putrid undesirable off flavors. Anomalous high contents in biogenic amines in fermented sausages are also indicative of high counts of undesirable microbiota (e.g. *Enterobacteriaceae*), consequence of poor quality of the raw materials and lack of hygiene and care during manufacture.

456 *Fermented Meat Products: Health Aspects*

3 FORMATION OF BIOGENIC AMINES IN FERMENTED MEAT PRODUCTS

3.1 Effect of the Environmental Conditions on the Biogenic Amine Formation

Differences in quantities and nature of biogenic amines in the different fermented meat products are consequence of environmental parameters that influences the microbial growth and the microbial decarboxylase activities. These activities are influenced by both extrinsic and intrinsic factors operating during fermentation/ripening such as pH, Eh, NaCl concentration, temperature, use of starter cultures, hygienic quality of raw materials and manufacturing practices, and even the size (diameter) of the sausage (Silla Santos 1996, Bover-Cid et al. 1999b, Gardini et al. 2001, Komprda et al. 2004, Latorre-Moratalla et al. 2008, Stadnik and Dolatowski 2010).

The pH is probably the most important factor influencing the decarboxylase activities and therefore the biogenic amine formation. Amino acid decarboxylases have generally optimum activities at acid pH values (4.0–5.5) (Sinell 1978), and higher amine concentrations were found in foods having low pH values (see Silla Santos 1996). Several factors influence and module the pH values in sausages: amounts and nature (easily to be metabolized) of fermentable carbohydrates in the mix, load of lactic acid bacteria, etc. However, very quick and intense acidifications can dramatically reduce the growth of biogenic amine producers (Maijala et al. 1995b); in fact the use of acidifying starter cultures or the addition of acidifying compounds such as GDL (glucono-δ-lactone) result in an important decrease in the levels of some biogenic amines.

The redox potential seems to affect the formation of some biogenic amines by some microorganisms, but contradictory effects were reported in literature. Anaerobic conditions seem to reduce the formation of concrete biogenic amines such as puterscine or cadaverine (Halász et al. 1994). In contrary, reducing the redox potential stimulates the histamine production (Halász et al. 1994). Histidine decarboxylase activity in concrete species of bacteria is inhibited in atmospheres having 80% CO_2 (Watts and Brown 1982), however, modified atmosphere packaging only provides slight inhibition of histidine-forming bacteria in fish (Oka et al. 1993).

The effect of the NaCl concentration on biogenic amine formation has been addressed in numerous scientific works. In general, increasing NaCl concentrations seem to decrease the formation and accumulation of biogenic amines, since high NaCl concentrations strongly inhibit the proteolytic bacteria and above all the amine-forming microorganisms

Biogenic Amines in Fermented Meat Products

(Silla Santos 1996, Suzzi and Gardini 2003). However, when the effect was studied on specific decarboxylase activities, contrary effects were reported; Korovicová and Kohajdová (2005) informed that NaCl inhibits histidine decarboxylase, but it activates tyrosine decarboxylase activity.

Regarding the effect of the temperature, it seems that ripening and storage temperature affects the formation of biogenic amines in sausages. However, contradictory effects were reported in literature in some cases (see Silla Santos 1996). In general, high temperatures, close to the optimum growth values of biogenic amine producing bacteria, increase the biogenic amine content in foods. In any case, low storage temperatures are not sufficient to inhibit the formation of some specific biogenic amines (Ababouch et al. 1991). On the other hand, biogenic amines are highly resistant to heat treatments, with the exception of spermine that decreased after treatment at 200°C for 2 hours (Silla Santos 1996).

Since composition and activity of indigenous microbiota of the sausages are very variable, starter cultures are used in order to control the ripening process, to ensure safety, contribute to color and flavor, extend shelf-life, and standardize production, maintaining the typical and distinctive characteristics obtained in artisanal production (Fonseca et al. 2013).The use of starter cultures, mainly lactic acid bacteria, notably decreases the accumulation of biogenic amines in the fermented sausages. However, in order to achieve optimal results, starter cultures should be carefully chosen regarding their metabolic abilities. The use of amine-producing strains can favor the formation and accumulation of biogenic amines, with contrary and negative effects.

A relationship was established between biogenic amine formation and the size (diameter) of the dry-fermented sausages (Bover-Cid et al. 1999b, Parente et al. 2001). Size notably influences the dehydration processes during ripening, the diffusion of O_2 from the environmental atmosphere, and therefore the microbial growth during fermentation/ ripening. Larger diameters result in higher moisture contents, higher a_w values, and lower salt shares, consequently favor microbial growth and biogenic amine formation. In general, thicker sausages have higher concentration of biogenic amines, and of tyramine and putrescine in particular, and for the same reasons external peripheral portions have less amine concentrations than the internal portions within the same sausage (Suzzi and Gardini 2003, Ruiz-Capillas and Jiménez-Colmenero 2004).

Despite the fact that spices (paprika, garlic, pepper, etc.) are normally used as ingredients in the manufacture of most of the dry-fermented sausages, and in spite of the generally recognized antimicrobial effect of some of these spices, scarce attention was given to the effect of the spices on the biogenic amine formation. Komprda et al. (2004) reported that sausages manufactured with low amounts of red pepper in the mix and 80 mm of diameter showed at the end of the ripening process

458   *Fermented Meat Products: Health Aspects*

higher contents of total biogenic amines, and of putrescine and tyramine in particular, than sausages manufactured with high amounts of red pepper in the mix and 55 mm of diameter. Unfortunately, the effect of the amount of red pepper added and the effect of the size were not studied in an independent way, and in such circumstances it is difficult to establish if the reduction of the biogenic amines is a consequence of the lower size of the sausages or of the inhibitory effect of the higher amounts of pepper added. In order to establish the effect of the spices on the biogenic amine production it is necessary to carry out specific works in that this effect is studied in an independent way of other factors influencing biogenic amine formation.

Finally, decarboxylation reactions seems to have a fed-back regulation. When histamine is accumulated in the medium, histidine decarboxylase activity is repressed (Omura et al. 1978); this same decarboxylase activity seems to be inhibited not only by histamine, but also by other biogenic amines as reported by some investigations reviewed by Silla Santos (1996).

3.2   Microbial Groups and Species Involved in the Biogenic Amine Formation

Microorganisms and their decarboxylases are, by far, the most responsible for the formation of biogenic amines in fermented meat products. Being the biogenic amine formation a defensive mechanism against the acidic environment, the capability of microorganisms to form biogenic amines should be necessarily correlated with their sensitivity to acid environments. In the last years, many studies were devoted to the ability of specific microbial species isolated from fermented sausages to produce biogenic amines "in vitro". Lactic acid bacteria, as highly diverse microbial group, show a strong variability in biogenic amine production. While Silla Santos (1998) did not show any significant amino acid decarboxylase activity in the strains of *Lactobacillus* studied, other authors (Bover-Cid and Holzapfel 1999, Bover-Cid et al. 2001c) described considerable production of biogenic amines, primarily tyramine, by enterococci, carnobacteria and lactobacilli. Quantities produced of each biogenic amine widely varied between species and even within the same species. Enterococci (*E. faecalis* and *E. faecium*), *Carnobacterium* (*C. divergens, C. gallinarum* and *C. piscicola*), and *Lb. curvatus* produced high quantities of tyramine; some strains of *Lb. curvatus*, in addition, were also able to produce high quantities of phenylethylamine, tryptamine, putrescine and cadaverine (Bover-Cid and Holzapfel 1999, Bover-Cid et al. 2001c).

Enterobacteriaceae are generally recognized as the microbial group displaying the higher amino acid decarboxylase activity (Silla Santos 1998). Several authors reported a frequent histidine, lysine, ornithine

Biogenic Amines in Fermented Meat Products 459

and tyrosine decarboxylase activity in enterobacteria from fermented sausages (Maijala et al. 1993, Silla Santos 1998, Bover-Cid and Holzapfel 1999, Bover-Cid et al. 2001c, Durlu-Özkaya et al. 2001, Pircher et al. 2007, Lorenzo et al. 2010). Regarding the amounts of biogenic amines produced by *Enterobacteriaceae,* quantities reported are very high, being common levels over 1000 mg/L for putrescine and cadaverine (Lorenzo et al. 2010). In any case, there is a considerable variability among species and even among strains within the same species.

Coagulase-negative staphylococci have also been considered as potential biogenic amine producers. Silla Santos (1998) reported a high frequency of histidine, tyrosine, ornithine, and lysine decarboxylase activity in strains of *Staphylococcus xylosus* and *St. saprophyticus* isolated from Spanish fermented sausages. Furthermore, Martín et al. (2006) observed ornithine and lysine decarboxylase activities in 57% of the strains of *St. xylosus* isolated from Iberian dry-cured sausages.

However, Drosinos et al. (2007) observed that only a low proportion of strains displayed amino acid decarboxylase activity; the species with the highest proportion of strains that displayed histidine, tyrosine, ornithine, or lysine decarboxylase activity were *St. saprophyticus, St. simulans,* and *St. xylosus,* but within each species the proportion of strains that were positive for a specific amino acid decarboxylase activity never surpassed the 50%. Martín et al. (2006) reported that only the 14.6% of the strains studied (35 strains) were able to decarboxylate one or more amino acids. In accordance with these results, Even et al. (2010) working with 129 strains of coagulase-negative staphylococci isolated from various environments including cheeses and fermented sausages, observed that only 5 strains (around 6%) were able to produce detectable amounts of biogenic amines. Bonomo et al. (2009) did not find any tyrosine or ornithine decarboxylase activity in any of the 37 staphylococci strains tested, and only observed histidine decarboxylase activity in two strains of *St. warneri.* Lysine was the amino acid most frequent decarboxylated, and 62% of strains, belonging mainly to the *St. equorum* and *St. xylosus* species, were able to decarboxylate this amino acid. In the latter study the highest proportion of strains capable to decarboxylate lysine were in the *St. pasteuri* and *St. succinus* species. Masson et al. (1996) did not observe histidine decarboxylase activity in any of the tested strains of *St. carnosus, St. xylosus, St. warneri,* and *St. saprophyticus* isolated from sausages. They observed tyrosine decarboxylase activity in all these strains, but the amounts of tyramine produced never achieved 40 µg/mL. Bover-Cid et al. (2001c) did not observe any decarboxylase activity in any of the staphylococci strains tested.

In a more recent study, Bermúdez et al. (2012) studied 38 strains of *Staphylococcus* isolated from Spanish sausages and they observed that the four decarboxylase activities studied (histidine, tyrosine, ornithine and

Table 2 Biogenic amine content (mg/kg) of some fermented sausages at the end of the ripening process (range of data reported in literature)

Sausage type	Agmatine	Tryptamine	2-Phenyle-thylamine	Putrescine	Cadaverine	Histamine	Tyramine	Spermidine	Sper-mine	Reference
Androlla	ND-48.69	39.90–145.88	0.85–103.88	97.73–346.34	22.96–693.13	18.54–246.36	65.12–338.91	7.07–20.25	14.87–45.79	Lorenzo et al. 2008
Belgian sausage*	–	–	ND-60.6	31.3–395.7	Tr-55.7	Tr-197.4	101.8–1506.3	–	–	Vandekerckhove 1977
Bologna sausage	–	–	–	ND-29	ND-57	ND-6	ND-29	1.5–4	15–36	Wortberg and Woller 1982
Botillo	ND-19.24	23.63–173.30	2.57–21.49	103–385.44	25.30–762.74	7.90–220.82	11.09–190.76	7.86–15.58	21.07–31.05	Lorenzo et al. 2008
Chinese sausage*	–	13.20–19.87	46.96–51.11	64.17–189.52	1.29–214.69	0.16–17.56	52.87–318.79	18.22–18.55	12.80–12.88	Lu et al. 2010
Chinese sausage*	–	ND-32.76	ND-2	70.03–564.5	232.51–644.1	28.93–177.4	268.1–502.8	–	101.2–328.6	Xie et al. 2015
Chorizo	–	ND-87.8	ND-51.5	2.6–415.6	ND-658.1	ND-314.3	29.2–626.8	1.9–10	13.8–43.5	Hernández-Jover et al. 1997a
Chorizo	–	ND-50	ND-5	78–223	ND-19	ND	62–132	ND-4	ND-2	González-Fernández et al. 2003
Dacia	–	39.38–48.66	79.83–88.17	26.33–49.94	37.59–90.65	15.14–21.45	57.69–141.35	4.85–8.32	28.16–30.68	Ciciu Simion et al. 2014
Danish sausage	–	<10–91	<1–4	<1–450	<1–790	<1–56	5–110	3–9	23–47	Eerola et al. 1998
Egyptian sausage	–	12.70	33.25	38.62	19.20	5.25	19.25	2.30	1.75	Shalaby 1993
Felino sausage	–	–	2–30	90–170	80–230	55–115	195–260	–	–	Tabanelli et al. 2012
Finnish dry-sausage	–	–	<5–67	–	–	6–69	50–211	–	–	Eerola et al. 1996

Product										Reference
Finnish sausage	—	<10–43	2–48	<1–230	<1–270	<1–180	4–200	2–7	19–46	Eerola et al. 1998
French sausage	—	ND–9	ND–8	61–410	31–192	15–151	84–268	2–6	61–119	Montel et al. 1999
Fuet	—	ND–67.8	ND–33.7	2.2–222.1	5.4–53.1	ND–357.7	31.8–742.6	0.9–11	9.4–30.1	Hernández-Jover et al. 1997a
Fuet*	—	0.35–1.39	ND–0.5	25–35	0.61–1.76		26–85			Bover-Cid et al. 1999a
Fuet*	0.5	4.8	3.7	41.5	10.8	2.4	133.1	7.7	30.1	Bover-Cid et al. 1999b
Fuet	—			7.5–24	3.75–31.5		81–96			Bover-Cid et al. 2008
Lubeck	—	<10–20	<1–7	<1–220	<1–8	<1–40	9–150	3–7	20–40	Eerola et al. 1998
Meetwurst	—	<10–54	<1–5	2–580	<1–16	<1–170	5–320	<1–14	22–38	Eerola et al. 1998
Milano sausage	—		ND–30	50–90	40–60	ND–23	85–130			Tabanelli et al. 2012
Norwegian salami	—	ND–2.38	ND–2.10	9.29–23.23	ND–1.66	ND–2.76	0.01–15.74	ND–3.8	ND–25.18	Hagen et al. 1996
Pepperoni	—	<10–91	<1–48	<1–580	<1–790	<1–200	3–320	<1–14	19–48	Eerola et al. 1998
Poličan*	3	5		54	6	25–32	86–92	2–3	2	Komprda et al. 2001
Russian sausage	—	<10–43	1–33	3–310	3–18	<1–200	6–240	2–8	23–40	Eerola et al. 1998
Salami	—	<10–51	<1–8	<1–210	<1–71	<1–9	3–200	1–8	19–45	Eerola et al. 1998
Salchichón	—	ND–65.1	ND–34.7	5.5–400	ND–342.3	ND–150.9	53.3–513.4	0.7–13.8	6.9–42.5	Hernández-Jover et al. 1997a
Salchichón*	0.3	4.4	7.2	119.8	11	6.3	144.7	4.3	27.5	Bover-Cid et al. 1999b

Table 2 (*Contd.*).

Table 2 Biogenic amine content (mg/kg) of some fermented sausages at the end of the ripening process (range of data reported in literature) (*Contd.*).

Salsiccia	–	–	ND	ND-77.74	ND-38.98	ND	ND-338.85	ND-57.17	ND-28.08	Perente et al. 2001
Sobrasada	–	–	–	–	–	3.10–14.25	14.15–77.55	–	–	Vidal-Carou et al. 1990
Sobrasada	–	ND-64.8	0.2–38.5	1.8–500.7	3.0–41.6	2.8–143.1	57.6–500.6	2.1–7	10.3–17.8	Hernández-Jover et al. 1997a
Soppressata	–	–	ND-19.90	ND-416.14	ND-271.36	ND-100.88	ND-556.88	ND-90.57	ND-97.86	Parente et al. 2001
Sucuk	–	ND-37	3.5–12.5	0–270	0–383	0–3.7	164–295	3.7–4.3	–	Kurt and Zorba 2010
Thin Fuet*	ND	ND	1.5	21.3	16.0	0.8	86.7	7.5	37.7	Bover-Cid et al. 1999b

*Expressed as mg/kg of dry matter
ND = Not detected
Tr = Traces
– = Not quantified

Biogenic Amines in Fermented Meat Products

lysine decarboxylase) were present in most of the strains. Accumulation of putrescine and cadaverine was quantified in the culture media of the strains that displayed ornithine and lysine decarboxylase activities. The aminogenic capability of the strains was very low, accumulating putrescine and cadaverine quantities lower than 25 and 5 mg/L, respectively.

Finally, some *Bacillus* species could be found in sausage formulations during the early stages of fermentation/ripening. Information regarding the amino acid decarboxylase activity of *Bacillus* strains isolated from meat products is very scarce. Roig-Sagués et al. (1996) found in salchichón (a Spanish traditional sausage) several *Bacillus* spp. strains capable to produce histamine, although in very low quantities (about 0.5 µg/mL). Bermúdez et al. (2012) studied 19 strains of *Bacillus* (13 of *B. subtilis* and 6 of *B. amyloliquefaciens*) isolated from Spanish traditional sausages and observed a very high percentages of strains displaying histidine, tyrosine, ornithine and lysine decarboxylase activities; however, putrescine and cadaverine amounts produced by these strains growing in synthetic culture media were very low, never higher than 20 mg/L.

In order to design strategies for specific inhibition of the production of biogenic amines, it is essential not only to obtain information about the potential production of biogenic amines by the microorganisms present in fermented meat products, but also about the microbial growth phase during which maximum production takes place. The existing information in literature on this subject is not conclusive. Halász et al. (1994) reported that production mainly took place in the stationary phase of microbial growth, while Lorenzo et al. (2010) reported that the production and accumulation of putrescine and cadaverine in *Enterobacteriaceae* takes place primarily in the exponential phase of the growth. At this respect, Lorenzo et al. (2010) concluded that the phase of growth in that the maximum production of biogenic amine takes place seems to be variable depending on the species and amine considered.

3.3 Biogenic Amine Content in Fermented Meat Products

The content of biogenic amines in fermented meat products is very variable in amounts and nature of the determined amines, even within the same type of sausage, this reflecting differences in quality of the raw materials (microbial loads and nature of the biota present) and in technological manufacturing processes (ingredients, environmental conditions in ripening rooms), as well as in the hygiene of processors, equipment and environment.

Table 2 shows data on the biogenic amine content of several fermented sausages reported in literature. Tyramine, putrescine and cadaverine are the main biogenic amines, a fact common in all the studied sausages. The concentration of cadaverine is the most variable. Some authors (Suzzi and Gardini 2003) indicated that the presence of putrescine and tyramine in sausages is attributable to the activity of lactic acid bacteria, while high quantities of cadaverine should be related to the poor hygienic practice and high microbial contamination of raw materials used in the production of sausages. Bover-Cid et al. (2001c) also suggested the relationship of the cadaverine content to the poor quality of the raw materials; in other studies, the content of this biogenic amine was related to high *Enterobacteriaceae* counts (Suzzi and Gardini 2003, Bover-Cid et al. 2006). Although *Enterobacteriaceae* are normally present in low numbers, or even absent in the ripened sausages, incorrect handling and storage of the meat used in the manufacture, as well as an incorrect fermentation, can provoke their proliferation and release of decarboxylases in the first stages of sausage manufacture (Bover-Cid et al. 2001a).

Several authors (see Table 2) have reported histamine as a common biogenic amine in sausages. Differences in the histamine content in the different sausage varieties and even within the same sausage variety may be due to variations in the duration of the ripening process, to the type and quality of meat used as raw material, and to the relation of lean meat/fat in the sausage mix (Shalaby 1993, 1996).

Spermidine and spermine are reported generally in low quantities. Both biogenic amines are natural amines that always appear in fresh meat (Hernández-Jover et al. 1997a). They are normal components both in the lean meat and in the fat used as raw materials for sausage production, and their levels are usually not affected by the fermentation/ripening processes. Their content generally remain constant throughout fermentation/maturation, and the spermine content may even decrease in comparison to levels in raw materials, as this amine is used by some microorganisms as a source of nitrogen (Bardócz 1995, Hernández-Jover et al. 1997a). Later, during storage, a decrease in the spermidine and spermine contents may also occur due to utilization by microorganisms grown at higher storage temperatures (Bover-Cid et al. 2001d).

Phenylethylamine is common in high quantities in cheese (Shalaby 1996), but rarely reported in sausages in noticeable amounts (see Table 2). This biogenic amine is generally formed when high concentrations of tyramine are present in the products (Lorenzo et al. 2008).

Toxic levels for biogenic amines were investigated in animal models (rats) in laboratory conditions. LD_{50} were higher than 2000 mg/kg body weight for tyramine and cadaverine, 2000 mg/kg for putrescine, and 600 mg/kg body weight for spermidine and spermine (Bacellar Rivas Rodríguez et al. 2014). The toxic levels of biogenic amines for humans are not well established. Toxicity of amines notably varies among individuals, depending on each particular metabolic capacity of detoxification conditioned by the physiological state and the concomitant intake of some drugs acting as potentiators of the toxic activity. According to some authors (Halász et al. 1994), the toxic levels are 10, 10 and 3 mg/100 g of food for histamine, tyramine and phenylethylamine, respectively; however, these quantities notably decrease in patients with asthma, gastric ulcer or taking drugs inhibiting monoamineoxidase. Histamine is the better studied amine regarding the toxic effects. An intake of 5–10 mg/kg of histamine can already negatively affect sensitive people, 10 mg/kg is considered the tolerable limit, 100 mg/kg produces medium toxicity and 1000 mg/kg results very toxic (Silla Santos 1996, Korovicová and Kohajdová 2005); based on data from food intoxication, a legal upper limit of 100 mg histamine/kg of food has been suggested (Silla Santos 1996).

Regarding the influence of the biogenic amine content on the food quality, the total amount of vasoactive biogenic amines (tyramine, histamine, tryptamine, and 2-phenylethylamine) of 200 mg/kg has been proposed as a possible indicator of the Good Manufacturing Practices in sausage production (Eerola et al. 1996). In any case, the relevance of the biogenic amine content as a quality index of foods strongly depends on the nature of food considered. This seems to be a satisfactory index in fresh meat and fish and in heat-treated products, but its usefulness notably decreases in fermented products (Ruíz-Capillas and Jiménez-Colmenero 2004) where a considerable microbial growth is normal and even desirable to achieve high quality products. A "biogenic amine index" (BAI) was established for fish and seafoods (Mietz and Karmas 1977), and also for meat to evaluate degree of spoilage (Wortberg and Woller 1982, Hernández-Jover et al. 1996). However, due to the reasons previously indicated, the development and use of similar indexes for fermented meat products is risky and of very doubtful practical utility.

3.4 Control of Biogenic Amine Formation in Fermented Meat Products

Being the biogenic amines a product of the microbial metabolism, measures to reduce the formation of these metabolites during the manufacture of fermented meat products are focused to the control of the microbial groups that produce these compounds. Concrete measures can be divided in three groups: (i) Control quality of the raw materials in order to minimize the microbial load of meat and other ingredients and additives; (ii) Use of appropriate starter cultures for the control of spoilage microbiota, (iii) Use of spices and/or additives, and control of the environmental conditions during fermentation/ripening.

Meat after animal slaughtering is contaminated with microorganisms of diverse nature and origin (Iacumin and Carballo 2016). Microorganisms come firstly from skin and accumulated dirt in the hooves, which often contain faecal debris. Microorganisms from the skin are, above all, psychrotrophs. Microorganisms from skin also come from deposited dust particles that provide soil flora, mainly esporulated bacteria. Dirt from the hooves in particular provides microorganisms of intestinal origin (*Enterobacteriaceae*, enterococci and lactic acid bacteria [LAB]). Microorganisms of intestinal origin can also access the carcass as a result of defective evisceration, which causes the breakdown of intestinal loops and the output of intestinal content that contaminates meat. Traditional scalding operations consisting of immersion in water baths can increase the microbial load of the skin surface due to the intense contamination of the scalding water. Later, carcasses and meat cuts can be contaminated by handlers (hands and clothes). Hands can transfer microorganisms of nasal or faecal origin to meat as a result of handlers' poor personal hygiene; additionally, dirty clothes are important sources of contamination. In both cases, enterobacteria and psychrotrophs are the main microbial contaminants involved. The equipment used in the quartering rooms (knives, saws, tables, baskets, etc.) as well as the contaminated surfaces (floors and walls) are also a not negligible source of microorganisms; air can also be a source of contamination. Contamination from the environment in quartering rooms usually consists of psychrotrophic microorganisms.

As a global result of all these contaminations, the microbial load (viable counts) on the surface of pork cuts usually ranges from 3 to $5 \log cfu/cm^2$; among these, coliforms and psychrotrophs usually represent more than $1–2 \log cfu/cm^2$. Augustin and Minvielle (2008) reported mean counts ranging from 0.6 to 2.2 $\log cfu/cm^2$ for *Enterobacteriaceae* and from 1.1 to 4.4 $\log cfu/cm^2$ for *Pseudomonas* in pork meat cuts from cutting plants; the observed counts depended on the year of processing, the type of meat cut and mainly the cutting plant.

As indicated in previous sections of this Chapter, most of these microbial groups contaminating pork are effective producers of biogenic amines. Thus, increasing microbial loads in meat used as raw material must have a negative effect by increasing the biogenic amine formation in the manufactured sausages. In the case of fermented sausages, high contents of some biogenic amines may be present as a result of the use of raw materials having poor quality, contamination and inappropriate conditions during manufacture and storage (Brink et al. 1990, Bover-Cid et al. 2001a). In fact, biogenic amines were suggested as chemical indicators of the hygienic quality of fermented foods (Slemr 1981). The appropriate selection of meat raw materials and the control of thawing and storage conditions (time and temperature) are critical in reducing biogenic amines accumulation in dry sausages (Maijala et al. 1995a). In this line, it has been demonstrated that freezing meat destined to sausage manufacture for few days before sausage production helps to reduce enterobacteria development and cadaverine production during the ripening process (Bover-Cid et al. 2006).

Starter cultures are widely used in the manufacture of fermented sausages with the aim of improving quality, safety and acceptability of the manufactured products. These cultures generally comprise selected lactic acid bacteria (LAB) that "start" or initiate the production of lactic acid from carbohydrates. Starter cultures also contain adjunct cultures that include bacteria and other microorganisms that do not produce lactic acid as their main metabolic activity, but they are involved in the development and stability of the typical red colour of such products, via nitrate reductase activity, and they enhance both texture and flavour development due to the release of low molecular weight compounds, including peptides, amino acids, aldehydes, amines and free fatty acids. In addition to the expected decrease in pH, starter LAB may also inhibit the growth of undesirable microbiota via the production of bacteriocins and other substances, such as hydrogen peroxide. These bacteria can also yield a number of flavour compounds derived from carbohydrate and citrate metabolism, such as acetaldehyde, diacetyl and acetoin. Proteases and peptidases from starter and adjunct cultures participate in the proteolysis that takes place during ripening of meat products, thus contributing to texture development and to the formation of flavour compounds (e.g. aldehydes, alcohols, sulphur compounds) derived from amino acid catabolism. In short, starter cultures enhance shelf life and microbial safety, improve texture and contribute to the pleasant sensory profiles of the cultured meat products (Centeno and Carballo 2015).

Lactic acid bacteria produce lactic acid and other antimicrobial compounds with well-known roles in fermented sausages (hydrogen peroxide, carbon dioxide, acetaldehyde, diacetyl and bacteriocins), as well as other antimicrobial substances with minor roles (ethanol, formic acid, benzoic acid, bacteriolytic enzymes, free fatty acids and ammonia). All of them contribute to the increase of the shelf-life and safety of the manufactured products through the inhibition of the spoiling and pathogenic bacteria. Due to their inhibiting effect on the spoilage bacteria, starter cultures are effective in controlling the biogenic amine generation during sausage fermentation and ripening (Buncic et al. 1993, Hernández-Jover et al. 1997b, Bover-Cid et al. 2000, Bover-Cid et al. 2001a, 2001b, González-Fernández et al. 2003, Komprda et al. 2004, Latorre-Moratalla et al. 2010, Lu et al. 2010, Ciuciu Simion et al. 2014). Starter cultures decrease the biogenic amine formation, but do not prevent it. However, sometimes the increased proteolysis due to the use of starter cultures may increase the availability of amino acids and results in a higher biogenic amine formation. In other cases, concrete strains potentially usable as starter cultures have an important amino acid decarboxylase activity (Cachaldora et al. 2013). The use of spices

468 *Fermented Meat Products: Health Aspects*

or acidulants (deltagluconolactone), and the control of temperature and time of fermentation can also help to reduce the biogenic amines formation in fermented sausages (Buncic et al. 1993, Maijala et al. 1993, Maijala et al. 1995b, Paulsen and Bauer 1997, Komprda et al. 2004).

4 CONCLUSION

The formation of biogenic amines in fermented meats is a real concern affecting both the quality and safety of the final products. Studies carried out on this subject allowed to identify the main biogenic amines concerned and the microbial groups producers of these amines, and to stablish the environmental conditions in that these metabolites are preferably produced in order to implement control measures and strategies. However, more studies are needed to achieve an effective control of the production of biogenic amines in these products.

Key words: *biogenic amines, fermented sausages, health aspects, quality aspects, biogenic amine formation, biogenic amine control*

REFERENCES

Ababouch, L., M.E. Afilal, H. Benabdeljelil and F.F. Busta. 1991. Quantitative changes in bacteria, amino acids and biogenic amines in sardine (*Sardina pilchardus*) stored at ambient temperature (25–28°C) and in ice. Int. J. Food Sci. Tech. 26: 297–306.

Augustin, J.-C. and B. Minvielle. 2008. Design of control charts to monitor the microbiological contamination of pork meat cuts. Food Control 19: 82–97.

Bacellar Ribas Rodríguez, M., C. da Silva Carneiro, M. Barreto da Silva Feijoo, C.A. Conte Junior and S. Borges Mano. 2014. Bioactive amines: Aspects of quality and safety in food. Food Nutr. Sci. 5: 138–146.

Bardócz, S. 1995. Polyamines in food and their consequences for food quality and human health. Trends Food Sci. Tech. 6: 341–346.

Bartsch, H. and R. Montesano.1984. Relevance of nitrosamines to human cancer. Carcinogenesis 5: 1381–1393.

Bermúdez, R., J.M. Lorenzo, S. Fonseca, I. Franco and J. Carballo. 2012. Strains of *Staphylococcus* and *Bacillus* isolated from traditional sausages as producers of biogenic amines. Front. Microbiol. 3: Article no. 151.

Bonomo, M.G., A. Ricciardi, T. Zotta, M.A. Sico and G. Salzano. 2009. Technological and safety characterization of coagulase-negative staphylococci from traditionally fermented sausages of Basilicata region (Southern Italy). Meat Sci. 83: 15–23.

Bover-Cid, S. and W.H. Holzapfel. 1999. Improved screening procedure for biogenic amine production by lactic acid bacteria. Int. J. Food Microbiol. 53: 33–41.

Bover-Cid, S., M. Izquierdo-Pulido and M.C. Vidal-Carou. 1999a. Effect of proteolytic starter cultures of Staphylococcus spp. on biogenic amine formation during the ripening of dry fermented sausages. Int. J. Food Microbiol. 46: 95–104.

Bover-Cid, S., S. Schoppen, M. Izquierdo-Pulido and M.C. Vidal-Carou. 1999b. Relationship between biogenic amine contents and the size of dry-fermented sausages. Meat Sci. 51: 305–311.

Bover-Cid, S., M. Hugas, M. Izquierdo-Pulido and M.C. Vidal-Carou. 2000. Reduction of biogenic amine formation using a negative amino acid-decarboxylase starter culture for fermentation of Fuet sausages. J. Food Protect. 63: 237–243.

Bover-Cid, S., M. Izquierdo-Pulido and M.C. Vidal-Carou. 2001a. Effectiveness of *Lactobacillus sakei* starter culture in the reduction of biogenic amine accumulation as a function of the raw material quality. J. Food Protect. 64: 367–373.

Bover-Cid, S., M. Izquierdo-Pulido and M.C. Vidal-Carou. 2001b. Effect of the interaction between a low tyramine-producing *Lactobacillus* and proteolytic staphylococci on biogenic amine production during ripening and storage of dry sausages. Int. J. Food Microbiol. 65: 113–123.

Bover-Cid, S., M. Hugas, M. Izquierdo-Pulido and M.C. Vidal-Carou. 2001c. Amino-acid decarboxylase activity of bacteria isolated from fermented pork sausages. Int. J. Food Microbiol. 66: 185–189.

Bover-Cid, S., M. Izquierdo-Pulido and M.C. Vidal-Carou. 2001d. Changes in biogenic amine and polyamine content in in slightly fermented sausages manufactured with and without sugar. Meat Sci. 57: 215–221.

Bover-Cid, S., M.J. Miguélez-Arrizado, L.L. Latorre Moratalla and M.C. Vidal Carou. 2006. Freezing of meat raw materials affects tyramine and diamine accumulation in spontaneously fermented sausages. Meat Sci. 72: 62–68.

Brink, B. ten, C. Damirik, H.M.L.J. Joosten and J.H.J. Huis in't Veld. 1990. Occurrence and formation of biologically active amines in foods. Int. J. Food Microbiol. 11: 73–84.

Buncic, S., L. Paunovic, V. Teodorovic, D. Radisic, G. Vojinovic, D. Smiljanic and M. Baltic. 1993. Effects of gluconodeltalactone and *Lactobacillus plantarum* on the production of histamine and tyramine in fermented sausages. Int. J. Food Microbiol. 17: 303–309.

Butler, A. 2015. Nitrites and nitrates in the human diet: Carcinogens or beneficial hypotensive agents? J. Ethnopharmacol. 167: 105–107.

Cachaldora, A., S. Fonseca, I. Franco and J. Carballo. 2013. Technological and safety characterization of Staphylococcaceae isolated from Spanish traditional dry-cured sausages. Food Microbiol. 33: 61–68.

Centeno, J.A. and J. Carballo. 2015. Starter and adjunct microbial cultures used in the manufacture of fermented and/or cured or ripened meat and dairy products. pp. 35–54. *In*: V.R. Rai and J.A. Bai (eds). Beneficial microbes in fermented and functional foods. CRC Press, Boca Raton, Florida.

Chowdhury, S. 2014. N-Nitrosodimethylamine (NDMA) in food and beverages: A comparison in context to drinking water. Hum. Ecol. Risk Assess. 20: 1291–1312.

470 *Fermented Meat Products: Health Aspects*

Ciuciu Simion, A.M., C. Vizireanu, P. Alexe, I. Franco and J. Carballo. 2014. Effect of the use of selected starter cultures on some quality, safety and sensorial properties of Dacia sausage, a traditional Romanian dry-sausage variety. Food Control 35: 123–131.

Crampton, R.F. 1980. Carcinogenic dose-related response to nitrosamines. Oncology 37: 251–254.

Drosinos, E.H., S. Paramithiotis, G. Kolovos, I. Tsikouras and I. Metaxopoulos. 2007. Phenotypic and technological diversity of lactic acid bacteria and staphylococci isolated from traditionally fermented sausages in Southern Greece. Food Microbiol. 24: 260–270.

Durlu-Özkaya, F., K. Ayhan and N. Vural. 2001. Biogenic amines produced by Enterobacteriaceae isolated from meat products. Meat Sci. 58: 163–166.

Eerola, S., R. Maijala, A.-X. Roig Sagués, M. Salminen and T. Hirvi. 1996. Biogenic amines in dry-sausages as affected by starter culture and contaminant amine-positive *Lactobacillus*. J. Food Sci. 61: 1243–1246.

Eerola, H.S., A.X. Roig Sagués and T.K. Hirvi. 1998. Biogenic amines in Finnish dry-sausages. J. Food Safety 18: 127–138.

Even, S., S. Leroy, C. Charlier, N. Ben Zakour, J.-P. Chacornac, I. Lebert, E. Jamet, M.-H. Desmonts, E. Coton, S. Pochet, P.-Y. Donnio, M. Gautier, R. Talon and Y. Le Loir. 2010. Low occurrence of safety hazards in coagulase negative staphylococci isolated from fermented foodstuffs. Int. J. Food Microbiol. 139: 87–95.

Fonseca, S., L.I.I. Ouoba, I. Franco and J. Carballo. 2013. Use of molecular methods to characterize the bacterial community and to monitor different native starter cultures throughout the ripening of Galician Chorizo. Food Microbiol. 34: 215–226.

Gardini, F., M. Martuscelli, M.C. Caruso, F. Galgano, M.A. Crudele, F. Favati, M.E. Guerzoni and G. Suzzi. 2001. Effects of pH, temperature and NaCl concentration on the growth kinetics, proteolytic activity and biogenic amine production of *Enterococcus faecalis*. Int. J. Food Microbiol. 64: 105–117.

González-Fernández, C., E.M. Santos, I. Jaime and J. Rovira. 2003. Influence of starter cultures and sugar concentrations on biogenic amine contents in chorizo dry sausage. Food Microbiol. 20: 275–284.

Hagen, B.F., J.-L. Berdagué, A.L. Holck, H. Naes and H. Blom. 1996. Bacterial proteinase reduces maturation time of dry-fermented sausages. J. Food Sci. 61: 1024–1029.

Halász, A., A. Báráth, L. Simon-Sarkadi and W. Holzapfel. 1994. Biogenic amines and their production by microorganisms in food. Trends Food Sci. Tech. 5: 42–49.

Hernández-Jover, T., M. Izquierdo-Pulido, M.T. Veciana-Nogués and M.C. Vidal-Carou. 1996. Biogenic amine sources in cooked cured shoulder pork. J. Agr. Food Chem. 44: 3097–3101.

Hernández-Jover, T., M. Izquierdo-Pulido, M.T. Veciana-Nogués, A. Mariné-Font and M.C. Vidal-Carou. 1997a. Biogenic amine and polyamine contents in meat and meat products. J. Agr. Food Chem. 45: 2098–2102.

Hernández-Jover, T., M. Izquierdo-Pulido, M.T. Veciana-Nogués, A. Mariné-Font and M.C. Vidal-Carou. 1997b. Effect of starter cultures on biogenic amine formation during fermented sausage production. J. Food Protect. 60: 825–830.

Biogenic Amines in Fermented Meat Products

Hotchkiss, J.H. 1987. A review of current literature on N-nitroso compounds in foods. Adv. Food Res. 31: 53–115.

Huis in't Veld, J.H.J., H. Hose, G.J. Schaafsma, H. Silla and J.E. Smith. 1990. Health aspects of food biotechnology. pp. 273–297. *In*: P. Zeuthen, J.C. Cheftel, C. Ericksson, T.R. Gormley, P. Link and K. Paulus (eds.). Processing and quality of foods. Vol. 2. Food biotechnology: avenues to healthy and nutritious products. Elsevier Applied Science, London and New York.

Iacumin, L. and J. Carballo. 2016. Microbiological ecology of pork meat and pork products. In press. *In*: A. De Souza Sant'Ana (ed.). Quantitative Microbiology in Food Processing. Wiley-Blackwell, London.

Jansen, S.C., M. van Dusseldorp, K.C. Bottema and A.E.J. Dubois. 2003. Intolerance to dietary biogenic amines: a review. Ann. Allergy Asthma Immunol. 91: 233–241

Komprda, T., J. Neznalovà, S. Standara and S. Bover-Cid. 2001. Effect of starter culture and storage temperature on the content of biogenic amines in dry fermented sausage poličan. Meat Sci. 59: 267–276.

Komprda, T., D. Smelà, P. Pechovà, L. Kalhotka, J. Stencl and B. Klejdus. 2004. Effect of starter culture, spice mix and storage time and temperature on biogenic amine content of dry fermented sausages. Meat Sci. 67: 607–616.

Kordiovská, P., L. Vorlová, I. Borkovcová, R. Karpísková, H. Buchtová, Z. Svobodobá, M. Krizek and F. Vácha. 2006. The dynamics of biogenic amine formation in muscle tissue of carp (*Cyprinus carpio*). Czech. J. Anim. Sci. 51: 262–270.

Korovicová, J. and Z. Kohajdová. 2005. Biogenic amines in food. Chem. Pap. 59: 70–79.

Kurt, S. and Ö. Zorba. 2010. Biogenic amine formation in Turkish dry fermented sausage (sucuk) as affected by nisin and nitrite. J. Sci. Food Agr. 90: 2669–2674.

Larsson, S.C., L. Bergkvist and A. Wolk. 2006. Processed meat consumption, dietary nitrosamines and stomach cancer risk in a cohort of Swedish women. Int. J. Cancer 119: 915–919.

Latorre-Moratalla, M.L., S. Bover-Cid, R. Talon, M. Garriga, E. Zanardi, A. Ianieri, M.J. Fraqueza, M. Elías, E.H. Drosinos and M.C. Vidal-Carou. 2010. Strategies to reduce biogenic amine accumulation in traditional sausage manufacturing. LWT-Food Sci. Technol. 43: 20–25.

Latorre-Moratalla M.L., T. Veciana-Nogués, S. Bover-Cid, M. Garriga, T. Aymerich, E. Zanardi, A. Ianieri, M.J. Fraqueza, L. Patarata, E.H. Drosinos, A. Lauková, R. Talon and M.C. Vidal-Carou. 2008. Biogenic amines in traditional fermented sausages produced in selected European countries. Food Chem. 107: 912–921.

Lorenzo, J.M., A. Cachaldora, S. Fonseca, M. Gómez, I. Franco and J. Carballo. 2010. Production of biogenic amines "in vitro" in relation to the growth phase by Enterobacteriaceae species isolated from traditional sausages. Meat Sci. 86: 684–691.

Lorenzo, J.M., S. Martínez, I. Franco and J. Carballo. 2008. Biogenic amine content in relation to physico-chemical parameters and microbial counts in two kinds of Spanish traditional sausages. Arch. Lebensmittelhyg. 59: 70–75.

Martín, B., M. Garriga, M. Hugas, S. Bover-Cid, M.T. Veciana-Nogués and T. Aymerich. 2006. Molecular, technological and safety characterization of

Gram-positive catalase-positive cocci from sligtly fermented sausages. Int. J. Food Microbiol. 107: 148–158.

Lu, S., X. Xu, G. Zhou, Z. Zhu, Y. Meng and Y. Sun. 2010. Effect of starter cultures on microbial ecosystem and biogenic amines in fermented sausages. Food Control 21: 444–449.

Maga, J.A. and I. Katz. 1978. Amines in foods. Crit. Rev. Food Sci. Nutr. 10: 373–403.

Maijala, R.L., S.H. Eerola, S. Aho and J. Hirn. 1993. The effect of GDL-induced pH decrease on the formation of biogenic amines in meat. J. Food Protect. 56: 125–129.

Maijala, R., S. Eerola, S. Lievonen, P. Hill and T. Hirvi. 1995a. Formation of biogenic amines during ripening of dry sausages as affect by starter cultures and thawing time of raw materials. J. Food Sci. 60: 1187–1190.

Maijala, R., E. Nurmi and A. Fischer. 1995b. Influence of processing temperature on the formation of biogenic amines in dry sausages. Meat Sci. 39: 9–22.

Masson, F., R. Talon, and M.C. Montel. 1996. Histamine and tyramine production by bacteria from meat products. Int. J. Food Microbiol. 32: 199–207.

McCabe-Sellers, B.J., C.G. Staggs and M.L. Bogle. 2006. Tyramine in foods and monoamine oxidase inhibitor drugs: A crossroad where medicine, nutrition, pharmacy, and food industry converge. J. Food Compos. Anal. 19 (Special Issue): S58–S65.

Mietz, J.L. and E. Karmas. 1977. Chemical quality index of canned tuna as determined by high-pressure liquid chromatography. J. Food Sci. 42: 155–158.

Montel, M.-C., F. Masson and R. Talon. 1999. Comparison of biogenic amine content in traditional and industrial French dry sausages. Sci. Aliment. 19: 247–254.

Oka, S., K. Fukunaga, H. Ito and K. Takama. 1993. Growth of histamine producing bacteria in fish filets under modified atmospheres. Bull. Fac. Fisher. Hokkaido Univ. 44: 46–54.

Omura, Y., R.J. Price and H.S. Olcott. 1978. Histamine-forming bacteria isolated from spoiled skipjack tuna and jack mackerel. J. Food Sci. 43: 1779–1781.

Parente, E., M. Martuscelli, F. Gardini, S. Grieco, M.A. Crudele and G. Suzzi. 2001. Evolution of microbial populations and biogenic amine production in dry sausages produced in Southern Italy. J. Appl. Microbiol. 90: 882–891.

Paulsen, P. and F. Bauer. 1997. Biogenic amines in fermented sausages: 2. Factors influencing the formation of biogenic amines in fermented sausages. Fleischwirtshaft Int. 77: 32–34.

Pircher, A., F. Bauer and P. Paulsen. 2007. Formation of cadaverine, histamine, putrescine and tyramine by bacteria isolated from meat, fermented sausages and cheeses. Eur. Food Res. Technol. 226: 225–231.

Roig-Sagués, A. X., M. Hernández-Herrero, E.I. López-Sabater, J.J. Rodríguez-Jerez and M.T. Mora-Ventura. 1996. Histidine decarboxylase activity of bacteria isolated from raw and ripened Salchichón, a Spanish cured sausage. J. Food Protect. 59: 516–520.

Ruiz-Capillas, C. and F. Jiménez-Colmenero. 2004. Biogenic amines in meat and meat products. Crit. Rev. Food Sci. Nutr. 44: 489–499.

Shalaby, A.R. 1993. Survey on biogenic amines in Egyptian foods: Sausage. J. Sci. Food Agr. 62: 291–293.

Shalaby, A.R. 1996. Significance of biogenic amines to food safety and human health. Food Res. Int. 29: 675–690.

Silla Santos, H. 1996. Biogenic amines: their importance in foods. Int. J. Food Microbiol. 29: 213–231.

Sinell, H.J. 1978. Biogene Amine als Risikofaktoren in der Fischhygiene. Arch. Lebensmittelhyg. 29: 206–210.

Slemr, J. 1981. Biogene Amine als potentieller chimischer Qualitätsindikator für Fleisch. Fleischwirtschaft 61: 921–926.

Stadnik, J. and Z.J. Dolatowski. 2010. Biogenic amines in meat and fermented meat products. Acta Sci. Pol., Technol. Aliment. 9: 251–263.

Stratton, J.E., R.W. Hutkins and S.L. Taylor. 1991. Biogenic amines in cheese and other fermented foods: A review. J. Food Protect. 54: 460–470.

Suzzi, G. and F. Gardini. 2003. Biogenic amines in dry fermented sausages: a review. Int. J. Food Microbiol. 88: 41–54.

Tabanelli, G., F. Coloretti, C. Chiavari, L. Grazia, R. Lanciotti and F. Gardini. 2012. Effects of starter cultures and fermentation climate on the properties of two types of typical Italian dry fermented sausages produced under industrial conditions. Food Control 26: 416–426.

Taylor, S.L. 1986. Histamine food poisoning, toxicology and clinical aspects. Crit. Rev. Toxicol. 17: 91–128.

Tricker, A.R. and R. Preussmann. 1991. Carcinogenic N-nitrosamines in the diet: occurrence, formation, mechanisms and carcinogenic potential. Mut. Res. 259: 277–289.

Vandekerckhove, P. 1977. Amines in dry fermented sausage. J. Food Sci. 42: 283–285.

Vidal-Carou, M.C., M.L. Izquierdo, M.C. Martín and A. Mariné. 1990. Histamina y tiramina en derivados cárnicos. Rev. Agroquím. Tecnol. Aliment. 30: 102–108.

Watts, D.A. and W.D. Brown. 1982. Histamine formation in abusively stored Pacific mackerel: Effect of CO_2-modified atmosphere. J. Food Sci. 47: 1386–1387.

Wendakoon, C.N. and M. Sakaguchi. 1992. Effects of spices on growth and biogenic amine formation by bacteria in fish muscle. pp. 306–313. In: H.H. Huss et al. (eds.). Quality assurance in the fish industry. Elsevier Science Publisher B.V., Amsterdam.

Wortberg, B. and R. Woller. 1982. Zur Qualität und Frische von Fleisch und Fleischwaren im Hinblick auf irhen Gehalt an biogenen Aminen. Fleischwirtschaft 62: 1457–1460, 1463.

Xie, C., H.H. Wang, H.K. Nie, L. Chen, S.L. Deng and X.L. Xu. 2015. Reduction of biogenic amine concentration in fermented sausage by selected starter cultures. CYTA-J. Food 13: 491–497.

Chapter 20

Fat Content of Dry-Cured Sausages and Its Effect on Chemical, Physical, Textural and Sensory Properties

José M. Lorenzo*, Daniel Franco and Javier Carballo

1 INTRODUCTION

Dry-cured sausages are meat products with high fat content. Commercial sausages have fat contents around 32% immediately after manufacture, but as a result of drying this rises to about 45–60% (Gómez and Lorenzo 2013). The fat is responsible for various properties of dry-fermented sausages: (i) fat contributes to the flavour, texture, mouthfeel, juiciness and lubricity, which determine the quality and acceptability of dry sausages and (ii) the granulated fat has a technological function in the manufacture of dry-fermented sausages as it helps to loosen up the sausage mixture to facilitate the continuous release of moisture from the inner layer of the sausage; a process necessary for undisturbed fermentation and flavour development (Wirth 1988). In addition, fat has an important role in reducing manufacturing costs because it generally has less economic value than lean meat.

On the other hand, fat has an important role in the nutritional quality of meat products because it is the source of liposoluble vitamins and

---

*For Correspondence: Centro Tecnológico de la Carne de Galicia, Rua Galicia no. 4, Parque ecnológico de Galicia, San Cibrao das Viñas, 32900 Ourense, Spain. Tel: +34-988-548277, Fax: +34-988-548277, Email: jmlorenzo@ceteca.net

essential fatty acids. However, the most commonly used fat in fermented sausages has a high level of saturated fatty acids due to its capacity to remain solid at room temperature. The high level of saturated fatty acids constitutes a major nutritional problem because governmental agencies recommend reduced saturated fat consumption as a way to reduce risk factors in the development of cardiovascular diseases (British Nutrition Foundation 2009, Joint WHO/FAO Expert Consultation 2003), which are the principal cause of death in developed countries (World Health Organization 2009).

Dietary fat impacts, both positively and negatively, on human health. The role of fat and fatty acid composition has been frequently reviewed, and proposals were made for optimum fat profiles and contribution of various foods to achieve a balance of individual fatty acid appropriate for our health. The FAO paper no. 91 on fats and fatty acids in human nutrition (Food and Agriculture Organization of the United Nations 2010) provides a relatively recent consensus on acceptable guidelines. Whilst a maximum of 35% of energy intake from fat is the recognized upper limit, a minimum dietary contribution of 15–20% is also important for health to ensure sufficient energy, fat soluble antioxidants and vitamins and essential fatty acid intake.

The high fat content of dry-fermented sausages (40–50%) is essential for sensory properties, such as hardness, juiciness and flavour, and for technological functions (Wirth 1988). However, from a health point of view, as comment previously, excessive fat intake is not recommended. For this reason, in recent years, increased concerns about the potential health risks associated with the consumption of high fat foods has led the food industry to develop new formulations or modify traditional food products to contain less fat (Mendoza et al. 2001). The Regulation (EC) No. 1924/2006 of the European Parliament and the claims of the Council on Nutrition and Health made on foods consider a product as reduced fat when the reduction in fat content is at least 30% compared to a similar product. Likewise, this regulation considers a food as energy-reduced when the total energy value is reduced by at least 30%.

One of the strategies for the development of low-fat fermented sausages was the reduction of fat content and the simultaneous addition of non-lipid fat replacers to minimize texture defects (Muguerza et al. 2004). In this regard, the addition of inulin, cereal and fruit fibres, and short-chain fructooligosaccharides gave satisfactory results for the reduction of fat content in dry-fermented sausages (dos Santos et al. 2012, Menegas et al. 2013, Petersson et al. 2014a, b, Salazar et al. 2009, Tomaschunas et al. 2013). Other strategies were focused on the replacement of pork back fat by olive oil in order to have a positive effect on consumer health (Beriain et al. 2011, del Nobile et al. 2009,

476   *Fermented Meat Products: Health Aspects*

Jiménez-Colmenero et al. 2013, Koutsopoulos et al. 2008, Triki et al. 2013a, b, Utrilla et al. 2014).

2 EFFECT OF FAT CONTENT ON PHYSICO-CHEMICAL PROPERTIES OF DRY-CURED SAUSAGES

The influence of fat content on pH values, chemical composition and TBARS values is presented in Table 1. The fat content of sausages has an effect on final pH values. Gómez and Lorenzo (2013) observed pH values ranging from 6.07 to 5.65 for "chorizo" (a Spanish dry-cured sausage) manufactured with 10, 20 and 30% of fat, and the "chorizo" manufactured with high fat content presented lower pH values compared to those manufactured with low fat content. A similar behavior was observed by Fonseca et al. (2015) who found the lowest pH values in "salchichón" manufactured with high fat content (5.74, 5.67 and 5.53, for 10, 20 and 30% fat, respectively) and by Beriain et al. (2000) who reported lower pH values in "salchichón" manufactured with high fat content. In addition, Bovolenta et al. (2008) also noticed lower pH values in "Pitina" sausages manufactured with 30% of fat compared to those manufactured with 10% of fat. However, Olivares et al. (2010) showed higher pH values in "salchichón" manufactured with high fat content (4.50 *vs.* 4.60 *vs.* 4.66, for 10, 20 and 30% of fat, respectively). Finally, Lorenzo and Franco (2012) also noticed higher pH values in foal "salchichón" manufactured with high fat content.

Regarding chemical composition, fat content also have a significant ($P<0.05$) effect. Fonseca et al. (2015) found that moisture content was not significantly different in the 10% and 20% fat batches (26.08% and 27.71%, respectively), while they were significantly ($P<0.05$) lower in the 30% fat batch (23.14%). Lorenzo and Franco (2012) also observed that low fat foal sausages had higher water content than the high fat batches (33.76% *vs.* 31.71%, $P<0.001$). To this regards, moisture content showed significant ($P<0.001$) differences among the three batches following the relationship low fat > medium fat > high fat (Gómez and Lorenzo 2013).

On the other hand, the fat content of "salchichón", as expected, was significantly different ($P<0.05$) among batches, with mean values of 42.35 in the 10% fat batch, 50.48 in the 20% fat batch and 58.34 in the 30% fat batch, expressed as g/100 g of dry matter (Fonseca et al. 2015). Gómez and Lorenzo (2013) also observed significant ($P<0.001$) differences on fat content among batches, with mean values of 47.4%, 57.3% and 60.9% for "chorizos" manufactured with 10%, 20% and 30% of fat, respectively. The same trend was showed by Beriain et al. (2000) who found higher fat content in "salchichón" manufactured with 30%

Fat Content of Dry-Cured Sausages and Its Effect on Chemical,....		477

of fat (58.7%) compared to those manufactured with 10% of fat (21.1%) and by Olivares et al. (2010) who observed lower fat content in sausages manufactured with less fat (34.9% *vs.* 37.1% *vs.* 42.7% for "salchichón" manufactured with 10%, 20% and 30%, respectively). Finally, Bovolenta et al. (2008) also noticed significant differences ($P<0.01$) on fat content between sausages (27.5% *vs.* 54.20% for sausages manufactured with 10% and 20% of pork lard, respectively).

Protein content was also different ($P<0.01$), presenting mean values of 40.12% in the 10% fat batch, 33.95% in the 20% fat batch and 28.07% in the 20% fat batch, expressed as g/100 g of dry matter (Fonseca et al. 2015). A similar behavior was observed by Beriain et al. (2000) who found higher protein content in sausages manufactured with low fat content (33.6% *vs.* 33.6% *vs.* 28.9% for "salchichón" manufactured with 20%, 25% and 30% of fat, respectively) and by Gómez and Lorenzo (2013) who reported lower protein content in "chorizo" manufactured with high fat content (37.6% *vs.* 33.3% *vs.* 28.0%; $P<0.001$, for sausages manufactured with 10%, 20% and 30%, respectively). Bovolenta et al. (2008) also showed higher protein content in "Pitina" sausages manufactured with 10% of fat compared to those manufactured with 30% of fat.

The effect of fat content on lipid oxidation (evaluated by measuring the TBARS index) is summarized in Table 1. Lorenzo and Franco (2012) observed significantly ($P<0.001$) higher TBARS values in foal sausages manufactured with high fat content (0.98 *vs.* 1.40 *vs.* 1.66 mg MDA/kg meat for "salchichón" manufactured with 10%, 20% and 30%, respectively). In addition, these authors suggested a positive correlation between fat content and TBARS values ($r = 0.826$; $P<0.001$). A similar trend was reported by Olivares et al. (2011) who found higher TBARS values in "salchichón" elaborated with 30% of fat (1.7 mg MDA/kg meat) compared to those manufactured with 10% of fat (1.3 mg MDA/kg meat) and by Gómez and Lorenzo who detected higher TBARS levels in "chorizo" manufactured with 30% of fat (0.92 mg MDA/kg meat) compared to those elaborated with 10% of fat (0.43 mg MDA/kg meat). In opposite, Fonseca et al. (2015) observed higher TBARS values in low fat sausages, a fact that the authors attributed to the higher oxidation status of intramuscular fat whose percentages increased with the reduction in fat. Finally, Bovolenta et al. (2008) reported similar TBARS values for sausages manufactured with 10 and 30% of fat (1.79 and 1.77 mg MDA/kg meat).

Table 2 shows the influence of fat content on color parameters and TPA traits. In general, the color parameters (L*, a* and b* values) of the sausages are affected by fat content. Gómez and Lorenzo (2013) observed that L* values increased with fat content (28.7 *vs.* 35.2 *vs.* 38.2 for sausages manufactured with 10, 20 and 30% of fat, respectively).

Table 1 Effect of fat content on pH values and chemical composition of dry-cured sausages

Reference	Gómez and Lorenzo 2013			Fonseca et al. 2015			Bovolenta et al. 2008		Beriain et al. 2000			Olivares et al. 2010, 2011			Lorenzo and Franco 2012		
Sausage variety	"Chorizo"			"Salchichón"			"Pitina"		"Salchichón"			"Salchichón"			Dry-cured foal sausage		
Fat content (%)	10	20	30	10	20	30	10	30	20	25	30	10	20	30	10	20	30
pH	6.07	5.84	5.65	5.74	5.67	5.53	5.52	5.37	4.87	4.62	4.66	4.50	4.60	4.66	5.97	5.94	6.00
Moisture (%)	20.19	18.07	16.42	26.08	27.71	23.14	29.20	26.30	39.49	38.52	37.29	36.98	35.17	33.56	33.76	33.52	31.71
Fat content (% of d.m.)	47.44	57.27	60.88	42.35	50.48	58.34	27.50	54.20	51.11	54.70	58.66	34.91	37.11	42.70	–	–	–
Protein content (% of d.m.)	37.60	33.28	28.00	40.12	33.95	28.07	55.20	32.8	33.59	33.64	28.92	56.57	55.31	49.92	–	–	–
TBARS (mg MDA/kg)	0.43	0.49	0.92	0.85	0.75	0.58	1.79	1.77	–	–	–	1.3	1.4	1.7	0.98	1.40	1.66

d.m. = dry matter

Table 2 Effect of fat content on colour parameters and texture traits of dry-cured sausages

Reference	Gómez and Lorenzo 2013			Fonseca et al. 2015			Olivares et al. 2010, 2011			Lorenzo and Franco 2012		
Sausage variety	"Chorizo"			"Salchichón"			"Salchichón"			Dry-cured foal sausage		
Fat content (%)	10	20	30	10	20	30	10	20	30	10	20	30
Color parameters												
Lightness (L*)	28.7	35.2	38.2	31.2	33.2	45.9	44.0	45.7	47.2	29.2	32.1	34.3
Redness (a*)	20.5	23.2	25.5	17.47	17.40	15.58	16.1	16.0	16.1	15.7	14.8	13.9
Yellowness (b*)	18.2	24.7	28.7	10.88	10.58	11.23	6.0	5.8	6.6	5.1	5.8	6.8
TPA test												
Hardness (N)	126.6	65.9	34.0	417.7	356.6	199.8	242.3	223.6	210.6	539.1	525.3	446.9
Springiness	0.41	0.36	0.34	0.49	0.50	0.52	0.78	0.80	0.76	0.65	0.55	0.53
Cohesiveness	0.21	0.23	0.21	0.39	0.39	0.33	0.78	0.77	0.77	0.45	0.44	0.42
Chewiness (N x mm)	11.3	5.5	2.5	78.5	69.5	34.10	147.3	137.6	124.2	243.1	126.5	98.9

Fat Content of Dry-Cured Sausages and Its Effect on Chemical,.... 479

The same trend was reported by Fonseca et al. (2015) in "salchichón", since the highest lightness values were found in 30% fat sausages (31.2 *vs.* 33.2 *vs.* 45.9 for "salchichón" manufactured with 10%, 20% and 30%, respectively), and by Lorenzo and Franco (2012) who notice the lowest lightness values in foal sausages manufactured with low fat. Regarding redness (a* values), Fonseca et al. (2015) and Lorenzo and Franco (2012) observed higher a* values in sausages manufactured with low fat content compared to those manufactured with high fat content. However, Gómez and Lorenzo (2013) noticed the opposite conclusion when studying the redness in "chorizo". In addition, Olivares et al. (2010) reported similar a* values among sausages manufactured with different fat content. Finally, Gómez and Lorenzo (2012) and Lorenzo and Franco (2012) found the highest yellowness (b* values) in sausages manufactured with high fat content. However, Fonseca et al. (2015) and Olivares et al. (2010) did not show significant differences on b* values among sausages manufactured with different amount of fat.

On the other hand, fat content has an influence on the texture properties of sausages. Fonseca et al. (2015) observed significant ($P<0.01$) differences when comparing the hardness and chewiness of 10%, 20% and 30% fat sausages at the end of the ripening, showing these latter the lowest values for both parameters, representing around half of those exhibited by 10% and 20% fat batches. These results are in agreement with those found by Lorenzo and Franco (2012) and Gómez and Lorenzo (2013), who reported higher hardness in low fat sausages compared to high fat ones, probably due to a more pronounced moisture loss in sausages with higher proportions of lean meat. However, springiness and cohesiveness were in general not significantly ($P>0.05$) affected by fat content (Fonseca et al. 2015, Olivares et al. 2010).

3 EFFECT OF FAT CONTENT ON FATTY ACID COMPOSITION OF DRY-CURED SAUSAGES

The influence of fat content on the fatty acid profile of dry-cured sausages is shown in Table 3. The relative percentages of most fatty acid differed significantly ($P<0.05$) as a function of the amount of added fat. Bovolenta et al. (2008), Leite et al. (2015) and Romero et al. (2013) observed that low fat sausages presented the highest levels of C14:0, C14:1 and C18:0, and the lowest levels of C18:1n-9 and C18:2n-6. So, the increase of fat content modified the total fatty acid profile, prompting a significant drop in the relative percentages of C14:0, and C18:0, together with a marked increase in C18:1n-9 and C18:2n-6 acids, as it has been expected once they are the predominant fatty acids in pork backfat (Lorenzo et al. 2012a).

Table 3 Effect of fat content on fatty acid profile of dry-cured sausages

Reference	Leite et al. 2015						Bovolenta et al. 2008		Romero et al. 2013	
Sausage variety	Goat meat sausages			Sheep meat sausages			"Pitina"		"Salami"	"Chorizo"
Fat content (%)	5	10	30	5	10	30	10	30	30	20
Fatty acid										
C12:0	0.14	0.13	0.13	0.10	0.11	0.12	0.12	0.14	–	–
C14:0	2.56	1.94	1.68	2.38	2.07	1.85	1.74	1.67	1.69	2.53
C14:1	0.42	0.12	0.07	0.17	0.09	0.06	0.20	0.06	0.06	0.28
C16:0	24.21	23.15	22.42	22.70	22.95	22.75	23.00	24.98	22.45	25.71
C16:1	2.57	2.20	2.07	2.31	2.14	2.03	1.98	2.16	2.43	3.05
C17:0	1.21	0.67	0.48	1.34	1.00	0.61	–	–	0.76	0.84
C17:1	0.99	0.46	0.34	0.73	0.51	0.36	0.00	0.38	0.43	0.53
C18:0	18.97	15.43	12.96	18.50	16.51	13.50	12.66	12.27	13.08	14.72
C18:1n-9t	0.61	0.73	0.39	0.55	0.76	0.43	0.39	0.06	0.63	0.53
C18:1n-9c	38.91	41.36	42.48	40.93	41.28	42.00	43.49	43.93	40.69	39.90
C18:1n-11t	1.47	0.86	0.58	2.25	1.72	0.84	–	–	–	–
C18:1n-7c	0.94	1.52	2.38	1.14	1.66	2.25	–	–	–	–
C18:2n-6c	3.74	8.22	10.60	3.91	6.62	9.96	10.03	10.40	13.90	8.02
C18:3n-3c	0.52	0.61	0.68	0.76	0.75	0.75	0.31	0.16	1.65	1.19
C20:0	0.09	0.22	0.24	0.14	0.18	0.23	–	–	0.06	0.08
C20:1n-9	0.12	0.74	0.99	0.13	0.55	0.53	–	–	–	–
SFA	47.79	41.86	38.12	45.82	43.29	39.35	38.17	39.24	39.08	45.60
MUFA	46.01	47.69	49.29	48.23	48.48	48.84	46.35	46.78	44.31	44.34
PUFA	6.20	10.45	12.59	5.94	8.22	11.80	15.48	13.98	16.56	9.94
MUFA+PUFA	52.21	58.14	61.87	54.18	56.70	60.65	61.83	60.76	60.87	54.28
PUFA/SFA	0.13	0.25	0.33	0.13	0.19	0.30	0.41	0.36	0.46	0.22
PUFA-n-3	1.08	1.35	0.97	1.11	1.07	0.98	1.62	0.94	1.82	1.26
PUFA-n-6	5.12	9.47	11.62	4.83	7.47	10.83	11.31	11.00	14.04	8.16
PUFA-n-6/ PUFA-n-3	4.78	7.43	11.96	4.38	7.08	11.09	7.44	11.68	8.07	4.19

SFA: saturated fatty acids; MUFA: monounsaturated fatty acids;
PUFA: polyunsaturated fatty acids

The relative percentages of most fatty acids differed significantly as a function of the percentage of fat content. The percentage of saturated fatty acid (SFA) and polyunsaturated fatty acid (PUFA) was affected by fat content of dry-cured sausages. The total amount of SFA in sausages decreased by approximately 2–9% with the increase of the percentage of fat included in the sausage formulation compared to those low fat sausages. Contrarily, the amount of PUFA increased a 40–50% with the fat level in relation to the low fat sausages. To this regards, dietary

Fat Content of Dry-Cured Sausages and Its Effect on Chemical,.... 481

guidelines have recommended avoiding saturated fat in order to prevent cardiovascular disease (Krauss et al. 2000). The mechanisms through which SFA exert pejorative effects in cardiovascular and general metabolic health are diverse. In this line, Kennedy et al. (2009) have proposed that an excessive consumption of SFA could promote white adipose tissue expansion and hypertrophy leading to apoptosis. These phenomena would promote the release of inflammatory proteins like cytokines and chemokines inducing inflammation and insulin resistance, thus increasing the risk of cardiovascular disease and metabolic syndrome (Haffner 2006, Willerson and Ridker 2004). In addition, the dietary intake of the long chain n-3 PUFAs, EPA and DHA has been positively associated with a number of health benefits including prevention and reduction of cardiovascular diseases (Roth and Harris 2010), inflammation and some inflammatory diseases (Calder 2006) as well as prevention or reduction of some types of cancer (Larsson et al. 2004).

4 EFFECT OF FAT CONTENT ON VOLATILE COMPOUNDS OF DRY-CURED SAUSAGES

The fat content has a significant ($P<0.05$) influence on volatile compounds. To this regards, higher amount of total volatile compounds was observed in the low fat sausages compared to high fat sausages (Gómez and Lorenzo 2013). Esters are the most abundant volatile compounds detected in dry-cured sausages showing higher values in low fat sausages (Gómez and Lorenzo 2013, Lorenzo et al. 2012b, Olivares et al. 2011). Esters are formed through the enzymatic esterification of fatty acids and alcohols during curing, mostly by the action of microorganisms such as lactic acid bacteria and *Micrococcaceae* (Purriños et al. 2011). This chemical family can modulate the global flavour due to their low odour thresholds, imparting fruity notes, mainly those formed from short-chain acids (Théron et al. 2010), whereas esters with long-chain acids have a slight fatty odor (Narváez-Rivas et al. 2012). Esters strongly affect the flavor of dry-cured meat products such as the typical aged meat products; in particular, the methyl branched short-chain esters have been associated with *"ripened flavor"* (Montel et al. 1996).

The second most important volatile compounds are the alcohols, which showed a greater content in high fat sausages (Lorenzo et al. 2012b, Olivares et al 2011). However, Gómez and Lorenzo (2013) observed higher amount of alcohols in sausages manufactured with low fat content. Alcohols originate from lipid oxidation, from the reduction of the corresponding aldehydes and methyl-ketones through the activity of lactic acid bacteria dehydrogenases and biochemical reactions or, in the case of short branched-chain alcohols, from the Strecker degradation

482   *Fermented Meat Products: Health Aspects*

of amino acids (Leroy et al. 2009, Rivas-Cañedo et al. 2011, Théron et al. 2010). Alcohols, because of their low odor threshold, contribute to the aroma of sausages, with fatty, woody and herbaceous notes (García and Timón 2001).

The fat content also induced significant ($P<0.05$) differences on ketones, since the lowest amounts were detected in sausages manufacture with low fat content (Gómez and Lorenzo 2013). Aliphatic ketones are known to originate from lipid oxidation and their methylated forms from the β-oxidation of unsaturated fatty acids (Poligné et al. 2001) while cyclopentones are typical volatiles of wood smoke (Yu et al. 2008).

On the other hand, aliphatic and aromatic hydrocarbons are not affected by fat content (Gómez and Lorenzo 2013). In general, significant differences ($P>0.05$) on hydrocarbons amounts were not found when compared low, medium and high fat sausages, except for hexane and octane, which showed a higher and a lower ($P<0.001$) content, respectively, with the increase of fat. Hydrocarbons in meat products come from fat degradation and chemical auto-oxidation (Leroy et al. 2009), and they are generally considered to have no substantial impact on flavour because of their high odour threshold values.

The amount of fat content has a significant ($P<0.05$) effect on aldehyde content. Aldehydes are known as the major contributors to the unique flavour of dry-cured meat products due to their rapid formation during lipid oxidation and their low odour thresholds (Ramírez and Cava 2007). Linear aldehydes are known to originate from the auto-oxidation of unsaturated fatty acids (Leroy et al. 2009), and branched-chain aldehydes from the Strecker degradation of amino acids (Théron et al. 2010). Among linear saturated aldehydes, hexanal was the most abundant compound, showing a lower amount in sausages manufactured with low fat content compared to those manufactured with high fat content (Gómez and Lorenzo 2013). To this regards, hexanal is generally considered as a good indicator of the oxidation level, and high concentrations of hexanal indicate flavour deterioration in meat products often resulting in a rancid aroma (Pham et al. 2008, Ramirez and Cava 2007).

5   EFFECT OF FAT CONTENT ON SENSORY PROPERTIES OF DRY-CURED SAUSAGES

The amount of fat content has a significant ($P<0.05$) effect on sensory properties of dry-cured sausages. Within appearance, sausages manufactured with high fat content showed higher scores for fat distribution (7.6 *vs.* 6.2 *vs.* 5.6, $P>0.05$, for sausages manufactured with 30, 20 and 10% of fat, respectively), and fat/lean ratio (7.3 *vs.* 4.7 *vs.* 2.1, $P<0.001$, for sausages manufactured with 30, 20 and 10% of fat,

Fat Content of Dry-Cured Sausages and Its Effect on Chemical,....   483

respectively), while they showed lower scores for fat/lean cohesiveness (6.1 *vs.* 6.0 *vs.* 7.1, $P>0.05$, for sausages manufactured with 30, 20 and 10% of fat, respectively) (Lorenzo and Franco 2012).

On the other hand, the reduction of fat in fermented sausages gave controversial results in relation to flavour. While Mendoza et al. (2001) and Olivares et al. (2010) obtained higher aroma scores in high fat dry-fermented sausages compared to low fat ones, other authors reported no differences in flavour (Liaros et al. 2009, Muguerza et al. 2002). Muguerza et al. (2003) reported an increase in oxidation and total volatile compounds in fat reduced sausages that was attributed to the higher intramuscular fat content of reduced fat products. Also, Chevance et al. (2000) indicated that fat reduction in salami increased the release of odour compounds.

Fat level significantly affected the texture attributes. Low fat dry-cured sausages showed high scores for hardness (5.1 *vs.* 3.2 *vs.* 2.1, $P<0.01$, for sausages manufactured with 10, 20 and 30% of fat, respectively). This outcome was in agreement with Papadima and Bloukas (1999) and Gómez and Lorenzo (2013) who reported that sausages with 30% of fat had the lowest scores for hardness when sausages were produced with three different fat levels (10%, 20% and 30% of fat). The observed differences in the perception of hardness attribute provided evidence that the lower abundance of fat content within the ground lean meat results in a higher force required to penetrate the sausage, and a lower degree of deformation of the dry-cured sausage before breaking. Contrary, the effect of fat content on juiciness exhibited an inverse trend, since the higher values were observed for the high fat sausages (4.8 *vs.* 6.8 *vs.* 7.4, $P>0.05$, for sausages manufactured with 10, 20 and 30% of fat, respectively). To this regards, juiciness is linked to the lubrication degree of food during chewing and swallowing and, due to the contribution of fat on saliva secretion, high fat foods increase juiciness by saliva coating tongue, teeth and other parts of the mouth (Fonseca et al. 2015).

In general, studies for sensorial analysis comparing dry-ripened sausages with different fat levels reported better sensory characteristics for low and medium fat sausages (Lorenzo and Franco 2012).

6 CONCLUSION

Results from this study showed that the fat content had an effect on physico-chemical properties, nutritional value and sensory characteristics of dry-cured sausages. Individual fatty acids have diverse effects on human health. Heart diseases have been reported to be favorably affected by the consumption of certain unsaturated fatty acids. Unsaturated fatty acids lower plasma total chlolesterol and "bad" cholesterol levels when

484 Fermented Meat Products: Health Aspects

substituted for saturated fatty acids. However, *trans* monounsaturated fatty acids were found to be intermediate between cis-monounsaturated fatty acid and long-chain saturated fatty acids in their effects on plasma total and "bad" cholesterol concentrations.

Key words: *fat level, fatty acid profile, nutritional value, chemical composition*

REFERENCES

Beriain, M.J., J. Chasco and G. Lizaso. 2000. Relationship between biochemical and sensory quality characteristics of different commercial brands of salchichón. Food Control 11: 231–237.

Beriain, M.J., I. Gómez, E. Petri, K. Insausti and M.V. Sarriés. 2011. The effects of olive oil emulsified alginate on the physico-chemical, sensory, microbial, and fatty acid profiles of low-salt, inulin-enriched sausages. Meat Sci. 88: 189–197.

Bovolenta, S., D. Boscolo, S. Dovier, M. Morgante, A. Pallotti and E. Piasentier. 2008. Effect of pork lard content on the chemical, microbiological and sensory properties of a typical fermented meat product (Pitina) obtained from Alpagota sheep. Meat Sci. 80: 771–779.

British Nutrition Foundation. 2009. Nutrient requirements and recommendations. http://www.nutrition.org.uk/upload/CVD%20pdf%20for%20website.pdf.

Calder, P.C. 2006. N-3 polyunsaturated fatty acids, inflammation, and inflammatory diseases. Am. J. Clin. Nutr. 83: 1505S–1519S.

Chevance, F.F., L.J. Farmer, E.M. Desmond, E. Novelli, D.J. Troy and R. Chizzolini. 2000. Effect of some fat replacers on the release of volatile aroma compounds from low-fat meat products. J. Agric. Food Chem. 48: 3476–3484.

Del Nobile, M.A., A. Conte, A.L. Incoronato, O. Panza, A. Sevi and R. Marino. 2009. New strategies for reducing the pork back-fat content in typical Italian salami. Meat Sci. 81: 263–269.

dos Santos, B.A., P.C.B. Campagnol, M.T.B. Pacheco and M.A.R. Pollonio. 2012. Fructooligosaccharides as a fat replacer in fermented cooked sausages. Int. J. Food Sci. Tech. 47: 1183–1192.

EU. 2006. Regulation (EC) No 1924/2006 of the European Parliament and of the Council of 20 December 2006 on nutrition and health claims made on foods. OJ L 404, 30 of December: pp. 9–25.

Fonseca, S., M. Gómez, R. Domínguez and J.M. Lorenzo. 2015. Physicochemical and sensory properties of Celta dry-ripened "salchichón" as affected by fat content. Grasas y Aceites 66: e059.

Food and Agriculture Organization of the United Nations. 2010. Fats and fatty acids in human nutrition. Report of an expert consultation. FAO food and nutrition, Vol. 91, Rome: Food and Agriculture Organization of United Nations.

García, C. and M.L. Timón. 2001. Los compuestos responsables del "flavour" del jamón. Variaciones en los distintos tipos de jamones. pp. 391–418. *In*: J. Ventanas (ed.), Tecnología del jamón Ibérico: de los sistemas tradicionales a la explotación racional del sabor y el aroma. Ediciones Mundi-Prensa, Madrid.

Fat Content of Dry-Cured Sausages and Its Effect on Chemical,.... 485

Gómez, M. and J.M. Lorenzo. 2013. Effect of fat level on physicochemical, volatile compounds and sensory characteristics of dry-ripened "chorizo" from Celta pig breed. Meat Sci. 95: 658–666.

Haffner, S.M. 2006. The metabolic syndrome: inflammation, diabetes mellitus, and cardiovascular disease. Am. J. Cardiol. 97: 3A–11A.

Jiménez-Colmenero, F., M. Triki, A.M. Herrero, L. Rodríguez-Salas and C. Ruiz-Capillas. 2013. Healthy oil combination stabilized in a konjac matrix as pork fat replacement in low-fat, PUFA-enriched, dry fermented sausages. LWT Food Sci. Technol. 51: 158–163.

Joint WHO/FAO Expert Consultation. 2003. Diet, nutrition and the prevention of chronic diseases. http://www.who.int/hpr/NPH/docs/who_fao_expert_report.pdf.

Kennedy, A., K. Martinez, C. Chuang, K. Lapoint and M. Mcintosh. 2009. Saturated fatty acid-mediated inflammation and insulin resistance in adipose tissue. Mechanisms of action and implications. J. Nutr. 139: 1–4.

Koutsopoulos, D.A., G.E. Koutsimanis and J.G. Bloukas. 2008. Effect of carrageenan level and packaging during ripening on processing and quality characteristics of low-fat fermented sausages produced with olive oil. Meat Sci. 79: 188–197.

Krauss, R.M., R.H. Eckel, B. Howard, L.J. Appel, S.R. Daniels, R.J. Deckelbaum, J.W. Erdman, P. Kris-Etherton, I.J. Golberg, T.A. Kotchen, A.H. Lichtenstein, W.E. Mitch, R. Mullis, K. Robinson, J. Wylie-Rosett, S. St. Jeor, J. Suttie, D.L. Tribble and T.L. Bazzarre. 2000. AHA dietary guidelines: Revision 2000: A statement for healthcare professionals from the Nutrition Committee of the American Heart Association. Circulation 102: 2284–2299.

Larsson, S.C., M. Kumlin, M. Ingelman-Sundberg and A. Wolk. 2004. Dietary long-chain n-3 fatty acids for the prevention of cancer: a review of potential mechanisms. Am. J. Clin. Nutr. 79: 935–945.

Leite, A., S. Rodrigues, E. Pereira, K. Paulos, A.F. Oliveira, J.M. Lorenzo and A. Teixeira. 2015. Physicochemical properties, fatty acid profile and sensory characteristics of sheep and goat meat sausages manufactured with different pork fat level. Meat Sci. 105: 114–120.

Leroy, F., C. Vasilopoulos, S. Van Hemelryck, G. Falcony and L. De Vuyst. 2009. Volatile analysis of spoiled, artisan-type, modified-atmosphere-packaged cooked ham stored under different temperatures. Food Microbiol. 26: 94–102.

Liaros, N.G., E. Katsanidis and J.G. Bloukas. 2009. Effect of the ripening time under vacuum and packaging film permeability on processing and quality characteristics of low-fat fermented sausages. Meat Sci. 83: 589–598.

Lorenzo, J.M, R. Montes, L. Purriños and D. Franco. 2012b. Effect of pork fat addition on the volatile compounds of foal dry-cured sausage. Meat Sci. 91: 506–512.

Lorenzo, J.M. and D. Franco. 2012. Fat effect on physico-chemical, microbial and textural changes through the manufactured of dry-cured foal sausage. Lipolysis, proteolysis and sensory properties. Meat Sci. 92: 704–714.

Lorenzo, J.M., R. Montes, L. Purriños, N. Cobas, and D. Franco. 2012a. Fatty acid composition of Celta pig breed as influenced by sex and location in the carcass. J. Sci. Food Agric. 92: 1311–1317.

Mendoza, E., M.L. García, C. Casas and M.D. Selgas. 2001. Inulin as fat substitute in low fat, dry fermented sausages. Meat Sci. 57: 387–393.

Menegas, L.Z., T.C. Pimentel, S. Garcia and S.H. Prudencio. 2013. Dry-fermented chicken sausage produced with inulin and corn oil: Physicochemical, microbiological, and textural characteristics and acceptability during storage. Meat Sci. 93: 501–506.

Montel, M.C., J. Reitz, R. Talon, J.L. Berdagué and S. Rousset-Akrim. 1996. Biochemical activities of Micrococcaceae and their effects on the aromatic profiles and odours of a dry sausage model. Food Microbiol. 13: 489–499.

Muguerza, E., D. Ansorena, J.G. Bloukas and I. Astiasarán. 2003. Effect of fat level and partial replacement of pork backfat with olive oil on the lipid oxidation and volatile compounds of Greek dry fermented sausages. J. Food Sci. 68: 1531–1536.

Muguerza, E., G. Fista, D. Ansorena, I. Astiasarán and J.G. Bloukas. 2002. Effect of fat level and partial replacement of pork backfat with olive oil on processing and quality characteristics of fermented sausages. Meat Sci. 61: 397–404.

Muguerza, E., O. Gimeno, D. Ansorena and I. Astiasaran. 2004. New formulations for healthier dry fermented sausages: a review. Trends Food Sci. Tech. 15: 452–457.

Narváez-Rivas, M., E. Gallardo and M. León-Camacho. 2012. Analysis of volatile compounds from Iberian hams: a review. Grasas y Aceites 63: 432–454.

Olivares, A., J.L. Navarro and M. Flores. 2011. Effect of fat content on aroma generation during processing of dry fermented sausages. Meat Sci. 87: 264–273.

Olivares, A., J.L. Navarro, A. Salvador and M. Flores 2010. Sensory acceptability of slow fermented sausages based on fat content and ripening time. Meat Sci. 86: 251–257.

Papadima, S.N. and J.G. Bloukas. 1999. Effects of fat level and storage conditions on quality characteristics of traditional Greek sausages. Meat Sci. 51: 103–113.

Petersson, K., O. Godard, A.C. Eliasson and E. Tornberg. 2014a. The effects of cereal additives in low-fat sausages and meatballs. Part 1: Untreated and enzyme-treated rye bran. Meat Sci. 96: 423–428.

Petersson, K., O. Godard, A.C. Eliasson and E. Tornberg. 2014b. The effects of cereal additives in low-fat sausages and meatballs. Part 2: Rye bran, oat bran and barley fibre. Meat Sci. 96: 503–508.

Pham, A.J., M.W. Schilling, W.B. Mikel, J.B. Williams, J.M. Martin and P.C. Coggins. 2008. Relationships between sensory descriptors, consumer acceptability and volatile flavour compounds of American dry-cured ham. Meat Sci. 80: 728–737.

Poligné, I., A. Collignan and G. Trystram. 2001. Characterization of traditional processing of pork meat into boucané. Meat Sci. 59: 377–389.

Purriños, L., R. Bermúdez, D. Franco, J. Carballo and J.M. Lorenzo. 2011. Development of volatile compounds during the manufacture of dry-cured "lacón" a Spanish traditional meat product. J. Food Sci. 76: C89–C97.

Ramírez, M.R. and R. Cava. 2007. Effect of Iberian × Duroc genotype on dry-cured loin quality. Meat Sci. 76: 333–341.

Rivas-Cañedo, A., C. Juez-Ojeda, M. Núñez and E. Fernández-García. 2011. Effects of high pressure processing on the volatile compounds of sliced cooked pork shoulder during refrigerated storage. Food Chem. 124: 749–758.

Romero, M.C., A.M. Romero, M.M. Doval and M.A. Judis. 2013. Nutritional value and fatty acid composition of some traditional Argentinean meat sausages. Food Sci. Tech. 33: 161–166.

Roth, E.M. and W.S. Harris. 2010. Fish oil for primary and secondary prevention of coronary heart disease. Curr. Atheroscler. Rep. 12: 66–72.

Salazar, P., M.L. García and M.D. Selgas. 2009. Short-chain fructooligosaccharides as potential functional ingredient in dry fermented sausages with different fat levels. Int. J. Food Sci. Tech. 44: 1100–1107.

Théron, L., P. Tournayre, N. Kondjoyan, S. Abouelkaram, V. Santé-Lhoutellier and J.L. Berdagué. 2010. Analysis of the volatile profile and identification of odor-active compounds in Bayonne ham. Meat Sci. 85: 453–460.

Tomaschunas, M., R. Zörb, J. Fischer, E. Köhn, J. Hinrichs and M. Busch-Stockfisch. 2013. Changes in sensory properties and consumer acceptance of reduced fat pork Lyon-style and liver sausages containing inulin and citrus fiber as fat replacers. Meat Sci. 95: 629–640.

Triki, M., A.M. Herrero, F. Jiménez-Colmenero and C. Ruiz-Capillas. 2013a. Storage stability of low-fat sodium reduced fresh merguez sausage prepared with olive oil in konjac gel matrix. Meat Sci. 94: 438–446.

Triki, M., A.M. Herrero, L. Rodríguez-Salas, F. Jiménez-Colmenero and C. Ruiz-Capillas. 2013b. Chilled storage characteristics of low-fat, n-3 PUFA-enriched dry fermented sausage reformulated with a healthy oil combination stabilized in a konjac matrix. Food Control 31: 158–165.

Utrilla, M.C., A.G. Ruiz and A. Soriano. 2014. Effect of partial replacement of pork meat with an olive oil organogel on the physicochemical and sensory quality of dry-ripened venison sausages. Meat Sci. 97: 575–582.

Willerson, J.T. and P.M. Ridker. 2004. Inflammation as a cardiovascular risk factor. Circulation 109: II2–II10.

Wirth, F. 1988. Technologies for making fat-reduced meat products. Fleischwirtschaft 68: 1153–1156.

World Health Organization. 2009. Preventing chronic diseases: A vital investment. http://www.who.int/features/factfiles/global_burden/en/index.html.

Yu, A.N., B.G. Sun, D.T. Tian and W.Y. Qu. 2008. Analysis of volatile compounds in traditional smoke-cured bacon (CSCB) with different fiber coatings using SPME. Food Chem. 110: 233–238.

Chapter 21

Lipid Oxidation of Fermented Meat Products

Slavomír Marcinčák

1 INTRODUCTION

Fermented meat products (salami and sausage) are very popular within consumers worldwide. They are produced as local specialties in small local companies as well as manufactured industrially in large companies. The basic ingredient is pork, beef, mutton and fat, or poultry meat with a lower or higher proportion of adipose tissue. Lipids generally form the major fraction in fermented sausages that ranges from 25–55% (Karslioglu et al. 2014). Higher fat content makes these products a subject of fat decomposition (hydrolysis and oxidation) with following manifestations of fat rancidity.

Oxidation is considered one of the main causes for functional, sensory and nutritional quality deterioration in meat and meat products due to insolubilization of proteins, off-flavour development and formation of free radicals and other oxidized compounds, such as cholesterol oxidation products (Ventanas et al. 2006). Nevertheless, lipid oxidation also has positive implications, since some of the volatile compounds which show pleasant flavour notes in dry-cured meat products arise from oxidation of unsaturated fatty acids (Carrapiso et al. 2002).

For Correspondence: University of Veterinary Medicine and Pharmacy in Košice, Department of Food Hygiene and Technology, Komenského 73, 041 81 Košice; Email: slavomir.marcincak@uvlf.sk

Fermented Meat Products—An Overview 489

2 HYDROLYSIS OF LIPIDS

Hydrolysis is one of the most common reactions in spoilage of fats. Fat hydrolysis is a process at which fatty acids are cleaved from glycerol and it is catalysed by lipases (Velíšek 1999). The result products are free fatty acids (FFAs), which can be readily oxidized. Lipases involved in the hydrolysis of triacylglycerols in both adipose tissues and muscles are obtained from adipose tissue cells, or are of microbial origin. Lipases in adipose tissue release FFAs and are produced due to the metabolic needs in the body. Regulation of these enzymes is hormonal and their production is increased in stressful situations in the organism such as before slaughter. Therefore it is necessary, that the meat and fat used in the manufacture of fermented products are obtained from animals slaughtered in terms of animal welfare standards.

Lipases of microbial origin also have an impact on fat hydrolysis. They are involved in fat hydrolysis of fermented meat products as a product of starter culture intentionally added for fermentation. Bacterial lipases are produced during the exponential growth phase and their production is greatly influenced by growth conditions with maximum amounts formed at optimum temperature and pH for growth of bacteria (Bingol et al. 2014). Time of meat processing (Karslioglu et al. 2014, Visessanguan et al. 2006, Chizzolini et al. 1998), water activity, and presence of cation catalysts such as magnesium and calcium salts (Wojciak and Dolatowski 2012) are additional parameters affecting the enzymatic hydrolysis and production of FFAs. One of the factors influencing production of FFAs in fermented products after applying starter cultures is a temperature of maturation. The influence of temperature on production of FFAs depends on the microbial species used in starter culture. Pikul et al. (1989) reported that 35–40°C is the optimal temperature for enzymatic hydrolysis. Bingol et al. (2014) also stated that the majority of bacterial lipases showed highest activity in the range of 30–40°C and at neutral or slightly alkaline pH range. On the other hand, Jin et al. (2011) declared that enzymatic activity was increasing by raising the temperature up to 35°C and then it declined rapidly. In the industrial production of fermented products, lower temperature regulated in the range of 5–27°C is used. At 8°C, the highest lipolytic activity was detected in samples with addition of *Lactobacillus sakei* L 110 and *Staphylococcus carnosus* M 72, while the temperature of 16°C increased lipase activity in samples with autochthonous microbes (Navarro et al. 1997). Starter cultures could be responsible for lipolysis in the early stages of ripening, when conditions of temperature, pH and NaCl concentration would be more favourable.

The pH of fermented products also affects the hydrolysis of fats. Lowering pH below 5 due to the activities of starter cultures causes a decrease in enzyme activity. Conversely, if the pH is 6–9, the activity of the lipolytic enzymes is increased (Jin et al. 2011).

Karslioglu et al. (2014) recorded a significant increase of FFAs content in products after the fermentation process. They are inclined to support the view of other authors (Casaburi et al. 2007, Gandemer 2002, Chizzolini et al. 1998) that the fermentation process has the effect on increasing of FFAs content in fermented products compared to raw material. The results of their measurements are satisfied the lowest FFAs value was measured in fermented products having the lowest pH and highest titratable acidity value after drying stage. This could be explained by the decreasing effect of low pH value on lipolytic activity (Zanardi et al. 2004).

Conversely, Molly et al. (1996, 1997) and Zanardi et al. (2004) stated that lipolysis is almost exclusively brought about by muscle and fat tissue endogenous enzymes and that *Micrococcaceae* do not increase overall lipolytic activity. Other authors also confirmed (Gandemer et al. 2002, Molly et al. 1997) that the contribution of bacteria to lipolysis in dry fermented sausages is weak because the medium conditions are far from the optimal conditions of bacterial lipases. The highest intensity of FFAs production in fermented products was recorded during the first days of their ripening period, depending on the pH (Mendoza et al. 2001, Balev et al. 2005). In other stages of production (drying, storage), there is a decrease of FFAs in products. This could prove a theory that the impact of starter cultures on fat hydrolysis is significant only in the first phase of production, unless there is a fermentation activity and a decrease in pH.

Olivares et al. (2011) offered the contrary argument by stating that the total FFAs levels were increased during processing as a results of lipolysis. The highest content of FFAs was recorded in final products. For them, the FFAs content in the final product at the end of processing proportion depends on the fat content in the products. The content of FFAs also increases with increasing amount of fat (Marco et al. 2006).

Zanardi et al. (2004) observed the extent of lipolysis in 4 types of fermented meat products (2 products from Mediterranean Europe and 2 products from North Europe) in relation to raw material, processing conditions and additives. After ripening process, the percentage of free poly- and mono-unsaturated fatty acids in the products was higher than of saturated fatty acids. Lipases preferentially attack the fatty acids placed on the outer positions of triacylglycerol molecules, and position 3 more than position 1, where unsaturated fatty acids are predominantly placed. Zanardi et al. (2004) concluded that an increase of FFAs during maturation appears independent from processing conditions and that the differences in polyunsaturated fatty acids increment found among

Fermented Meat Products—An Overview

the formulation appear to be due inherent variations of raw material. Lipolysis in dry fermented products is mainly due to endogenous lipases. The lipolytic activity was primarily attributed (about 70%) to meat tissue lipases whereas lipolytic starters, such as micrococci/staphylococci, were mainly playing the role in reducing nitrates and nitrites.

The quality of raw material is an important factor affecting the size of hydrolytic changes in fermented products. Only fresh meat and fat components, not microbially contaminated, should be used for the production. Adipose tissue should be separated from pig halves immediately after the slaughter and it should be stored in the refrigerator for 2–3 days before freezing. Thereby water will be partially evoporated. The fatty tissue with lower water content has the better processing characteristics and a longer shelf life (Kameník 2011). The use of older raw material, although without visible signs of hydrolytic changes, can cause a rapid increase in oxidative processes and fast oxidative spoilage in fermented products limits their shelf life.

3 LIPID OXIDATION

Lipid oxidation is one of the main causes of deterioration in the quality of meat and meat products during storage and processing (Asghar et al. 1988, Morrissey et al. 1998, Gandemer 2002) resulting in a variety of breakdown products which produce off-odours and flavours (Broncano et al. 2011). It is an autocatalytic process occurring in food and biological membranes, which leads to significant damage of food quality (Eriksson and Na 1995). Oxidation of fatty component in fermented meat products is a serious problem that can be overcome only with difficulty and leads to losses in their production in terms of shorter shelf life, worse sensory, functional and nutritional properties of manufactured products (Schmidt and Pokorný 2005). Loss of sensory properties is due to the formation of undesirable substances, particularly in the degradation of unsaturated fatty acid during oxidation. In terms of sensory properties, this process does not lead only to the creation of products with adverse effects on the smell and taste of stored products (off flavour) but the breakdown products at low concentrations are often the basis of typical aromas and flavours of many fermented meat products (Gandemer 2002).

The most common type of oxidation in food is autoxidation (Velíšek 1999). Autoxidation is oxidative degradation of unsaturated fatty acids by autocatalytic process based on the mechanism of radical chain reactions (Schmidt 2011). Fat autoxidation in food is usually initiated with hydroperoxides formed either by photosynthetic oxidation or by the action of the enzyme lipoxygenase. The principle of lipid autoxidation is generally known and is carried out in three phases:

initiation, propagation and termination. The initiation step involves removing a hydrogen atom from a free polyunsaturated fatty acid. The hydrogen atom most commonly removed is located between two double bounds (Wojciak and Dolatowski 2012). This will provide a free hydrogen radical (H$^\bullet$) and free fatty acid radical (R$^\bullet$), which are products of homolytic cleavage of a covalent bond (C–H) of hydrocarbon chain. This reaction may be catalysed by light, heat, radiation, metals, enzymes, singlet oxygen and free radicals and also, in particular, by the hydroxyl radical OH$^\bullet$. It has a strong oxidation capacity and is regarded as a dominant initiating oxidant. In the initiation phase the free fatty acid radical (R$^\bullet$) is produced, and then it reacts in a propagation step of chain radical reaction.

In propagation stage, peroxyl radical (ROO$^\bullet$) is developed due to the reaction of fatty acid radical with molecular oxygen (O$_2$). This reaction is very fast and requires almost no activation energy. Peroxyl radical is highly reactive and abstracts a hydrogen atom from another unsaturated fatty acid molecule to form a hydroperoxide (ROOH) and another free fatty acid radical (R$^\bullet$) is formed. The resultant free fatty acid radical immediately reacts with oxygen and the process is repeated cyclically until all free fatty acids are oxidized in the environment. The reaction 21.3 (Figure 1) proceeds slowly, its speed is limited by the ability of lipids to provide hydrogen from the fatty acid.

Initiation:
$$In^\bullet + R\text{-}H \rightarrow H^\bullet + R^\bullet \qquad (21.1)$$

Propagation:
Formation of peroxyl radical
$$R^\bullet + O_2 \rightarrow R\text{-}O\text{-}O^\bullet \qquad (21.2)$$

Formation of hydroperoxide
$$R\text{-}O\text{-}O^\bullet + R\text{-}H \rightarrow R\text{-}O\text{-}OH + R^\bullet \qquad (21.3)$$

Termination:
$$2R^\bullet \rightarrow R\text{-}R \qquad (21.4)$$

$$R^\bullet + R\text{-}O\text{-}O^\bullet \rightarrow R\text{-}O\text{-}O\text{-}R \qquad (21.5)$$

$$2R\text{-}O\text{-}O^\bullet \rightarrow R\text{-}O\text{-}O\text{-}R + {}^1O_2 \qquad (21.6)$$

Figure 1 The course of lipid autoxidation (Velíšek 1999)

If the concentration of free radicals in the environment is high, there is a probability that two free radicals react together to form a stable non-

Fermented Meat Products—An Overview

radical product. This phase is called termination and it terminates radical chain reactions of lipid oxidation. A flow diagram of autooxidation consists of the chemical reactions shown in Figure 1.

The overall level of oxidation is affected by the ability to separate the hydrogen from double bonds of polyunsaturated fatty acids (PUFA). The reactivity of the double bonds of unsaturated fatty acids makes them the primary objectives of free radicals. The double bonds separated each other by only one methylene group are much more reactive than those that are separated by more methylene groups. During lipid oxidation, shifting of double bonds in the chain of ester-linked fatty acids can occur. The methylene groups adjacent to the double bonds are particularly highly reactive. If the number of double bonds in a molecule is higher, the amount of the methylene hydrogen is increased, which will increase the level of oxidation. The reactivity of unsaturated fatty acids increases with their chain length and a number of double bounds (Guillevic et al. 2009). The number of double bonds dramatically increases production of hydroperoxides, which further significantly increase the number of degradation reactions and the overall production of secondary products (German 1999). Beside the number of double bonds in fatty acids, their oxidative reactivity is also determined by the position of the double bond and the type of fatty acid. N-3 polyunsaturated fatty acids (PUFA) undergo oxidative changes more rapidly as n-6 PUFA, and these are oxidized more rapidly than n-9 PUFA (Cosgrove et al. 1987). The most stable fatty acids (FA) are mono-unsaturated (MUFA) and saturated fatty acids (SFA).

In all raw material and food with fat content, there are also products of lipid oxidation, either to a greater or lesser degree. Lipid oxidation products are divided into two categories: primary and secondary. The primary products of the oxidation of unsaturated fatty acids are hydroperoxides. These products are quite volatile and can undergo are subject of further decomposition changes. The effect of thermal or catalytic disintegration leads to the formation of secondary oxidation products (radical and non-radical) from hydroperoxides. Radical products re-enter the autooxidation reactions and non-radical adversely affect the sensory and hygienic value of food. Results of these reactions are dihydroperoxides, epoxides, diols, unsaturated ketones, aldehydes, volatile products, etc. (Takácsová et al. 1998). These products change the sensory properties of food (colour, taste and smell) which leads to the fact that consumers refuse to eat it. Malondialdehyde is the dominant product of hydroperoxide degradation arising from fatty acids with three and more double bonds, which makes it important to determine the amount of oxidative degradation products of lipids (Marcinčák et al. 2003).

Products formed during the process of fat degradation must be considered very carefully, because of their negative impact on human health. They are included into mechanisms of many diseases such as arteriosclerosis, cancer, suppression of immunity, Parkinson's disease, Alzheimer's disease and numerous cardiovascular diseases (Patthy et al. 1995, Dave and Ghaly 2011).

4 FACTORS INFLUENCING OXIDATION OF FERMENTED MEAT PRODUCTS

The most important factors affecting oxidation of fatty components in fermented meat products are fat content in products, its fatty acid composition, processing steps throughout the entire preparation of meat products, fermentation and drying conditions, duration of storage, as well as application of additives into meat products.

Lipids are important components of food and they have insufficient status in human nutrition. Fat content of meat products has an effect on oxidation processes. Several authors investigated the effect of fat content on the amount of oxidation products in final products (Fuentes et al. 2014, Kouba and Mourot 2011, Olivares et al. 2011). In general, we can say that the rate of fat oxidation and hydrolysis increases with increasing fat content in meat products. As fermented meat products belong among the products with a higher proportion of fat (about 50%), the oxidation can be a serious problem in the production of these products. Selection of raw materials is therefore an important factor affecting stability of products. The most common raw materials for production of fermented meat products are pork, beef and pork bacon. They contain a higher proportion of fats with a high content of saturated fatty acids, which are not subject to rapid oxidative changes. Therefore, besides the fat content in meat products, a composition of essential fatty acids in fat of the raw material is also of great importance due to oxidation changes. SFA and MUFA are more stable and do not undergo oxidative changes in normal conditions. On the other hand, PUFA react at lower temperatures (4°C) and undergo oxidative changes. As for the quality of fermented meat products, it is desirable when fats contain mostly SFA and MUFA, but PUFA only in smaller amounts.

Fatty acid composition in final products is determined by the type of raw material. The fatty acid composition depends on the animal species from which the raw material is obtained. The highest content of PUFA is in fish and poultry fat, where MUFA and PUFA predominate, whereas SFA portion is low. A lower proportion of PUFA is found in the pork, beef and mutton fat. The meat and fat of those species show relative stability against oxidation. Therefore, these raw materials are most commonly used in the production of fermented meat products. The proportion of individual fatty acids in the fat of animal species

Fermented Meat Products—An Overview

is not constant and depends on several factors (age, sex, breeding, feeding). Fat composition of individual animals can be also affected by feeding. In monogastric animals, FA profile in muscle and fat can be easily influenced by nutrition because FA are absorbed unchanged in their small intestine (Kouba and Mourot 2011). In ruminants, the effect of feeding on FA composition is limited due to the process of biohydrogenation in the rumen, where unsaturated fatty acids are changed to saturated fatty acids by the action of microorganisms. Thus only a minority of PUFA from fodder becomes part of phospholipids and triacylglycerols in various tissues (Jalč and Čertík 2005).

Intracellular fat in muscle is another factor which can affect the oxidation of meat products. Its amount is relatively constant and the proportion of the fat components depends on the structures involved in intracellular membranes. These structures mainly include a membrane-bound lipoproteins and phospholipids with a large proportion of unsaturated fatty acids. Their oxidation after muscle tissue disruption is very rapid and results in the formation of undesirable sensory characteristics of meat products. Increase of stress burden in animals shortly before the slaughter accelerates oxidative processes, in particular of this fatty component.

Fermented products with higher levels of SFA are not considered healthy due to adverse effects of SFA on consumer health. Relatively high cholesterol level and low PUFA/SFA ratio represent a risk of developing a pathological condition such as coronary heart disease (Rubio et al. 2008, Muguerza et al. 2001). For this reason, several authors have focused on the reduction and partial substitution of fat in fermented meat products (Olivares et al. 2011, Liaros et al. 2009). Products with a lower fat content (10%), even if they are oxidatively stable, had lower sensory properties (they were tough due to weight loss and also flavour was weaker) than products with a higher fat content (30%). Besides lowering the saturated fat content, there is an attempt to influence the fatty acid composition to reduce portion of SFA and, on the other hand, to increase proportion of PUFA in the products. The trend is to produce foods enriched with n-3 PUFA, which are deficient in a diet. There is a tendency to add olive oil (Severini et al. 2003, Muguerza et al. 2001) or linseed oil (de Ciriano et al. 2010) into fermented meat products to increase the proportion of PUFA, especially α-linolenic acid. Lipid alterations of raw materials through lipid alterations of feed is another way to increase the proportion of PUFA in fermented meat products. The latest studies try to modify the composition of the animal raw matter (pigs) through a change in the diet (Muguerza et al. 2004, Kouba and Mourot 2011). N-3 PUFA, the least oxidatively stable FA, are liable to oxidative changes which will affect the quality of final product. However, if dietary manipulation is performed to increase PUFA, then antioxidant technologies must also be employed to minimize oxidative deterioration. Inhibition of oxidative deterioration could be accomplished by increasing muscle antioxidants by diet (Cardenia et al. 2011, Petrovič et al. 2012, Marcinčák et al. 2008a) or food ingredients (Bowser et al. 2014, Berasategi et al. 2011, de Ciriano et al. 2010), decreasing storage temperature and/or oxygen exclusion by vacuum or modified atmosphere packaging (Decker and Park 2010).

Besides increased oxidation due to the higher amount of PUFA in meat products, there is also a change in sensory properties. Raw materials and fermented meat products with a higher proportion of MUFA and PUFA have softer consistency and their fat can drip off at room temperature as a consequence of a lower melting point of MUFA and PUFA compared to SFA, which are solid at room temperature.

Other factors influencing the fat oxidation are technology and processing of fermented meat products. Shredding, mincing, and mixing of the meat during the manufacture of the sausage have increased the surface area of the meat and exposed it to oxygen and oxidation catalysts (Visessanguan et al. 2006). Processes of preparing raw materials necessary for the production of meat products (cutting, grinding, mixing raw materials) are responsible for breakdown of muscle membrane system and may cause oxidation of fatty components and consequently an increase in the conversion of hydroperoxides of PUFA into secondary products (Rubio et al. 2008, Raharjo et al. 1992). The disintegration of organelles in animal cells causes the release of fat, oxidation catalysts and enzymes responsible for the oxidation (Ladikos and Lousgovois 1990). To reduce the impact of these processes, the raw material used for producing should be frozen. Immediately before production, it should be thawed at temperature of approximately –6°C, and then should be cut into smaller pieces. Grinding and mixing of raw materials should be performed at temperatures of about 0°C. At these temperatures, the oxidation process takes place more slowly.

Even the temperatures of fermentation (maturation), drying and storage of the products can affect the oxidation process. In general, the oxidation process is faster at higher temperatures. Fermentation process is most commonly performed at temperatures of 17–25°C, during which the products are also smoked. Since the smoke is a good antioxidant, FFAs are protected from oxidation during this phase. Drying and storage of products have even more significant impact on the oxidation process. There is the largest increase of degradation products during the drying phase (Kameník et al. 2012, Rubio et al. 2008). The increase of degradation products of fat expressed as TBARS (thiobarbituric acid reactive substances) in fermented products was also observed in the first 2 months of storage of fermented products by several authors (Šojić et al. 2014). Following storage period for 3–12 months mostly caused a decline of TBARS in the products (Wojciak et al. 2015,

Kameník et al. 2012, Šojić et al. 2014). A decrease of TBARS during storage may be caused by the fact that malondialdehyde, the main product of PUFA oxidation, reacts with amino acids, sugars and nitrates forming stable complexes. This may be the cause of lower amount of TBARS at the end of the storage period of fermented products. Despite this fact, we can say, that the storage time significantly affects the lipid oxidation parameters at the end of storage. The deterioration of oxidation stability is also accompanied by deterioration in sensory characteristics of the products at the end of the storage.

Kameník et al. (2012) investigated the effect of two different storage temperatures (5 and 15°C) on the content of oxidation products in the vacuum sealed fermented product (sausage Poličan) during fermentation (30 days) and storage period (120 days). Fermentation for 30 days has increased degradation products of fat (TBARS). The results in changes of fat degradation (TBARS) during the storage supported the claims that the increase in degradation was recorded during first 60 days at both storage temperatures (5 and 15°C). Subsequently, after 90 and 120 days, a decrease of TBARS was recorded at both temperatures. Lower TBARS values were recorded at 15°C during the entire storage period. The results obtained in this study indicate that different temperatures in the course of storage did not have a significant effect on the progression of oxidation changes. This observation corresponds to those of other authors who have come to conclusion that the method of packaging has a much more significant influence on lipid oxidation in meat products than the storage temperature (Zanardi et al. 2002).

Packaging in modified atmospheres (vacuum and gas packaging) is being introduced as a commercial way for the retail selling of dry fermented sausages. Both methods increased shelf life of products, and thus the attractiveness of the commercial sale of sliced meat products. Another advantage of these packaging methods from the producer's and seller's point of view is that the product does not dry further after packaging and there is no weight loss during storage, transport and sale. After fermentation and drying process fermented meat products should be packed by one of the methods, under vacuum or under protective atmosphere (with different ratios of N_2/CO_2, most commonly 20–25% of CO_2 and 75–80% of N_2), to slow down the oxidation process. Scientific opinions on the effect of packaging on the oxidation rate in fermented products are unclear and divergent. On one hand, several authors confirm that the packaging affects slowing down of oxidation in the products. On the other hand, others point to a higher degree of oxidation in vacuum packaged products than in unpacked products. Rubio et al. (2008) investigated the effect of the two packaging methods on oxidation stability of fermented products during their storage up to 210 days at 6°C. Gas packaged samples showed lower TBARS values than those found in vacuum packaged samples. Also Zanardi et al. (2002) stated that there is lower oxidative stability of vacuum packaged Milano-type sausages stored for 28 days, compared to those wrapped in a protective atmosphere containing 100% N_2. Summo et al. (2006) investigated the influence of vacuum-packaging on the sensory properties and the degradation level of lipid fraction of some ripened sausages (30 days). Subsequently, the sausages were further stored for 40 days under vacuum-packaging and compared to those kept unpackaged for the same period of time. On the basis of their results, Summo et al. (2006) found that the vacuum packed products had lower oxidation stability, and higher amounts of oxidation products than unpacked products.

As for the sensory properties, the unpackaged sausages showed significantly higher mean scores of red intensity and global flavour after storage than vacuum packed ones indicating a brighter red colour of meat and a greater overall appreciation by the panellists.

5   THE IMPACT OF OXIDATION PRODUCTS ON PROPERTIES OF MEAT PRODUCTS

Fermented meat products contain a high percentage of fat (Šojić et al. 2014). Intramuscular lipids of raw material play an important role in the development of chemical and sensory characteristics of fermented meat products. During processing intramuscular lipids are gradually degraded through both lipolysis and oxidation (Jin et al. 2010, Veiga et al. 2003). Products, formed during lipolysis and lipid oxidation, have an important role in the formation of odour, taste, and texture of the final product (Visessanguan et al. 2006). The first step in the conversion from lipids to flavour compounds is lipolysis. In this process, free fatty acids are released from the triacylglycerols, and phospholipids of meat and adipose tissue used for the production of fermented meat products. Some authors have studied the lipolysis in fermented meat products (Zanardi et al. 2004, Fuentes et al. 2014, Olivares et al. 2011) and dry-cured ham (Jin et al. 2010, Huang et al. 2014, Yang et al. 2005) and most of them found that phospholipids were the most important fractions in intramuscular lipids because free fatty acids originated mainly from phospholipids. PUFA having three and more double bonds are primarily tied to phospholipids and are important for development of the specific flavour state of meat products (Šojić et al. 2014). FFAs themselves do not directly affect the sensory quality of fermented products. Their main benefit to the sensory properties is that FFAs are fat components, which are susceptible to oxidative changes that result in the formation of a large variety of volatile compounds (aldehydes, hydrocarbons, esters, ketones). These components, even in low concentrations, are responsible

Fermented Meat Products—An Overview

for specific taste and aroma of fermented meat products. Resolution, whether the aroma and flavour of products are acceptable and when they are not dependent on the type of product. Milder flavour and aroma are accepted in the fermented products, which mature in a shorter time, a few days or weeks. On the contrary, the products maturing for several months are allowed to have even stronger taste and aroma and a higher concentration of oxidation products responsible for pleasant nutty to fruity aroma and flavour of the product.

In the products with the intensive oxidation, oxidation products are generated in such an amount, that there is a change in the organoleptic characteristics, and rancidity of fats. Foods with such significant changes in taste, aroma and colour are unacceptable for consumers and represent a serious problem in the production of foods with a high fat content. Rancid aroma is correlated to oxidation products, mainly to aldehydes such as nonanal and 2-hexenal which exhibit a strong odour (Gandemer 2002, Ruiz et al. 1999). Malondialdehyde (MDA) is also an important product of PUFA oxidation (Grau et al. 2000). According to the international standards, threshold for the appearance of rancidity off-flavour in fresh meat is lower than 0.5 mg MDA per kg (Lanari et al. 1995) and in cooked meat 1.0 mg/kg (Igene et al. 1979). Higher concentrations of secondary oxidation products arising from the oxidation of free PUFA in meat products promote unpleasant rancid flavour and aromas, known as warmed-over flavour, and the colour of muscle and adipose tissues also changes (Balev et al. 2005).

Oxidative processes in meat and meat products lead to the degradation of lipids and proteins (including pigments) which in turn contribute to the deterioration in colour (Wojciak et al. 2015). Meat discolouration compromises its appearance and is due to the conversion of red oxymyoglobin to brownish metmyoglobin. It was concluded that changes in oxymyoglobin and colour of meat appeared to be driven by lipid oxidation and correlated strongly with oxidation product content (Zakrys et al. 2008). The mechanisms by which lipid oxidation could enhance myoglobin oxidation have been explained primarily on the reactivity of primary and secondary oxidation products derived from PUFA (Faustman et al. 2010). Oxidation products could interact with haem pigment, resulting in increased susceptibility of haem iron to oxidation and meat product deterioration (Alderton et al. 2003). When ferrous iron in haem oxidizes to its ferric form, oxygen is released and replaced by a water molecule (Faustman et al. 2010). Factors that affect oxymyoglobin oxidation include temperature, pH, partial oxygen pressure and lipid oxidation (Faustman et al. 2010, Wojciak and Dolatowski 2012). Oxidation of oxymyoglobin is favoured by higher temperatures, low pH environment, and the presence of non-haem iron (Allen and Cornforth 2006, Faustman et al. 2010). This process results

500   *Fermented Meat Products: Health Aspects*

in a colour change from red to brown or meat products lose their red colour. Thus, the products lose their attractiveness and it is necessary to stabilize the colour. The effective additives used in fermented meat products that stabilize the colour of the products are nitrates, vitamin E, ascorbic acid (L-ascorbic acid) and erythorbic acid (D-isoascorbic acid) and their salts.

Fat is an important factor influencing the flavour of dry fermented meat products not only by the generation of flavour compounds but also as a solvent of aroma compounds (Leland 1997). Fermented meat products are included in the group of cured meat products where the reduction of fat is more difficult (Flores et al. 2013). The fermented meat products with a higher proportion of fat (> 30%) show better sensory characteristics than the products with a lower fat content in the sensory evaluation (Olivares et al. 2011). Conversely, fat reduction in the products causes the problems with the flavour and aroma deterioration of meat products. This should be considered in the current trend to reduce fat content in fermented meat products, which may contribute to the loss of the typical sensory characteristics in individual products. One of the problems that can occur in the production of dry fermented sausages with the reduced fat is that there are greater losses of moisture which can cause problems with firmness and on surface of produced products.

6   METHODS PREVENTING OXIDATION IN FERMENTED MEAT PRODUCTS

In the present, great attention is being paid to lipid protection against oxidation. Oxidative stability in foods depends on the concentration, composition and activity of reactants, prooxidants and antioxidants. Shifting the balance between these three factors is the key for the development of oxidatively stable foods (Decker 1998). Although fermented products belong to the relatively stable oxidation products due to their composition, a high proportion of fat and relatively long time of processing and storing designate them to protect them against oxidation.

Reducing the amount of unsaturated fatty acids in fats and reducing the oxygen concentration are known as methods of inhibiting fat oxidation. Choosing appropriate raw material for the meat production is an important factor in minimizing the process of hydrolysis and oxidation. For the production of fermented products fresh raw material, without microbial contamination and with the optimal ratio of saturated, monounsaturated and polyunsaturated acids should be used. It is desirable if the basic raw materials such as pork and beef meat and fat component are derived from older animals (sows, cows). The meat is

Fermented Meat Products—An Overview 501

darker, it contains less water and has a more appropriate ratio of SFA/PUFA. Thus the fat is more stable and it is not subject to significant oxidative changes. The current trend to increase the proportion of PUFA, especially n-3, in meat and subcutaneous fat of animals, by changing the diet of animals, complicates the choice of materials. Feeding animals with a higher proportion of PUFA will also reflect in the composition of raw materials, i.e. in an increase of PUFA. Pork enriched with PUFA was already used for manufacturing of meat products (Rubio et al. 2008, Guillevic et al. 2009, Haak et al. 2008). All researchers concluded that an elevated proportion of PUFA have increased the amount of oxidation products in stored meat and meat products. When using such raw material, it is necessary to use other methods of protection against degradation products of fats.

Minimizing access of oxygen using packaging under vacuum or modified atmosphere packaging can reduce oxidative reactions. Storage of fermented meat products in a controlled atmosphere of gas seems like a good way to protect fats from oxidation. This method is particularly important if protected products are sliced and then stored. Combination of low temperatures and the use of suitable antioxidants ensures that the products are oxidatively stable; there is no increase of their oxidation products and no change in their organoleptic characteristics (flavour and colour). Several authors (Rubio et al. 2008, Zanardi et al. 2002) suggested that the vacuum-packing of fermented products had a lower effect on the protection of fats from oxidation compared to modified atmosphere packaging (CO_2/N_2).

Limiting other factors, such as the use of lower temperatures during ripening and storage and access of light, in particular UV radiation, also significantly helps to prevent auto-oxidation of fatty acids in lipids. It is necessary to store the fermented meat products at low temperatures (15°C) and in dark rooms.

Prooxidants in food can be either kept in an inactive form (by reducing the light activation in photosensitive compounds) or removed (reduce metal content), or inactivated (denaturation of lipoxygenases). However, most prooxidant food components and food additives contribute to the desired properties of foods, because they act as nutritional factors (e.g. Fe, Cu and riboflavin) and pigments (myoglobin, chlorophyll). In these cases, it is necessary to minimize the prooxidant activity by the addition of substances with the ability to scavenge free radicals in the medium and thereby reducing the oxidation of the fatty acids. Nowadays, the most effective way of inhibition of oxidative processes is with the use of antioxidants (Marcinčák et al. 2005). Antioxidants are used to extend the shelf life of food, and to ensure health safety. Their use in the production of fermented products is one of the main ways of protecting fats from oxidation during fermentation and storage.

6.1 Antioxidants

Antioxidants are used to preserve food products by retarding discolouration and deterioration as a result of oxidation (Decker et al. 2005). Antioxidants are those ingredients that can prevent, slow down or hinder the oxidation process in the stored food and raw materials, or during food processing. They delay the onset of oxidation of lipids and other oxidatively labile compounds or reduce the rate at which the reaction proceeds. These components are either a natural part of the food, or are deliberately added to the product, or are formed during the processing. Although their role is not to directly improve the quality of food, they help to maintain it (especially the original nutritional value) and prolong their shelf life.

According to the mechanism of their action which prevents or retards oxidation, antioxidants are classified into: primary (reacting with free radicals and interrupting the chain reaction) and secondary (reducing already formed hydroperoxides, antioxidants, binding to the complexes with the catalytically active metals, and antioxidants eliminating present oxygen).

Primary antioxidants prevent auto-oxidation when reacting with lipid and peroxyl free radicals (electron donors) and convert them into the more stable non-radical products (Figure 2). In the reaction, a fatty acid and a free radical of antioxidant of low reactivity are developed (21.7, 21.8). As primary antioxidants are hydrogen donors, they have higher affinity to the peroxyl radicals than to the lipids (Porter et al. 1995). Antioxidants may also react directly with lipid radicals (21.9).

Figure 2 The effect of primary antioxidants on free radicals (Schmidt 2011)

In the termination phase (Figure 3), radical of antioxidant reacts with peroxyl or alkoxyl radical of fatty acid or other radical of antioxidant (21.10, 21.11, 21.12) to form non-radical products, thus stopping the chain reaction of fatty acid oxidation (Maestri et al. 2006). The group of primary antioxidants includes phenolic antioxidants: tocopherols, propyl gallate, butylhydroxytoluene (BHT), butylhydroxyanisole (BHA), tertiary butylhydroxyquinone (TBHQ), flavonoids, phenolic acids.

Secondary antioxidants inhibit lipid autoxidation primarily by deceleration and braking of oxidation. These antioxidants slow down the oxidation rate by a number of different mechanisms, but do not convert free radicals to more stable products. They have broad-spectrum effects, and may also act as synergists (increase effect) of primary antioxidants.

Fermented Meat Products—An Overview

The secondary antioxidants can bind metal ions, remove the oxygen, decompose hydroperoxides to non-radical components, absorb UV light and deactivate singlet oxygen.

$$ROO^{\cdot} + A^{\cdot} \longrightarrow ROOA \qquad (21.10)$$

$$RO^{\cdot} + A^{\cdot} \longrightarrow ROA \qquad (21.11)$$

$$A^{\cdot} + A^{\cdot} \longrightarrow AA \qquad (21.12)$$

Figure 3 Reactions of primary antioxidants in the termination phase of oxidation (Schmidt 2011)

Complex-forming agents bind to the ions of some prooxidant metals (Fe, Cu, Mn, Cr) forming inactive complexes. These transition metals, especially iron, play an important role in the initiation and promotion of lipid oxidation. Iron is found in fermented products mainly in bound form in the blood pigment hemoglobin, muscle myoglobin and cytochromes. The most common chelating agents are citric acid and lipophilic monoacylglycerol ester, phosphoric acid and its derivatives (Schmidt and Rodrick 2003, Haahr and Jacobsen 2008).

Oxygen scavengers are preferably oxidized with oxygen acting as reducing agents and thereby protect the lipids in foods. Oxygen adsorption is therefore most effective in foods packaged in containers where oxygen is consumed. The most effective oxygen scavengers in foods are L-ascorbic acid, and its stereoisomer D-isoascorbic acid and their salts and sulphites.

Singlet oxygen quenchers inactivate the oxygen chemically and physically. They take away the energy excess from oxygen. Effective quenchers are carotenoids, in particular β-carotene, lycopene and lutein (Schmidt 2011).

Hydroperoxide-degrading agents are substances which convert hydroperoxides derived from fatty acids to alcohols or other non-radical products. These particularly include various phenols, tocopherols and trolox (Schmidt and Pokorný, 2005).

Antioxidants can be also classified by their origin to:

- synthetic antioxidants,
- natural antioxidants.

6.1.1 Synthetic Antioxidants

Although the use of synthetic antioxidants is quite widespread, their amount is limited due to possible health risks (Fasseas et al. 2007). Maximum levels (mg/kg) of synthetic antioxidants are laid down by Regulation No 1333/2008 of the European Parliament and of the Council (Anonymous 2008). The following substances can be used in a limited

504 *Fermented Meat Products: Health Aspects*

way as antioxidants in meat products: propyl gallate (E 310), octyl gallate (E 311), dodecyl gallate (E 312), erythorbic acid (isoascorbic acid, E 315), sodium erythorbate (E 316), tertiary butylhydroxyquinone (E 319), butylhydroxyanisole (E 320, BHA) and butylhydroxytoluene (E 321, BHT). Although BHT and BHA are one of the most effective antioxidants in practice, currently, ascorbic acid and its salts, erythorbic acid and sodium erythorbate are mainly used. Erythorbic and ascorbic acids stabilize the colour, flavour and aroma of smoked meat products (Schmidt and Rodrick 2003). The role of both acids as antioxidants in meat products is to protect and stabilize pink dye N-nitrosomyochromogen against oxidation by atmospheric oxygen. In the food, ascorbic acid acts mainly as a secondary antioxidant with more features: it recaptures oxygen; it shifts redox potential of a food matrix into the reduction zone; it is synergistic with the complex-forming agents and regenerates primary antioxidants. Ascorbyl palmitate, synthetic derivative of ascorbic acid, is preferably used in fat-rich foods. Esters of ascorbic acid are more soluble in the fat phase than ascorbic acid (Pokorný and Schmidt 2003). Ascorbic and erythorbic acids are also added to fermented meat products because of preventing the formation of N-nitro compounds by microorganisms during salting with nitrite salting mixture.

Nitrites (sodium nitrite and potassium nitrite) can be considered as the additive substances having antioxidant effect added to fermented meat products. Nitrites are added to meat products with the salt in the form of nitrite salting mixture. Nitrite added to meat products contributes to the stability of the colour of meat products, to the formation of flavour and has preservative and antioxidant effects (Dave and Ghaly 2011). The antioxidant effect of nitrite is based on the principle of the oxygen consumption, while nitrite is oxidized to nitrate, and also on the formation of a stable complex with iron bound to the haem (chelating effects), thus preventing the release of iron ions. The iron ions then are not able to initiate lipid autoxidation.

6.1.2 *Natural Antioxidants*

Natural antioxidants are either a natural part of raw materials or added to foods as additives. The oil-soluble vitamins (vitamins A and E), or their analogues are the most commonly used. Vitamin A is usually applied as a provitamin of β-carotene. Vitamin E is the collective noun of eight derivatives of 6-hydroxy-chromane soluble in fat (tocopherols, tocotrienols) with significant biological and antioxidant activity (Watson and Preedy 2009). Vitamin E belongs to the primary antioxidants. Its mechanism of action is based on the direct interruption of the propagation phase of oxidation. Antioxidant effect primarily depends on the substrate, temperature, concentration, the presence of synergists and light (Kamal-Eldin and Appelqvist 1996).

In the body, vitamin C contributes to the protection of lipids indirectly by regenerating other antioxidants, but a relatively significant chelating effect on iron ions has been also described (Dognin and Crichton 1975). β-carotene scavenges oxygen and vitamin C withdraws oxygen.

In general, natural antioxidants are preferable because unlike synthetic compounds do not represent a health risk (Pokorný and Schmidt 2003). Currently, there is an increased interest in the food industry to use natural antioxidants obtained from plants, primarily because of the improved safety of the plant extracts (Aziza et al. 2010). Spices, used only for improving the sensory properties of foods until recently, are becoming increasingly important also in storing and prolonging shelf life of animal products.

A large group of plants and spices freely available in nature contains substances with antioxidant properties. Rosemary, thyme, marjoram, sage and oregano, of the family *Lamiaceae*, have high antioxidant activity (Barros et al. 2010, Marcinčák et al. 2008b). Potent antioxidant properties were also found in spices such as clove, cinnamon and ginger (Suhaj 2006). Extracts from the leaves of green and black tea (Toschi et al. 2000) but also components of some oil plants (Niklová et al. 2000) are also recommended as antioxidants. Most extracts with antioxidant effects contain flavonoid and phenolic compounds whose effect is to interrupt the chain of propagation of autoxidation (Ou et al. 2003). The amount of compounds in the plant material acting as antioxidants may vary and depends on several factors such as the quality of the plant, the location of its origin, the time of collection, processing and storage conditions (Pokorný et al. 1998).

A number of studies have compared BHA and BHT with plant extracts for their antioxidant activity in various meat products. Generally, plant extracts exhibit better lipid oxidation inhibition in cooked, fermented and irradiation processed meat products in comparison with synthetic antioxidants (Hygreeva et al. 2014). The use of plant extracts finds application in the production of fermented meat products, especially when used components contain a higher proportion of PUFA, which must be protected from oxidation during manufacture and storage. Extracts of spices such as paprika, black pepper, white pepper and garlic are a common part of fermented meat products and participate in anti-oxidative stability of manufactured products. Rosemary (Bowser et al. 2014) as well as lemon balm (de Ciriano et al. 2010, Berasategi et al. 2011) have been successfully tested as potent antioxidants in fermented meat products.

7 CONCLUSION

Fermented meat products belong to the relatively stable oxidation products due to their composition, a high proportion of fat (about 50%) and relatively long time of processing and storing designate them to protect them against oxidation. Choosing appropriate raw material for the meat production is an important factor in minimizing the process of hydrolysis and oxidation. For the production of fermented products fresh raw material, without microbial contamination and with the optimal ratio of saturated, monounsaturated and polyunsaturated acids should be used. The use of older raw material, although without visible signs of hydrolytic changes, can cause a rapid increase in oxidative processes and fast oxidative spoilage in fermented products limits their shelf life. Inhibition of oxidative deterioration during processing and storage could be accomplished by adding of antioxidants, decreasing storage temperature and/or oxygen exclusion by vacuum or modified atmosphere packaging.

Key words: *antioxidants, fat content, hydrolysis, fermented meat products, lipid oxidation, polyunsaturated fatty acids, oxidation products*

REFERENCES

Alderton, A.L., C. Faustman, D.C. Liebler and D.W. Hill. 2003. Induction of myoglobin redox instability by adduction with 4-hydroxynonenal. Biochemistry 42: 4398–4405.

Allen, K.E. and D.P. Cornforth. 2006. Myoglobin oxidation in a model system as affected by nonheme iron and iron chelating agents. J. Agric. Food Chem. 54: 10134–10140.

Anonymous. 2008. Regulation (EC) No. 1333/2008 of the European Parliament and of the Council of 16 December 2008 on food additives.

Asghar, A., J.I. Gray, D.J. Buckley, A.M. Pearson and A.M. Booren. 1988. Perspectives on warmed-over flavor. Food Technol. 42: 102–108.

Aziza, A.E., N. Quezada and G. Cherian. 2010. Antioxidative effect of dietary Camelina meal in fresh, stored or cooked broiler chicken meat. Poultry Sci. 89: 2711–2718.

Balev, D., T. Vulkova, S. Dragoev, M. Zlatanov and S. Bahtchevanska. 2005. A comparative study on the effect of some antioxidants on the lipid and pigment oxidation in dry-fermented sausages. Int. J. Food Sci. Technol. 40: 977–983.

Barros, L., S.A. Heleno, A.M. Carvalho and I.C.F.R. Ferreira. 2010. Lamiaceae often used in Portuguese folk medicine as a source of powerful antioxidants: Vitamins and phenolics. LWT – Food Sci. Technol. 43: 544–550.

Fermented Meat Products—An Overview

Berasategi, I., S. Legarra, M.G. de Ciriano, S. Rehecho, M.I. Calvo, R.Y. Cavero, I. Navarro-Blasco, D. Ansorena and I. Astiasarán. 2011. High in omega-3 fatty acids bologna-type sausages stabilized with an aqueous-ethanol extract of *Melissa officinalis*. Meat Sci. 88: 705–711.

Bingol, E.B., G. Ciftcioglu, F.Y. Eker, H. Yardibi, O. Yesili, G.M. Bayrakal, and G. Demirel. 2014. Effect of starter cultures combinations on lipolytic activity and ripening of dry fermented sausages. Ital. J. Anim. Sci. 13: 776–781.

Bowser, T.J., M. Mwavita, A. Al-Sakini, W. McGlynn and N.O. Maness. 2014. Quality and shelf life of fermented lamb meat sausage with rosemary extract. The Open Food Sci. J. 8: 22–31.

Broncano, J.M., M.L. Timón, V. Parra, A.I. Andrés and M.J. Petrón. 2011. Use of proteases to improve oxidative stability of fermented sausages by increasing low molecular weight compounds with antioxidant activity. Food Res. Int. 44: 2655–2659.

Cardenia, V., M.T. Rodriguez-Estrada, F. Cumella and L. Sardi. 2011. Oxidative stability of pork meat lipids as related to high-oleic sunflower oil and vitamin E diet supplementation and storage conditions. Meat Sci. 88: 271–279.

Carrapiso, A.I., J. Ventanas and C. Garcia. 2002. Characterization of the most odor active compounds of Iberian ham headspace. J. Agric. Food Chem. 50: 1996–2000.

Casaburi, A., M.C. Aristoy, S. Cavella, R. Di Monaco, D. Ercolini, F. Toldra and F. Villani. 2007. Biochemical and sensory characteristics of traditional fermented sausages of Vallo di Diano (Southern Italy) as affected by the use of starter culture. Meat Sci. 76: 295–307.

Chizzolini, R., E. Novelli and E. Zanardi. 1998. Oxidation in traditional Mediterranean meat products. Meat Sci. 49: 87–99.

Cosgrove, J.P., D.F. Church and W.A. Pryor. 1987. The kinetics of the autoxidation of polyunsaturated fatty-acids. Lipids 22: 299–304.

Dave, D. and A.E. Ghaly. 2011. Meat spoilage, mechanisms and preservation techniques: A critical review. Am. J. Agric. Biol. Sci. 6: 486–510.

de Ciriano, M.G.I., E Larequi, S. Rehecho, M.I. Calvo, R.Y.Cavero, I. Navaro-Blasco, I. Astiasarán and D. Ansorena. 2010. Selenium, iodine, ω-3 PUFA and natural antioxidant from Melisaa officinalis L.: A combination of components from healthier dry fermented sausages formulation. Meat Sci. 85: 274–279.

Decker, E.A. 1998. Strategies for manipulating the prooxidative/antioxidative balance of foods to maximize oxidative stability. Trends Food Sci. Tech. 9: 241–248.

Decker, E.A. and Y. Park. 2010. Healthier meat products as functional foods. Meat Sci. 86: 49–55.

Decker, E.A., K. Warner, M.P. Richards and F. Shahidi. 2005. Measuring antioxidant effectiveness in food. J. Agric. Food Chem. 53: 4303–4310.

Dognin, J. and R.R. Crichton. 1975. Mobilization of iron from ferritin fractions of defined iron content by biological reductants. FEBS Lett. 54: 234–236.

Eriksson, C.E. and A. Na. 1995. Antioxidant agents in raw materials and processed foods. Biochem. Soc. Symp. 61: 221–234.

Faustman, C., Q. Sun, R. Mancini and S.P. Suman. 2010. Myoglobin and lipid oxidation interactions: Mechanistic bases and control. Meat Sci. 86: 86–94.

508   *Fermented Meat Products: Health Aspects*

Fasseas, M.K., K.C. Montzouris, P.A. Tarantilis, M. Polissiou and G. Zervas. 2007. Antioxidant activity in meat treated with oregano and sage essential oils. Food Chem. 106: 1188–1194.

Fuentes, V., M. Estévez, J. Ventanas and S. Ventanas. 2014. Impact of lipid content and composition on lipid oxidation and protein carbonylation in experimental fermented sausages. Food Chem. 147: 70–77.

Flores, M., A. Olivares and S. Corral. 2013. Healthy trends affect the quality of traditional meat products in Mediterranen area. Acta Agric. Slov. 4: S183–S188.

Gandemer, G. 2002. Lipids in muscles and adipose tissues, changes during processing and sensory properties of meat products. Meat Sci. 62: 309–321.

German, B.J. 1999. Food processing and lipid oxidation. pp. 23–50. *In*: L.S. Jackson, M.G. Knize, J.N. Morgan (eds.). Impact of processing on food safety. Kluwer Academic, Plenum Publishers, New York.

Grau, A., F. Guardiola, M. Boatella, A. Barroeta and R. Codony. 2000. Measurement of 2-thiobarbituric acid values in dark chicken meat through derivative spectrophotometry: influence of varios parameters. J. Agric. Food Chem. 48: 1155–1159.

Guillevic, M., M. Kouba and J. Mourot. 2009. Effect of a linseed diet on lipid composition, lipid peroxidation and consumer evaluation of French fresh and cooked pork meat. Meat Sci. 81: 612–618.

Haak, L., S. DeSmet, D. Fremaut, K. Van Walleghem and K. Raes. 2008. Fatty acid profile and oxidative stability of pork as influenced by duration and time of dietary linseed or fish oil supplementation. J. Anim. Sci. 86: 1418–1425.

Haahr, A.M. and C. Jacobsen. 2008. Emulsifier type, metal chelation and pH effect oxidative stability of n-3 enriched emulsions. Eur. J. Lipid Sci. Technol. 110: 949–961.

Huang, Y., H. Li, T. Huang, F. Li and J. Sun. 2014. Lipolysis and lipid oxidation during processing of Chinese traditional smoke-cured bacon. Food Chem. 149: 31–39.

Hygreeva, D., M.C. Pandey and K. Radhakrishna. 2014. Potential applications of plant based derivatives as fat replacers, antioxidants and antimicrobials in fresh and processed meat products. Meat Sci. 98: 47–57.

Igene, J.O., A.M. Pearson, R.A. Merkel and T.H. Coleman. 1979. Effect of frozen storage time, cooking and holding temperature upon extractable lipids and TBA values of beef and chicken. J. Anim. Sci. 49: 701–707.

Jalč, D. and M. Čertík. 2005. Effect of microbial oil, monensin and fumarate on rumen fermentation in artificial rumen. Czech J. Anim. Sci. 50: 467–472.

Jin, G., J. Zhang, X. Yu, Y. Lei and J. Wang. 2011. Crude lipoxygenase from pig muscle: Partial characterization and interactions of temperature, NaCl and pH on its activity. Meat Sci. 87: 257–263.

Jin, G., J. Zhang, X. Yu, Y Zhang and Y. Lei. 2010. Lipolysis and lipid oxidation in bacon during curing and drying-ripening. Food Chem. 123: 465–471.

Kamal-Eldin, A. and L.A. Apelqvist. 1996. The chemistry and antioxidant properties of tocotrienols. Lipids. 31: 671–701.

Kameník, J. 2011. Fermented meat products (in Czech). VFU, Brno.

Kameník, J., A. Salakova, G. Borilova, Z. Pavlik, E. Standarova and L. Steinhauser. 2012. Effect of storage temperature on the quality of dry fermented sausage Poličan. Czech J. Food Sci. 30: 293–301.

Karslioglu, B., U.E. Cicek, N. Kolsarici and K. Candogan. 2014. Lypolytic changes in fermented sausages produced with turkey meat: Effect of starter culture and heat treatment. Korean J. Food Sci. An. 34: 40–48.

Kouba, M. and J. Mourot. 2011. A review of nutritional effects on fat composition of animal products with special emphasis on n-3 polyunsaturated fatty acids. Biochimie 93: 13–17.

Ladikos, D. and V. Lougovois 1990. Lipid oxidation in muscle foods: A review. Food Chem. 35: 295–314.

Lanari, M.C., D.M. Schaefer, R.G. Cassens and K.K. Scheller. 1995. Atmosphere blooming time affect colour and lipid stability of frozen beef from steers supplemented with vitamin E. Meat Sci. 40: 33–44.

Leland, J.V. 1997. Flavor interactions: The greater whole. Food Technol. 51: 75–80.

Liaros, N.G., E. Katsanidis and J.G. Bloukas. 2009. Effect of ripening time under vacuum and packaging film permeability on processing and quality characteristics of lowfat fermented sausages. Meat Sci. 83: 589–598.

Maestri, D.M., V. Nepote, A.L. Lamarque and J.A. Zygadlo. 2006. Natural products as antioxidants. pp. 105–135. *In:* F. Imperato (ed.) Phytochemistry: advances in research. Research Signpost, Kerala, India.

Marcinčák, S., P. Bystrický, P. Turek, P. Popelka, D. Máté, J. Sokol, J. Nagy and P. Ďurčák. 2003. Effect of natural antioxidants on the oxidative processes in pork meat. Folia Vet. 47: 215–217.

Marcinčák, S., R. Cabadaj, P. Popelka and L. Šoltýsová. 2008a. Antioxidative effect of oregano supplemented to broilers on oxidative stability of poultry meat. Slov. Vet. Res. 45: 61–66.

Marcinčák, S., P. Popelka and L. Šoltysová. 2008b. Polyphenols and antioxidative activity of extracts from selected slovakian plants. Acta Sci. Pol. - Medicina Veterinaria 7: 9–14.

Marcinčák, S., P. Popelka, P. Bystrický, K. Hussein and K. Huḍecová. 2005. Oxidative stability of meat and meat products after feeding of broiler chickens with additional amounts of vitamin E and rosemary. Meso 7: 34–39.

Marco, A., J.L. Navarro and M. Flores. 2006. The influence of nitrite and nitrate on microbial, chemical and sensory parameters of slow dry fermented sausage. Meat Sci. 73: 660–673.

Mendoza, E., M.L. Garcia, C. Casas and M.D. Selgas. 2001. Inulin as fat substitute in low fat, dry fermented sausages. Meat Sci. 57: 387–393.

Molly, K., D. Demeyer, T. Civera, and A. Verplaetse. 1996. Lipolysis in a Belgian sausage: relative importance of endogenous and bacterial enzymes. Meat Sci. 43: 235–244.

Molly, K., D. Demeyer, G. Johansson, M. Raemaekers, M. Ghistelinck and I. Geenen. 1997. The importance of meat enzymes in ripening and flavour generation in dry fermented sausages. First results of a European project. Food Chem. 59: 539–545.

Morrissey, P.A., P.J.A. Sheehy, K. Galvin, J.P. Kerry and D.J. Buckley. 1998. Lipid stability in meat and meat products. Meat Sci. 49: S73–S86.

Muguerza, E., O. Gimeno, D. Ansorena, J.G. Bloukas and I. Astiasarán. 2001. Effect of replacing pork backfat with pre-emulsified olive oil on lipid fraction and sensory quality of Chorizo de Pamplona – A traditional Spanish fermented sausage. Meat Sci. 59: 251–258.

510   *Fermented Meat Products: Health Aspects*

Muguerza, E., O. Gimeno, D. Ansorena, I. Astiasara´n. 2004. New formulations for healthier dry fermented sausages: a review. Trends Food Sci. Tech. 15: 452–457.

Navarro, J.L., M.I. Nadal, L. Izquierdo and J. Lores. 1997. Lipolysis in dry cured sausages as affected by processing conditions. Meat Sci. 45: 161–168.

Niklová, I., Š. Schmidt and S. Sekretár. 2000. Antioxidative active compounds in oilseeds (In Slovak). Bulletin Food Res. 39: 101–116.

Olivares, A., J.L. Navarro and M. Flores. 2011. Effect of the fat content on aroma generation during processing of dry fermented sausages. Meat Sci. 87: 264–273.

Ou, B., D. Huang, M. Hampsch-Woodili and J.A. Flanagan. 2003. When the east meets west: the relationship between yin-yang and antioxidation–oxidation. The FASEB J. 17: 127–129.

Petrovič, V., S. Marcinčák, P. Popelka, J. Šimková, M. Mártonová, J. Buleca, D. Marcinčáková, M. Tučková, L. Molnár and G. Kováč. 2012. The effect of supplementation of clove and agrimony or clove and lemon balm on growth performance, antioxidant status and selected indices of lipid profile of broiler chickens. J. Anim. Physiol. Anim. Nutr. 96: 970–977.

Patthy, M., I. Király and I. Sziráki. 1995. Separation of dihydroxybenzoates, indicators of *in-vivo* hydroxyl free radical formation, in the presence of transmitter amines and some metabolites in rodent brain, using high performance liquid chromathography with electrochemical detection. J. Chromatogr. B 664: 247–252.

Pikul, J., D.E., Leszczyński and F.A. Kummerow. 1989. Evaluation of three modified TBA methods for measuring lipid oxidation in chicken meat. J. Agr. Food Chem. 37: 1309–1313.

Pokorný, J., Z. Réblová and W. Janitz. 1998. Rosemary ans sage extracts as antioxidants in fats and oils. Czech J. Food Sci. 16: 227–234.

Pokorný, J. and Š. Schmidt. 2003. The impact of food processing in phytochemicals: the case of antioxidants. pp. 298–314. *In:* I. Johnson and G. Williamson (eds.). Phytochemical functional foods. Woodhead Publishing, Cambridge.

Porter, N.A., S.E. Caldwell and K.A. Mills. 1995. Mechanisms of free radical oxidation of unsaturated lipids. Lipids. 30: 277–290.

Raharjo, S., J.N. Sofos and G.R. Schmidt. 1992. Improved speed, specificity, and limit of determination of an aqueous acid extraction thiobarbituric acid $-C_{18}$ method for measuring lipid peroxidation in beef. J. Agric. Food Chem. 40: 2182–2185.

Rubio, B., B. Martínez, M.D. García-Cachán, J. Rovira and I. Jaime. 2008. Effect of the packaging method and the storage time on lipid oxidation and colour stability on dry fermented sausage salchichón manufactured with raw material with a high level of mono and polyunsaturated fatty acids. Meat Sci. 80: 1182–1187.

Ruiz, J., J. Ventanas, R. Cava, A. Andres and C. Garcia. 1999. Volatile compounds of dry-cured Iberian ham as affected by the length of the curing process. Meat Sci. 52: 19–27.

Severini, C., T. De Pilli and A. Baiano. 2003. Partial substitution of pork backfat with extra-virgin olive oil in salami products: Effects on chemical, physical and sensory quality. Meat Sci. 64: 323–331.

Fermented Meat Products—An Overview 511

Schmidt, Š. 2011. Antioxidants and oxidation changes of lipids in food (In Slovak), STU, Bratislava.

Schmidt, Š. and J. Pokorný 2005. Potential aplication of oil seeds as sources of antioxidants for food lipids – a rievew. Czech J. Food Sci. 23: 93–102.

Schmidt, R.H. and G.E. Rodrick. 2003. Food safety handbook. John Wiley & Sons, Inc., Hoboken, New Jersey.

Šojić, B.V., L.S. Petrović, A.I. Mandić, I.J. Sedej, N.R. Džinić, V.M. Tomović, M.R. Jokanović, T.A. Tasić, S.B. Škaljac and P.M. Ikonić. 2014. Lipid oxidative changes in tradititional dry fermented sausage Petrovská klobása during storage. Hem. Ind. 68: 27–34.

Suhaj, M. 2006. Spice antioxidants isolation and their antiradical activity: a review. J. Food Comp. Anal. 19: 531–537.

Summo, C., F. Caponio and A. Pasqualone. 2006. Effect of vacuum-packaging storage on the quality level of ripened sausages. Meat Sci. 74: 249–254.

Takácsová, M., N.D. Vinh, D.M. Nhat and A. Príbela. 1998. Antioxidative properties of some spices (In Slovak). Bulletin of Food Res. 37: 1–10.

Toschi, T.G., A. Bordoni, S. Hrelia, A. Bendini, G. Lercker and P.L. Biagi. 2000. The protective role of different green tea extracts after oxidative damage is related to their catechin composition. J. Agric. Food Chem. 48: 3973–3978.

Veiga, A., A. Cobos, C. Ros and O. Diaz. 2003. Chemical and fatty acid composition of "Lacón gallego" (dry-cured pork foreleg): Differences between external and internal muscles. J. Food Compos. Anal. 16: 121–132.

Velíšek, J. 1999. Chemie potravin (In Czech) OSSIS, Pelhřimov.

Ventanas, S., M. Estévez, J.F. Tejeda and J. Ruiz. 2006. Protein and lipid oxidation in longissimus dorsi and dry cured loin from Iberian pigs as affected by crossbreeding and diet. Meat Sci. 72: 647–655.

Visessanguan, W., S. Benjakul, S. Riebroy, M. Yarchai and W. Tapingkae. 2006. Changes in the lipid composition and fatty acid profile of Nham, a Thai fermented pork sausage, during fermentation. Food Chem. 94: 580–588.

Watson, R.R. and V.R. Preedy. 2009. Tocotrienols: vitamin E beyond tocopherols. CRC Press, Boca Raton, Florida.

Wojciak, K.M., M. Karwowska and Z.J. Dolatowski. 2015. Fatty acid profile, color and lipid oxidation of organic fermented sausage during chilling storage as influenced by acid whey and probiotic stains addition. Sci. Agric. 72: 124–131.

Wojciak, K.M. and Z.J. Dolatowski. 2012. Oxidative stability of fermented meat products. Acta Sci. Pol., Technol. Aliment. 11: 99–109.

Yang, H.J., C.W. Ma, F.D. Qiao, Y. Song and M. Du. 2005. Lipolysis in intramuscular lipids during processing of traditional Xuanwei ham. Meat Sci. 71: 670–675.

Zanardi, E., V. Dorigoni, A. Badiani and R. Chizzolini. 2002. Lipid and colour stability of Milano-type salamis: effect of packing conditions. Meat Sci. 61: 7–14.

Zanardi, E., S. Ghidini, A. Battaglia and R. Chizzolini. 2004. Lipolysis and lipid oxidation in fermented sausages depending on different processing conditions and different antioxidants. Meat Sci. 66: 415–423.

Zakrys, P.I., S.A. Hogan, M.G. O'Sullivan, P. Allen and J.P. Kerry. 2008. Effects of oxygen concentration on the sensory evaluation and quality indicators of beef muscle packed under modified atmosphere. Meat Sci. 79: 648–655.

Chapter 22

HACCP in Fermented Meat Production

Igor Tomašević* and Ilija Djekić

1 INTRODUCTION

Hazard Analysis and Critical Control Points (HACCP) is a conceptually simple, science based system whereby the food safety hazards can be identified and evaluated by the meat establishment, which then institute controls necessary to prevent those hazards from occurring or keeping them within acceptable limits, monitor the performance of controls and maintain records routinely. Its established control systems are focused on prevention rather than relying on end-product testing (Radovanović et al. 2003a). Today, HACCP is internationally recognized resource of managing food safety. Its sole direct objective is to ensure food production and processing control in a manner that the final product does not represent a risk to the consumer. It is a distinctly structured and defined system that implies disciplined approach with a focus upon preventive actions and monitoring and control of critical points only if necessary. All the procedures are defined and executed in a way that any loss of control upon food safety hazards is detected instantly and corrective actions are followed so the reoccurrence of the same problem is discouraged. When HACCP is based upon strong foundations of Good Manufacturing Practices (GMP's) and Standard Sanitation Operating

---

*For Correspondence: Animal Source Food Technology Department, University of Belgrade – Faculty of Agriculture Nemanjina 6, 11080 Belgrade, Republic of Serbia. Tel: + 381 11 2615315/272, cell: + 381 60 4299998, Email: tbigor@agrif.bg.ac.rs

HACCP in Fermented Meat Production

Procedures (SSOP's) it enables good support for the achievement of consumer confidence in the safety of the food produced in such facility, regardless of its size. It clearly demonstrates producer's dedication to food safety and improves comprehension of duties and responsibilities to all employees, starting with a top management, regarding consumer health protection (Tomić and Tomašević 2007). Any HACCP system is capable of accommodating changes, such as advances in equipment design, processing procedures, as well as technological developments. It is the best system currently available for maximizing the safety of meat and meat products, as well as food in general, which is why it has been recommended for use in the food industry and promoted by the governments and scientific groups for the last 40 years.

2 LEGISLATION

Since the year 1959, when as a part of NASA laboratories "space food" program it had only two objectives (to produce the food for astronauts totally free of crumbs and free water and without presence of pathogenic microorganisms and biological toxins) until today, HACCP has traveled a long way. From a food safety concept for the production of food for 3–5 astronauts up to a food safety system for the industry that provides food for people around the world.

Once a visionary approach and today almost 60 years old, HACCP is an integral part of American food industry but before all the food safety inspection system. The US Food and Drug Administration (FDA) used HACCP-based principles in the development of low-acid food canning regulations in the 1970s. In 1995, the FDA issued regulations that made HACCP mandatory for fish and seafood products, and in 2001 they issued regulations for mandatory HACCP in juice processing and packaging plants. In addition, a voluntary HACCP program was implemented in 2001 for Grade A fluid milk and milk products under the cooperative federal/state National Conference on Interstate Milk Shipments (NCIMS) program. The Food and Drug Administration (FDA) has also implemented pilot HACCP programs for a variety of other food processing segments as well as for retail foods. HACCP has also been implemented by the United States Department of Agriculture (USDA). In 1998, USDA's Food Safety and Inspection Service (FSIS) mandated HACCP for the nation's meat and poultry processing plants. Currently, HACCP systems are utilized for pathogen reduction in over 6,500 raw meat and poultry plants. The US food processing industry will inevitably be faced with more mandatory HACCP programs under FDA and USDA/FSIS regulations in the future. In fact, the Food Safety Modernization Act of 2010 (FSMA) will incorporate the mandatory

514 *Fermented Meat Products: Health Aspects*

use of preventative food safety programs (such as HACCP) in several segments of the food industry, including produce.

In the European Union (EU) the HACCP system is laid down in Regulation 852/2004 on the hygiene of foodstuffs (Regulation (EC) no 852/2004 of the European Parliament and of the Council of 29 April 2004 on the hygiene of foodstuffs, Official Journal of European Union). According to EU regulations, food business operators are responsible for safety of their products and have to follow the principles of HACCP system. The only exception for the implementation of HACCP is the producers of the primary products, such as animal farmers, although they have to follow the principles of hygiene practices. HACCP system has also been implemented under regulation in other countries (e.g., Canada, Australia, and New Zealand) and is a high priority program under Codex Alimentarius, the world food standards authority.

3 PREREQUISITE PROGRAMMES

Before implementing a HACCP system, a food (meat) business should already have in place various practices that may be collectively termed 'prerequisite programmes' (PRPs). There is a big confusion between PRPs and HACCP plan, their relations and how they should be managed. It seems to be very difficult to implement a HACCP based system in meat industry, when a high proportion of employees is not familiar with this reality and does not participate in PRPs (Gomes-Neves et al. 2011).

Prerequisites include cleaning and sanitation, maintenance, personal hygiene and training, pest control, plant and equipment, premises and structure, services (compressed air, ice, steam, ventilation, water etc.), storage, distribution and transport, waste management and physical separation of activities to prevent potential food contamination. HACCP system cannot be imagined and properly executed if any of those previously mentioned activities are missing.

These activities are usually divided in two large sub-groups of practices according to the nature of problems they are dealing with. The first group is called "Good manufacturing practices" (GMPs) and is sometimes referred to as 'control points' and is identified as the correct processes and procedures to be followed in the preparation of food to prevent microbial, chemical and physical contamination of the finished product. In other words, GMPs define what has to be done to prevent contamination, when it has to be done and by whom. GMPs do not address specific hazards, and lack of control would not necessarily result in an unacceptable health hazard to the consumer. Areas covered by the GMP programme include: personnel, including task and hygiene

HACCP in Fermented Meat Production 515

training, job description and organizational structure, premises, including location and structure, equipment, including design, maintenance and calibration, services, including sanitary services, disposal of waste materials, the provision of electricity, water, refrigeration and steam, raw materials, including live animals, packaging, food ingredients and chemical product traceability, documentation.

The second group is called "Good hygiene practices" (GHPs). This may be defined as those operations involved in providing a clean sanitary environment for the preparation, handling and storage of meat and meat products. In other words, the GHPs define what has to be done in relation to cleaning and hygiene, when it has to be done and by whom (Arvanitoyannis 2009). The importance of GHPs is perceived contrarily in different countries. Cleaning and disinfection procedures were in place in 100% Polish, 96.1% Serbian and 85.7% Chinese meat producing facilities, while it was the case with only 71.7% Spanish and 29.6% of Turkish food business operators (Tomašević et al. 2013, Baş, et al. 2006, Wang et al. 2010, Konecka-Matyjek et al. 2005, Celaya et al. 2007).

4 HACCP PRINCIPLES AND IMPLEMENTATION IN A PRODUCTION OF "UŽICE GOVEÐA PRŠUTA" (DRY-CURED BEEF HAM)

The raw materials and the autochthonous microorganisms found in traditional fermented meat products are not just responsible for the expression of their typical flavor but they also affect the safety of food. Although the microbiota that cause the production of organic acids, ethanol, hydrogen peroxide and diacetyl, antifungal compounds and bacteriocins during the fermentation process help protect these kinds of foods, they are not necessarily enough to guarantee that there will not be unwanted microorganisms in the final mixture. This is why traditional fermented meat products safety is due to the combined action of several preservative factors/hurdles. Example of such traditional meat fermented product, called "Užice Goveđa Pršuta" is processed in a remote area of mountain Zlatibor (the southwestern part of the Republic of Serbia). It largely contributes to the local economy and Serbian gastronomic heritage. Being made of the most valuable parts of beef carcass (round muscles, loin muscles and tenderloin) originating from well-fed, 3- to 5-year-old cattle, this product conforms fully with its national Protected Designation of Origin (PDO) mark (Tomic et al. 2008).

When the HACCP concept was first presented to the public, at the 1971 National Conference on Food Protection, it consisted of three principles (Bernard and Stevenson 1999):

516   Fermented Meat Products: Health Aspects

1. Identification and assessment of hazards associated with growing/ harvesting to marketing/preparation.
2. Determination of the critical control points to control any identifiable hazard.
3. Establishment of systems to monitor critical control points.

Along with these principles, the system identified a critical control point (CCP) as a point in the manufacture of a product whose loss of control would result in an unacceptable food safety risk.

The preventive nature of the HACCP system is readily apparent when these principles are paraphrased, as follows:

1. Identify any safety-related problems associated with the ingredients, product and process.
2. Determine the specific factors that need to be controlled to prevent these problems from occurring.
3. Establish systems that can measure and document whether or not these factors are being controlled properly.

The Codex Alimentarius Committee on Food Hygiene (Codex) has been actively involved in the development of HACCP guidelines for use in international trade. This is a committee of the United Nations WHO/ FAO Codex Alimentarius Commission (CAC) that have revised and refined explanations of the HACCP principles and guidelines for use in applying the HACCP principles to various food production operations. Codex adopted the latest revision of the HACCP guidance document a more than a decade ago (CAC 2003).

HACCP implementation is a logical process that needs to be followed step by step in order for it to work correctly. It is recommended by CAC that the practical application of HACCP should be carried out in twelve steps that must include all of its seven principles. Their twelve-stage logic sequence must be followed in a specific order:

Step 1: Assemble HACCP Team

Step 2: Describe the product

Step 3: Identify intended use

Step 4: Construct flow diagram

Step 5: On-site confirmation of flow diagram

Step 6: (Principle 1) Conduct a hazard analysis

Step 7: (Principle 2) Determine the CCPs

Step 8: (Principle 3) Establish critical limit(s)

Step 9: (Principle 4) Establish a system to monitor control of the CCP

Step 10: (Principle 5) Establish the corrective action to be taken when monitoring indicates that a particular CCP is not under control

HACCP in Fermented Meat Production 517

Step 11: (Principle 6) Establish procedures for verification to confirm that the HACCP system is working effectively

Step 12: (Principle 7) Establish documentation concerning all procedures and records appropriate to these principles and their application.

4.1 HACCP Team

There is no doubt that the appropriate product-specific knowledge must be available for the development of an effective HACCP plan. If it is developed by one person it is likely to fail. Although colleagues like to think of themselves as "HACCP experts" they should also acknowledge that no single person can poses all the multi-disciplinary knowledge and expertise necessary to establish a solid HACCP system. This is why the size and composition of the team will vary with the size and complexity of the meat processing facility.

Every HACCP team must have its "Leader" who is responsible for ensuring that the HACCP team training and expertise meets minimum standards. The team should include representatives from every aspect of the meat operation, including management, maintenance, sanitation, operations, and receiving. The team members must understand the biological, chemical or physical hazards connected with a particular meat or meat product group. They should be responsible for, or are closely involved with, the technical process of manufacturing the products. They have a working knowledge of the hygiene and operation of the meat plant and equipment. The predominant professions among meat plants HACCP team members are food technologists followed by veterinarians and food chemists although it is not unusual to include economists in it (Tomašević et al. 2013).

4.2 Product Description and Its Intended Use

In order to properly describe the product, the HACCP team must have a full understanding of its manufacturing process and technology. Product description must include detailed information about its composition, its raw materials, its processing and handling methods, and its distribution conditions (Table 1). The list of all incoming ingredients and especially the ones that are restricted, such as phosphate, nitrite or nitrate compounds, and potential allergens such as soy, cereals containing gluten, sulphites, milk, etc., must be clearly identified. The product description also provides basic information on the products shelf life, packaging, labeling, display and storage requirements, and its intended use.

HACCP team must have a full understanding of the product's intended use and its likely consumers. Weather the target audience

518 *Fermented Meat Products: Health Aspects*

is general population or the consumers of the product may have a particular food safety requirement. It is often the case with so called "vulnerable groups" that include allergy sufferers, infants and young children, elders, pregnant women and people that have an impaired immune response (for instance those undergoing chemotherapy or have AIDS) or any other health related problem that makes them unsuitable as a consumers of such a product (Table 1).

Table 1 Product Description of "Užice Goveđa Pršuta"

Product name	Užice Goveđa Pršuta
Composition	Beef meat, Sodium nitrite (E 250)
Important characteristics	(pH < 5.6) (a_w < 0.88) (moisture < 50%) (NaCl > 6%)
Intended use	Not suitable for babies, elders or adults suffering from hypertension or kidney problems
Preparation	Ready to eat
Shelf life	Shelf stable product
Packaging	Sliced, vacuum packed (plastic bags of 125 g and 250 g net weight) Bulk package in cardboard boxes (boxes of 2.5 kg, 5 kg and 10 kg net weight)
Labeling instructions	Keep under 10°C "High salt content" warning
Distribution instruction	Distributed under 10°C

4.3 Flow Diagram and On-Site Confirmation

HACCP team should construct a flow diagram. It is an important document in a HACCP plan. Codex defines a flow diagram as "A systematic representation of the sequence of steps or operations used in the production or manufacture of a particular food item." It usually begins with receiving meat, ingredients, and packaging – then continues through the steps of making a product and concludes with packaging, storage and distribution. Essentially, the flow diagram continues until the product is no longer in the control of the establishment, nor the establishment's responsibility. It should be a simple block diagram (not a complex engineering drawing) that is easily understood by all the employees involved. Ideally, it should be in a simple linear format, from top to bottom of the page. Avoiding the lines that cross over the flow diagram makes it easier to read and understand. Although it should contain sufficient technical data, it should be kept in mind that its main purpose is to accurately represent all of the steps in the production process from raw materials to the end product.

For the production of "Užice Goveđa Pršuta" (in further text – UGP) only selected beef cuts or individual muscles are used, like *m. quadriceps femoris, m. semitendinosus, m. longissimus thoracis et lumborum* or *m. gracialis* with a portion of *m. semimembranosus*, from well-fed, 3–5-year old cattle. The cuts are shaped in accordance to the traditional dimensions: length between 30 and 50 cm, width between 12 and 15 cm and height in a range of 8 to 10 cm. The whole process is relatively simple and all it takes besides beef is a sea salt, smoke, air and 4–5 weeks of winter time. It begins with salting (no nitrates/nitrites, no sugars or spices added) when all the muscles are covered with sea salt, one by one and by hand. The amount of salt added equals 3% of the weight of the meat. It is very important to use the right amount of salt according to the cut's weight and shape and to make sure it is evenly covered. This personalized care is essential in all stages of the traditional processing of UGP.

The removal of salt crystals, after 7 to 10 days of salting, from the surface of the beef cuts is done by hand, while rinsing with cold or lukewarm water was never perceived as a traditional practice. Meat cuts are then hung on sticks and the period of drying, smoking, ripening (maturing) and fermentation follows. During this phase, of 4 to 5 weeks, meat develops the typical flavor, color and texture. Since the traditional production takes place on Mt. Zlatibor and during winter months of November till February, the average air temperature is between 3.5°C and –2.2°C with its relative humidity (RH) in a range of 81 to 85% (Radovanović 1990).

The average height of Mt. Zlatibor is about 1000m and it is not prone to strong winds. This allows drying, smoking and fermentation to occur in natural drying areas (rooms), where temperature and humidity are controlled through ventilation and burning of wood during the frequent periods of cold (extensive) smoking. The favorite type of wood used for that purpose is European beech (*Fagus sylvatica*), while European hornbeam (*Carpinus betulus*), English (*Quercus robur*) and Durmast oak (*Quercus petraea*) are less preferred. Temperature, in this natural drying/smoking/fermentation chambers, ranges from 8°C to 10°C (never exceeds 12°C) and the RH is between 70 and 75%. By the end of a production process, a UGP will lose more than 40% of its weight through moisture loss (Stamenković et al. 2003), helping to concentrate the flavor and leaving the final product with the heavy salt content of more than 5% (Radovanović et al. 2003). Flow diagram of operations used in the production of UGP is given in Figure 1.

It is essential that the flow diagram matches the process that is actually carried out. To do this, the HACCP team must observe the operation at all stages of manufacture, from start up to shut down, across shift patterns, by 'walking the line'. It is vital that both the process and diagram match at all times. If they do not, then it is because non-existent or out-of-date process is still present in the diagram or because a newly introduced production step(s) were not included in it. In either case, flow diagram that is not regularly updated, whenever the production line was altered, completely diminishes all the subsequent and efforts made beforehand to make a proper HACCP plan.

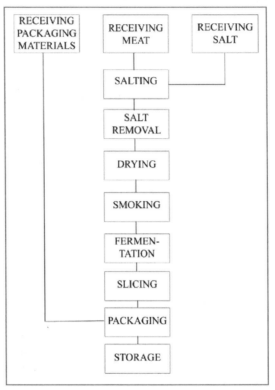

Figure 1 Flow diagram of "Užice Goveđa Pršuta" production.

4.4 Hazard Analysis

Potential hazards include all hazards (biological, chemical, physical) that may be reasonably expected to occur at a particular process step. On occasions when products are, or could be, apparently allergen-free, or free from a particular allergen, allergens can be considered as a separate hazard category.

Raw materials, the manufacturing process, equipment, product storage and distribution should all be considered when identifying hazards. This is exactly why flow diagram is used when conducting a hazard analysis and each step in it must be assessed for potential hazards. When conducting a hazards analysis it is necessary to consider all the hazards introduced by any of the inputs or as a consequence of applying the process step itself (e.g. metal fragments from processing

HACCP in Fermented Meat Production 521

equipment), hazards carried over in the product from the previous step and the adverse impacts of process steps on existing hazards (e.g. growth of microorganisms). The HACCP Team must carry out a hazard analysis to identify for the HACCP plan which hazards are of such nature that their elimination or reduction to acceptable levels is essential to the production of safe food. A consideration of the significance of the hazard at this stage helps with preparation of the CCP determination, since there is no point in taking non-significant hazards through to the next stage.

The HACCP team will focus on hazards that can be prevented, eliminated or controlled by the HACCP plan. When other types of hazards are identified during the process, then effective control by the relevant PRP should be confirmed, and the results recorded. A justification for including or excluding the hazard is reported and possible control measures are identified.

While receiving cuts of beef meat in the production of UGP we must be aware of the potential biological and physical hazards introduced at this processing step. Raw cuts of meat generally have the same level of contamination as the carcass and the occurrence of pathogens such as *Salmonella*, *Escherichia coli* O157:H7 on beef carcasses, today represent a common knowledge. *Listeria monocytogenes* and *Staphylococcus aureus* are also reasonably likely to occur. There must be a subsequent processing step(s) that will reduce the risk of these biological hazards to an acceptable level. Foreign materials such as broken needles, although regarded as severe physical hazard, have never been recorded during the reception of beef meat. Current regulatory programs in animal (beef) production reduce the risk of antibiotic residue presence.

Packaging materials used do not introduce any of the biological, chemical or physical hazards because only food grade materials are used. On each reception certificates of conformity are checked from all suppliers. Similar should be said regarding receiving salt, because specifications for chemical, physical and biological issues are mandatory on each and every delivery from salt suppliers.

The health hazards associated with smoked foods are related to the presence of carcinogenic components in wood smoke and smoked meats, polycyclic aromatic hydrocarbons (PAH) (Pleadin and Bogdanović, Chapter 18 in this book). Wood smoke contains different PAH with a wide range of molecular weights (MW). The low-molecular members of this group, below MW 216, are not regarded as carcinogenic, contrary to many heavy MW PAHs. Very mutagenic and carcinogenic is benzo(a) pyrene (BaP). Therefore it was recognized as an indicator of carcinogenic PAHs in wood smoke and smoked products. Regulation (EU) of 2011, a maximum concentration of BaP in smoked meat and meat products has been maintained for the present (5 ng/g until 31 August 2014 and 2 ng/g after this date). However, additionally the upper limit for PAH4 has been introduced (30 ng/g from 1 September 2012 until 31 August 2014, and 12 ng/g after this date). PAH isolated from smoked products are mainly compounds with a MW less than 216 (Sikorski and Sinkiewicz 2014).

The Serbian meat industry was already interested in studying PAHs in smoked meat and meat products, in order to improve food safety of products from protected geographic areas and their compliance with the EU regulations, especially because smoked meat products present a significant part of the human diet in Serbia (Djinovic et al. 2008). The study revealed that the distance between the firing place (smoke generator) and the meat product in a smokehouse plays a role in preventing the heavy MW PAH reaching the final product. Higher distance meant less content of the PAHs examined in observed meat products. In the case of UGP smoked with beech wood at the distance of 2m highest detected BaP concentrations were 1.09 ng/g and maximum PAHs concentrations were 8.62 ng/g, all well below the limits set by EU (Djinovic et al. 2008). According to Radovanović (1990), the distances between beef meat and firing place used in traditional smoking chambers of Mt. Zlatibor is around 2.5 m. Therefore, the concentrations of chemical hazards (carcinogenic components) introduced during the smoking of UGP are not reasonably likely to exceed acceptable levels.

All the other processing steps like salting, drying, fermentation or storage are used only to control biological hazards introduced previously. If all the procedures defined in GMP and GHP protocols are in place and working as planned then we should not consider that any of the biological, chemical or physical hazards are reasonably likely to occur at these processing steps. For example, someone could argue that slicing of UGP may introduce the potential chemical hazard(s) and promote cross-contamination with cleaner/sanitizer used for cleaning the slicing machine. This can be done only if cleaning and disinfection procedures within GHP have not addressed this matter earlier. In that case modifications should be applied to an inadequate PRP and by no means addressed within a hazard analysis.

Finally, the temperature in a meat handling room of a licensed Serbian meat facility shall not exceed 12°C. Usually the GMP procedures deal with this type of meat safety requirement. In a case of UGP this does not have to be controlled particularly, since the air temperature during the traditional production period does not go above this value. The only exception is a temperature of a storage room since the shelf life of a product can extend beyond the winter months.

4.5 CCP Determination

The information established during the hazard analysis should allow the identification of CCPs. During the process of CCP determination

HACCP team must ask themselves a few important questions for every processing step presented in a flow chart. There are two possible scenarios on how a processing step becomes a CCP.

The first one includes all operations that, according to hazard analysis, have introduced any kind of safety hazards and are designed in a way that allow them to eliminate or reduce the likely occurrence of those hazards to an acceptable level. The second one involves processing steps that have not introduced new safety hazards but are able to control or eliminate any of them introduced in operations that precede them. If a hazard has been identified at a step where control is necessary for safety and no control measure exists at that step, or any other, then the process should be modified at that step, or at any earlier or later stage, to include a control measure. That modified step becomes a CCP by default.

The correct determination of CCPs is vital to ensure that there is effective management of food safety. The number of CCPs in a process will depend on the complexity of the process itself. Therefore, it is defined as a step at which control can be applied and is essential to prevent or eliminate a food safety hazard or reduce it to an acceptable level. Not surprisingly, the emphasis is on the necessity of the measure involved, because the HACCP team frequently tends to be too cautious and designates too many CCPs, rather than correctly identifying only the essential ones. This means that HACCP plan has lost its focus and credibility and that commitment of people that should execute it on every day basis will also be lost over time. The attitude of a HACCP team that there are always steps that can further enhance the safety of our product is a wrong one. On the other hand, having too few CCPs could jeopardize product safety.

The CCP determination is perhaps the most difficult part of making a HACCP plan. This is especially true for the fermented meat products where safety is achieved by multiple "hurdles" that control microbial growth. Salt, water activity, pH and smoke are examples of hurdles that can be used for preservation. No single preservation method alone would create a stable product; when combined, however, these methods result in a desirable, stable, and safe product. Therefore, when· conducting a CCP determination for fermented meat product we must have in mind that there is no single, essential processing step that will prevent or eliminate a food safety hazards or reduce them to an acceptable level. Instead, the combination of processing steps (hurdles) is vital for assuring food safety of traditional fermented meat products. It is also important to remember that HACCP is a food safety system, not a food quality system. Thus, CCPs specifically deal with food safety issues, not food quality issues. For example, if the UGP at the end of the production becomes too salty or too dry in order to be accepted by the consumer, it only means that its microbial safety is stronger.

The first major preservative added to meat was salt. Today, there is a strong demand of the consumers, based on medical advices, to decrease sodium content in meat products. For traditional meat products sodium still plays a role in reducing the growth of pathogens and organisms that spoil products and reduce their shelf life. Salt is effective as a preservative because it reduces the water activity of foods. Salt's ability to decrease water activity is thought to be due to the ability of sodium and chloride ions to associate with water molecules (Damodaran et al. 2008). Loss of water from the microbial cells as a result of a osmotic shock suffered because of the addition of salt causes their death or retards their growth (Doyle and Buchanan 2013). It has also been suggested that for some microorganisms, salt may limit oxygen solubility, interfere with cellular enzymes, or force cells to expend energy to exclude sodium ions from the cell, all of which can reduce the rate of growth (Leora and Julie 2005). For all the before mentioned reasons the processing step of salting in a production of UGP must be identified as a CCP. Because it is a first determined CCP and because it controls biological hazards it should be labeled as a CCP-1B.

Drying is one of the oldest methods of meat preservation, where water activity is reduced by separating out water. This is achieved by evaporation of water from the peripheral zone of the meat to the surrounding air and the continuous migration of water from the deeper meat layers to the peripheral zone. Drying of UGP is achieved in traditional meat-drying facilities where the design of a drying room allows natural air convection. During the 4 weeks period, more than 40% of the initial weight of beef muscles is lost and the water activity of the final product decreases sufficiently to prevent microbial spoilage. Therefore drying is identified as a CCP-2B.

Smoking of meat can be regarded as "hurdle technology". Preservative effect of smoking is related to the concentration of salt due to pre-smoking treatment, the time–temperature regime and loss of water during processing, as well as the composition and quantity of smoke deposited on the meat. Numerous components of wood smoke have antimicrobial activity in concentrations similar to those found in smoked meats. UGP is extensively smoked meat product and smoking should be identified as a CCP-3B.

Traditional fermentation relies on indigenous microorganisms that are present on the meat ingredients, the spices and especially the environment. When safety of traditional meat products is discussed, fermentation and drying are usually elaborated together because in practice they are impossible to separate. These preservation methods are several thousand years old and early meat manufacturers did not originally realize that in drying meat more or less quickly they actually subjected the products to some degree of fermentation. The amount

HACCP in Fermented Meat Production 525

of acid produced during fermentation and the lack of moisture in the finished product after drying have been shown to cause the death of pathogenic bacteria. This is why fermentation is designated as a CCP-4B.

Finally, although the end product is shelf stable, when it is sliced and vacuum packed, the moister interior can permit staphylococcal multiplication and thus sliced dried and cured hams should be refrigerated (FSIS-USDA 2015). Therefore storage of UGP must become a CCP-5B.

4.6 Critical Limits

A maximum and/or minimum value to which a biological, chemical or physical parameter must be controlled at a CCP to prevent, eliminate or reduce to an acceptable level the occurrence of a food safety hazard is a critical limit (NACMCF 1998). A critical limit is defined by the Codex Alimentarius Commission as "a criterion which separates acceptability from unacceptability". By unacceptable, Codex is referring to unsafe products; those likely to cause harm if the product reached the consumer (CAC 2003).

All CCPs must have critical limits and they must be scientifically based, easy and quick to measure by test or observation. This is why factors like temperature, time, pH, water activity and salt concentration are commonly used as critical limits. Microbiological critical limits should be avoided because microbiological factors can usually only be monitored by growing the organism of concern in the laboratory. This process is usually too slow and would prevent instant reaction when a failure to meet a critical limit (deviation) occurs.

The critical limits may be established by the HACCP team using information from different sources like regulatory standards and guidelines, surveys of published research, experimental results and experts (e.g. academics, consultants, food scientists, microbiologists, equipment manufacturers etc.). As it is an absolute value it must be a single value and cannot be a range.

Changes in salt content of beef meat used for the production of UGP are most intensive in the first 7 to 9 days of salting. It was previously established that, due to the uniform size and weight of meat cuts, it takes no more than 10 days for salt to penetrate to central parts of the muscles reaching concentrations above 3% (Radovanović et al. 1993, Tomic et al. 2008). Salting followed with drying will lead to a final product with more than 6% of salt and less than 50% of moisture (Stamenković et al. 2003). Water phase salt content of UGP, when calculated with these values, exceeds 10.5%. This is an important limiting condition for the growth of almost all major pathogenic microorganisms like *Clostridium botulinum*, *Bacillus cereus*, *Cl. perfringens*, *Escherichia coli*, *L. monocytogenes*

526   *Fermented Meat Products: Health Aspects*

and *Salmonella* spp. (FDA 2011). Therefore, for the CCP-1B and the operation of salting a minimum duration of 10 days must be set as a critical limit.

By the end of the drying process, UGP will lose more than 40% of the initial weight of the muscles with a maximum yield of 60%. This yield will relate to approximately 45% of the moisture, 40% of the meat proteins and $a_w \le 0.88$ of the final product (Radovanović et al. 2003b). Before, quantity of moisture in a product in relationship to the quantity of meat protein, expressed as X parts (or percent) of moisture for each part (or percent) of meat, called moisture: protein ratio (MPR) was used to define the shelf stability of meat products. Today it is only used to define the standard of identity of meat products. Since the water activity of 0.88 is a typical value for the most common dry cured hams, successfully limiting the microbial growth of all major pathogens with the exception of *L. monocytogenes* (Grau et al. 2015), the maximum yield of 60% during drying of UGP (CCP-2B) must be set as a critical limit.

Phenols formed during cold smoking may decrease the population of *L. monocytogenes* on the surface of the products while formaldehyde arrests the growth of *Cl. botulinum* in the concentration of 40 mg/g. However, it is not possible to predict exactly the concentration of total smoke components that are necessary for an inhibitory effect but we do know that smoke components present in a lightly smoked meats are not effective enough (Grau et al. 2015). For the production UGP 8 hours per day of cold smoking is applied (Rajkovic 2012). In order to assure extensive smoking and maintain its preservative effect on the final product, minimum duration of 8 hours of smoking a day must be set as a critical limit for the CCP-3B.

When using fermentation as a main safety hurdle, most fermented meat products are microbiologically stable when pH 5.3 or lower is obtained within a relatively short period of time. Traditional meat fermentation is anything but fast because it relies on lactic acid bacteria which naturally occur in fresh meat and the environment to initiate the process. While this practice today is perceived as an "art form" itself, it is highly unreliable as a single meat safety measure. The UGP has a final pH $\le$ 5.6 which coupled with the $a_w \le 0.88$ achieved during drying ensures unlikely growth of *L. monocytogenes* and *St. aureus* on this product according to the Shelf Stability Predictor Developed by the Center for Meat Process Validation at the University of Wisconsin–Madison. Therefore, for the CCP-4B and the operation of fermentation a maximum pH of 5.6 must be set as a critical limit.

It was previously mentioned that sliced dried and cured hams should be refrigerated but what is the maximum temperature allowed for the storage of sliced UGP? It was previously established that these products should not be stored at temperatures higher than 12°C when outgrowth

HACCP in Fermented Meat Production 527

and toxin production of *St. aureus* is concerned (Rajkovic 2012). Yet, since the Serbian Meat Products Regulation demands the storage of sliced dried meat products in facilities with the temperature below or equal to 10°C, this legal requirement defines the critical limit for the CCP-5B.

4.7 Monitoring

Monitoring must provide a measurement or observation at a given processing step, identified as a CCP, that serves a purpose of testing whether process is operating within the previously established critical limits or not. Monitoring procedures detect if deviation occurs. The frequency of monitoring will depend on the nature of the CCP and the type of monitoring procedure. These may be continuous where data are constantly recorded or discontinuous where measurements/observations are made at specified time intervals during the process (e.g. every 5 minutes, every hour, once a day). It is vital that a suitable frequency is determined for each monitoring procedure. Since the importance of monitoring is enormous and because it is usually assigned to a meat handler rather than to a member of a HACCP team, it is of paramount significance that this employee is adequately trained.

The differences between the appropriate monitoring procedures and their frequencies are easily observed in the HACCP Plan for the UGP. Monitoring of the CCP-1B will include the observation of the time passed since the beginning of the salting of meat cuts for all different lots of products in a salting chamber. The procedure is discontinuous and its frequency is once a day at the beginning of the shift. When the critical limit of 10 days of salting is achieved the appropriate lot will be sent to the next production step of salt removal. The weight measurement and calculation of yield for 3 different beef UGP in every lot is used as a monitoring procedure for drying (CCP-2B). This should be done at the end of each day of drying. Monitoring of the smoking (CCP 3-B) is also discontinuous but the nature of critical limit involved demands higher frequency. Therefore, the operator will check the time passed since the beginning of the smoking every 4 hours. When the limit of 8 hours per day of smoking is reached operator will stop the process and its monitoring until the next day. The employee responsible for the monitoring of fermentation (CCP 4-B) will test pH of 3 UGP for each lot by probe every other day and until the critical limit of pH ≤ 5.6 is reached when the production ends. Finally, for the purpose of monitoring the storage of the final product (CCP 5-B) a continuous procedure is used. This is achieved with a temperature data logger, also called temperature monitor with warning alerts. The temperature of the storage room is measured non-stop and if the deviation from the critical limit happens visual and sound alerts are set-off immediately.

4.8 Corrective Actions

There are two main types of corrective actions. First type of action adjusts the process to maintain control and prevent a deviation at the CCP. This first type of corrective action normally involves the use of operational limits within the critical limits. Monitoring systems where the CCP monitor takes action when the operational limits are approached or exceeded, and thus prevents a CCP deviation are good examples of this type of corrective actions (Mortimore and Wallace 2013).

In the HACCP Plan for UGP this kind of corrective action is taken every time when a CCP monitor finds out that a critical limit (time) necessary for salting (10 days) or smoking (8 hours per day) is not achieved. By prolonging the time of salting/smoking until critical limits are achieved the monitor actually prevents deviations from occurring. The same applies for the monitoring of drying and fermentation where corrective actions represent the continuation of the operations until the yield of at least 60% or pH ≤ 5.6 is accomplished.

Second type of corrective action is taken after the deviation of a CCP has already occurred. This requires swift reactions which will include adjusting the process to bring it back under control and dealing with the material that was affected during the deviation period (Mortimore and Wallace 2013).

This kind of corrective actions are taken after the alert was received about the deviation from the critical limit during the continuous monitoring of the CCP 5-B. Information that the temperature of the cold storage has exceeded the limit of 10°C forces the HACCP team to take corrective action in order to bring back the temperature to acceptable level (e.g. repair the compressor, adjust the door of the chamber, bring back electricity etc.) If the time necessary to complete this corrective action was too long and there is a reasonable doubt that safety of the final product was affected, then another corrective action is needed. The HACCP team must make a decision about additional microbiological or toxicological testing in order to reassure itself that all hazards remained below acceptable levels. This could be perceived as an additional verification of the final products.

4.9 Verification

Verification is the principle which confirms that the HACCP plan if followed will produce safe food for the final consumer. According to Codex, verification is "the application of methods, procedures, tests and other evaluations, in addition to monitoring, to confirm compliance with the HACCP plan" (CAC 2003).

HACCP in Fermented Meat Production 529

Depending on the type of product and scale of business, verification activities may include: internal audits, customer audits and evaluation of customer feedback analysis including complaints, external audits of suppliers, observation of operations at CCPs, their monitoring and the effectiveness of corrective actions, review of the HACCP system and its records, prior shipment review, instrument calibration, etc. Undertaking microbiological testing of the raw materials, food contact surfaces or end products is generally too slow for monitoring purposes, but it can be of great value in verification, since many of the identified hazards are likely to be microbiological. Verification should be scheduled on a regular basis. The HACCP Team must ensure that documentation of verification activities is undertaken.

It should be noted that verification and validation of HACCP plans are often confused, sometimes even equaled. Therefore, it is of use to repeat that validation focuses on the collection and evaluation of scientific, technical and observational information to determine whether control measures are capable of achieving their specified purpose in terms of hazard control. The validation should ask whether the HACCP plan will ensure that safe food will be produced, while verification should ask if the HACCP plan is working and producing safe food. It should be obvious that a HACCP plan that was not validated in the first place must not get a chance to be verified.

In a case of a HACCP Plan for UGP this is best explained for a production step of drying and CCP-2B. Critical limit is maximum yield of 60% and it was validated in-house by weighing 24 different beef cuts at the start and at the end of the process. When the yield was calculated they were sent to a testing laboratory for a determination of water activity. To ensure safety, the water activity must be 0.88 or lower. The highest yield for any of the beef cuts tested that met the applicable water activity of 0.88 was 60%. This is how the critical limit for the CCP-2B was established and validated.

From this time on, after the validation, the yield for 3 different UGP is determined on each lot to monitor the critical limit of CCP-2B. When the production process without deviations is over, only one final product randomly chosen from the lot is sent for microbiological testing, salt content and/or water activity analysis in order to accomplish the verification of the HACCP Plan.

4.10 Documentation

The HACCP plan and all the activities and operations arising from it, must be fully documented at all stages. It is important that this documentation is prepared and stored systematically, and is readily available to all relevant members of staff. Hand written and computer records are equally acceptable, but documentation and record keeping does need to be appropriate to the nature and size of the operation. The record of a HACCP system should include but are not limited to the records for critical control points, establishments of limits, corrective actions, results of verification activities, and the HACCP plan including hazard analysis.

In the unfortunate event of a food safety incident that is connected to the products whose safety was managed within related HACCP plan, it might be asked from the producers to demonstrate that all reasonable precautions have been taken to produce food safely. When the principles of HACCP have been correctly applied as required by law and when documentation and records are kept, they provide important evidence of due diligence in the event of legal action.

5 HACCP CERTIFICATION

Producers have to make sure that food is prepared, stored and sold in a hygienic way, to identify food safety hazards and to ensure that safety controls are in place within the HACCP based food safety system (Smigic et al. 2015). In most countries, official inspections check the level of implementation of prerequisites and HACCP plans by analyzing various indicators to verify the effectiveness of the food safety (Doménech et al. 2011). In spite of the fact that assessment of HACCP is under the jurisdiction of local inspection services, mistrust occurred regarding the competence of personnel performing these assessments (Lee and Hathaway 2000, Gagnon et al. 2000, Barnes and Mitchell 2000). As a result of lack of competence of the official institutions, certification bodies started providing third party audits and HACCP certification as an added value for food exporters. HACCP audits fall under various food safety schemes performed by certification bodies (Djekic et al. 2014). These schemes are either unaccredited with self-made guidelines for auditing HACCP based food safety systems or in line with accreditation protocols issued by accreditation bodies such as the Dutch Accreditation Council (RvA 2014).

Food producers seeking certification consider a certificate as a proof of an implemented and effective system (Djekic et al. 2011, Djekić 2006). There are researchers that criticize the certification process emphasizing that certification is a paper-driven process of limited value for the company performance and is used as a marketing tool. However, an independent assessment by an expert provides additional value to the industry it serves as well as supporting and complementing the role of the food-law enforcement agencies (Tanner 2000). Audits may provide audited organizations with a unique opportunity to receive advice, new ideas and help from the outside (Djekic et al. 2011). In that way, the role of audits is not only to verify for the effectiveness of a food safety systems, but also to serve for organizational learning and continual improvement.

6 CONCLUSION

By strict application of the HACCP plan and all of the prerequisite programmes, stable and safe fermented meat products can be guaranteed. For the production of "Užice Goveđa Pršuta" (dry-cured beef ham) following 5 biological CCP's and their critical limits were proposed: CCP-1B for the processing step of salting with a minimum duration of 10 days; CCP-2B for drying with the maximum yield of 60% as a critical limit; CCP-3B for smoking with a minimum duration of 8 h of smoking per day; fermentation was designated as a CCP-4B and maximum pH of 5.6 was set as a critical limit and finally CCP-5B for storage of sliced dried meat products in facilities with the temperature below or equal to 10°C as a legal requirement.

HACCP has taken center stage as food manufacturers seek the best ways to stay compliant, avoid costly recalls and provide an easier audit experience. However, in a year of its 56th anniversary, it is tempting to look forward and envision further improvements that can be accomplished. Since its birth the HACCP system has grown from three to seven principles. The future of HACCP may be exciting and we could see the addition of more principles, perhaps related to training, validation, prerequisite programs etc. It may also be bleak and disappointing if we do not have the courage to take some bold steps. We need to address the urgent need for knowledge and leadership across the global food industry (every single person), we need to continue the work aimed at having common standards and science-based regulations, and we need a global infrastructure to provide global strategy and oversight (Wallace et al. 2010).

One of expected breakthroughs is the development of specific product based quality management systems deployed in terms of quality of the products, of the processes and of the systems. Deploying the generic criteria from food industry to specific food sectors becomes a challenge in the 21st century (Djekic and Tomasevic 2015). Perhaps we should look forward to a creation of such a specific fermented meat quality management system and support its integration with HACCP.

Key words: *fermented meat production, food safety, HACCP, Užice Goveđa Pršuta*

REFERENCES

Arvanitoyannis, I.S. 2009. HACCP and ISO 22000: application to foods of animal origin. Chichester, U.K.; Ames, Iowa: Wiley-Blackwell.

Barnes, J. and R.T. Mitchell. 2000. HACCP in the United Kingdom. Food Control 11: 383–386.

Baş, M., A.Ş. Ersun and G. Kıvanç. 2006. Implementation of HACCP and prerequisite programs in food businesses in Turkey. Food Control 17: 118–126.

Bernard, D.T. and K.E. Stevenson. 1999. Association National Food Processors, and Institute Food Processors. HACCP: a systematic approach to food safety: a comprehensive manual for developing and implementing a hazard analysis and critical control point plan. Washington, D.C.: Food Processors Institute.

CAC. 2003. Recommended international code of practice general principles of food hygiene. In: CAC/RCP 1–1969, Rev.4–2003. Roma: FAO/WHO.

Celaya, C., S.M. Zabala and P. Pérez. 2007. The HACCP system implementation in small businesses of Madrid's community. Food Control 18: 1314–1321.

Damodaran, S., K.L. Parkin and O.R. Fennema. 2008. Fennema's food chemistry. Boca Raton: CRC Press/Taylor & Francis.

Djekić I. 2006. Quality and safety of food – problems and dilemas. Quality 16 (9–10): 71–74.

Djekic, I., I. Tomasevic and R. Radovanovic. 2011. Quality and food safety issues revealed in certified food companies in three Western Balkans countries. Food Control 22: 1736–1741.

Djekic, I., V. Zaric and J. Tomic. 2014. Quality costs in a fruit processing company: a case study of a Serbian company. Qual. Assur. Saf. Crop. 6: 95–103.

Djekic, I. and I. Tomasevic. 2015. How to improve the meat chain: overview of assessment criteria. Meso 4: 371–377.

Djinovic, J., A. Popovic and W. Jira. 2008. Polycyclic aromatic hydrocarbons (PAHs) in different types of smoked meat products from Serbia. Meat Sci. 80: 449–456.

Doménech, E., J.A. Amorós, M. Pérez-Gonzalvo and I. Escriche. 2011. Implementation and effectiveness of the HACCP and pre-requisites in food establishments. Food Control 22: 1419–1423.

Doyle, M.P. and R. Buchanan. 2013. Food microbiology fundamentals and frontiers. Available from http://www.asmscience.org/content/book/10.1128/9781555818463.

FDA. 2011. Fish and fishery products hazards and controls guidance – Fourth Edition. Available from http://www.fda.gov/downloads/Food/GuidanceRegulation/UCM251970.pdf.

FSIS-USDA. 2015. Ham and Food Safety. Available from www.fsis.usda.gov/wps/portal/fsis/topics/food-safety-education/get-answers/food-safety-fact-sheets/meat-preparation/ham-and-food-safety/ct_index/.

Gagnon, B., V. McEachern and S. Bray. 2000. The role of the Canadian government agency in assessing HACCP. Food Control 11: 359–364.

Gomes-Neves, E., C.S. Cardoso, A.C. Araújo and J.M. Correia da Costa. 2011. Meat handlers training in Portugal: A survey on knowledge and practice. Food Control 22: 501–507.

HACCP in Fermented Meat Production

Grau, R., A. Andres and J.M. Barat. 2015. Principles of drying. pp. 31–39. *In:* F. Toldrá (ed). Handbook of fermented meat and poultry. 2nd edition. Wiley-Blackwell, Chichester, UK.

Konecka-Matyjek, E., H. Turlejska, U. Pelzner and L. Szponar. 2005. Actual situation in the area of implementing quality assurance systems GMP, GHP and HACCP in Polish food production and processing plants. Food Control 16: 1–9.

Lee, J.A. and S.C. Hathaway. 2000. New Zealand approaches to HACCP systems. Food Control 11: 373–376.

Leora, A.S. and S. Julie. 2005. Indirect and miscellaneous antimicrobials. pp. 573–598. *In:* P.M. Davidson (ed.). Antimicrobials in food, Third edition. CRC Press.

Mortimore, S. and C. Wallace. 2013. HACCP a practical approach. Springer. Available from http://public.eblib.com/choice/publicfullrecord.aspx?p=1030814.

NACMCF. 1998. Hazard analysis and critical control point principles and application guidelines. Adopted August 14, 1997. National Advisory Committee on Microbiological Criteria for Foods. J. Food Protect. 61: 1246–1259.

Radovanović, R. 1990. "Užice beef prshuta" protected designation of origin – major elaboration. Belgrade: The Intellectual Property Office of the Republic of Serbia.

Radovanović, R., J. Adamović, T. Šakota and I. Tomašević. 2003a. HACCP plan for the "Serbian Sausage" traditional semi-dry product. *In:* J.N. Beraquet (ed.) Proceedings of 49th International Congress of Meat Science and Technology (ICoMST). São Paulo – Brazil: Centro de Tecnologia de Carnes – CTC.

Radovanović, R., P. Bojović, Z. Arbutina and L. Vukajlović. 1993. Changes of selected quality factors during processing and storage of "Uzice Beef Prshuta" – traditional dry meat product. *In:* Proceedings of 39th International Congress of Meat Science and Technology. Calgary, Canada.

Radovanović, R., T. Stamenković and S. Saičić. 2003b. Sensory properties and chemical composition of beef prshuta. Tehnologija mesa 44: 212–219.

Rajkovic, A. 2012. Incidence, growth and enterotoxin production of Staphylococcus aureus in insufficiently dried traditional beef ham "govedja pršuta" under different storage conditions. Food Control 27: 369–373.

RvA. 2014. Specific accreditation protocol for certification of food safety management systems. edited by R. v. Accreditatie. Utrecht, The Netherlands: Dutch Accreditation Council (Raad voor Accreditatie).

Sikorski, Z.E. and I. Sinkiewicz. 2014. SMOKING | Traditional. pp. 321–327. *In:* M. Dikeman, and C. Devine (eds). Encyclopedia of meat sciences (Second Edition), Oxford: Academic Press.

Smigic, N., A. Rajkovic, I. Djekic and N. Tomic. 2015. Legislation, standards and diagnostics as a backbone of food safety assurance in Serbia. Brit. Food J. 117: 94–108.

Stamenković, T., N. Šušnjarac, V. Jovanović and S. Jovanović. 2003. Weight loss, sensory properties and chemical parameters of beef prshuta obtained by traditional and changed smoking procedures. Tehnologija mesa 44: 79–84.

Tanner, B. 2000. Independent assessment by third-party certification bodies. Food Control 11: 415–417.

Tomašević, I., N. Šmigić, I. Djekić, V. Zarić, N. Tomić and A. Rajković. 2013. Serbian meat industry: A survey on food safety management systems implementation. Food Control 32: 25–30.

Tomic, N., I. Tomasevic, R. Radovanovic and A. Rajkovic. 2008. "Uzice Beef Prshuta": Influence of different salting processes on sensory properties. J. Muscle Foods 19: 237–246.

Tomić, N. and I. Tomašević. 2007. Quantitative and qualitative properties of selected beef muscles during production of "Uzice beef prshuta" – traditional Serbia dry-cured meat product. In: Đ. Okanović (ed.). Proceedings of International Congress "Food technology, quality and safety" – Technology, quality and safety in pork production and meat processing. Novi Sad, Srbija.

Wallace, C.A., W.H. Sperber and S.E. Mortimore. 2010. The future of food safety and HACCP in a changing world. pp. 48–61. In: Food safety for the 21st century. Wiley-Blackwell.

Wang, D., H. Wu and X. Hu. 2010. Application of hazard analysis critical control points (HACCP) system to vacuum-packed sauced pork in Chinese food corporations. Food Control 21: 584–591.

Chapter 23

Official Controls of Raw Meat Fermented Sausage Production

Milorad Radakovic* and Slim Dinsdale

1 INTRODUCTION

There is no clear agreed legal international definition of fermented sausages. The Food and Agriculture Office of the United Nations (FAO) document on Raw-Fermented Sausages notes that *"Raw-fermented sausages receive their characteristic properties (tangy flavour, in most cases chewy texture, intense red curing colour) through **fermentation processes**, which are generated through physical and chemical conditions created in raw meat mixes filled into casings. Typical raw-fermented sausages are **uncooked meat products** and consist of coarse **mixtures of lean meats and fatty tissues** combined with salts, nitrite (curing agent), sugars and spices as non-meat ingredients. In most products, uniform fat particles can clearly be distinguished as white spots embedded in dark-red lean meat, with particle sizes varying between 2–12 mm depending on the product. In addition to fermentation, **ripening phases** combined with moisture reduction are necessary to build-up the typical flavour and texture of the final product. The need for moisture reduction requires the utilization of **water-vapour permeable casings**. The products are not subjected to any heat treatment during processing and are in most cases distributed and consumed raw"* (Heintz and Hautzinger 2010).

---

*For Correspondence: University of Cambridge, Department of Veterinary Medicine, Madingley Road, Cambridge CB3 0ES, UK. Tel: +44 1223 765048 Cell: +44 7917 228664, Email: mr412@cam.ac.uk

The term raw meat fermented sausages (RMFS) as such may be misleading since the final product is ready-to-eat (RTE), although may also be consumed cooked. The fact that it is 'raw meat' and RTE has been known to cause confusion amongst consumers and regulators with a limited understanding of the processes involved. Legal definitions, if available for different type of products within each country's legal framework, may differ in definitive wording, but the real meaning should be equivalent and in line with internationally agreed standards.

For consistency, we will be referring throughout the chapter to the definitions that are available within the EU legal framework. RMFS clearly fall into a definition of "Meat products" as *"processed products resulting from the processing of meat or from the further processing of such processed products, so that the cut surface shows that the product no longer has the characteristics of fresh meat"* (Anonymous 2004a). "Processing" is defined in EU legislation, as *"any action that substantially alters the initial product, including heating, smoking, curing, maturing, drying, marinating, extraction, extrusion or a combination of those processes"* (Anonymous 2004b).

Responsibility for the organisation of Official Controls (OCs) of fermented meat products, as for any other product of animal origin (POAO), is a political one. At the country central level this responsibility is given to the so-called Central Competent Authority (CCA), which may be within the Ministry of the Agriculture and/or Food, or Department of Health. In addition to being responsible for the OCs, the CCA is responsible for international negotiations, law creation and standard setting for food production and official controls.

Practical delivery of OCs is usually delegated to the official controllers at the regional and local level with their performance being monitored by the CCA. OCs of fermented meat production may be carried out by a number of professionals such as Veterinarians, Environmental Health Practitioners, Sanitary Inspectors and others. Each of these professionals may be authorised to enforce compliance with either food safety or food quality laws or both. In order to carry out the tasks required within the scope of the applicable legislation, it is essential that official controllers have adequate competencies. There are a number of definitions of competency. For the purpose of this chapter 'competency' means that an authorised professional has an ability, in line with the requirements of his profession, to effectively carry out audits and inspections of the fermented meat production process. It is also essential that a professional has an adequate level of knowledge and understanding of the food production chain and processes from farm-to-fork, as the safety of the final product may be affected at any stage of the production process.

Effective official control procedures also need to be considered in the context of the state of 'development' of each country, in accordance with internationally agreed FAO/OIE/ WHO standards. This is because the successful application of controls is dependent on a range of factors, including those of a socioeconomic, cultural, and religious nature, all of which have the same aims: safe food production. The responsibility for the safety and quality of fermented meat products clearly rests with the Food Business Operators (FBOs). Official controllers, on the other hand, are required to have adequate knowledge and understanding of fermented meat production in order to discharge their statutory function (inspection and auditing tasks), and should adopt a flexible approach to enforcement, particularly with small and traditional businesses (Anonymous 2010), whilst also bearing in mind the proper application of the precautionary principle where a need is demonstrated.

2 FROM ANIMALS TO MEAT

Fresh meat from any species can be used as a raw material for fermented sausages production. A number of fermented sausages, subject to the recipe, may contain meat from more than one species with different spices being added. It is beyond the scope of this chapter to go into the details of fresh meat production.

Any official controller, however, should be familiar with the basic slaughter process, the production of fresh meat as the main ingredient, and the fact that at any stage of the production process there is the potential for cross contamination with non-visible pathogens and chemicals that may affect the safety of raw material used for RMFS, and consequently the final product. The simplified flow diagrams "From animals to meat" for pigs and cattle are as follows:

Pigs

Reception, lairage → Ante-mortem inspection → Stunning and bleeding → Dressing (scalding, hair removal or skinning) → Evisceration → Post mortem inspection → Chilling → Cutting and boning → Meat for consumption and further processing.

Cattle

Reception, lairage → Ante-mortem inspection → Stunning and bleeding → Dressing (head, hoof and hide removal) → Evisceration → Post mortem inspection → Chilling → Cutting and boning → Meat for consumption and further processing.

2.1 Official Controls at Abattoirs

Ante and post mortem inspection official controls at abattoirs are usually carried out by Official Veterinarians (OVs), assisted by Official

538   *Fermented Meat Products: Health Aspects*

Auxiliaries (OAs) who are primarily responsible for practical post mortem inspection. Both ante and post-mortem inspection tasks have their limitations and neither activity guarantees the 'absolute safety' of fresh meat, and absence of non-visible pathogens and chemicals that may be present on, or in, meat.

Table 1 outlines the OC activities in abattoirs and the relevance of ante and post-mortem decisions taken, in particular, for the safety of meat intended to be used in fermented food production. These basic common principles also apply to any other product that is made from fresh meat. A judgement of the carcase and offal fitness for human consumption can only be made after the post mortem inspection is completed and the satisfactory results of any additional tests, if required and performed, are available.

Table 1   Official controls at abattoirs that are relevant to fermented meat production

Tasks	Relevance for fermented food production
Ante-mortem inspection: Routine observation, clinical inspection or clinical examination including, if required, sample taking. Food Chain Information (FCI) is a part of the ante-mortem inspection procedure.	Removal from the food chain of diseased animals, specifically those known to be affected with zoonotic agents. Verification of the welfare of animals, in particular determining (and minimising) the level of stress which may affect the quality and suitability of meat for fermented meat production.
Post-mortem inspection: Visual, Palpation, Incision, Sample taking (e.g. Trichinella or residue testing).	Removal of abnormal and diseased carcases and offal which may have an impact on human and/or animal health. Includes verification of Specified Risk Material (SRM) and Animal-by-Products (ABP) removal.
Health marking	Verification that, at the time of inspection, the meat is fit and safe for human consumption.

2.2   Quality Issues with Fresh Meat Relevant to RMFS

There are two key quality defects of fresh meat used as raw material for fermented sausages. These are pale, soft, exudative (PSE) meat in pig carcases and dark, firm and dry (DFD) meat in beef and sheep carcases. Both PSE and DFD fresh meat, if used on its own for fermented sausages, may affect the quality of the final product. If this meat is used in RMSF production, it is often mixed with non-PSE and non-DFD meat.

PSE in pig carcases results from an accelerated rate of postmortem glycolysis. This causes rapid production of lactic acid in the muscle,

Official Controls of Raw Meat Fermented Sausage Production   539

and a fall in pH to 5.4–5.6. This occurs whilst the carcass is still warm causing muscle proteins to denature leaving the proteins with a reduced ability to bind to water. The cause is believed to be severe stress just before slaughter (poor transport and manhandling) with faster growing, leaner, pigs being more susceptible. Quiet handling and letting pigs rest in the lairage for some time before slaughter will significantly reduce the risk of PSE.

DFD can occur in beef and sheep carcases (and sometimes pigs) soon after slaughter. The muscle glycogen has been depleted during the pre-slaughter period (from stress, injury or disease) resulting in less lactic acid production, and with a consequently higher pH of 6.4–6.8. The meat is darker in colour, appears drier and has a firmer texture due to the muscle's increased water binding capacity, and the high pH may limit shelf life as it has a greater ability to support the growth of spoilage bacteria.

2.3   Safety Issues with Fresh Meat Used as Raw Material for RMFS

Raw meat from which fermented sausages are produced may be contaminated with a number of biological, chemical (e.g. dioxin, Veterinary Medicine residues) and physical hazards (Chapters 10 and 18 in this book). Some important food safety bacterial pathogens are:

2.3.1   *Verocytotoxic Escherichia coli (VTEC)*

VTEC (such as *E. coli* O157) are of particular importance because they have a very low infective dose for humans, and can produce a devastating illness. They are associated with ruminants and their meat, but may also found on raw pork meat and, although it may not multiply during the production process of RMFS, they are capable of surviving. Since the infective dose is very low, survival in the product represents a major risk to the consumer.

To assist the assessment of safety of RMFS, Quinto et al. (2014) developed a modelling approach, and found that the inactivation of VTEC was greatest at higher temperatures and produced a 6 log reduction following 66 days storage at 25°C. This was in contrast to a reduction of only around 1 log when the storage temperature was 12°C. The model found that the greatest inactivation of VTEC was predicted in dry-fermented raw meat sausages with long drying periods, whilst the smallest reduction was predicted in semi-dry fermented raw meat sausages with short drying periods. Because of this, semi-dry fermented raw meat sausages should not be consumed by high-risk groups, such as children, the elderly, pregnant women and those who are immunocompromised.

2.3.2 *Clostridium botulinum*

Endospores of *Clostridium botulinum* are highly resistant to destruction and are likely to be found on, or in, raw meat fermented sausages. Following germination of the endospores the resulting vegetative cells go on to multiply to produce botulin, a powerful neurotoxin. Preventing germination of the endospores is, therefore, of critical importance. The only effective material that prevents germination is nitrite (normally sodium nitrite or potassium nitrite), although the mechanism has not been elucidated. The European Food Safety Authority (EFSA 2003) notes that: "The panel is of the opinion that the ingoing amount of nitrite, rather than the residual amount, contributes to the inhibitory activity against *Cl. botulinum*. Therefore, control of nitrite in cured meat products should be via the input levels rather than the residual amounts." The panel also notes that: "Nitrates have no direct activity against *Cl. botulinum*. In particular traditional products (e.g. dry ripened fermented sausages of high pH, dry cured ham), nitrates act as reservoirs for nitrites generated by microbial activity." Nitrites also contribute to the pink/red colour of cured meats by combining with myoglobin to form nitrosomyoglobin (and also with haemoglobin to form nitrosohaemoglobin).

There is still some controversy over the safety of nitrite when consumed at normal levels used in the production of fermented sausages. In the body, nitrates are converted to nitrites and these react with haemoglobin to produce methaemoglobin, which reduces the ability of haemoglobin to carry oxygen around the body. Infants are at risk if they consume fermented sausages where there has been an error in the formulation and too much nitrite has been added. Nitrites can also combine with myoglobin to form nitrosamines in RMFS and in the stomach and concern has been raised due their ability to cause cancer (Sindelar and Milkowski 2012).

2.3.3 *Staphylococcus aureus* and *Salmonella* **spp**.

Pork meat may contain *Staphylococcus aureus* and some strains have the ability to produce toxins, which are secreted into food. Although *St. aureus* is a poor competitor, it still has the potential to grow and produce emetic toxins in fermented sausages during the early stages of production if acid development is slow, and the temperature is sufficient to support growth. *Salmonella* may be found on raw pork and, although a poor competitor, the bacteria may survive unless acid development from lactic acid bacteria (LAB) is rapid.

3 RAW MEAT FERMENTED SAUSAGE PRODUCTION

Full details of the production of fermented sausages are dealt elsewhere in this book (Chapter 1 and 3) and other reference books (Toldrá 2010). An outline account is provided below to assist the understanding of RMFS production in the context of official controls. This simplified type of description is particularly important, for example in cases when official controllers need to explain to the courts (lay persons) the whole process in terms they are more likely to understand.

The process of RMFS production takes raw meat, naturally contaminated with spoilage bacteria, such as *Pseudomonas* spp. and *Acinetobacter* spp. and pathogens such as, *Staphylococcus aureus* and *Salmonella* spp. Fermented sausage production renders the product safe to eat by:

- Inactivating and preventing multiplication of pathogens, and,
- Preventing spoilage by bacteria which could reduce quality of the product, and,
- Preventing surviving endospores of, for example, *Clostridium* spp. from germinating and, growing, and in the case of *Cl. botulinum*, production of a lethal neurotoxin.

In essence, the following hurdles (see also Chapters 7 and 15 in this book) are used:

- The addition of salt, which reduces the water activity to a level where bacteria are unable to multiply;
- The addition, or use of naturally occurring, lactic acid bacteria, which will produce lactic acid and lower the pH of the meat;
- The addition of nitrite, a powerful anti-clostridial agent, and which is also important in the development of nitrosomyoglobin which provides the red colour;
- Dehydration during the maturing phase, which lowers the water activity even further.

A typical process for preparing dry and semi-dry fermented sausage is as follows. Minced pork is mixed with sodium chloride, a nitrate/nitrite mixture (the nitrate acts as a 'reservoir' for nitrite, as it can be broken down by bacteria in the meat into nitrite), sugars, and flavourings such as pepper. A starter culture of LAB may also be added rather than relying on the bacteria naturally present. The mixture is filled into casings of varying lengths and diameters, commensurate with the type of sausage being produced.

The fermentation (or ripening) process takes place initially at a temperature of around 15–30°C, depending on the nature of the sausage (semidry sausages are held at higher temperatures than dry sausages), following which the temperature is reduced. The water activity declines progressively during the ripening (or maturation) period as water is lost through dehydration. Salt increases in concentration for the same reason, as do nitrates and nitrites (although the concentration of nitrates and nitrites will fall as nitrites are used up during the fermentation process). Maturation continues until the moisture content and flavour development has reached the optimum for that product.

The drying process changes the casings from being water permeable to reasonably airtight, and a white covering of mould can develop. Typical pH values for traditionally produced fermented sausages are in the range of 5.5–5.7 in the final product. Total bacterial counts at the end of the maturation period can be high. Faecal enterococci may also reach high numbers and are believed to contribute to the organoleptic characteristics of the product.

4   RELEVANT LEGISLATION

Within the country's legal framework, each official controller is authorised to enforce compliance with specific legislation. Broadly speaking, legislation applicable to RMFS can be divided into food safety and quality. In some countries the same officials may be authorised to enforce the compliance with both sets of legislation, or there may be a clear separation of the responsibilities between different officials. The legislation on quality of RMFS generally covers labelling and nutritional information (Anonymous 2011), traditionally protected food schemes (Anonymous 2012) and organic food (Anonymous 2007). The main European quality schemes are: Protected Designation of Origin (PDO), Protected Geographical Indication (PGI), Traditional Speciality Guaranteed (TSG) and Organic farming schemes.

The decision as to who does what in terms of official controls, and at which level (e.g. wholesale, retail), varies from country to country. Regardless of these decisions and divisional responsibilities, the CCA should ensure that there is effective communication, cooperation, collaboration and coordination of all official activities at the national, regional and local level (Anonymous 2004c).

In the real world, however, it is not always easy to separate food safety and quality, as, for example, some deterioration of quality such as rancidity may not pose a food safety risk to consumers, but still the products may not be desirable for consumers. A final judgement is often clouded by unnecessary amplification, or diminishment, of the food safety risks.

4.1 Definitions

As a starting point to any type of official controls, or to any case when there is a need to make judgments on product, process or structure and equipment where processing takes place it is necessary to become familiar with the list of legal definitions that are available within the country's legal framework.

In addition to the already quoted definition of "Meat products" from Regulation (EC) 853/2004, Regulation (EC) No 882/2004, on official controls, provides relevant wide-ranging definitions. Further relevant definitions are provided in Regulation (EC) 178/2002 laying down the general principles and requirements of food law, establishing the European Food Safety Authority and laying down procedures in matters of food safety (Anonymous 2002).

Is there a legal definition of safe food and quality?

Safe food is not legally defined as such. Instead, the main provisions for safe food are laid down in Regulation (EC) 178/2002 with the overriding purpose to protect human health. It applies to all stages of commercial production, processing and distribution of food, and lays down procedures to ensure that food and drink businesses only place safe products on the market. More specifically, Article 14 deals with Food Safety Requirements, the relevant parts as follows:

- Food shall not be placed on the market if it is unsafe.
- Food shall be deemed to be unsafe if it is considered to be:
  - injurious to health;
  - unfit for human consumption.
- In determining whether any food is unsafe, regard shall be had:
  - to the normal conditions of use of the food by the consumer and at each stage of production, processing and distribution, and
  - to the information provided to the consumer, including information on the label, or other information generally available to the consumer concerning the avoidance of specific adverse health effects from a particular food or category of foods.
- In determining whether any food is injurious to health, regard shall be had:
  - not only to the probable immediate and/or short-term and/or long-term effects of that food on the health of a person consuming it, but also on subsequent generations;

544   *Fermented Meat Products: Health Aspects*

- – to the probable cumulative toxic effects;
- – to the particular health sensitivities of a specific category of consumers where the food is intended for that category of consumers.
- • In determining whether any food is unfit for human consumption, regard shall be had to whether the food is unacceptable for human consumption according to its intended use, for reasons of contamination, whether by extraneous matter or otherwise, or through putrefaction, deterioration or decay.

This legislation provides a wide-ranging definition of 'unsafe' which needs to be interpreted with caution as outlined later when issues of 'safe or unsafe' are considered (see below). Quality is also not defined as a such, but is implied with the existence of recognised Quality Schemes and nutritional food labelling, each with a legislative background. Food quality may deteriorate as a result of the growth of spoilage organisms and, whilst not harmful, may be caught by Regulation 178/2002 if it is considered to be "unacceptable for human consumption".

4.2 Auditing Tasks

In terms of food safety, the main purpose of the official controllers task is to check and verify whether or not the FBO applies procedures that guarantee the safe production of food, and that systems exist to identify and respond to food safety problems. In reality official controllers audit FBO's continuous compliance with Good Hygiene Practice (GHP) and Hazard Analysis and Critical Control Points (HACCP) based procedures (Anonymous 2004d, Table 2). In terms of quality, the main purpose of the OCs task is to verify the compliance with specific Quality assurance schemes and the compliance with legislation under which these schemes have been developed.

As a part of the auditing process, there may be a specific inspection i.e. a reality check, consisting of the examination of the whole establishment or part of it, or its equipment, or documents. Overall, though, it is an examination of the whole system for which the FBO has primary responsibility, but it does not have to be completed in one go. An audit of the whole system is usually completed on a risk-based frequency and this very much depends on the assessment carried out by the CCA.

For both food safety and quality official controls it is essential to fully understand the traceability requirements of Article 18 of Regulation (EC) 178/2002. Traceability in meat processing establishments is a complex issue that needs to be considered from farm-to-fork. Relevant factors in determining traceability requirements are the level of product

Official Controls of Raw Meat Fermented Sausage Production   545

throughput (industrial *versus* small/traditional) and the type of final product. Depending on the size of production, i.e. large industrial or small/traditional, the fermented sausages may be traced back to the day of production, but, in reality, not always to particular animals. The legal concept of traceability requires external but not internal traceability to be established. It is up to FBOs (not official controllers) to establish an internal level of traceability within a particular process and to decide what constitutes a unit (lot or batch) of their internal production. In cases when there is a need to withdraw and recall the final product from the market because of food safety risks, those FBOs with limited internal traceability procedures may suffer a bigger financial loss as more product than would otherwise be necessary may need to be withdrawn or recalled.

Table 2   Auditing tasks

Auditing tasks	Relevance to fermented food production
Checks on food chain information	See Table 1 – Part of ante-mortem inspection
The design and maintenance of premises and equipment	Should be of a suitable design and quality to prevent, eliminate, or reduce the risk to an acceptable level. Some derogations exists for Traditional food production
Pre-operational, operational and post-operational hygiene	As above, the objective is to prevent, eliminate or reduce the risk to an acceptable level
Personal hygiene	As above, the objective is to prevent, eliminate or reduce the risk to an acceptable level
Training in hygiene and in work procedures	Adequate training tailored to the type of production
Pest control	As above, the objective is to prevent, eliminate or reduce the risk to an acceptable level
Water quality	Potable water is required
Temperature control	Some flexibilities may be permitted with no compromise to food safety
Controls on food entering and leaving the establishments and any accompanying documentation	This includes the source of raw material/additives spices, etc. Identification mark – FBO's responsibility Traceability – FBO's responsibility
Verify continuous compliance with animal-by-products (ABP) and Specified Risk Material (SRM) legislative requirements	More relevant for slaughterhouse operations, although fermented food production may also generate ABP
Audit of HACCP based procedures	Dealt with in a separate chapter.

5 EFFECTIVE OFFICIAL CONTROLS

The approach to official controls, with the same aim of safe food production, may vary from country to country. These differences may also be noted in terminology being used when describing official controls. For the purpose of this chapter, in very general terms, the UK approach, and the terminology in use, is described within the context of the EU and UK applicable laws.

Effective official controls of fermented meat production depend on a number of interrelated factors such as having effective legislation (including enforcement powers), an adequate competency level of official controllers and coordinated enforcement activities at national, regional and local level. The first step following adequate training, is that official controllers must be authorised under specific national legislation to discharge their statutory duties.

Practical, day-to-day, enforcement may simply be described as informal and formal. Informal enforcement consists of verbal and written communication through, for example, advisory letters and official reports. In the majority of cases this is sufficient to achieve full compliance. However, in more serious cases, or when an informal approach fails to achieve the desired effect, other formal enforcement powers are available. These may range from:

- Detention of either animals or meat for further examination;
- Seizures of either animals or meat with a view for further examination, or making a recommendation for condemnation;
- In urgent cases, such as a serious risk to public health, immediate action may be required;
- In less urgent cases, such as structural deficiencies, formal notices may be served to require certain actions to be carried out by the FBO within the specified time scale;
- Suspension of the approval or registration;
- Withdrawal of approval;
- In some or all cases, depending on the case, a recommendation for prosecution.

The UK enforcement officers in the meat sector in the course of their duties are required to apply the principles of an "Enforcement Concordat" (Anonymous 1998) and to adhere to its main principles, which include proportionate action with regard to risk, and consistency of enforcement. This has been reinforced by the 2014 Regulators Code (Anonymous 2014) which notes that "Food Authorities should ensure that enforcement action taken by their authorised officers is reasonable, proportionate, risk-based and consistent with good practice."

5.1 Gathering and Preserving Evidence – Objective Evidence

An official controller's opinion of why something is non-compliant or 'not right' must be based on factual and objective evidence. To be able to produce this evidence an official controller should keep accurate records of his/her observations, organoleptic examination of food, structure and equipment, any samples taken and results obtained. These records may be kept in the form of a contemporaneous notebook or on specially designed forms and official reports. Samples may be taken from meat and from the premises, together with photographs and video clips, the latter also being useful with live animals. This can be helpful in all cases, but particularly when more serious breaches are alleged.

Furthermore, any type of communication with the FBO as well as the results of any immediate action that has been taken by an official controller should also be recorded. In fact, all records kept by official controllers may be used as evidence in court. If it comes to that stage then the official is legally obliged to give a statement for the court to detailing the facts, and may also be required to give oral evidence if the facts are challenged.

5.2 Classification of Non-Compliances Depend on the Task Performed

Non-compliances following ante and post- mortem inspection can simply be divided into "pass" and "fail". Animals and carcases that have "passed" simply receive health or identification marks demonstrating that, at the time of inspection, the meat and offal were fit for human consumption.

On the other hand the classification of non-compliances and corrective actions, following an audit, are much more complex because the officials must verify the FBO's food safety management system with regard to its legislative responsibilities. In this case, minor non-compliances with no immediate risk to public or animal health may be dealt with within an agreed time scale, whilst more serious non-compliances must be dealt more urgently.

Decisions and actions taken after both inspection and audit, will therefore largely depend on the assessment of any risk to public health or, where appropriate, animal health and animal welfare. What, in many cases, poses challenges and sometimes difficulties for FBOs and enforcement officers is the interpretation of words such as, "adequate", "sufficient", "as reasonably practicable", "abnormal", "fit", "unfit", "safe", "unsafe", etc. Inevitably, there is a level of subjectivity in judging whether meat is "normal or abnormal", "fit or unfit" and "safe or unsafe". A competent professional approach, however, should lead to

548   *Fermented Meat Products: Health Aspects*

sound decisions being made with minimal risk of official controllers being challenged for their judgements.

5.2.1 Normal or Abnormal?

Organoleptic examination of meat, or fermented meat product may reveal some deviations from "normal" colour, consistency and smell. This may be due to factors, such as age, breed, feed, sex, and stress (live animals), or storage conditions of product which may have no relevance with regard to food safety. For example, the yellow colour of a fresh meat carcase may be caused by disease and classed as abnormal. Cattle fed on grass, on the other hand, may have only yellow fat which is considered to be normal. Careful post mortem examination would lead to the rejection of the first carcase, whilst the latter would be passed as fit for human consumption. Some other examples, such as PSE meat and DFD meat, are considered abnormal, but in the absence of other abnormal post mortem changes, should not lead to the rejection of the meat. However, although fit for human consumption, PSE meat and DFD meat may not suitable on its own be for fermented sausage production for quality reason as noted earlier in this chapter.

5.2.2 Fit or Unfit?

If meat is fit, it is suitable according to the safety and quality required. The list of legislative reasons why meat should be rejected relates well with the extreme definition of abnormal. However, the available list also indicates (although not in exact words), that in some cases visibly "normal" meat must be rejected as indicated by the following categories:

- animals which have received no ante mortem inspection;
- meat containing residues or contaminants in excess of the level laid down in Community legislation;
- animals affected by some notifiable diseases listed on the OIE list. For some specific hazards, which are diagnosed by the officials, the legal requirement is to reject all meat for human consumption if there is evidence of, for example, *TSE, Trichinella, Glanders*. On the other hand for some conditions such as *Cysticercus bovis, Tuberculosis* the official controller is required to perform an examination and make a judgement for a partial or total rejection of the carcase.
- meat that has been produced illegally.

5.2.3 Safe or Unsafe?

The scope for deciding whether a food is unsafe is extremely broad and care must be used when invoking it. For example, RMFS may be

Official Controls of Raw Meat Fermented Sausage Production 549

"contaminated" with large numbers of bacteria, some of them faecally derived, but this does not necessarily make them unsafe. Further, RMFS may show signs of rancidity over time, which could be described as 'deterioration' but would also not necessarily make them unsafe.

The literal meaning of, "safe" means without risk, which is a state that does not exist in the food (or any other) industry. For example, the absence of common human pathogens such as *Salmonella* and *Campylobacter* in raw meat, based on the results from a small number of samples cannot be guaranteed. The qualification of "safe" should, therefore, include the level of risk, which is "acceptable", and for this to be properly applied requires a high level of competence of both the FBOs and the official controllers. After the examination of all relevant information, including that provided by the FBO (whose duty it is to provide safe meat) a judgement needs to be made. Clearly, if the meat (or meat product) is likely to constitute a risk to public or animal health, or is unsuitable for human consumption for any other reason, the official controller should consider rejecting the meat/meat product.

5.3 Actions Following Auditing of Good Hygiene Practices and HACCP Based Procedures

Corrective actions and follow up activity depend on the risk to public health and, in slaughterhouses where appropriate, animal health and animal welfare. Immediate action is required if the evidence suggests there is a serious risk to public health, and this could involve stopping production, and the rejection of product already produced, or detaining it for further examination.

Immediate action on the other hand is unlikely to be necessary in the majority of non-compliances such as some structural deficiencies and shortcomings in dressing and cutting procedures, through to failures to keep and maintain adequate HACCP or pre-requisite records. In these cases "reality checks" are important to verify whether or not public or animal health, or animal welfare (slaughterhouses), is compromised.

In all cases where the results of tests on RMFS (as ready-to-eat product) show the presence of specific pathogens, or chemicals, that are harmful to human health there should be no argument with regard to safety and such products, in accordance with applicable legislation, must be rejected for human consumption.

5.3.1 Follow Up to Non-compliances

Follow up of non-compliances will largely depend on the risk and type of non-compliances, as each case is different. The auditing system allows the official controllers to employ the option that they consider

550   *Fermented Meat Products: Health Aspects*

to be the best according to the circumstances, so as to achieve FBO compliance with the legislation. More frequent auditing, sometimes through unannounced visits, often suffices to achieve desired results.

5.3.2 Real Life Scenario with RMFS

The following real life example involved fermented meat products. A small speciality retailer imported RMFS (and other cured meat products) from Italy, sliced and vacuum packed the products ready for display and sale. The retailer also vacuum packed other non-cured cooked meats, and cheeses, similarly, for display and sale. The retailer had a single vacuum packing machine which he used for all of his products, all of which were ready-to-eat.

An official controller carried out a routine inspection (also known as an Intervention) and noticed the RMFS being vacuum packed by the retailer. The OC invoked Guidance produced by the UK's Food Standards Agency (2014) which recommended that raw foods and ready to eat foods should each have their own dedicated vacuum packing machine. Detailed explanations were given to the OC with regard to the RMFS, including the product specification and extensive microbiological and chemical analysis results, and how they were in a ready-to-eat form.

The OC determined that a separate vacuum packing machine was required, even though the RMFS products were ready to eat, as the products contained "raw meat". In this case, the OC showed a gross lack of competence in the ability to evaluate the process of fermented meat production and an inability to understand the product specification and microbiological and chemical results. In due course, the OC was reluctantly persuaded that the term "raw meat" in this case could not be equated with, for example, raw fresh meat.

6   CONCLUSION

Both Food Business Operators (FBOs) and Official Controllers share the same common goals; safe food and safe food production. These common goals can only be achieved with an effective interpretation and implementation of the legislation by competent people and through an effective communication, collaboration, cooperation and coordination process at all levels.

An approach to Official Controls of RMFS production (as with any other food) can only occur through the consideration of the whole food chain, from primary producers to retailers. Official controllers contribute to the safe production of RMFS though inspection and auditing, but have no direct responsibility for the day-to-day safe food production.

Official Controls of Raw Meat Fermented Sausage Production 551

Their primary role is to verify that FBOs are compliant with their legal responsibilities such as the structure and hygiene of establishments by carrying out regular audits. This verification should be carried out in the accordance with national laws and prescribed frequencies.

It is neither possible nor necessary to have over-prescriptive legislation, guidance and instructions for officials as what to do in each case because differences in individual production processes exist, which is particularly true for RMFS. Any non-compliance raised by official controllers must be based on objective evidence, and action taken must be proportionate to the risk. In some cases an official controller, in addition to raising a non-compliance, may decide to instigate more formal legal proceedings against an FBO in a court of law. The actual legal procedures depend very much on the legal framework of a particular country and it is beyond the scope of this chapter.

In any case, official controllers should carry out their statutory tasks by avoiding unnecessary amplification or diminishment of risk to public health. In those (probably rare) cases where a serious risk to food safety is suspected, with full evidence lacking, the OC should not hesitate to invoke the precautionary principle and support the decision with a fully reasoned explanation as why the action was taken. This can only occur, of course, if the OC has the necessary competencies.

Key words: *official controls, legislation, raw meat fermented sausages, audit, pathogens, animals, meat*

REFERENCES

Anonymous. 1998. ENFORCEMENT CONCORDAT. The Principles of Good Enforcement: Policy and Procedures, Available from http://www.parliament.uk

Anonymous. 2002. Regulation (EC) No 178/2002 of the European Parliament and of the Council of 28 January 2002 laying down the general principles and requirements of food law, establishing the European Food Safety Authority and laying down procedures in matters of food safety.

Anonymous. 2004a. Regulation (EC) No 853/2004 of the European Parliament and of the Council of 29 April 2004 laying down specific hygiene rules for on the hygiene of foodstuffs.

Anonymous. 2004b. Regulation (EC) No 852/2004 of the European Parliament and of the Council of 29 April 2004 on the hygiene of foodstuffs.

Anonymous. 2004c. Regulation (EC) No 882/2004 of the European Parliament and of the Council of 29 April 2004 on official controls performed to ensure the verification of compliance with feed and food law, animal health and animal welfare rules.

552 *Fermented Meat Products: Health Aspects*

Anonymous. 2004d. Regulation (EC) No 854/2004 of the European Parliament and of the Council of 29 April 2004 laying down specific rules for the organisation of official controls on products of animal origin intended for human consumption.

Anonymous. 2007. Council Regulation (EC) No 834/2007 of 28 June 2007 on organic production and labelling of organic products and repealing Regulation (EEC) No 2092/91.

Anonymous. 2010. The European Commission working document. Commission staff working document on the Understanding of certain provisions on flexibility provided in the Hygiene Package Frequently Asked Questions Guidelines for the competent authorities. Available from: http://ec.europa. eu/food/food/biosafety/hygienelegislation/docs/faq_all_public_en.pdf

Anonymous. 2011. Regulation (EU) No 1169/2011 of the European Parliament and of the Council of 25 October 2011 on the provision of food information to consumers, amending Regulations (EC) No 1924/2006 and (EC) No 1925/2006 of the European Parliament and of the Council, and repealing Commission Directive 87/250/EEC, Council Directive 90/496/EEC, Commission Directive 1999/10/EC, Directive 2000/13/EC of the European Parliament and of the Council, Commission Directives 2002/67/EC and 2008/5/EC and Commission Regulation (EC) No 608/2004.

Anonymous. 2012. Regulation (EU) No 1151/2012 of the European Parliament and of the Council of 21 November 2012 on quality schemes for agricultural products and foodstuffs.

Anonymous. 2014. Regulators' Code. Available from https://www.gov.uk/government/publications/regulators-code

EFSA. 2003. Opinion of the Scientific Panel on Biological Hazards on the request from the Commission related to the effects of nitrites/nitrates on the microbiological safety of meat products. The EFSA Journal 14: 1–31.http://www.efsa.europa.eu/en/efsajournal/doc/14.pdf

Food Safety Agency. 2014. Guidelines on re-commissioning vacuum packers. Available from: http://www.food.gov.uk/sites/default/files/re-commissioning-vacuum-packer-guidance.pdf

Heintz, G. and P. Hautzinger. 2010. Meat processing technology for small- to medium- scale producers. Food and Agriculture Organization of the United Nations, Regional Office for Asia and the Pacific, Bangkok.

Quinto, E.J., P. Arinder, L. Axelsson, E. Heir, A. Holck, R. Lindqvist, M. Lindblad, P. Andreou, H.L. Lauzon, V.Þ. Marteinsson, and C. Pin. 2014. Predicting the concentration of verotoxin-producing Escherichia coli bacteria during processing and storage of fermented raw-meat sausages. Appl. Environ. Microbiol. 80: 2715–2727.

Sindelar, J.J. and A.L. Milkowski. 2012. Human safety controversies surrounding nitrate and nitrite in the diet. Nitric Oxide 26: 259–266.

Toldrá, F. (ed.) 2010. Handbook of meat processing. Blackwell Publishing Professional, Ames, Iowa.

Index

A

Active packaging, 253, 254, 255, 268

Animals, 2, 24, 28, 30, 41, 43, 50, 58, 59, 60, 61, 62, 63, 66, 67, 78, 79, 82, 89, 98, 111, 207, 228, 231, 303, 319, 322, 324, 329, 335, 361, 391, 402, 406, 417, 418, 419, 420, 421, 423, 424, 425, 428, 429, 430, 431, 432, 433, 442, 489, 495, 500, 501, 515, 537, 538, 545, 546, 547, 548

Antimicrobial agents, 88, 321, 323, 335, 337

Antimicrobial resistance, 54, 88, 239, 279, 319, 320, 321, 322, 323, 325, 326, 332, 333, 335

Antioxidants, 3, 33, 83, 346, 398, 399, 406, 407, 408, 409, 475, 496, 500, 501, 502, 503, 504, 505, 506, 511

Audit, 531, 544, 547

Autochthonous starter cultures, 173, 270, 272, 274, 275, 280, 282, 288, 289, 291, 292

B

Bacteria, 2, 3, 6, 16, 19, 21, 22, 24, 28, 29, 30, 31, 32, 33, 37, 38, 41, 43, 44, 51, 52, 54, 59, 60, 65, 66, 67, 68, 80, 81, 87, 88, 90, 95, 96, 97, 98, 99, 100, 103, 104, 105, 106, 107, 108, 109, 110, 111, 112, 113, 114, 117, 119, 120, 121, 122, 127, 128, 129, 140, 158, 168, 169, 173, 174, 175, 176, 179, 180, 182, 196,
207, 208, 228, 229, 231, 232, 234, 236, 239, 240, 244, 245, 246, 247, 249, 251, 255, 256, 270, 274, 275, 280, 281, 282, 283, 284, 287, 289, 290, 292, 295, 297, 298, 308, 310, 311, 319, 322, 323, 324, 325, 326, 329, 332, 333, 335, 343, 344, 348, 352, 359, 379, 391, 397, 398, 402, 404, 425, 426, 439, 440, 450, 452, 456, 457, 458, 464, 466, 467, 468, 481, 489, 490, 525, 526, 539, 540, 541, 549, 552

Bacteriocins, 68, 69, 87, 129, 139, 140, 168, 214, 215, 216, 217, 229, 230, 231, 236, 242, 244, 245, 246, 247, 251, 253, 255, 256, 257, 271, 272, 274, 280, 283, 284, 286, 304, 329, 335, 348, 467, 515

Basic chemical properties, 380

Beef, 18, 19, 23, 24

Biodiversity, 168, 169, 170, 172, 174, 185, 236, 281

Biogenic
—amines, 33, 65, 66, 89, 90, 140, 168, 229, 236, 239, 240, 253, 271, 272, 275, 279, 332, 396, 397, 434, 437, 438, 439, 450, 451, 452, 453, 454, 455, 456, 457, 458, 459, 463, 464, 465, 466, 468, 473
—amine control, 468
—amine formation, 456, 457, 458, 466, 467, 468, 473

Biological hazards, 29, 61, 62, 68, 521, 522, 524

C

Chemical
—composition, 95, 291, 362, 364, 476, 478, 533
—hazards, 418, 442, 522
Chicken, 24
Cholesterol, 41, 43, 275, 295, 296, 304, 361, 364, 369, 371, 380, 390, 391, 399, 400, 401, 402, 403, 404, 405, 407, 483, 488, 495
Colour, 2, 3, 6, 81, 82, 83, 86, 101, 102, 105, 107, 119, 120, 128, 139, 149, 150, 155, 168, 229, 236, 271, 346, 364, 372, 373, 379, 380, 434, 440, 467, 478, 493, 498, 499, 500, 501, 504, 535, 539, 540, 541, 548
Contaminants of meat processing, 442
Culture-dependent and culture-independent molecular methods, 171, 183, 190

D

Denaturing gradient gel electro-phoresis (DGGE), 171, 186
Dried, 17, 24
Dry
—cured ham, 28, 37, 38, 39, 43, 47
—cured meat, 128, 141, 151, 152, 154, 155, 156, 158, 364, 366, 432, 433, 440, 481, 482, 488, 534
—fermented sausage, 28, 45
Dynamics and changes of microbial ecology, 186

E

E. coli O157:H7, 69, 197, 198, 200, 202, 203, 204, 205, 212, 216, 217, 249, 251, 256
Encapsulation, 169, 253, 256, 258
Environmental contaminants, 30, 181, 418, 420, 427, 430, 442

F

Fat, 2, 3, 8, 9, 17, 24, 27, 28, 30, 32, 41, 42, 43, 44, 52, 59, 61, 65, 67, 68, 79, 80, 81, 83, 85, 86, 114, 115, 116, 117, 200, 210, 211, 213, 235, 238, 296, 304, 311, 335, 360, 361, 362, 364, 365, 366, 367, 368, 369, 371, 373, 380, 389, 390, 391, 392, 393, 398, 399, 400, 401, 402, 403, 404, 405, 407, 409, 419, 421, 422, 424, 425, 426, 428, 431, 436, 464, 474, 475, 476, 477, 478, 479, 480, 481, 482, 483, 488, 489, 490, 491, 493, 494, 495, 496, 497, 498, 499, 500, 501, 504, 506, 535, 548
—content, 2, 28, 41, 80, 86, 114, 115, 116, 200, 238, 304, 335, 367, 371, 373, 390, 392, 399, 401, 404, 409, 424, 426, 436, 474, 475, 476, 477, 478, 479, 480, 481, 482, 483, 488, 490, 493, 494, 495, 499, 500, 506, 510
—level, 81, 401, 480, 486
—oxidation, 65, 86, 494, 496, 500
Fatty acid, 28, 38, 39, 42, 43, 44, 61, 113, 129, 149, 231, 275, 279, 300, 360, 361, 367, 368, 369, 370, 375, 380, 390, 398, 401, 404, 405, 406, 467, 475, 479, 480, 481, 482, 483, 488, 489, 490, 491, 492, 493, 494, 495, 498, 500, 501, 503, 506, 510
—composition, 380, 406
—profile, 41, 42, 79, 296, 364, 368, 369, 399, 400, 479, 480, 511
Fermentation, 2, 3, 4, 5, 6, 7, 8, 9, 15, 16, 17, 19, 20, 21, 22, 24, 27, 28, 33, 34, 38, 41, 50, 51, 52, 55, 58, 59, 66, 69, 80, 81, 82, 83, 86, 88, 90, 98, 105, 106, 107, 108, 111, 112, 114, 115, 117, 118, 127, 128, 129, 139, 140, 141, 149, 150, 155, 167, 168, 170, 171, 172, 173, 174, 181, 182, 183, 184, 185, 197, 199, 201, 207, 208, 209, 210, 211, 212, 213, 214, 215, 216, 217, 228, 229, 230, 231, 236, 238, 242, 243, 250, 251, 252, 253, 270, 271, 272, 273, 274, 279, 281, 282, 284, 285, 286, 287, 289, 290, 291, 307, 310, 311, 332, 333, 335, 344, 345, 347, 349, 350, 351, 352, 361, 365, 372, 380, 389, 393, 400, 426, 430, 432, 434, 438, 441, 452, 456, 457, 463, 464, 465, 467, 468, 474, 489, 490, 494, 496, 497, 501, 515, 519, 522, 524, 525, 526, 527, 528, 531, 535, 541, 542
Fermented, 15, 16, 17, 19, 20, 21, 22, 24, 26

Index 555

—meat, 16, 19, 20, 21, 23, 24, 27, 29, 49, 50, 52, 53, 54, 55, 61, 80, 87, 88, 90, 96, 97, 98, 113, 127, 167, 168, 169, 170, 173, 174, 179, 181, 182, 183, 184, 185, 196, 197, 198, 199, 201, 214, 217, 228, 229, 230, 231, 236, 238, 239, 240, 241, 242, 251, 252, 253, 256, 257, 279, 283, 286, 288, 290, 291, 292, 296, 305, 307, 308, 310, 311, 319, 320, 327, 329, 332, 333, 334, 335, 344, 345, 346, 347, 348, 350, 352, 359, 360, 361, 362, 363, 364, 367, 369, 370, 371, 372, 373, 374, 375, 376, 377, 378, 379, 380, 393, 396, 398, 402, 403, 404, 417, 431, 432, 433, 434, 438, 439, 442, 450, 451, 452, 456, 458, 463, 465, 489, 490, 491, 494, 495, 496, 497, 498, 499, 500, 501, 504, 505, 506, 515, 523, 526, 531, 536, 537, 538, 546, 548, 550

— meat production, 50, 55, 327, 333, 531, 536, 537, 538, 546, 550

—meat products, 16, 19, 20, 21, 23, 24, 27, 29, 49, 50, 52, 53, 54, 80, 87, 88, 90, 96, 97, 98, 113, 127, 167, 168, 169, 170, 173, 179, 181, 182, 184, 185, 196, 197, 198, 199, 201, 214, 217, 228, 229, 230, 231, 236, 238, 239, 241, 242, 251, 252, 253, 256, 257, 279, 283, 286, 288, 291, 292, 296, 305, 307, 308, 310, 311, 319, 320, 329, 332, 333, 334, 344, 345, 346, 347, 348, 350, 352, 360, 362, 363, 367, 369, 370, 371, 372, 373, 374, 375, 376, 377, 378, 379, 380, 393, 396, 402, 403, 404, 417, 431, 432, 433, 434, 438, 439, 442, 450, 451, 452, 456, 458, 463, 465, 489, 490, 491, 494, 495, 496, 497, 498, 499, 500, 501, 504, 505, 506, 515, 523, 526, 531, 536, 537, 550

—products, 15, 16, 17, 20, 24, 52, 61, 69, 79, 81, 86, 89, 90, 129, 167, 168, 173, 245, 252, 271, 272, 273, 274, 281, 310, 325, 330, 359, 371, 375, 393, 465, 489, 490, 491, 496, 497, 498, 499, 500, 501, 503, 506

—sausages, 1, 2, 4, 5, 7, 8, 11, 17, 18, 24, 27, 28, 33, 36, 37, 43, 44, 51, 52, 53, 54, 55, 61, 64, 65, 66, 67, 68, 79, 80, 81, 82, 83, 86, 96, 97, 98, 99, 100, 101, 102, 103, 105, 106, 107, 108, 110, 111, 112, 113, 114, 115, 116, 117, 118, 119, 122, 128, 129, 130, 131, 132, 133, 134, 136, 138, 139, 140, 141, 142, 143, 144, 145, 146, 147, 148, 149, 150, 151, 152, 154, 155, 156, 158, 169, 172, 173, 174, 180, 181, 182, 183, 184, 185, 214, 215, 216, 232, 233, 234, 236, 237, 238, 239, 240, 242, 248, 251, 254, 256, 257, 270, 272, 274, 275, 276, 277, 278, 283, 284, 285, 286, 287, 289, 290, 291, 292, 308, 310, 320, 321, 327, 328, 329, 330, 331, 332, 333, 334, 335, 359, 360, 363, 364, 365, 366, 368, 369, 371, 372, 373, 374, 375, 376, 377, 378, 379, 390, 393, 397, 398, 400, 432, 433, 437, 439, 440, 441, 455, 457, 458, 459, 460, 462, 463, 466, 467, 468, 474, 475, 483, 488, 490, 497, 500, 506, 535, 536, 537, 538, 539, 540, 541, 542, 545

Flavour compounds, 380, 467, 498, 500

Food-borne agents, 102, 112, 113, 120, 122

Food safety, 24, 28, 29, 38, 44, 50, 52, 55, 61, 79, 87, 90, 173, 175, 238, 254, 255, 270, 272, 282, 287, 288, 360, 398, 409, 419, 435, 512, 513, 514, 516, 518, 522, 523, 525, 530, 531, 536, 539, 542, 543, 544, 545, 547, 548

Formulation, 24

Functional foods, 80, 274, 309, 311, 397, 398, 399, 404, 405, 510

G

Game meat, 58, 59, 60, 61, 62, 63, 64, 65, 66, 69, 292

General and selection criteria, 288

Goat, 28, 78, 79, 80, 81, 82, 83, 84, 85, 86, 88, 89, 90, 148, 149, 233, 428, 485

Gram-positive catalase-positive cocci, 127, 141, 272, 292, 472

H

HACCP, 24, 29, 44, 87, 418, 440, 512, 513, 514, 515, 516, 517, 518, 519, 520, 521, 523, 525, 527, 528, 529, 530, 531, 544, 545, 549

Hams, 2, 4, 9, 10, 11, 19, 27, 28, 33, 37, 38, 39, 40, 41, 42, 43, 44, 47

Health

—aspects, 53, 61, 468

—aspects of production, 45

—benefits, 80, 173, 174, 294, 295, 296, 297, 298, 299, 305, 306, 307, 310, 311, 407, 481

Hepatitis E, 62, 64, 67, 76

Heritage, 49, 55, 515

High hydrostatic pressure (HPP), 119, 256, 257, 258

Home-made fermented meat products, 56

Hurdle concept, 196, 208, 214, 335, 344, 353

Hydrolysis, 111, 273, 279, 363, 367, 488, 489, 490, 494, 500, 506

I

Identification and characterization, 181, 182, 183, 186

In vitro test, 311

In vivo test, 311

L

Lactic acid bacteria, 2, 3, 6, 16, 24, 43, 52, 65, 67, 80, 87, 106, 108, 122, 127, 168, 173, 208, 228, 232, 234, 251, 282, 283, 289, 290, 292, 295, 319, 321, 323, 335, 343, 398, 450, 452, 456, 457, 464, 466, 467, 468, 481, 526, 540, 541

Lamb, 24

Legislation, 59, 252, 283, 311, 436, 536, 542, 544, 546, 548, 549, 550

Lipid

—content, 45

—oxidation, 41, 139, 280, 349, 369, 375, 376, 390, 391, 407, 408, 477, 481, 482, 488, 493, 497, 498, 499, 503, 505, 506, 511

Listeria monocytogenes, 4, 28, 31, 62, 64, 65, 66, 69, 88, 89, 102, 103, 104, 113, 120, 196, 197, 200, 202, 203, 204, 205, 207, 208, 209, 210, 215, 216, 217, 241, 242, 247, 248, 249, 250, 251, 252, 254, 255, 256, 257, 275, 308, 521, 525, 526

M

Meat, 15, 17, 19, 20, 22, 23, 24, 26

Minerals, 296, 361, 372, 380, 390, 398, 400, 404, 407, 446

Molds, 20, 22, 30, 32, 44, 53, 97, 108, 109, 110, 114, 127, 128, 150, 155, 158, 207, 252, 272, 281, 284, 343, 344, 348, 349, 350, 397, 423, 432, 433, 434

Mycotoxins, 20, 30, 32, 44, 53, 88, 158, 272, 290, 291, 397, 418, 423, 427, 432, 433, 434, 448

N

NaCl, 31, 65, 83, 97, 99, 100, 102, 114, 119, 120, 122, 213, 281, 282, 287, 344, 345, 372, 378, 390, 391, 392, 393, 398, 456, 457, 489, 518

Next generation sequencing (NGS), 175, 186

Nitrite, 3, 6, 9, 16, 17, 18, 21, 28, 32, 33, 38, 43, 59, 64, 65, 66, 67, 68, 69, 81, 83, 100, 101, 102, 103, 105, 108, 114, 122, 208, 214, 215, 230, 235, 286, 310, 333, 344, 345, 346, 347, 379, 389, 390, 391, 393, 394, 395, 398, 403, 434, 440, 441, 455, 504, 517, 518, 535, 540, 541, 552

Nitrosamine, 346, 394, 395, 441, 455

Non-volatile compounds, 375, 380

Nutritional value, 2, 96, 122, 257, 364, 371, 400, 404, 435, 483, 502

O

Official controls, 536, 537, 541, 542, 543, 544, 546, 552

Ostrich, 24, 59, 68, 69, 220

Oxidation products, 390, 391, 488, 493, 494, 497, 498, 499, 500, 501, 506

P

PAHs, 395, 396, 403, 419, 434, 435, 436, 437, 521, 522, 532

Parasites, 29, 30, 55, 62, 427

Index

Pathogens, 2, 8, 9, 24, 28, 29, 33, 43, 51, 54, 63, 66, 69, 90, 98, 129, 140, 196, 199, 200, 201, 207, 208, 214, 217, 229, 241, 242, 246, 251, 254, 256, 257, 272, 283, 299, 300, 308, 320, 325, 332, 333, 335, 344, 425, 521, 524, 526, 537, 538, 539, 541, 549

pH, 2, 3, 4, 5, 6, 7, 8, 9, 15, 16, 17, 18, 19, 20, 21, 22, 24, 28, 29, 32, 33, 37, 44, 52, 53, 59, 60, 61, 62, 63, 64, 65, 66, 67, 68, 69, 80, 81, 87, 88, 89, 95, 96, 97, 99, 101, 102, 103, 104, 106, 107, 108, 110, 111, 112, 113, 114, 118, 119, 120, 122, 127, 128, 129, 150, 168, 199, 207, 208, 209, 210, 211, 212, 213, 214, 215, 230, 231, 232, 233, 234, 235, 242, 243, 244, 245, 248, 249, 251, 279, 280, 281, 282, 283, 286, 290, 296, 310, 333, 344, 345, 346, 363, 365, 366, 367, 378, 379, 393, 398, 402, 404, 408, 432, 439, 441, 452, 453, 456, 467, 476, 478, 489, 490, 499, 518, 523, 525, 526, 527, 528, 531, 539, 540, 541, 542

Phenotypic and genotypic methods, 271, 275, 335

Physico-chemical characterization, 380

Polyunsaturated fatty acids, 28, 42, 43, 61, 367, 369, 370, 404, 405, 406, 480, 490, 493, 506, 510

Pork, 23, 24

Probiotics, 80, 90, 140, 169, 184, 185, 256, 294, 295, 296, 297, 298, 299, 303, 304, 305, 306, 307, 308, 309, 310, 311, 335, 398

Production technology, 45

Prophylactic agents, 419, 442

Protective cultures, 239, 251, 252, 253, 257, 272

Q

Quality aspects, 43, 468

R

Raw meat fermented sausages, 536, 540

Redoxpotential, 122

Reduced pH, 24

Resistance genes, 54, 88, 89, 239, 240, 320, 321, 322, 323, 325, 327, 332, 333, 335, 341

Resistance transfer, 322, 335, 340

Ripened meats, 1, 2, 4, 9, 10, 11

Ruminants, 3, 9, 63, 64, 67, 68, 82, 401, 405, 406, 495, 539

S

Safety requirements, 229, 238, 258, 518, 522, 543

Salmonella spp., 28, 40, 62, 64, 69, 102, 122, 196, 199, 202, 204, 205, 207, 211, 217, 241, 247, 256, 344, 526, 540, 541

Salt, 2, 3, 9, 16, 17, 21, 27, 28, 30, 32, 37, 38, 43, 44, 50, 51, 52, 65, 79, 80, 81, 89, 98, 99, 100, 101, 102, 105, 113, 114, 115, 119, 120, 122, 140, 149, 155, 158, 215, 232, 235, 238, 279, 281, 286, 301, 333, 335, 360, 361, 364, 365, 367, 378, 379, 380, 389, 390, 391, 392, 393, 396, 398, 402, 403, 439, 457, 504, 518, 519, 521, 524, 525, 527, 529, 541

Salt cured meat, 56

Sausage, 18, 24, 25

Semi-dried, 24

Sensorial properties, 53, 179, 211, 213, 329, 349, 360, 380, 470

Sheep, 3, 6, 8, 11, 68, 78, 79, 80, 81, 82, 83, 84, 85, 86, 88, 89, 90, 233, 426, 428, 538, 539

Shelf-life, 52, 60, 61, 81, 168, 174, 179, 251, 254, 288, 294, 308, 329, 440, 457, 467

Specialties, 11

Spoilage, 2, 19, 24, 32, 38, 43, 51, 80, 95, 96, 97, 98, 99, 105, 129, 168, 174, 179, 208, 229, 231, 236, 243, 246, 248, 249, 251, 254, 256, 272, 283, 332, 344, 345, 347, 348, 349, 350, 351, 352, 434, 438, 451, 452, 465, 467, 489, 491, 506, 524, 539, 541, 544

Starter culture, 17, 19, 20, 65, 66, 67, 86, 101, 108, 111, 112, 155, 158, 173, 208, 209, 212, 216, 238, 239, 243, 252, 274, 275, 279, 287, 290, 291, 306, 307, 308, 310, 311, 348, 397, 430, 432, 489, 541

St. aureus, 52, 66, 69, 88, 114, 141, 142, 144, 145, 147, 148, 149, 203, 204, 207, 208, 214, 217, 241, 247, 332, 526, 527, 540
STEC, 102, 197, 346, 357
Substances having anabolic effect, 442

T
Therapeutic agents, 398, 419, 442
Tradition, 3, 22, 78, 283, 359, 413

U
Užice goveđa pršuta, 515, 518, 520, 531

V
Varieties, 7, 8, 9, 11, 46
Veterinary drugs, 30, 417, 418, 419, 420, 442, 449
Veterinary public health, 335

Vitamins, 296, 361, 371, 380, 390, 400, 404, 405, 474, 475, 504
Volatile compounds, 39, 109, 140, 149, 155, 229, 368, 374, 375, 377, 380, 481, 483, 488, 498

W
Water activity, 4, 5, 6, 8, 9, 11, 16, 18, 19, 20, 22, 29, 40, 43, 44, 51, 52, 64, 81, 95, 96, 97, 100, 113, 114, 116, 122, 127, 129, 196, 230, 251, 369, 398, 432, 436, 489, 523, 524, 525, 526, 529, 541, 542
Wild boar, 55, 61, 62, 63, 64, 65, 66, 233

Y
Yeasts, 6, 32, 97, 109, 114, 117, 127, 150, 155, 158, 207, 229, 241, 272, 281, 283, 284, 289, 295, 343, 344, 348, 349, 356